David N. Cooper

The Molecular Genetics of Lung Cancer

David N. Cooper

The Molecular Genetics of Lung Cancer

With 34 Figures, 13 in Colour

Springer

David N. Cooper
Professor of Human Molecular Genetics
Institute of Medical Genetics
Cardiff University
Heath Park
Cardiff, CF14 4XN, UK

Library of Congress Control Number: 2004111936

ISBN 3-540-22985-X Springer Berlin Heidelberg New York

Springer is a part of Springer Science+Business Media
springeronline.com

Printed in Germany

Editor: Dr. Rolf Lange, Heidelberg, Germany
Desk editor: Hiltrud Wilbertz, Heidelberg, Germany
Production editor: Helmut Schwaninger, Heidelberg, Germany

Cover design: Erich Kirchner, Heidelberg, Germany
Typesetting: Mitterweger & Partner, Plankstadt, Germany

Printed on acid-free paper 27/3150 hs – 5 4 3 2 1 0

To Paul, Catrin and Duncan

Preface

Lung cancer, the second most common cancer in men and the third most common in women, is one of the most studied and hence probably one of the best understood of the 'non-heritable' human cancers. Familial aggregation of lung cancer is comparatively rare and, as yet, no gene has been described whose mutation is pathognomonic for, or even largely confined to, lung cancer, as found for example in cancers of the colon, kidney or breast. This should not however be taken as implying that genes are not involved in the etiology of lung cancer. Indeed, a considerable number of inherited genetic variants are now thought to confer susceptibility to the disease. An even larger number of genes have been shown to harbour somatic mutations in lung tumour cells or tissue. The role that these somatic mutations play in lung tumorigenesis can increasingly be understood in terms of their ability to promote cellular growth, to interfere with DNA repair, to confer resistance to apoptosis or to induce cellular transformation, tumour growth, invasiveness, angiogenesis, evasion of host immunity and finally, metastasis.

Perhaps the most characteristic feature of lung cancer is however the very strong association, evidenced by a multitude of epidemiological studies, with cigarette smoking. Although the precise underlying biological mechanisms and pathways responsible for this association are not yet understood, it is nevertheless clear that in lung tumorigenesis we are witnessing a highly complex interplay between genes and environment. Unravelling this complexity promises to be a very substantial undertaking. This is reflected in the fact that the molecular genetics of lung cancer is a burgeoning subject with a very large and widely dispersed literature that is difficult to access. This volume is primarily intended to summarize the current state of knowledge regarding the molecular and genetic mechanisms underlying lung cancer, and focuses specifically on the proximal genetic causes (as evidenced by the different types of somatic and germline mutations in a variety of different genes) and consequences (as adduced by mRNA expression and protein profiling) of this condition.

The introductory chapter is intended to provide an overview of the molecular basis of cancer and discusses the different categories of gene known to be involved in tumorigenesis as well as the underlying mechanisms of mutagenesis. Chapter 2 seeks to set the scene, with short sections on the history of lung cancer research, clinical aspects of the disease, disease classification and the major contribution made by both classical cytogenetic and molecular cytogenetic analysis to the identification of the genes involved. In the absence of a familial predisposition to the disease, the molecular genetic analysis of lung cancer (both small cell and non-small cell) has been largely confined to the study of the tumour cells themselves. Somatic mutations involving some 120 different human genes have so far been characterized in primary lung tumour tissue or in cells derived from lung tumours. Such mutations may be gross and affect entire chromosomal regions or may instead be more subtle and serve to alter the fine structure of specific genes. A growing number of genes have also been shown to be inactivated by promoter methylation ('epimutation') during lung tumorigenesis. Genes compromised by mutation or epimutation include oncogenes and tumour suppressor genes as well as genes encoding proteins that perform functions in DNA repair, telomerase activity, apoptosis, and cell cycle regulation. This topic is reviewed in Chapter 3. Chapter 4 focuses on recent advances in our understanding of signal transduction pathways, apoptosis and cell cycle control mechanisms that are beginning to allow us to account for the emerging mutation data in terms of, for example, either a growth advantage accruing to the cells and/or an escape from apoptosis. The determination of the temporal order in which these lesions occur and the elucidation of the functional consequences of different combinations and permutations of lesions are extremely important. Chapter 4 attempts to address these topics as well as the identification of new prognostic indicators and the potential for early clinical detection.

Without straying into either epidemiology or toxicology, Chapter 5 attempts to review current knowledge about the genetic damage in the lung cells of smokers and smoking lung cancer patients. The cytogenetic, genetic and epigenetic (methylation) changes that have been associated with smoking are discussed, as are reported differences in mutation frequency or lung cell gene expression between smokers and non-smokers. Finally, the ongoing debate and controversy surrounding the potential role of benzo[*a*]pyrene as a mutagen (specifically with respect to the *TP53* gene) is described and discussed in detail.

A variety of inherited variants in genes encoding xenobiotic metabolising enzymes have been described that may influence lung cancer risk by playing a role in determining levels of cellular exposure to potential exogenous mutagens and carcinogens; these are reviewed in Chapter 6. Also covered are inherited variants in genes encoding DNA repair enzymes that can confer inter-individual differences in DNA repair capacity which may, at least in principle, indirectly modulate lung cancer risk.

New approaches to studying the lung tumour cellular phenotype (e.g. cDNA and oligonucleotide microarrays) are explored in Chapter 7. These approaches promise to greatly increase our knowledge of the consequences of mutation at the level of gene expression and are vital not only for obtaining a better understanding of lung cancer pathogenesis, but also for improving existing tumour diagnostics (identification/classificatory system) and for providing the novel prognostic markers and indicators that are a prerequisite for optimal patient management (Chapter 8). Advances in our understanding of the molecular biology of lung cancer have guided the design and application of a variety of new gene therapy approaches that are currently being explored in the search for the next generation of treatments (Chapter 8).

This volume therefore attempts to describe how the new techniques, methods and approaches of molecular genetics have been used in an attempt to unravel the complexities of lung tumorigenesis by analysis at DNA, RNA and protein levels with potentially important implications for diagnosis, prognosis and treatment as well as providing new insights into how lung tumours arise from precancerous lesions and eventually progress to metastasis. In relation to the prevalence of the disease, research into lung cancer has historically been under-subscribed and poorly resourced by comparison with the common 'heritable cancers'. It is hoped that by stimulating interest in lung cancer research, this volume will help to reverse this trend.

Cardiff, July 2004

David N. Cooper

Acknowledgements

I am most grateful to Anna-Lisa Fisher who originally conceived the idea of a volume on the molecular genetics of lung cancer. Grateful thanks are also due to Karl-Friedrich Baetz, Nadia Chuzhanova, Sunil Dolwani, Michael Krawczak, Paul Lewis, Ann Procter, Nick Thomas, Ray Thornton and Meena Upadhyaya for their help and advice with the manuscript, to Hester Wain and Elspeth Bruford of the *Human Gene Nomenclature Committee* for providing gene symbols, and to Rolf Lange and Julia Heidelmann of Springer-Verlag for their much appreciated practical support during the production process.

Table of Contents

CHAPTER 1

An Introduction to the Molecular Basis of Cancer

> "Tumors destroy man in a unique and appalling way, as flesh of his own flesh which has somehow been rendered proliferative, rampant, predatory and ungovernable".
>
> Peyton Rous (1966) Nobel Lecture presentation speech

> "To understand cancer is to gain access to the logic of the system which imposes on cells the constraints of the organism"
>
> Francois Jacob (1973) *The Logic of Life*

Cancer has been defined as a "genetic disease, arising from an accumulation of mutations that promote clonal selection of cells with increasingly aggressive behaviour" (Fearon 1997). In other words, cells that become malignantly transformed have a tendency to exhibit uncontrolled growth and to acquire invasive properties. The development of malignancy is in most cases thought to be a multistep process involving sequential mutations in different genes that encode protein products which play a role in cell growth and differentiation (Fearon and Vogelstein 1990). Cells that possess such mutations may acquire a growth advantage (cellular 'selection') by comparison with other cells that lack these lesions (Tomlinson *et al.*, 1996; Tomlinson and Bodmer 1999; Michor *et al.*, 2004). Since cancer is an evolutionary process in which mutations accumulate in different cellular lineages, cellular heterogeneity is inevitable, a fact that serves to complicate greatly the molecular genetic analysis of tumour tissue.

Clonal expansion occurs through the selection of cells bearing mutations that confer an increased rate of cellular proliferation (either by providing a constitutive growth signal or through insensitivity to normal growth inhibitory signals), and/or a decreased cellular death rate (apoptosis). Over time, those cells that have a competitive growth advantage will come to predominate. As they then steadily acquire more mutations that increase this growth advantage, their growth can become uncontrolled. Some tumours acquire a 'mutator phenotype' that is responsible for greatly increasing the cellular mutation rate and with it the probability with which these cells can acquire further mutations that will provide them with a still greater competitive advantage (Cahill *et al.*, 1999; Loeb and Loeb 2000; Loeb *et al.*, 2003). In addition, the majority of tumours express molecules involved in telomere maintenance and which allow cells to escape from the normal processes of senescence and cell death. Growing tumours must also acquire the ability to induce the formation of new vasculature (angiogenesis) giving them increased access to oxygen and nutrients. Finally, cancer cells can acquire invasive properties that potentiate their spread into surrounding tissue (metastasis). This they do by detaching themselves from the extracellular matrix (ECM), destroying the ECM in order to migrate through tissue, and then being transported by lymph or blood to remote sites that they can colonize by virtue of their having lost certain cell-cell adhesion molecules or growth factors (Engbring and Kleinman 2003). These processes are rarely contemporaneous. Indeed,

> "tumour cells don't usually start out bad; they generally acquire their malignant properties over years or decades, going from benign to invasive to metastatic to lethal, in a process known as tumour progression".
>
> BR Zetter & J Banyard (2002) Cancer: The silence of the genes. *Nature* 419: 572-573.

From the above discussion, it will be clear that many workers have conceptualized cancer in terms of a process of neo-Darwinian selection with natural selection being the driving force behind cellular evolution during tumour progression (e.g. Cahill *et al.*, 1999; Gatenby and Vincent 2003). A struggle for existence occurs at the cellular level with cells competing with one another for growth advantage, jostling for *Lebensraum* in the microenvironment of the tumour (Frank 2003). Mutation, as the origin of genetic variation, is clearly a *sine qua non* for evolution to occur, and so it is with cellular evolution. The genomic and karyotypic instability that characterizes most if not all forms of cancer provides the somatic genetic variation from which the specific cellular phenotypes that confer a growth advantage may be selected (Sieber *et al.*, 2003). The result is progression along defined (albeit highly variable) pathways to tumorigenesis, yielding a tumour that is often extremely heterogeneous in terms of its constituent cell populations.

It should however be noted that an increased mutation rate does not automatically confer a selective advantage upon the cell. Indeed, any cell that manifests a dramatically increased mutation rate will necessarily experience a considerable amount of genetic damage and will consequently be more likely to be apoptosed than to develop into a tumour. Cahill *et al.* (1999) therefore put forward a 'just enough' genetic instability model of tumour progression which proposes that those cells which survive the tumult of the process of tumorigenesis will have successfully achieved a compromise between excessive genetic instability (which would have resulted in programmed cell death) and insufficient genetic instability (which would have yielded too few mutations to give rise to cancer).

What we eventually see clinically are the tumour masses or metastases that have come to clinical attention. Composed of clonal populations of those cells that have survived against all the odds, that have evaded the normal growth constraints most effectively, and which have succeeded in avoiding senescence and in eluding the clutches of the apoptotic machinery of programmed cell death, the cellular make-up of these tumours reflects the selective pressures experienced during their evolutionary history. What we witness is the end-result of the survival of the 'fittest cells', the inevitable consequence of a cellular struggle for existence. Tumour cell evolution has, however, no more foresight than organismal evolution. Thus, fitness is defined here not at the level of the organism, which the tumour will ultimately destroy, but at the level of cellular growth advantage within the tissues and organs of that organism; fitness in the sense of ruthlessly out-competing other cells that possess genotypes which are less able to exploit the niches available in the micro-environment of the tumour.

Cancer genes

> "Part of the revolution in cancer research in the 1980s was the discovery that......the same kinds of genes are mutated in all tumors".
> B Vogelstein and KW Kinzler (1993) The multistep nature of cancer. *Trends Genet.* 9: 138-141.

Although ~1% of cancers arise in individuals with an hereditary cancer syndrome resulting from the inheritance of a germline mutation, the vast majority of cancer-associated mutations are somatic. Such lesions are now known to occur in several different types of gene:

(a) Oncogenes
(b) Tumour suppressor genes
(c) Cell cycle control genes
(d) Mutator (DNA repair) genes
(e) Apoptosis regulatory genes
(f) Telomerase-associated genes

Oncogenes encode proteins whose normal function is to promote cellular proliferation. When activated by mutation, the oncogene products 'gain a function' by becoming over-active, inappropriately active, or by losing their potential to be inactivated. This serves to increase the rate of cellular proliferation. By contrast, tumour suppressor genes are characterized by loss-of-function mutations. The protein products of tumour suppressor genes often inhibit cell proliferation or alternatively participate in the process of programmed cell death known as 'apoptosis' that is designed to protect the organism from the consequences of proliferation of irreparably damaged cells. Loss of a tumour suppressor gene may thus also increase the rate of cellular proliferation. If we adopt a vehicular analogy, inactivated tumour suppressor genes may be regarded as being equivalent to broken brakes whilst activated oncogenes can be visualized as being jammed accelerators.

The above categories of gene are by no means mutually exclusive and there is some overlap depending upon the functions of the protein in question. Another way of classifying the different types of cancer susceptibility gene is to distinguish 'gatekeepers' (genes that encode protein products that control cellular proliferation directly) from 'caretakers' (genes which encode proteins that help to maintain the integrity of the genome). In cancer, an altered gatekeeper is likely to bring about tumour initiation whereas an altered caretaker may be expected to promote tumour progression. These different types of cancer susceptibility gene will now be described in more detail in order to provide the necessary background to understand the nature and significance of some of the genetic changes thought to play a role in lung tumorigenesis.

Oncogenes

Oncogenes encode proteins whose normal function is to promote cellular proliferation. When activated, usually by a solitary heterozygous missense mutation, the oncogene products become over-active, inappropriately active, or instead lose their potential to be inactivated. Such "gain-of-function" somatic mutations arising in only one allele of a gene may in some cases be sufficient to alter the phenotype of the cell. Oncogene protein products fulfil a variety of different cellular roles: signal transduction (e.g. Ras), transcriptional regulation (e.g. myc), cell surface receptors (e.g. epidermal growth factor receptor) or components of the cyclin/cdk system that controls the cell cycle (e.g. cyclin D1). Activated oncogene products therefore include transcription factors that are capable of up-regulating genes whose role is to promote cell growth, or growth factor receptors able to transmit stimulatory signals in the absence of their ligands*. The biological effects of an activated oncogene can be assessed by transforming cells *in vitro* with copies of that activated oncogene. By way of example, Ravi *et al.* (1998) transformed human small cell lung cancer cells with an activated Raf1 (*RAF1*) oncogene and reported marked induction of cyclin-dependent kinase inhibitor p27 and a decrease in cdk2 protein kinase activity.

Tumour suppressor genes

Tumour suppressor genes have been defined as "genes that sustain loss-of-function mutations in the development of cancer" (Haber and Harlow 1997) and their protein products often inhibit cell proliferation. Tumour suppressor genes are however involved in the regulation of a diverse array of different cellular functions including cell cycle checkpoint control, detection and repair of DNA damage, protein ubiquitination and degradation, mitogenic signaling, cell specification, differentiation and migration, and tumour angiogenesis (Sherr 2004). Thus, tumour suppressor genes often encode proteins with a regulatory role in cell cycle progression (e.g. Rb) but may also encode DNA-binding transcription factors (e.g. p53) and inhibitors of cyclin-dependent kinases required for cell cycle progression (e.g. p16).

This notwithstanding, the rule of thumb is that mutational inactivation of both tumour suppressor alleles is required to change the phenotype of the cell (reviewed by Levine 1993). This "two hit (Knudson) hypothesis" provides the basis for a mechanistic understanding of tumour suppressor gene mutagenesis: a first (inherited) mutation in one tumour suppressor allele is followed by the somatic loss of the remaining allele via a number of different mutational mechanisms (Knudson 2001). Whilst the inherited lesion is usually fairly subtle, the second (somatic) hit may often involve the deletional loss of the entire gene or even a substantial portion of the chromosome involved. Alternatively, both 'hits' may constitute somatic mutations: the end result is the same – the loss of both gene copies. As with all rules, however, there are exceptions: thus, it should be appreciated that there are certain types of tumour for which one 'hit' appears to be sufficient to promote tumour formation ('haploinsufficiency'; Largaespada 2001; Kemkemer *et al.*, 2002; Santarosa and Ashworth 2004) whereas in other cases, a 'third hit' may be required (Su *et al.*, 2000).

* It should be noted that the non-mutant cellular versions of activated oncogenes should properly be termed *proto-oncogenes*.

Gross deletions often serve to remove alleles in the vicinity of the tumour suppressor gene as well as the tumour suppressor gene itself, thereby bringing about a *loss of heterozygosity* (LOH) at numerous loci in the region (Tischfield 1997). Heterozygosity refers to the presence of discernibly different alleles derived from an homologous chromosome pair. Large-scale chromosomal deletions will lead to the loss of alleles at numerous linked loci - hence the term, loss of heterozygosity. In any given cancer, LOH can thus be used as a pointer to the location of tumour suppressor genes whose mutation is associated with the tumorigenic process.

Mutator genes and genetic instability

> "Although instability may be essential for neoplasia to develop, it may also prove to be its Achilles' heel when the tumour is attacked by the right [chemotherapeutic] agents".
>
> C. Lengauer, K.W. Kinzler & B. Vogelstein (1998) *Nature* 396: 643-649.

A common finding in a diverse array of cancers is genetic instability. Its relationship to tumorigenesis is particularly well understood in hereditary non-polyposis colorectal cancer (HNPCC), a syndrome characterized by predisposition to colorectal carcinoma and other cancers of the gastrointestinal and genitourinary tracts. In HNPCC, somatic genetic instability occurs at simple highly repetitive microsatellite sequences. This microsatellite instability (MI) has been linked to mutations in a number of mismatch repair (MMR) genes including *MLH1*, *MSH2*, *GTBP*, *PMS1* and *PMS2* (Kinzler and Vogelstein 1996; Papadopoulos and Lindblom 1997). Mutation rates in cells exhibiting MI are two to three orders of magnitude higher than in normal cells (Kinzler and Vogelstein 1996) such that tumour cells manifesting MI can carry >100,000 mutations (Ionov *et al.*, 1993). Since multiple mutations are necessary for malignancy, MMR deficiency greatly speeds up the process of accumulating mutations at those loci that are critical for tumour progression. Pursuing my earlier vehicular analogy of brakes and accelerators, a mismatch repair gene defect could perhaps be likened to a drunken mechanic attempting a routine automobile service!

The secondary target genes of the genome-wide somatic hypermutability evident in mismatch repair-deficient cells are now beginning to be identified (Duval and Hamelin 2002; Woerner *et al.*, 2003). One prime example is that of the gene (*TGFBR2*) encoding TGFβ receptor II that is intimately involved in cellular growth regulation. Colorectal tumours are generally insensitive to the growth suppressing hormone TGFβ and in colorectal tumours manifesting MI, this insensitivity is almost invariably due to frameshift mutations within a microsatellite sequence (polyadenine tract) embedded within the *TGFBR2* gene (Markowitz *et al.*, 1995; Lu *et al.*, 1995; Parsons *et al.*, 1995). Similarly, the target sequence of mutations within the transcription factor *E2F4* gene is a $(CAG)_{13}$ trinucleotide repeat within the putative transactivation domain (Yoshitaka *et al.*, 1996). Thus, secondary target genes bearing mutations that contribute to the development of colorectal neoplasia may be identified firstly through their involvement in the negative regulation of cell growth and secondly on the basis of their containing repetitive sequences which represent likely targets for mutation in cells exhibiting MI. Some of these secondarily mutated genes are directly involved in growth regulation [e.g. *APC* (Huang *et al.*, 1996) and *IGF2R* (Souza *et al.*, 1996)] or in promoting apoptosis [e.g. *BAX* (Rampino *et al.*, 1997] and are therefore likely to play a role in tumour progression. Further, the mismatch repair genes may themselves represent mutational targets (Perucho 1996; "the mutator that mutates the other mutator"), occurrences which increase genomic instability still further.

Cell cycle control genes

> "Cancer results from the accumulation of mutations in the genes that directly control cell birth or cell death".
>
> C. Lengauer, K.W. Kinzler & B. Vogelstein (1998) *Nature* 396: 643-649.

Genes encoding proteins of the cell cycle are common targets of mutagenesis in cancer (Molinari 2000). The cell cycle has a cell division phase (M phase) and a long interphase that comprises S phase (DNA synthesis), G_1 phase and G_2 phase

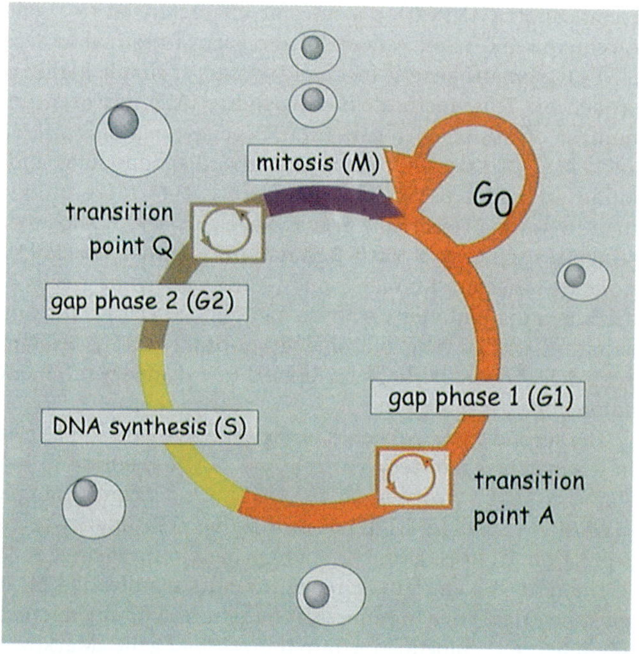

Fig. 1.1. Phases of the cell cycle. M phase constitutes chromosome segregation and mitosis. Phases G_1, S and G_2 comprise the interphase. DNA replication takes place in the S phase. A variety of checkpoints occur during the gap phases G_1 and G_2 that determine whether the cell proceeds to S or M phases. G_1: DNA integrity, cell size, presence of nutrients and growth factors are checked in preparation for synthesis. G_2: Integrity of replication checked. The transition points A and Q represent points of hesitation where progress is halted for an undefined period. When the conditions for growth are unfavourable, the cell cycle arrests in G_0 [Reprinted from Fig. 10.4 (page 233), Chapter 10, in *Signal Transduction*, by BD Gomperts, PER Tatham & IM Kramer, Copyright (2002) by kind permission from Elsevier]

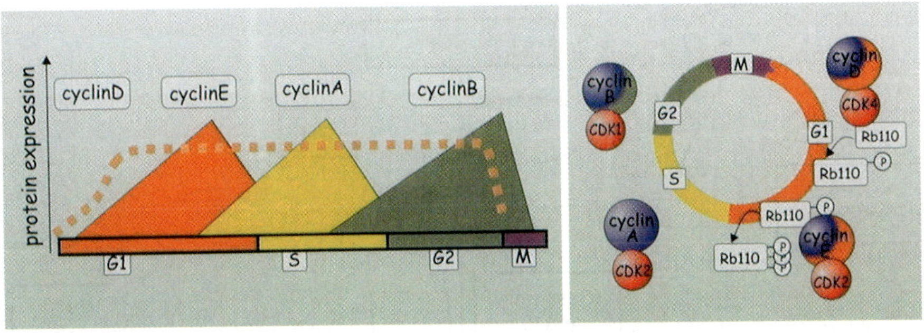

Fig. 1.2. Cyclin activity during the cell cycle. Progress through the cell cycle is controlled by the sequential generation of cyclins that associate with cyclin-dependent protein kinases (cdks). Cyclins D and E associate with cdk4 and cdk2 respectively and drive the cell through G_1 phase. They prepare the cell for DNA replication, a process that requires phosphorylation of Rb. Cyclin A associates with cdk2 and drives the cell through the S phase. Cyclin B associates with cdk1 and determines the time of onset of mitosis [Reprinted from Fig. 10.5 (page 234), Chapter 10, in *Signal Transduction*, by BD Gomperts, PER Tatham & IM Kramer, Copyright (2002) by kind permission from Elsevier]

(Figure 1.1). Cells normally only enter S phase if they are committed to mitosis whereas non-dividing cells remain quiescent in G_0 phase. Checkpoints serve to regulate the order of events in the cell cycle as well as ensuring that DNA repair is coordinated with cell cycle progression. These checkpoints are regulated by the cyclin-dependent kinases (cdks) which phosphorylate key substrates that allow the cell cycle transitions to occur, their activating proteins (the cyclins) and cdk inhibitors such as p15 and p16 (Hunter and Pines 1994; Figure 1.2). In G_1, cyclin D complexes with cdk4 or cdk6. Later in G_1, cyclin E forms a complex with cdk2 to promote the G_1 to S phase transition. During S phase, cyclin A complexes with cdk2 and regulates DNA replication and exit to the G_2 phase. During G_2 and M phases, cyclin B complexes with cdk2 to mediate the G_2 to M phase transition. Mammalian cell cycle control comprises an elaborate regulatory network of proteins that is only just beginning to be unravelled (Kohn 1999). Cancer cells often possess defective cell cycle checkpoints that allow cells to bypass quiescence (Malumbres and Barbacid 2001) thereby causing disruption to the DNA repair process (Hall and Peters 1996). Thus, cells that lack a specific checkpoint can go on to acquire further mutations that may confer a selective (growth) advantage (Paulovich *et al.*, 1997; Kaufman and Paules 1996).

Apoptosis regulatory genes

DNA repair processes are inherently error-prone and in any case may be inadequate to the task if genomic damage is extensive. Such damage may constitute a neoplastic risk if cell division is not arrested. An alternative option for the cell is suicide, the rationale of which is to prevent the clonal expansion of potentially malignant cells. To this end, both DNA damage and cellular injury initiate signaling pathways leading to programmed cell death or *apoptosis* (Zhou and Elledge 2000; see Figure 1.3). These signaling pathways involve either cytokine receptors (e.g. Fas) or transcrip-

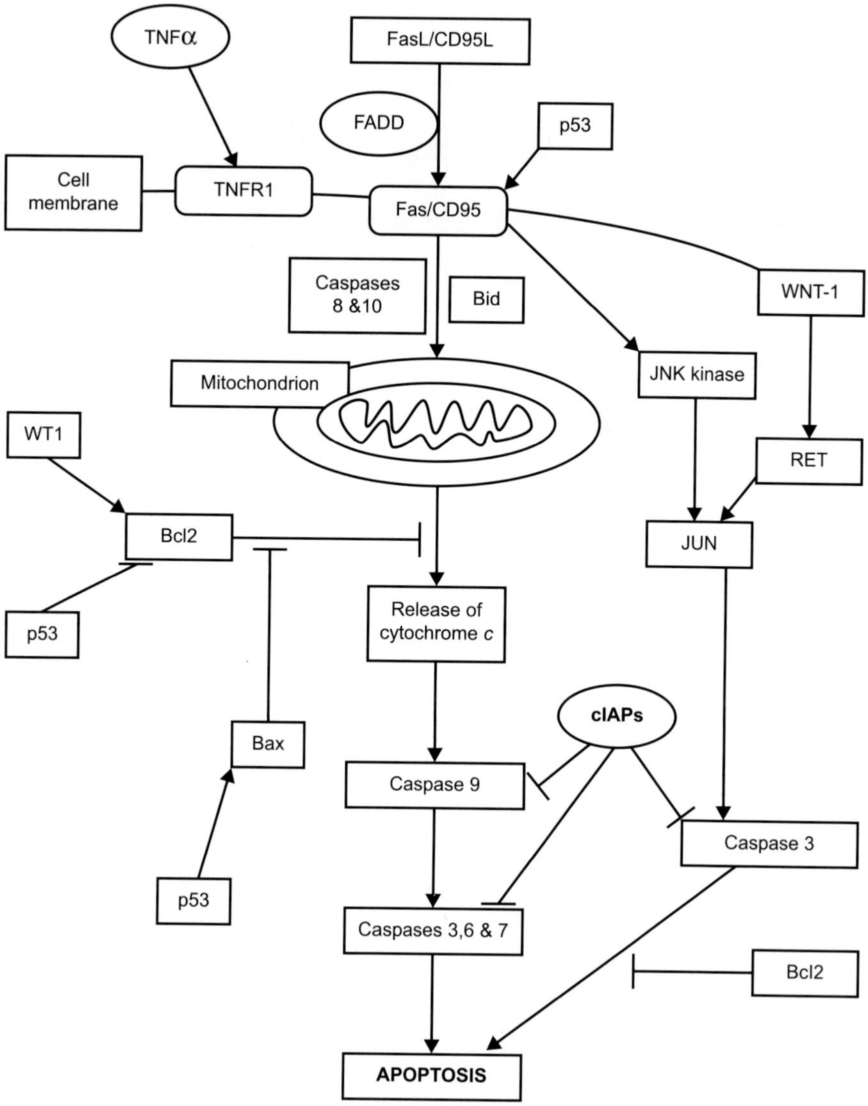

Fig. 1.3. Apoptosis signaling pathways

tion factors (e.g. myc, p53, Rb) whose transduction into caspase activation is regulated by Bcl2, a repressor of apoptosis, and a family of Bcl2-like proteins including Bax, a promoter of apoptosis (Wyllie 1997). Apoptosis probably plays a major role in limiting tumour progression and its inactivation represents an important mechanism in tumorigenesis (Cory and Adams 2002; Townson *et al.*, 2003).

Cancer as a disease of differentiation?

> "The fundamental problem of chemical physiology and of embryology is to understand why tissue cells do not all express, all the time, all the potentialities inherent in their genome. The survival of the organism requires that many, and in some cases most, of these potentialities be unexpressed, that is to say repressed".
>
> F Jacob & J Monod (1961) Genetic regulatory mechanisms in the synthesis of proteins. *J. Mol. Biol.* 3: 318-356.

Although it is often blithely assumed that somatic cells do not multiply unless stimulated to do so, there is another contrasting viewpoint to consider, and that is that proliferation is the cell's natural default state (Sonnenschein and Soto 1999). In this case, the question becomes not so much what induces the cells to proliferate but rather what it is that restrains them from so doing. It should be appreciated that the suppression of cellular proliferation is actually inherent in the process of differentiation which underlies the acquisition of tissue specificity. If, as suggested by Harris (2004), we envisage cancer as a disease of differentiation rather than as a disease of cellular proliferation, the situation becomes more complex. Mutations in oncogenes may still be viewed as key steps in the tumorigenic process by virtue of their impeding the function of the growth-restraining tumour suppressor genes. Whilst it has long been known that the inactivation of specific tumour suppressor genes gives rise to only certain types of cancer involving particular tissues, this tissue specificity still remains to be explained mechanistically in terms of pathways that relate the processes that normally restrain cellular proliferation, to differentiation. If, however, cancer is to be conceptualised as a disease of differentiation, then perhaps we should be moving away from the rather over-simplistic idea of competing checks and balances and seek instead to introduce our knowledge of the workings of these intricate control systems into the milieu of differentiation with its temporal notions of cellular determination, commitment, loss of diversification, and the acquisition of specialized functions. In so doing, we might also consider the sequential mutational changes of tumorigenesis as part of a ratchet-like process by which cells become channelled by acquiring new and specialized functions at the same time as experiencing the irreversible loss of others (Gan *et al.*, 2003).

Cancer, signaling and acquired capabilities

> "Although cancers are indeed extremely diverse and heterogeneous, ... underlying this variability lies a relatively small number of 'mission critical' events whose convergence is required for the development of any and all cancers".
>
> GI Evan & KH Vousden (2001) Proliferation, cell cycle and apoptosis in cancer. *Nature* 411: 342-348.

The protein products encoded by the above-mentioned oncogenes, tumour suppressor genes and apoptotic and cell cycle regulatory genes interact with each other to potentiate cellular growth signaling which allows extracellular signals in the form of hormones and growth factors to elicit multiple biological effects within the normal cell. Growth signaling comprises a variety of different inter-connected pathways whose protein components either promote or inhibit cellular growth, differentiation

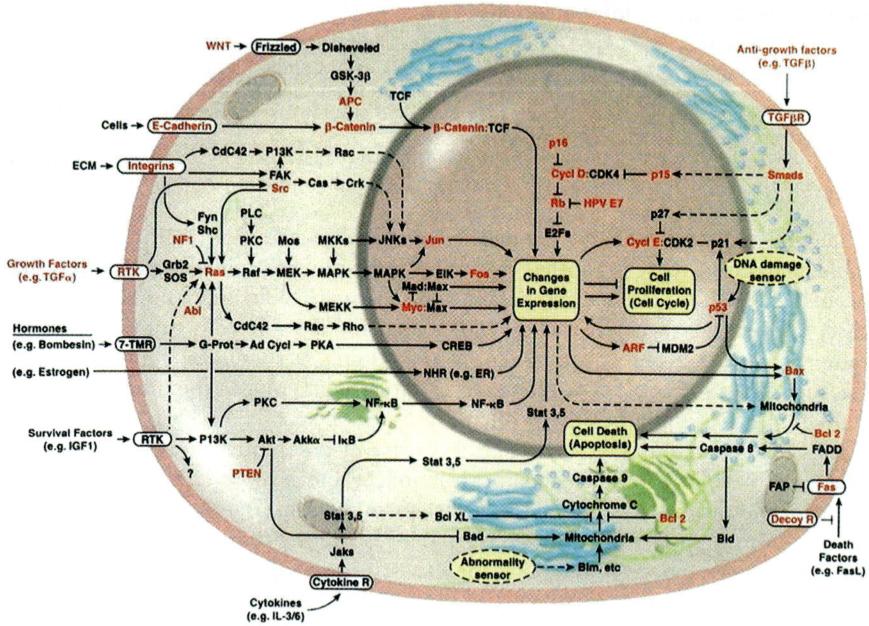

Fig. 1.4. Signaling pathways in the cell. Signaling pathways transduce extracellular stimuli to the interior of the cell, transmit growth promoting and inhibiting signals, and mediate pro- and anti-apoptotic instructions. Genes reported to be functionally altered in human tumorigenesis are highlighted in red. [Reprinted from *Cell*, 100: D Hanahan and RA Weinberg, The hallmarks of cancer, pp57-70, Copyright (2000) by kind permission from Elsevier]

and survival (Figure 1.4). We currently know of perhaps 20 different signal transduction pathways in human cells, each of which ends with a transcription factor that serves to up- or down-regulate the expression of a discrete set of genes (Nebert 2002; Gomperts *et al.*, 2002). By understanding that many of these growth signaling pathways become deregulated during tumorigenesis, we come a long way towards being able to establish the ground rules that govern the transformation of normal cells into malignant tumours. One of the best recent overviews of the molecular basis of cancer is to be found in Hanahan and Weinberg (2000). These authors have viewed tumorigenesis in terms of cells acquiring specific capabilities with these capabilities being properties shared by most, perhaps even all, types of human cancer. These capabilities will now be briefly outlined:

(i) Self sufficiency in growth signals

Normal cells require growth signals before they leave their default quiescent state and begin to proliferate. Such signals, for example in the form of growth factors, are transmitted into the cell via transmembrane receptors. By contrast, tumour cells exhibit a greatly reduced degree of dependency on exogenous growth stimuli. Cells may synthesize their own growth factors thereby creating a positive feedback loop (*autocrine* stimulation). The cell surface receptors that transduce the growth stimulatory signals may be over-expressed (allowing the cancer cell to be abnormally responsive to growth factors) or structurally altered (leading to constitutive ligand-independent signaling). Cellular autonomy with respect to growth signaling may also arise from mutations or changes in the expression of the downstream molecules that are responsible for receiving and transmitting the signals toward their end effectors.

We should not neglect the potential contribution from neighbouring ancillary cells present in a tumour. In normal tissue, cell-to-cell signaling occurs with the signals emanating either from neighbouring cells (*paracrine*) or systemically from a different cellular source (*endocrine*). Under some circumstances, tumour cells appear to be able to co-opt their neighbours into providing a supporting role during tumorigenesis by inducing them to release growth-stimulating signals.

(ii) Insensitivity to growth inhibitory signals

In normal tissue, growth inhibitory signals, acting through transmembrane surface receptors and coupled to intracellular signaling pathways, serve to maintain cellular quiescence. These signals block cellular proliferation either by temporarily removing cells from the cell cycle back into the quiescent (G_0) state or by permanently preventing cells from proliferating by inducing them to differentiate. Clearly, tumour cells must avoid growth inhibitory signals if they are to continue to proliferate. Much of the circuitry that allows normal cells to respond to inhibitory signals is associated with the cell cycle and its regulation; thus, the protein components of this circuitry represent mutational targets during tumorigenesis. Finally, tumour cells may also extinguish the expression of integrins and other cell adhesion molecules whose function is to transmit growth inhibitory signals.

(iii) Evasion of programmed cell death (apoptosis)

If tumour cells are to thrive, they must not only proliferate, but they must also avoid programmed cell death which can be triggered, for example, by over-expression of an oncogene. In tumorigenesis, resistance to apoptosis is acquired both by the loss of pro-apoptotic regulatory molecules and by up-regulation of the genes encoding proteins with anti-apoptotic potential.

(iv) Limitless replicative potential

In order to be able to proliferate indefinitely, tumour cells still need to overcome another barrier – cellular senescence. A variety of normal human cell types have the capacity for some 60-70 replicative generations, implying a kind of cellular clock. This clock takes the form of a natural limit to the progressive shortening experienced by the ends of chromosomes (*telomeres*) with every cycle of replication. This limit can however be circumvented by tumour cells that up-regulate or reactivate an enzyme, telomerase, which maintains telomere length by adding hexanucleotide repeats onto the ends of telomeric DNA.

(v) Sustained angiogenesis

Without adequate oxygen and nutrients, the mass of proliferating cells of newly formed tumour tissue would not survive and so the growth of new blood vessels, *angiogenesis*, becomes essential. Tumour cells may thus induce and sustain angiogenesis by activating an 'angiogenic switch', perhaps by altering the balance between inducers and inhibitors. The up-regulation of the expression of genes encoding inducers of angiogenesis and the down-regulation of the expression of genes encoding inhibitors of angiogenesis represents one strategy to achieve this.

(vi) Tissue invasion and metastasis

Tissue invasion and metastasis allows cancer cells to escape from the primary tumour and to colonize new sites distant from the original tumour where space is less limiting and access to nutrients less restricted. These processes are currently not understood in any detail, but the altered expression of a variety of molecules including extracellular proteases, integrins, cadherins and cell adhesion molecules, probably all play a role.

In their conceptualisation of the various general principles governing tumorigenesis and the processes underlying it, Hanahan and Weinberg (2000) have therefore provided us with a broad framework within which to understand the still fragmentary picture that is emerging of the molecular basis of tumorigenesis. Clearly, a variety of different processes serve to restrain cellular proliferation, including terminal differentiation, senescence and apoptosis, with tumorigenesis only proceeding when these growth inhibitory mechanisms are compromised or inactivated (Figure 1.5).

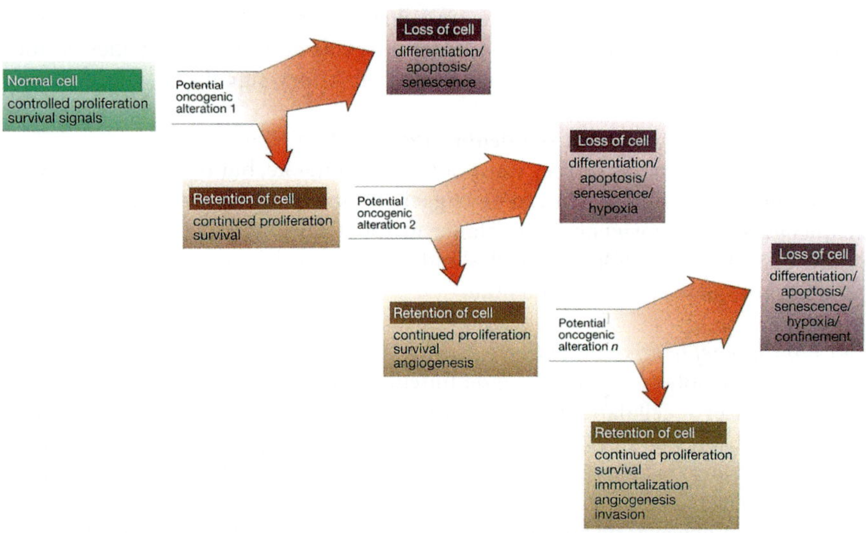

Fig. 1.5. Evolution of a tumour cell population. Potentially oncogenic proliferative signals are coupled to a variety of growth inhibitory processes (e.g. induction of apoptosis, differentiation or senescence) that restrict clonal expansion and neoplastic evolution. Tumour progression only occurs when these growth inhibitory mechanisms are abrogated. Reprinted with kind permission from *Nature* [GI Evan & KH Vousden (2001) Proliferation, cell cycle and apoptosis in cancer. *Nature* 411: 342-348. Copyright (2001), Macmillan Magazines Limited]

Mutational mechanisms in cancer

> "One of the most important developments in genetics over the past decade has been the proof that cancer is, in essence, a genetic disease."
> B Vogelstein and KW Kinzler (1993) The multistep nature of cancer. *Trends Genet.* 9: 138-141.

Cancer is now recognized to be a multi-gene, multi-mutation disorder. Somatic mutations typically occur at a rate of ~10^{-7} per gene per cell, a rate that would appear at first sight to preclude the emergence of multiple mutations in a given cell. However, since the average human possesses some 10^{14} cells and 30,000–40,000 genes in each cell (Venter *et al.*, 2001), it is perhaps surprising that cancer is not more common than it actually is. That such multiple mutations (Boland and Ricciardiello 1999) occur at all, and have the opportunity to promote cellular survival and uncontrolled growth, is due to a variety of different factors:

(i) Mutations that increase the rate of cellular proliferation also serve to increase the mutation rate. Mitogenesis increases mutagenesis because the probability of converting endogenous DNA damage into a stable mutation is increased when cells are dividing (Ames and Gold 1990) and because mutational changes become ever more likely as cells age with repeated divisions (DePinho 2000). It has been estimated that even those tumour cells which do not display microsatellite instability can accumulate as many as one non-synonymous somatic mutation per megabase (Mb) of coding DNA (Wang *et al.*, 2002f), a figure which would translate as ~3,000 mutations per haploid genome for any given neoplastic cell.

(ii) Mutations in 'mutator genes' greatly increase the genome-wide mutation rate (Loeb *et al.*, 2003; *see* Mutator Genes and Genetic Instability).

(iii) In some cell types (e.g. lung, colon), the mutation rate may be greatly increased by intimate and sustained exposure to exogenous carcinogens and/or mutagens (Hussain and Harris 2000).

(iv) Each human cell possesses a variety of potent anti-tumorigenic mechanisms. However, even once oncogenic mutations have occurred, tumours only arise when the mechanisms designed to suppress cellular proliferation or trigger their suicide, fail (Evan and Littlewood 1998).

It must however be remembered that whatever the underlying cause of an increased mutation rate, it is ultimately selection, at the cellular level, for advantageous (growth promoting) somatic mutations that is critical for clonal expansion and tumour growth (Tomlinson *et al.*, 1996). A positive feedback cycle ensues with the growth promoting lesions serving to propagate those cells bearing them, thereby increasing the number of target cells available to acquire further growth promoting mutations. As Evan and Vousden (2001) put it:

> "Cancers 'progress' for the same reason organisms seem to – we see only the successes, not the failures. This distorts our statistical view of cancer progression. No matter how rare the genesis and evolution of a cancer cell or how effective the anti-cancer therapy administered, our perception is only of the rare surviving clones that beat all the odds and appear as clinical disease. Our inability to discern the mechanisms that thwart the vast majority of inchoate tumours deprives us of great insight into how these mechanisms break down in cancer and, correspondingly, how we might best reactivate them".

Pathological mutations in inherited disease and cancer

> "We hear sometimes of cancer-cells and should, therefore, be justified in demanding such information respecting them as will place us in a condition to distinguish a cancer-cell from any other".
>
> C. Wedl (1855) *Rudiments of Pathological Histology*.

Human gene mutations have long been held to belong to one of two distinct categories: those that are of pathological importance on the one hand, and those that may be termed neutral polymorphisms on the other. This somewhat arbitrary distinction has however become blurred with the realization that a sizeable proportion of gene-associated polymorphisms serve to alter the structure, function or expression of their host genes. These "functional polymorphisms" may sometimes be of pathological significance but without in-depth analytical studies, such mutations are often difficult to distinguish from polymorphisms with little or no clinical significance (Brookes 1999; Cooper 2002; Hemminki and Shields 2002).

In terms of pathologically significant mutations, there are in general three types of mutation that give rise to an inherited disease: (i) mutations which lead to a loss of function, (ii) mutations which lead to a gain of function that is deleterious and (iii) dominant negative mutations that adversely affect protein subunit activity or assembly. Characterized gene mutations causing genetic disease have been found to occur within coding sequences, untranslated sequences, promoter and locus control regions, in splice junctions, within introns and in polyadenylation sites [see *Human Gene Mutation Database* at http://www.hgmd.org, an information resource which currently (July 2004) contains details of over 47,000 different inherited mutations in some 1870 different genes]. Indeed, they may interfere with any stage in the pathway of expression from gene to protein product. Mutations may be classified on the basis of whether they lead to the reduced synthesis of a gene product or the synthesis of a structurally/functionally abnormal gene product and secondarily on the basis of whether they affect promoter function, gene structure, RNA processing or translation. Some gene lesions may be placed into more than one category. For example, a missense mutation close to an intron/exon splice junction could affect mRNA splicing efficiency as well as protein structure.

Distinguishing missense mutations of potential pathological significance from neutral polymorphisms is not always straightforward. Evidence for pathological involvement of a given mutation may come from one or more different lines of evidence:

- Location in a protein region of structural or functional importance
- Location in an evolutionarily conserved nucleotide sequence and/or amino acid residue
- Non-conservative substitutions are more likely to disrupt protein function
- Previous independent occurrence in an unrelated patient
- Absence in normal controls
- Novel appearance and subsequent cosegregation of the lesion and disease phenotype through the family pedigree
- Demonstration that the mutant protein has the same properties *in vitro* as its *in vivo* mutant counterpart

- Reversal of the pathological phenotype in patient/cultured cells by gene replacement

In the context of human inherited disease, by far the most frequent genetic lesions in the genome are missense and nonsense mutations and micro-deletions. The remainder comprise a mixed assortment of insertions, deletions, duplications, inversions, sequence amplifications and complex rearrangements (Antonarakis *et al.*, 2001). The same types of lesion are found in association with cancer but their relative prevalence is rather different. For example, translocations and large deletions occur much more frequently in cancer than in inherited disease. Such gross rearrangements are non-randomly distributed in the human genome for a variety of reasons (Mitelman, 2000; Petes, 2001). Firstly, in order to come to clinical attention, somatic gene rearrangements associated with cancer must confer a growth advantage upon the affected cells or tissue, usually via gene deregulation or through the creation of a hybrid gene encoding a tumour-specific fusion protein. This contrasts with the spectrum of germ-line gene rearrangements causing inherited disease; in order to come to clinical attention, these must confer a disadvantage upon the individual, usually through haploinsufficiency. Secondly, gross gene rearrangements are often associated with recombination 'hotspots', DNA sequences that promote either homologous unequal recombination or non-homologous recombination, the two main pathways of double strand break repair. Double strand breaks arise as a consequence of DNA damage but may also occur in programmed fashion during V(D)J recombination or immunoglobulin heavy chain class switching. Whilst the maintenance of genomic integrity requires the accurate repair of double strand breaks, genomic rearrangements may arise through their mis-repair. Deletion breakpoints are AT-rich whereas by comparison, translocation breakpoints tend to be GC-rich (Abeysinghe *et al.*, 2003) implying the action of different mutational mechanisms. A number of recombination-associated motifs have been found to be over-represented at translocation breakpoints (including DNA polymerase pause sites/frameshift hotspots, immunoglobulin heavy chain class switch sites, heptamer/nonamer V(D)J recombination signal sequences, translin binding sites and the χ element) but, with the exception of the translin-binding site and immunoglobulin heavy chain class switch sites, none of these motifs are over-represented at deletion breakpoints (Abeysinghe *et al.*, 2003). This is consistent with a role for homologous unequal recombination in deletion mutagenesis and a role for non-homologous recombination in the generation of translocations.

Mutational spectra in cancer

Kinzler and Vogelstein (2002) posed the question: "Are the DNA alterations in cancer different from those in other genetically determined diseases?" Although these authors then proceeded to list five broad categories of cancer-associated alteration (subtle, chromosome number changes, chromosome translocations, amplifications, and the insertion of exogenous sequences), they rather disappointingly failed to make a formal comparison with inherited disease in terms of either mutation type or relative frequency.

The study of mutational spectra in different oncogenes and tumour suppressor genes is very important as a means to be able (i) to distinguish somatic mutations caused by the action of exogenous mutagens from somatic mutations that have arisen via endogenous mechanisms (Friedberg *et al.*, 2004) and (ii) to relate the type, frequency and location of somatic mutations in a particular gene to a specific gain or loss of function in the protein product (Hussain and Harris 1998; Davidson *et al.*, 2002; Sarasin 2003; Olivier *et al.*, 2004). Empirically, activating point mutations and gene amplification tend to occur in oncogenes whilst tumour suppressor genes tend either to harbour frameshift/truncating mutations or to manifest loss of heterozygosity due to gross deletions. As yet, however, few studies have systematically compared germline and somatic mutational spectra in inherited cancer syndromes. At least as far as the more subtle micro-lesions are concerned, these studies often tend to show that the observed germline and somatic mutational spectra display remarkable similarities in terms of mutation type, relative frequency of occurrence and putative underlying mechanisms of mutagenesis (Ali *et al.*, 1999; Upadhyaya *et al.*, 2004). It may be that the similarities between germline and somatic mutational spectra also extend to gross mutations (Kolomietz *et al.*, 2002; Kost-Alimova *et al.*, 2003).

Krawczak *et al.* (1995) demonstrated that the bulk of the spectrum of somatic single base-pair substitutions in the *TP53* gene strongly resembles that of their germline counterparts seen in other human genes. The latter set of mutations have, however, arisen in a tissue that is usually well protected against exogenous mutagens and carcinogens: the germ cells. Since spectral similarity is strongly suggestive of the involvement of similar mutational mechanisms, it would appear that many *TP53* mutations in the soma have arisen directly or indirectly as a consequence of endogenous cellular mechanisms (probably DNA repair and replication) rather than through the action of exogenous mutagens. The similarity noted between the cancer-associated mutational spectrum of *TP53* and germline gene mutations was consistent with the intriguing idea that cancer might be a critical mediator of negative selection against excessive germline mutation. Sommer (1994) speculated that for such a mediator function to work, there must be a correlation between germline and somatic mutation rates. If specific mutations were to occur that enhanced the rates of both germline and somatic mutation, a consequent increase in the incidence of cancer before the end of the normal reproductive period would serve to militate against their survival. It would then follow that p53 could act as a critical sensor that is built into the genome's molecular warning system and which, through carcinogenesis, "kills the individual and saves the species" (Sommer 1994).

Studies such as these have highlighted the fact that the CpG dinucleotide is a hotspot for mutation in the *TP53* gene, as it is in a wide range of different human genes (Krawczak *et al.*, 1998). Indeed, the rate at which CpG mutates to either TpG or CpA is some 5-fold higher than the basal mutation rate per nucleotide. The putative underlying mechanism is the error proneness of the G:T mismatch repair process that is designed to counteract the frequent methylation-mediated deamination of 5-methylcytosine. This represents a major endogenous cause of gene mutability leading to both human genetic disease and cancer (Zingg and Jones 1997).

Mutation in lung cancer

A wide variety of genes have been shown to harbour somatic mutations in human lung tumour cells or tissue. Genes compromised by such mutations include oncogenes and tumour suppressor genes as well as genes encoding proteins that perform key functions in DNA repair, telomerase activity, apoptosis, and cell cycle regulation. A variety of different types of somatic mutation have also become apparent including missense, nonsense and splicing mutations, micro-deletions and micro-insertions, promoter mutations, gene amplification events, gross deletions, translocations and promoter methylation. The role that these different types of somatic mutation play in lung tumorigenesis is increasingly being understood in terms of their ability to promote cellular growth, to interfere with DNA repair, to confer resistance to apoptosis or to induce cellular transformation, tumour growth, invasiveness, angiogenesis, evasion of host immunity and finally, metastasis. We are also now beginning to determine the temporal order in which these lesions commonly occur and to elucidate the functional consequences of different combinations and permutations of specific gene mutations.

In sharp contradistinction to the above somatic mutations are the considerable number of inherited genetic variants that have been proposed to confer an innate susceptibility to lung cancer (Wu *et al.*, 2004b). Thus, inherited variants in genes encoding xenobiotic metabolising enzymes may play an important role in determining levels of exposure to potential exogenous mutagens and carcinogens. Similarly, inherited variants in genes encoding DNA repair enzymes may confer inter-individual differences in DNA repair capacity that could indirectly modulate lung cancer risk. In the latter case, there is probably going to be some interplay between germline and somatic mutations in that the possession of certain inherited gene variants may well increase the likelihood that particular types of somatic gene mutation will subsequently occur in specific individuals. These inherited variants, which include missense mutations, promoter and untranslated region mutations, gene amplification events and gross gene deletions, occur polymorphically in the general population but, for the reasons briefly outlined, may be associated with an increased propensity to develop lung cancer.

In subsequent Chapters, these somatic and inherited gene variants will be reviewed together with the genes/proteins with which they are associated, and the possible mechanistic pathways through which they exert their influence on lung cancer risk, initiation and progression.

CHAPTER 2

Lung Cancer: Setting the Scene

Introduction

Lung tissue is highly complex with at least 40 different component cell types having been identified (Breeze and Wheeldon 1977). These include airway epithelial cells [e.g. ciliated, mucous (goblet), serous, Clara, brush, basal, intermediate], alveolar cells (type I and II pneumocytes, alveolar macrophages), connective tissue cells (e.g. fibroblasts, interstitial cells, mast cells, eosinophils, neutrophils and lymphocytes), chondrocytes, smooth muscle cells, as well as endothelial and mesothelial cells. By virtue of their function, the lungs are exposed to higher oxygen concentrations than most other tissues whilst increased oxidative stress represents part of the pathology of obstructive lung disease as well as lung cancer. Not surprisingly, lung tissue is protected against oxidants by a number of different antioxidant mechanisms including the superoxide dismutases (Kinnula and Crapo 2003), catalase, glutathione peroxidase, and the glutathione *S*-transferases (Rahman *et al.*, 1999; Kinnula *et al.*, 2004).

Lung cancer tends to arise in the epithelium lining the bronchi or in the fine air sacs at the periphery of the lungs. Lung tumours tend to metastasize widely to lymph nodes in the neck and chest and to the pleura, liver, adrenal glands and bone. Cancer Research UK lists lung cancer as the second most common cancer in men and the third most common in women. According to Cancer Research UK, there are more than 38,000 new cases of lung cancer in the UK every year (http://www.cancerresearchuk.org/aboutcancer/statistics/incidence). Of 154,460 cancer-related deaths in the UK in 1991, some 33,390 (22 %) were from lung cancer, making lung cancer the leading cause of cancer mortality (UK Office for National Statistics; http://www.statistics.gov.uk). Similar lung cancer incidence and mortality data are available from the USA (Jemal *et al.*, 2001; 2003; Greenlee *et al.*, 2001). Whilst the incidence of lung cancer in North American and Western European men may have peaked, comparable data from other parts of the world, and from women suggest that the epidemic of lung cancer is far from over (Janssen-Heijnen and Coebergh 2003; Patel *et al.*, 2004a; Bain *et al.*, 2004; Gasperino and Rom 2004). Nevertheless, the age-adjusted incidence rates of lung cancer in men worldwide still exceed by some two-fold, those among women. Furthermore, male gender appears to be an independent unfavourable prognostic indicator for survival in NSCLC (Visbal *et al.*, 2004).

Over the last 50 years, a plethora of epidemiological studies have (i) confirmed the relationship between tobacco smoke exposure and lung cancer and (ii) documented increases in lung cancer incidence that have paralleled the increase in smoking prevalence (Shields 2000; Proctor 2001; Thun *et al.*, 2002; Alberg and Samet 2003; Doll *et al.*, 2004; Vineis *et al.*, 2004). Although nowadays, most authorities regard tobacco smoking as the major cause of lung cancer responsible perhaps for up to 90 % of cases of the disease, epidemiological data suggest that exposure to coal products, radon gas (Darby *et al.*, 2001) and a number of other chemicals including arsenic, vinyl chloride, nickel chromate, mustard gas and asbestos are also associated with an increased risk of the disease (Schottenfeld 2000; Williams and Sandler 2001). However, air pollution (Cohen 2000; Zmirou *et al.*, 2000; Schottenfeld 2000; Pope *et al.*, 2002; Nafstad *et al.*, 2003), ozone (Palli *et al.*, 2004), radiation (Little 2002; Pierce *et al.*, 2003), diet (Gao *et al.*, 1996; Fortes *et al.*, 2003; Tsai *et al.*, 2003; Shen *et al.*, 2003b; Miller *et al.*, 2004), papillomavirus infection (Cheng *et al.*, 2001c; Zafer *et al.*, 2004), chronic inflammation (Ballaz and Mulshine 2003), a familial history of lung cancer (Ooi *et al.*, 1986; Sellers *et al.*, 1990) and a previous history of non-neoplastic lung disease (Schottenfeld 2000) may also be risk factors. Interestingly, the prevalence of co-morbidity, especially cardiovascular disease (23 %) and chronic obstructive pulmonary disease (22 %), among newly diagnosed lung cancer patients is about twice as high as in the general population (Janssen-Heijnen *et al.*, 1998).

Whilst racial differences in lung cancer incidence exist, these may be in part attributable to differences in socio-economic variables and smoking habits as well as perhaps, to an inherited predisposition (Schottenfeld 2000; Gadgeel and Kalemkerian 2003). In the absence of a clearly understood biological mechanism, the evidence for causation provided by the epidemiological data on smoking remains inherently probabilistic. Employing the categories of causation definition 'necessary and sufficient' and 'sufficient-component', allows one to cite smoking as a cause of lung cancer "only when the existence of unknown deterministic variables is assumed" (Parascandola and Weed 2001). It is therefore of the utmost importance to establish the nature of these deterministic variables with a view to being able to link risk factors mechanistically in a pathway of disease that connects initial mutational trigger to hyperplasia and metaplasia, through dysplasia to lung carcinoma *in situ*, and finally to metastasis. The task I have set myself here is to review the molecular genetics of lung cancer in order for us to see more clearly how we are progressing toward our eventual goal of placing our understanding of the molecular basis of lung cancer on a firm mechanistic footing.

Human lung development

The trachea and the lungs originate from the anterior foregut where an outpocketing of the endodermal epithelium forms the tracheal rudiment. This then buds to form the two primary bronchi that grow, surrounded by mesenchyme, and produce secondary branches that form the basic structural network underlying the lobes of the lung. Several further rounds of branching produce finer branches, termed *bronchioli*, that terminate in alveoli which function in gas exchange with the circulatory system.

The early development of the tubular epithelium that will eventually give rise to the lungs is dependent upon the action of a number of transcription factors that potentiate the 'cross-talk' between endodermally-derived epithelium and the mesoderm (reviewed by Warburton *et al.*, 2000; Minoo 2000; Costa *et al.*, 2001; Cardoso *et al.*, 2001). Fibroblast growth factor 10 (FGF10) expression in the mesenchyme surrounding the tracheal rudiment is required for bronchial bud formation (Bellusci *et al.*, 1997a) and may be under the control of sonic hedgehog (Shh) signaling (Watkins *et al.*, 2003). β-Catenin signalling is required for the formation of the distal airways which mediate gas exchange but not the proximal airways that conduct air (Mucenski *et al.*, 2003). Interestingly, the FGF signaling pathway is also involved in the development of the *Drosophila* respiratory (tracheal) system (Metzger and Krasnow 1999). Other key transcription factors in lung morphogenesis lie downstream of FGF10 and Shh and include thyroid transcription factor 1/Nkx2.1 (Archavachotikul *et al.*, 2002; Reynolds *et al.*, 2003), midkine (Reynolds *et al.*, 2004), the Gli-Krüppel family members Gli2 and Gli3 (Grindley *et al.*, 1997; Motoyama *et al.*, 1998), patched (Bellusci *et al.*, 1997b), the GATA zinc finger family member Gata6 (Yang *et al.*, 2002), the helix-loop-helix factors such as Id1, Id2 and Id3 (Jen *et al.*, 1996), achaete-scute homologue 1 (Borges *et al.*, 1997; Sriuranpong *et al.*, 2002) and Pod-1 (Quaggin *et al.*, 1998), the forkhead box transcription factors Foxf1 (Kalinichenko *et al.*, 2004), Foxp1 and Foxp2 (Shu *et al.*, 2001) and hepatocyte nuclear factor (HNF) 3β (Stahlman *et al.*, 1998), the ETS domain transcription factor ERM/Pea3 (Liu *et al.*, 2003c) and the glucocorticoid and retinoic acid receptors (Malpel *et al.*, 2000). Together, these proteins serve to integrate lung morphogenesis and cell lineage differentiation. Receptors, including fibroblast growth factor receptor (FGFR), epidermal growth factor receptor (EGFR), insulin-like growth factor receptor (IGFR), platelet-derived growth factor receptor (PDGFR) and c-met all play a role in lung morphogenesis (Mendelson 2000) whilst TGFβ receptor exerts an inhibitory influence (Bartram and Speer 2004). Bone morphogenetic protein 4 (BMP4), whose expression is induced by FGF10, contributes to branching by suppressing epithelial proliferation (Zhu *et al.*, 2004b).

A short history of lung cancer research

The first person to document a lung cancer from autopsied material is thought to have been the Italian anatomist, Giovanni Battisti Morgagni (1761), one of the founders of morbid anatomy, who described an *ulcus cancrosum* in the right lung of a 60-year-old man. Subsequent cases reported by others in the 18th Century tended however to be poorly described and may not actually have been lung cancer.

References to lung cancer in the 19th Century medical literature were still fairly sparse (Stokes 1837; Graves 1848), probably a consequence of the comparative rarity of the condition at this time coupled with its under-ascertainment (Arkin and Wagner 1936). Thus, Langhans (1871) was able to begin a research paper with the statement:

> "Primäre Krebse der Bronchien und Trachea sind jedenfalls eine der grössten Seltenheiten". ["Primary cancers of the bronchi and trachea are in any case one of the greatest rarities"].

For a long time, the distinction between primary and secondary tumours of the lung was simply not made. However, a certain J. Bell, publishing in the London *Monthly Journal of Medical Science* in 1846 (cited in Adler 1912), is credited with being the first to diagnose a primary lung tumour. Adler (1912) also credited Josefson and Pfannenstill, publishing in an obscure Stockholm *Festband*, with the realisation that although heredity

> "has always been considered a very potent factor in the etiology of malignant neoplasms in general,.....this [did] not apply to lung tumours".

Härting and Hesse (1879) documented a high frequency of "lymphosarcoma" (lung cancer) in autopsied miners in Schneeberg, Silesia, Germany, a consequence of occupational exposure to radon. Since some 75 % of the miners eventually died from the *Schneeberger Lungenkrebs*, and the disease was confined to those who worked directly in the mines, this was regarded, even at the time, as the first evidence for an exogenous influence on the origin of malignancy. In similar vein, Rottman (1898) later suggested that a cluster of cases of lung cancer amongst tobacco workers in Leipzig might be due to the inhalation of tobacco dust.

For his landmark book, *Primary Malignant Growths of the Lungs and Bronchi*, Adler (1912) painstakingly collated a total of 374 autopsied cases of lung cancer from the literature of the time. He observed a disproportionate number of male over female cases, noted a maximum incidence between the ages of 50 and 60, and discussed the question as to whether trauma and tuberculosis might be important in the etiology of lung cancer. Adler also noted that his cases occurred disproportionately in the right lung and he opined that since

> "the right bronchus is shorter and wider than the left, its course is considerably straighter, it seems natural enough that irritating substances, both chemical and mechanical, are aspirated more easily into the right than into the left bronchus".

Adler has been widely credited with being the first to suggest that lung cancer might be associated with tobacco smoking. However, although he undeniably showed great prescience, close examination of the text of his 1912 volume suggests that he was actually rather more reticent on this subject than some of his laudators have subsequently claimed. Thus whilst Adler was aware that Kurt Wolf had pointed out some 19 years earlier that

> "bronchial carcinomata are nearly always found in those places which are most subjected to slight, but chronic, irritations, especially on the right side and more particularly near the bifurcations".

he pointed out that

> "naturally, all the irritations of aspiration, of dust, tobacco, and so on, as well as coughs, are apt to centre about these points".

However, despite extremely careful documentation of his 374 cases in terms of detailed clinical histories and symptoms, patient age and gender, duration of the illness from diagnosis, autopsy notes, metastases, and family history of other malignant conditions, in only one case did he volunteer that a patient was an 'inveterate smoker'. Adler's reasons for reticence on the topic of smoking remain unknown. He nevertheless hints at his views on the need for careful scientific corroboration and verification in the following explanation for the preponderance of lung cancer cases in males:

> "The domestic life led by women, with their consequent retirement and immunity from the irritations and traumatisms which must be frequent in the more unprotected life of men (the abuse of tobacco and alcohol, the many trades and vocations which are accompanied by irritations of the respiratory organs, etc) has been adduced in explanation of this fact. The entire subject is not yet ready for final judgement".

William George Barnard (1926) performed one of the first histological studies, on an "oat-celled sarcoma of the mediastinum", showing it to be a primary cancer of the lung whilst Graham and Singer (1933) reported the first surgical removal of an entire lung in a case of carcinoma of the bronchus.

In 1926, Philippson associated pipe smoking with oral cancer, pointing the finger of blame at 'tobacco juice' and at one of its constituents in particular, nicotine:

> "In der Zeit, in welcher die lange Pfeife Mode war, es auch Sitte war, die Pfeife nicht ausgehen zu lassen, beobachtete man gelegentlich am Mundwinkel, wo die Pfeife gehalten wurde, ein kleines Geschwür, das sich an der Oberfläche und in die Tiefe verbreitete und als Krebsgeschwür entpuppte. Als rein mechanisches Geschwür wird es wohl niemand ansprechen können. Denn bei Trompetern, Flötisten, Glasbläsern ist dergleichen nie beschrieben worden......Es bleibt als einzige Ursache der Tabakdampf und der sogenannte Schmergel, der Tabaksaft. Man wird wohl nicht fehlgehen, wenn man als eigentliche Krebsursache das Nicotin und seine Zersetzungsprodukte ansieht".
> ["In the days when the long pipe was fashionable and when it was customary not to let the pipe go out, one would occasionally observe a little ulcer at the corner of the mouth where the pipe is held, an ulcer which was growing in width and depth, and which turned out to be cancer. It cannot be regarded as a purely mechanical ulcer since it has never been described in trumpeters, flautists or glassblowers......The only cause remaining is the tobacco smoke and the so-called *Schmergel*, the tobacco juice. It is certainly not wrong to regard nicotine and its decomposition products as the actual cause of cancer"].

Two years later, Lombard and Doering (1928) published a study which claimed that heavy smoking was more common among 217 cancer (including lung cancer) patients than among sex- and age-matched controls. However, the patient and control groups were 75% female and only 5 cases of lung cancer were considered (all of whom were heavy smokers, cf. 20% heavy smokers in the control group). In the following year, Lickint (1929) also noted a correlation between smoking and lung cancer whilst McNally (1932), in reviewing the topic, wrote:

> "Comparing the enormous consumption of cigarettes in 1925-1931 with the increase in pulmonary cancer, one is certainly led to believe that cigarette smoking is an important factor in the increase of cancer of the lungs. In England during the period 1901-1910 the death rate was 1 per 100,000, increasing to 2.3 per 100,000 in 1927".

However, Miller and Jones (1930) presented 32 cases of lung cancer from their own hospital and speculated that the disease might be due to bronchial irritation due to urban dust. Lickint (1935) nevertheless revisited the postulated association between smoking and lung cancer with some conviction and opined that some individuals might be more prone to the carcinogenic effects of inhaled tobacco smoke than others:

> "Abschliessend möchte ich daraus die Lehre ziehen, dass m. E. kein Zweifel mehr daran bestehen kann, dass der Tabakrauch auch eine erhebliche Bedeutung für die Entstehung der Bron-

chialkrebserkrankung im allgemeinen und die auffallende Zunahme dieser Krankheit beim männlichen Geschlecht im besonderen besitzt, und dass wir daraus auch den Schluss ziehen sollten, zum mindesten die Sitte des Tabakrauchinhalierens zu bekämpfen, soweit wir nicht sogar Angehörigen krebsdisponierter Familien vom Rauchen überhaupt abraten müssen!"
["Finally I would draw the conclusion from the aforesaid that in my view there can be no doubt that tobacco smoke is of considerable significance for the development of bronchial cancer in general and for the striking increase in incidence of this disease in the male sex in particular, and that from this we should conclude that we ought at least to fight the habit of inhaling tobacco if indeed we should not discourage the members of families with a predisposition for cancer from smoking altogether!"]

Ochsner and Debakey (1939) discussed the possibility that smoking might cause lung cancer by irritating the bronchial mucosa. In the same year, the earliest controlled study of the relationship between smoking and lung cancer was performed by Franz Hermann Müller (1939), who reported his findings in a now classic paper entitled *Tabakmißbrauch und Lungencarcinom* (Tobacco abuse and lung carcinoma). Müller compared the smoking histories of 86 male lung cancer cases and an identical number of poorly defined controls ("gesunden Männern"), and found that the first group contained a disproportionately greater number of heavy smokers as well as a rather smaller proportion of non-smokers (data on smoking habits were however obtained indirectly and retrospectively via a questionnaire sent to relatives of the deceased). Müller expressed his opinion that tobacco smoking was the single most important factor responsible for the increasing incidence of lung cancer:

"Wir haben somit an einem großeren Krankengut die Frage geprüft, welche Momente für die Zunahme des primären Lungencarcinoms verantwortlich zu machen sind und stellen als wichtigstes Moment das Tabakrauchen heraus".
["Using a larger sample of patients, we have examined the question of which factors can be held responsible for the development of primary lung carcinoma and have singled out tobacco smoking as the most important factor"].

Müller did not however entirely ignore the cases of lung cancer among the non-smokers:

"Daß unter den Lungenkrebskranken auch mäßige Raucher und Nichtraucher anzutreffen sind, hier etwa in 1/3 der Fälle, spricht dafür, daß es außer dem Tabakrauchen noch andere krebserzeugende Momente gibt. In Übereinstimmung hierzu steht die Tatsache, daß in unserem Krankengut die Beamtenberufe und in reiner Atemluft arbeitenden Berufe, wie Kaufleute, Angestellte, Wirte usw. sich fast ausnahmslos unter den starken Rauchern befinden, während sich in den meisten Fällen bei den maßigen Rauchern und Nichtrauchern sonstige, meist berufliche Einwirkungen auf die Atemwege nachweisen lassen".
["The fact that there are also non-smokers among the patients with lung cancer – here approximately 1/3 of all cases – indicates that there are cancer-producing agents other than tobacco. This is supported by the fact that in our patient sample there are civil servants and those working in a clean air environment such as salespeople, white-collar workers, pub landlords etc, a group of people who without exception can be regarded as heavy smokers, whereas there are also those among the patients who are moderate smokers or non-smokers but who are exposed to work-related influences on their respiratory systems"].

Schairer and Schöniger (1943) also performed a case-control study of lung cancer in smokers and non-smokers (using controls from both the general population and men from their own Jena hospital who had died of stomach cancer) demonstrating there to be a disproportionate number of cases of lung cancer in the smoking group. Although reports such as those described were not simply anecdotal, they lacked rigorous statistical support.

Some years were to elapse before the first steps were taken, by three different groups at around the same time, to use proper statistical analysis to put the association between cigarette smoking and lung cancer on a sounder, truly epidemiological footing: Doll and Hill (1950), Wynder and Graham (1950) and Levin, Goldstein and Gerhardt (1950). Wynder and Graham (1950) reported that only 1.3 % of some 605 men with lung cancer were non-smokers (defined as having smoked fewer than one cigarette per day over the preceding 20 years) whereas 51.2 % had smoked more than 20 cigarettes a day over the same period. By contrast, these authors found that 14.6 % of male hospital patients of the same age composition were non-smokers whilst only 19.1 % smoked more than 20 cigarettes a day. They concluded that

> "excessive and prolonged use of tobacco. . ..seems to be an important factor in the induction of bronchiogenic carcinoma".

Very similar results were reported four months later by Doll and Hill (1950) who compared 709 lung cancer patients with 709 non-cancer hospital patients selected so as to be comparable with respect to age and gender, allowing the authors to state that

> "smoking is a factor, and an important factor, in the production of carcinoma of the lung".

Sir Richard Doll went on, however, to study lung cancer in non-smokers (defined as persons who had not consistently smoked for as long as one year at the rate of as much as one cigarette or one grame of tobacco a day) and concluded that up to one in five lung cancer deaths in persons age 25-74 might be attributable to causes other than smoking (Doll 1953).

A study of death certificates in England and Wales from 1928-1945 provided evidence for a dramatic increase in mortality due to lung cancer (Kenneway and Kenneway 1936; 1947) whilst Ochsner and Debakey (1941) found that the increased prevalence of lung cancer correlated well with increased cigarette sales. Grosse (1953), in reviewing the statistics from the Dresden Pathological Institutes, noted a marked increase in the incidence of lung cancer from 0.3 % in 1852 to 5.7 % in 1951. Despite a rearguard action by that doyen of statistical genetics, R.A. Fisher (Fisher 1958a; 1958b), by 1964 the U.S. Surgeon General felt able to state that:

> "Cigarette smoking is causally related to lung cancer in men; the magnitude of the effect of cigarette smoking far outweighs all other factors".

Cytological studies of lung cancer were initiated in the late 1940s (Liebow *et al.*, 1948; Woolner and McDonald 1949; Farber *et al.*, 1951). In 1957, Auerbach *et al.* reported that the relative frequency of basal cell hyperplasia, squamous metaplasia and carcinoma *in situ* were all significantly higher in the lungs of smokers than in non-smokers. These initial findings were confirmed by the same authors (Auerbach *et al.*, 1961) who reported that the degree of cellular abnormality was related to the extent of cigarette smoking. They also reported that 'cells with disintegrating (atypical) nuclei' were found almost exclusively in bronchial epithelial sections from ex-smokers (Auerbach *et al.*, 1962) and that the number of such cells appeared to be related to the extent of the smoking habit. Auerbach *et al.* (1962) noted that the

> "histologic changes in bronchial epithelium in relation to cigarette smoking parallel[ed] epidemiologic findings on lung cancer in relation to cigarette smoke. [The] findings of an increase

in number of cells with atypical nuclei after exposure to cigarette smoke.....provide a reasonable explanation for the epidemiologic findings".

In the absence of cytogenetic techniques, however, the cytological diagnosis of lung cancer remained fairly crude, depending as it did upon the measurement of the size and shape of cells, the 'regularity of nuclear borders', the 'relative preservation of [the] nucleocytoplasmic ratio' and the 'clumping of chromatin' to differentiate between normal and malignant lung cells (Umiker 1957).

In the years prior to the advent of recombinant DNA technology, the genetics of lung cancer susceptibility appeared to be refractory to analysis by conventional methods. Tokuhata and Lilienfeld (1963) noted that lung cancer deaths among the non-smoking first degree relatives of 270 lung cancer cases were 4-fold higher than in the non-smoking first degree relatives of age-, race- and sex-matched controls. Noting this reported familial aggregation of cases, and making a number of assumptions, Burch (1964) estimated that 'the frequency of any particular allele predisposing to lung cancer is likely to be appreciably less than 3 per cent'.

An hereditary influence on murine lung tumorigenesis was first mooted by Lynch (1926) who noted that different mouse strains exhibited widely differing rates of incidence of lung cancer. Although an extensive study of both induced and spontaneous lung tumours in mice suggested to Heston (1942) that multiple genetic and non-genetic factors might play a role, Falconer and Bloom (1961) emphasized the importance of genetic factors over non-genetic factors in influencing susceptibility to induced lung tumours. Indeed, evidence was subsequently obtained, from the crossing of inbred mouse strains, for the existence of a recessive 'pulmonary tumor resistance' gene that appeared to account for most of the observed inter-strain difference in susceptibility (Bloom and Falconer 1964).

In 1976, Haig Kazazian, now Professor of Molecular Biology in Genetics at the University of Pennsylvania, put forward some ideas for studying the genetics of lung cancer, or bronchiogenic carcinoma as he then termed it (Kazazian 1976). Kazazian was much taken with an earlier paper by Kellermann *et al.* (1973) that explored the possible relationship between aryl hydrocarbon hydroxylase (AHH) inducibility and the risk of lung cancer: in a series of patients with lung cancer, 96% exhibited high or intermediate levels of inducibility as compared to 55% of controls whilst low levels of inducibility were noted in 4% of lung cancer patients and 45% of controls. AHH is today better known as cytochrome P450, subfamily 1, polypeptide 1, encoded by the *CYP1A1* gene. As detailed later on, associations between polymorphic variants in the *CYP1A1* gene and the risk of lung cancer are still being explored. Even back in 1976, however, it was apparent to Kazazian that genetic variation at one locus would be insufficient to explain why one person got lung cancer while another did not. He was aware of the epidemiological studies that pointed to smoking being a risk factor but emphasized that all diseases have both genetic and environmental components.

"Few persons [with lung cancer] will have single-gene or single environmental causes. Gradations of risk will be modulated by different genes in combination and different environmental factors. By studying......patients with [this] common disease, we should find common mutant genes that in the homozygous state are rare causes of the disease...... and in the heterozygous state are risk factors."

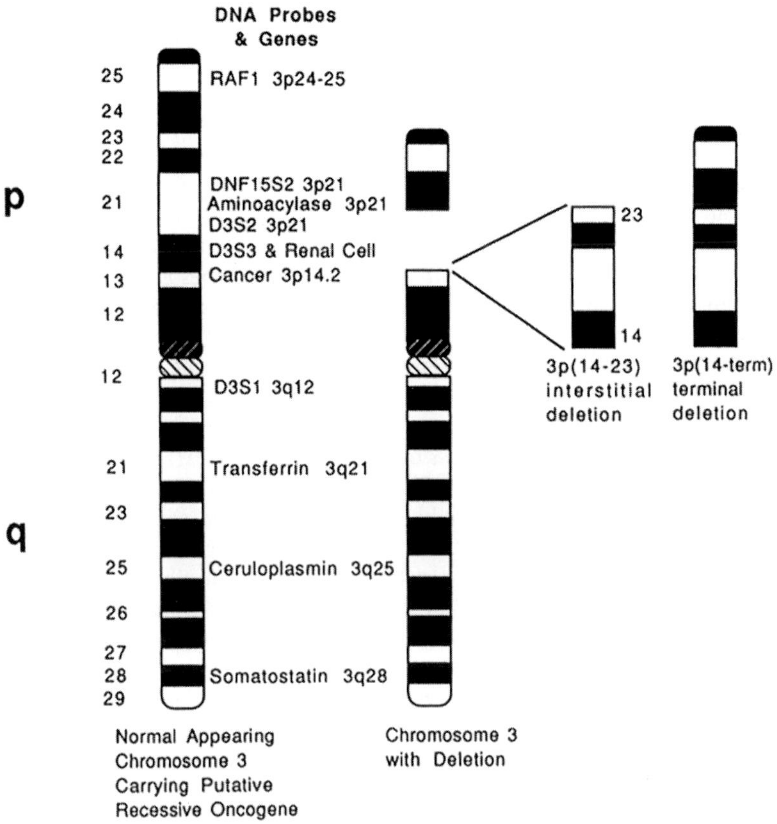

Fig. 2.1. Chromosome 3p deletions in lung cancer. Deletions of the short arm of chromosome 3 are frequently found in both SCLC and NSCLC. Reprinted from *Seminars in Oncology*, 15, MJ Birrer & JD Minna, Molecular genetics of lung cancer, pp226-235, Copyright (1988), with kind permission from Elsevier

The cytogenetic analysis of lung tumour tissue began to report gross karyotypic abnormalities at a very early stage (Davidson and Bulkin 1966; Miles 1967; Falor *et al.*, 1969). It was not, however, until 1982 that the application of higher resolution cytogenetic techniques succeeded in identifying a frequent deletion, at chromosome 3p14-p23, in both human small cell lung cancer cell lines and tumours (Whang-Peng *et al.*, 1982a; 1982b; Figure 2.1). This was held to imply the existence of 'anti-oncogenes', or as they would become known, 'tumour suppressor genes', in the vicinity. Also in 1982, the first missense mutations in a cellular oncogene ('proto-oncogene') were being described in bladder cancer. In the following year, copy number amplification of the *KRAS* oncogene was reported from a human lung cancer cell line (McCoy *et al.*, 1983) followed closely by the demonstration of activating missense mutations in the *HRAS* (Yuasa *et al.*, 1983), *NRAS* (Yuasa *et al.*, 1984) and *KRAS* (Capon *et al.*, 1983), oncogenes also in lung cancer cell lines. The RAS oncogene family was initially identified by virtue of its homology to the transforming elements of certain RNA tumour viruses and by their ability to transform mammalian cells in culture. Ras functions as a cytoplasmic guanine nucleotide binding protein that transduces external signals to the nucleus. It was soon realised that acquired missense mutations in codons 12, 13 and 61, in or near the guanine nucleotide binding domain, were able to confer transforming potential upon the oncogene by locking the Ras protein into a conformation that promoted the continued stimulation of cell growth (Nakano *et al.*, 1984; Santos *et al.*, 1984; Yamamoto and Perucho 1984). At around the same time, the amplification of the *MYC* oncogene was also being described in lung cancer (Little *et al.*, 1983; Saksela *et al.*, 1985) with Taya *et al.* (1984) reporting the amplification of both *KRAS* and *MYC* genes in the same giant cell lung carcinoma cell line. Reports of the amplification of several other oncogenes in human lung cancer cell lines soon followed, including *EGFR* (Hunts *et al.*, 1985), *MYCL1* (Nau *et al.*, 1985) and *MYCN* (Wong *et al.*, 1986; Nau *et al.*, 1986; Saksela *et al.*, 1986).

The first reports of mutated tumour suppressor genes in human lung cancer required however the demonstration of loss of heterozygosity (LOH) at specific chromosomal loci. LOH at chromosome 3p in both SCLC and NSCLC was first noted by Mooibroek *et al.* (1987), closely followed by Brauch *et al.* (1987) and Naylor *et al.* (1987). The subsequent discovery of LOH at 13q and 17p in lung cancer cells was key to identifying the first mutations in (or at least involving) the Rb retinoblastoma (*RB1*; Harbour *et al.*, 1988) and p53 (*TP53*; Takahashi *et al.*, 1989b) tumour suppressor genes. Thus, by 1989, eight of the most important genes involved in lung tumorigenesis had been identified, and the current paradigm that lung cancer is the consequence of the steady accumulation of mutations in multiple genes affecting key pathways governing cellular proliferation and apoptosis, had become established. For the last 15 years, it has been a question of steadily building upon these foundations. Showing considerable prescience, Birrer and Minna (1988) opined that the emerging complexity of the mechanisms underlying the disease suggested that there was

> "no *a priori* reason to believe that these mechanisms contribute equally to every lung tumour and probably vary considerably among patients".

Lung cancer classification, staging, treatment and prognosis

> "Clinical observations, classifications, and theories of cancer extend to the dawn of medical history".
>
> MB Shimkin (1985) *Cancer, Diagnosis, Treatment and Prognosis.*

The initial diagnosis of lung cancer usually depends on clinical evaluation of chronic cough, hoarseness, weight loss, dyspnoea, haemoptysis or chest pain (Kraut and Wozniak 2000). Clinically and histologically, lung cancer can be broadly divided into two distinct categories - small cell lung cancer (SCLC) and non-small cell lung cancer (NSCLC) - which require somewhat different management and treatment regimens (Spira and Ettinger 2004).

SCLC comprises some 20-25 % of lung cancers. It is characterized by a more rapid progression but is relatively sensitive to chemotherapy and radiotherapy although it has often metastasized by the time of diagnosis. In SCLC, metastasis occurs so early on that surgery is seldom employed even in cases with no signs of metastasis at the time of diagnosis. Treatment therefore relies upon combination chemotherapy and concomitant thoracic irradiation and is dependent upon stage. Limited stage disease is treated by chemotherapy and radiation therapy which achieves a cure rate of ~20 %, whilst extensive stage disease is treated primarily with chemotherapy leading to a response rate of 20-30 % (Simon and Wagner 2003). Despite this, however, the median survival time for patients with limited stage disease is ~18 months and for patients with extensive stage disease, ~9 months (Simon and Wagner 2003). In reality, progress in terms of treatment of lung cancer has been fairly minimal (Breathnach *et al.*, 2001) and age-related differences in patient management and survival are evident (Peake *et al.*, 2003).

SCLC tumours often show evidence of neuroendocrine differentiation upon ultrastructural (dense core secretory granules) and immunohistochemical [e.g. CYFRA 21-1 (a cytokeratin 19 fragment), neuron-specific enolase, PGP9.5, bombesin etc] analysis (Coulson *et al.*, 2003). Neuroendocrine properties are not however confined to SCLC. Thus, lung cancer showing characteristics of neuroendocrine differentiation can be roughly divided into four types: classic SCLC, variant SCLC, pulmonary carcinoid, and NSCLC with neuroendocrine features (Junker *et al.*, 2000; Huang *et al.*, 2002). The ectopic production of neuroendocrine peptides has given rise to the hypothesis that SCLC arises from the neuroendocrine cells of the bronchial mucosa. The alternative view, that both SCLC and NSCLC arise from the same pluripotent lung stem cells is however supported by (i) the occasional observation of cells with features of SCLC and NSCLC from within the same biopsy sample, (ii) the observation that some 10-15 % of SCLC cases do not express neuroendocrine products and (iii) that some 10-15 % of NSCLC cases exhibit neuroendocrine properties.

NSCLC, which comprises between 75 % and 80 % of lung cancer, is a morphologically more diverse group of lung cancers including squamous cell carcinoma, large cell lung cancer and adenocarcinoma. Ultrastructurally, these NSCLC tumour types are often quite heterogeneous (Bombi *et al.*, 2002). Squamous cell carcinomas are centrally located in the larger airways and are characterized by extensive keratinization and by intercellular bridging linked by desmosomes. Adenocarcinomas, a tumour type characterized by mucin production, arise peripherally in the smaller airways and constitute the majority of lung tumours in women and non-smokers. In NSCLC, the therapeutic approach is based around surgery because metastases occur later than in SCLC. For patients manifesting the most limited stage of NSCLC at the time of diagnosis, the 5-year cure rate is ~50 %. For NSCLC patients with nonresectable lung cancer who are treated by chemotherapy and irradiation, long term response rates are only 35 % and the median survival time is ~25 weeks.

A common classificatory problem, particularly with adenocarcinomas and squamous cell carcinomas, is intra-tumoral heterogeneity (Desinan *et al.*, 1996; Yesner 2001; Bombi *et al.*, 2002). Tumours are sometimes categorized as 'adenosquamous carcinoma' or as having combined SCLC with NSCLC features, and such assessments may be even more subjective in poorly differentiated tumours. The World Health Organization (WHO) has recently updated its classificatory scheme in order to ensure reproducibility but also to reduce the number of unclassifiable specimens (Travis *et al.*, 2000; Brambilla *et al.*, 2001). It recommends the use of quantitative criteria for tumour classification into histological classes. Thus, in the case of the intra-tumoral heterogeneity manifested by adenocarcinomas and squamous cell carcinomas, if the less frequently observed pattern exceeds 10 % of the tumour, it is classified as adenosquamous, otherwise it is classified by the more frequent pattern. Other changes to the pre-existing WHO classification have been the inclusion of two new types of *pre-invasive lesion* (atypical adenomatous hyperplasia and diffuse idiopathic pulmonary neuroendocrine cell hyperplasia), the recognition of large cell neuroendocrine carcinoma as a sub-class of large cell carcinoma, and the definition of a new class termed carcinoma with pleomorphic, sarcomatoid or sarcomatous elements that is characterized by a range of features of epithelial to mesenchymal differentiation (Brambilla *et al.*, 2001). In the longer term, it is likely that lung cancer classification will be greatly improved and extended by the recruitment of molecular (and genetic) markers. This will be of enormous benefit in helping to reduce the very significant levels of inter-observer variability still evident in histopathological subtyping of lung tumours (Sorensen *et al.*, 1993).

Pre-invasive (or pre-neoplastic) lesions constitute an intermediate morphological phenotype that is considered to link hyperplasia to invasive lung neoplasia (Kerr *et al.*, 2001; Greenberg *et al.*, 2002). Although to some extent plagued by arbitrarily made pathological distinctions, three different types of pre-invasive lesion have been traditionally recognized: atypical adenomatous hyperplasia (AAH) which could be a precursor to adenocarcinoma/bronchioloalveolar carcinoma, squamous dysplasia/carcinoma *in situ* (CIS) which may be a precursor to squamous cell carcinoma, and diffuse idiopathic pulmonary neuroendocrine cell hyperplasia (DIPNECH) which could progress to carcinoid. Ruffini *et al.* (2004) determined the occurrence of pre-invasive lesions in surgical specimens taken following lung resection and concluded that the occurrence of AAH is strongly correlated with adenocarcinoma, CIS with squamous cell carcinoma, and DIPNECH with well-differentiated neuroendocrine tumours. However, these authors did not find that the presence of these pre-invasive lesions had any impact either upon the long-term survival of the patients or on the occurrence of subsequent lung carcinomas in the remaining lobes.

The accurate staging of lung cancer is a prerequisite for the selection of the most appropriate treatment regimen in individual cases, for ensuring uniformity both within and between studies and clinical trials, and for making prognostic predictions. Indeed, the clinical outcome is critically dependent upon stage and tumour type. Staging in NSCLC can be clinical and non-invasive (clinical assessment, radiologic testing, positron emission tomography), or can be pathologically based and employ invasive (bronchoscopy, endoscopic ultrasound needle biopsy or thoracoscopic surgery) techniques. Staging in SCLC is achieved by means of computed tomographic or magnetic resonance imaging scans (Simon and Wagner 2003). The *International System for Staging Lung Cancer* defines the clinical/surgical/pathological stages (0, I-IV) that allow each patient to be classified with respect to the primary tumour (T), regional lymph nodes (N) and metastasis (M) [Mountain 2000; Beadsmoore and Screaton 2003]. Hoffman *et al.* (2000) have clearly summarised the definitions of the various stages of lung cancer as currently used; these will now be summarized briefly:

- Stage 0: carcinoma *in situ*.
- Stage IA: tumour ≤3 cm in greatest dimension (T1), but with no metastasis to regional lymph nodes (N0) and no distant metastasis (M0).
- Stage IB: tumour ≥3 cm in greatest dimension (T2), (N0) and (M0).
- Stage IIA: T1, metastasis to lymph nodes in the peribronchial and/or ipsilateral hilar region (N1), and M0.
- Stage IIB: T2, N1 and M0 or a tumour of any size with direct extension into the chest wall, diaphragm, mediastinal pleura or pericardium without involving the heart, great vessels, trachea or oesophagus (T3), N0 and M0.
- Stage IIIA: T3, N1 and M0 or T1/T2/T3, metastasis to ipsilateral mediastinal lymph nodes and subcarinal lymph nodes (N2), and M0.
- Stage IIIB: tumour of any size with invasion of the mediastinum or involving the heart, great vessels, trachea or oesophagus (T4), N0/N1/N2 and M0, or T1/T2/T3/T4, metastasis to contralateral mediastinal lymph nodes, contralateral hilar lymph nodes, ipsilateral or contralateral scalene or supraclavicular lymph nodes (N3), and M0.
- Stage IV: T1/T2/T3/T4, N0/N1/N2/N3, and distant metastasis present (M1).

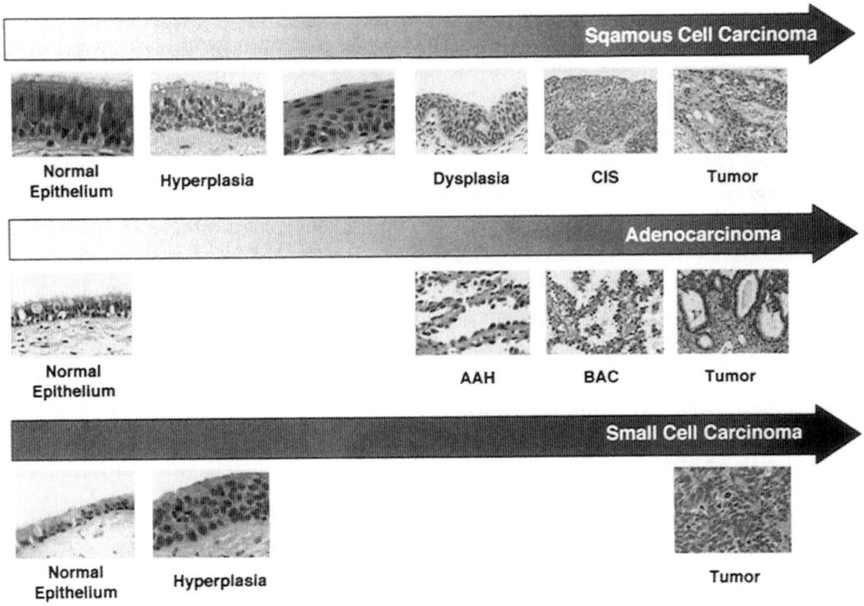

Fig. 2.2. Histopathological and molecular pathogenesis of lung cancer. Histopathological and molecular changes noted during lung tumorigenesis. Sequential changes are thought to be involved in the development of NSCLC (squamous cell carcinoma and adenocarcinoma) whilst SCLC may arise from histologically normal or mildly abnormal epithelium. CIS: Carcinoma *in situ*. AAH: Atypical adenomatous hyperplasia. BAC: bronchioalveolar carcinoma (non-invasive adenocarcinoma). Reprinted from II Wistuba *et al. Oncogene*, 21, 7298-7306 (2002) by kind permission of Nature Publishing Group

The average age of diagnosis of lung cancer is ~60 years suggesting that lung cancer arises and evolves through a series of pathological genetic changes in the respiratory mucosa. In squamous cell carcinoma of the lung, such mucosal changes include hyperplasia, metaplasia, dysplasia and carcinoma *in situ* (Figure 2.2). Hyperplasia and metaplasia are generally considered to be reactive and reversible changes whilst dysplasia and carcinoma *in situ* are the changes most frequently associated with squamous cell carcinoma. The temporal sequence of 'pre-neoplastic lesions' is less well worked out for the other types of lung cancer. Adenocarcinoma may be accompanied by various changes such as atypical adenomatous hyperplasia (Mori *et al.*, 2001), although the malignant potential of these lesions is unclear (Mori *et al.*, 2001; Greenberg *et al.*, 2002; Figure 2.2). It is still not known if SCLC and NSCLC are derived from the same cell lineages.

Familial aggregation of lung cancer

> "In discussing the question of heredity Dr. Adler quotes Josefsen and Pfannenstiel, who found only one case of accredited heredity among their 70 cases. Judging also from a large number of other cases recorded, it is evident that the incidence of malignant growths of the lungs is not seriously affected by hereditary strain."
>
> Review of Adler's *Primary Malignant Growths of the Lungs and Bronchi, The Lancet*, Dec. 14th 1912.

There are as yet no genes whose mutation is pathognomonic for, or even largely confined to, lung cancer, as found for example in cancers of the colon or breast. Perhaps the closest we come to an inherited lung cancer gene is the case of the rare familial cancer syndrome, Li-Fraumeni syndrome. Caused by inherited mutations in the *TP53* gene (Varley 2003), Li-Fraumeni syndrome is characterized by the early onset of various cancers including sarcoma, leukaemia, adrenal cortical carcinoma and cancers of the breast, brain, lung and larynx in affected family members. Nichols *et al.* (2001) have reported that lung cancer became evident at a significantly younger age in Li-Fraumeni syndrome patients bearing germline *TP53* gene lesions than in the general population. Indeed, being a *TP53* mutation carrier has been estimated to carry a 33-fold higher risk of lung cancer (95% CI=13.1-83.1) relative to non-carriers after adjustment for age, sex, smoking habits, and familial cancer (Hwang *et al.*, 2003). Moreover, those Li-Fraumeni syndrome patients carrying germline *TP53* mutations and who smoke, appear to have an increased risk of lung cancer as compared to their counterparts who do not smoke (OR=3.2; Hwang *et al.*, 2003).

Lung cancer has been reported in occasional individuals possessing germline mutations in the *BRCA1* (Johannsson *et al.*, 1996) and *BRCA2* (Boettger *et al.*, 2003) genes. However, this may be no more than coincidence since lung cancer has not been specifically noted in *BRCA1* and *BRCA2* mutation carriers. Indeed, in the lung adenocarcinoma of the patient shown by Boettger *et al.* (2003) to contain a germline frameshift micro-deletion in the *BRCA2* gene, it was the mutant rather than the wild-type allele that was somatically lost. This notwithstanding, Etzel *et al.* (2003) noted a 7-fold increased risk of breast cancer among daughters of lung cancer cases and a 5-fold increased risk of breast cancer among mothers of never smoking lung cancer cases. Carriers (N=144) of germline mutations in the *RB1* gene who survived retinoblastoma, have also been reported to exhibit a greatly increased risk of fatal lung cancer – standardized mortality ratio = 7.01; 95% confidence interval, 3.83-11.76 (Fletcher *et al.*, 2004) although no data on smoking status were available for this cohort. Similarly, carriers (N=240) of germline mutations in the Peutz-Jeghers syndrome (*STK11*) gene that encodes serine/threonine protein kinase 11 have been found to manifest a 7-fold increase in the risk of lung cancer (Lim *et al.*, 2004) although smoking status was again not taken into consideration when arriving at this estimate of risk. It must however be appreciated that although defects in these hereditary cancer syndrome genes can and do predispose towards lung cancer (among other cancers), none of these conditions are common and they are therefore unlikely to contribute significantly to the heritable component of lung cancer risk as far as the general population is concerned.

Familial aggregation of lung cancer is comparatively rare but there are a number of studies in the literature which have suggested that hereditary factors may influence lung cancer risk (Ghadirian *et al.*, 1997; Tomizawa *et al.*, 1998; Schwartz 2000). The first of these was the early study of Tokuhata and Lilienfeld (1963) that reported an excess of mortality due to lung cancer in the relatives of 270 lung cancer patients. Subsequent studies have tended to confirm this finding but few have properly allowed for smoking status in exploring the familial aggregation of lung cancer. In those studies that have been consistent with the existence of a familial association, this has been most apparent for lung cancer in non-smokers (Schwartz *et al.*, 1996; Amos *et al.*, 1999; Schwartz 2000).

In a study of 337 lung cancer patients, Ooi *et al.* (1986) found that the relatives of these patients were 2.4 times more likely to develop lung cancer than the controls (relatives of spouses). Other estimates of the relative risk of lung cancer in relatives of lung cancer patients have been 2.8 (Shaw *et al.*, 1991), 2.5 (Sellers *et al.*, 1988) and, using the Utah Population Database, 2.5 (Cannon-Albright *et al.*, 1994; Goldgar *et al.*, 1994; Thomas *et al.*, 1999). Wunsch-Filho *et al.* (2002) have claimed a mildly elevated risk of lung cancer among persons with a positive family history of lung cancer but the magnitude of a putative interaction between familial aggregation and smoking was small (OR=1.21; 95% CI 0.50-2.92). It has also been reported that the first degree relatives of non-smoking lung cancer cases have a slightly increased risk of cancers of other organs (Brownson *et al.*, 1997; Perkins *et al.*, 1997; Schwartz *et al.*, 1999). More recently, Wood *et al.* (2000) reported that some 11.6% of individuals with lung cancer have a positive family history of the disease. This claim should however be taken with a pinch of salt – chance co-occurrence represents a confounding factor in all studies of familial cancers and should not automatically be taken as evidence for the existence of an autosomal dominant lung cancer gene segregating in the family. The selection criteria used by Wood *et al.* (2000) [at least two first degree relatives with lung cancer, one of which must have been diagnosed before the age of 55] are also insufficiently stringent, and probably fail to allow sufficiently for a shared familial environment (e.g. with respect to smoking behaviour).

Other studies have claimed familial aggregation of lung cancer in relatives of early onset cases of lung cancer. Thus, Schwartz *et al.* (1996) reported that relatives of non-smokers with early onset lung cancer manifested a 6-fold increased risk of lung cancer whilst Kreuzer *et al.* (1998) noted that lung cancer in a first degree relative was associated with a 2.6-fold increase in risk of lung cancer in young adults (<46 years old). Similarly, Bromen *et al.* (2000) reported a 4.7-fold increase in lung cancer risk among the relatives of lung cancer patients <50 years of age.

A family history of lung cancer has been found to be a predictor of lung cancer in non-smoking females (OR=5.7; 95% CI 1.9-16.9) from Taiwan (Wu *et al.*, 2004a). The association was found to be stronger for individuals having a female relative with lung cancer (OR=14.4; 95% CI 2.7-75.5) and for individuals <60 years of age at lung cancer onset (OR=11.2; 95% CI 2.2-56.9). Wu *et al.* (2004a) postulated the existence of a rare autosomal codominant gene lesion that influences the risk of lung cancer in their non-smoking Taiwanese population. This population is extremely interesting because only ~50% of lung cancer cases are associated with cigarette smoking and fewer than 10% of Taiwanese women smoke (Wen Cheng and Lee 2003). The etiology of lung cancer in non-smoking Taiwanese women is therefore still enigmatic although it has been proposed that papillomavirus infection could provide an explanation for both the high prevalence and the familial aggregation of lung cancer in this group (Cheng *et al.*, 2001c; Chiou *et al.*, 2003).

Another large study on the familial aggregation of lung cancer, was that of Etzel *et al.* (2003) which involved 806 lung cancer patients from the M.D. Anderson Cancer Center and 663 controls from the same locality matched for age, sex, ethnicity and smoking status. Family history data were obtained for 6430 first degree relatives of the cases and 4936 first degree relatives of the controls. An increased risk (RR=1.08, p=0.02) for all cancers was observed among the first degree relatives of lung cancer cases. An increased risk of lung cancer was observed among the first degree relatives of lung cancer cases after adjustment for age and the smoking behaviour of the cases and their relatives (RR=1.33, p=0.03). However, the excess risk was confined to relatives of lung cancer cases whose age of onset was >55 (RR=1.6, p=0.006) and was not apparent in early onset (≤55 years) cases (RR=1.07, p=0.06). No evidence of increased risk of lung cancer was found among the relatives of never-smokers. Taken together, these data therefore make a strong case for there being genetic factors that both influence and modulate lung cancer susceptibility.

The largest study to date on the relative influence of environmental and heritable factors on the causation of lung cancer has been reported by Lichtenstein *et al.* (2000). Data on 44,788 pairs of twins (both monozygotic and dizygotic) from the Swedish, Danish and Finnish twin registries were combined in order to assess the risks of a range of cancers, including lung cancer, for the twin siblings of individuals with cancer. Heritable factors were estimated to account for 26% of the variation in susceptibility to lung cancer, shared environmental effects for 12%, and non-shared environmental effects for the remaining 62%. The authors' statistical model therefore predicted the involvement of major environmental factors plus relatively minor genetic components (the rather limited heritability estimated for lung cancer did not actually attain statistical significance) which may or may not interact with these environmental factors.

Although studies of loss of heterozygosity in tumour DNA have implicated certain regions of the genome as being closely associated with lung tumorigenesis, there is as yet little evidence for such regions also being involved in a familial predisposition to lung cancer through the inheritance of specific mutational lesions. One possible exception to this assertion is provided by the putative lung cancer susceptibility locus recently found on chromosome 6q23-q25 by means of a genome-wide linkage analysis performed in a total of 52 families with multiple affected individuals (Bailey-Wilson *et al.*, 2004). In general, however, whilst a positive family history of lung cancer represents a significant risk factor (Sellers *et al.*, 1990; Shaw *et al.*, 1991; McDuffie *et al.*, 1991; Gauderman and Morrison 2000; Bromen *et al.*, 2000), the genetic component still appears to be relatively weak when compared with the influence of environmental factors (Braun *et al.*, 1994; Yang *et al.*, 1997). This is not to say that genetics is unimportant in lung cancer research; rather, that the emphasis should perhaps be placed primarily upon elucidating the nature, underlying causes and timing of the mutational lesions that occur in the soma, and secondarily upon the germline mutations that may be present in the relatively small number of potential lung cancer susceptibility genes and which could confer heightened risk of the disease.

Clues to candidate genes from cytogenetic abnormalities and loss of heterozygosity studies

> "Multipolar mitoses lead to the origin of such cells which have lost their balance. Could this not throw some light on the riddle of tumours?"
>
> T. Boveri (1902) *Verh. Phys. Med. Ges. Würzburg* 35: 67-90.

Conventional cytogenetic analysis has been instrumental in implicating specific chromosomes or chromosomal regions in lung cancer pathogenesis (Lukeis *et al.*, 1990; Testa 1996; Testa *et al.*, 1997; Mitelman *et al.*, 1997; Balsara and Testa 2002; Mitsuuchi and Testa 2002). A complete and up-to-date listing of chromosomal aberrations (including translocations, inversions, deletions and aneuploidies) in lung cancer is available from the *Mitelman Database of Chromosomal Aberrations in Cancer* (http://cgap.nci.nih.gov/Chromosomes/Mitelman).

The consistent loss of a particular chromosomal arm in lung cancer tissue strongly supports the hypothesis that one or more tumour suppressor genes important in lung tumorigenesis reside within that region. Indeed, such cytogenetic changes may be considered as having unmasked an inactivated tumour suppressor allele on the remaining karyotypically-normal chromosome. As a result of such studies, a considerable number of different chromosomal regions have been implicated in either SCLC or NSCLC or sometimes both (Table 1). Conversely, allele gains could indicate the duplication or amplification of certain chromosomal regions and these could be indicative of the involvement of oncogenes, residing in these regions, in lung tumorigenesis. It must however be remembered that amplified chromosomal regions usually contain multiple genes and it is not always straightforward to determine which of these genes, by virtue of its amplification and consequent over-expression, has contributed to the functional and clonal selection of a given cell lineage (Racz *et al.*, 1999).

Loss of heterozygosity in NSCLC

In NSCLC, aneuploidy (DNA content abnormalities) is very common (Zimmerman *et al.*, 1987; Schmidt *et al.*, 1992; Smith *et al.*, 1996; Tosolini and Testa 2000; Choma *et al.*, 2001) and involves virtually all chromosomes. It is however not altogether clear whether these abnormalities are a cause or a consequence of tumorigenesis (Matzke *et al.*, 2003). Cytogenetic studies have nevertheless identified chromosomes 3p, 9p and 12p as those arms most often lost in cells derived from the primary tumour (Lukeis *et al.*, 1990; Neville *et al.*, 1995; Testa 1996; Testa *et al.*, 1997; Suzuki *et al.*, 1998b; Varella-Garcia *et al.*, 1998; Wistuba *et al.*, 1999b; Tosolini and Testa 2000; Figure 2.1). In addition, loss of heterozygosity (LOH) studies using highly polymorphic microsatellite markers have implicated a considerable number of additional regions including 1p, 2q, 4q, 5q, 8p, 9p, 11p13-p15, 11q23-q24, 13, 14, 15q, 16q24, 17q, 18q, 19, 21q, 22q and Yq (Center *et al.*, 1993; Lukeis *et al.*, 1993; Testa *et al.*, 1994; Bepler and Garcia-Blanco 1994; Fong *et al.*, 1995b; Lu *et al.*, 1996a; Shiseki *et al.*, 1996; Cooper *et al.*, 1996; Sato *et al.*, 1998b; Takei *et al.*, 1998; Kohno *et al.*, 1998c; Abujiang *et al.*, 1998; Wistuba *et al.*, 1999a; Wang *et al.*, 1999a; Lerebours *et al.*, 1999; Mendes da Silva *et al.*, 2000; Kurimoto *et al.*, 2001; Chizhikov *et al.*, 2001; Cho *et al.*, 2002; Bepler *et al.*, 2002; Li *et al.*, 2003a; Kee *et al.*, 2003; Lee *et al.*, 2003b;

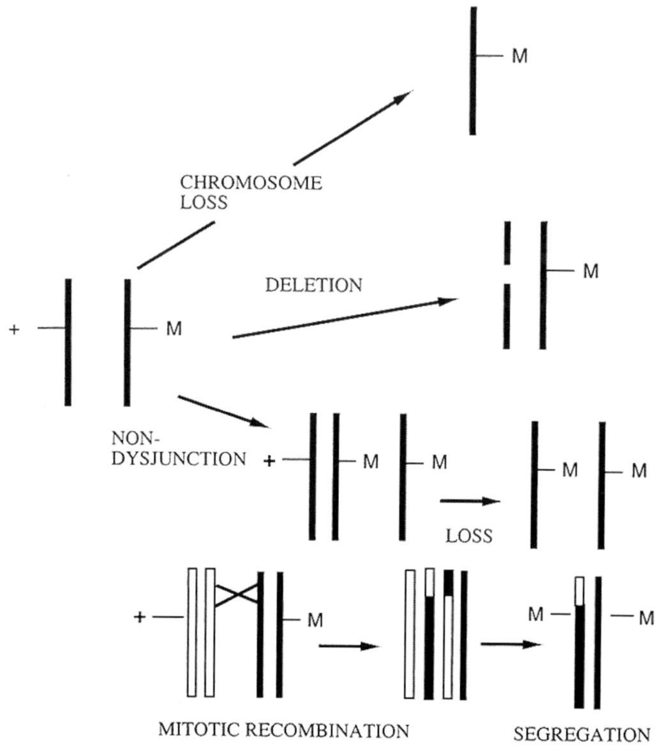

Fig. 2.3. Mechanisms that give rise to loss of heterozygosity (LOH). From a normally heterozygous cell carrying a recessive mutation (M) in one of the tumor suppressor gene (TSG) alleles, the loss of the entire chromosome carrying the normal wild-type gene (+) leads to the exposure of the mutation. Similarly, chromosome deletion, reduplication as a consequence of non-dysjunctional loss, or mitotic recombination between chromatids of homologous chromosomes, can lead to the production of cells lacking both copies of the TSG [Reprinted from Fig. 7 (page 12), Chapter 1, Basic principles in cancer genetics, by JK Cowell in *Molecular Genetics of Cancer* 2nd *Ed.*, (2001) Ed. JK Cowell by kind permission of Taylor & Francis/BIOS, Oxford]

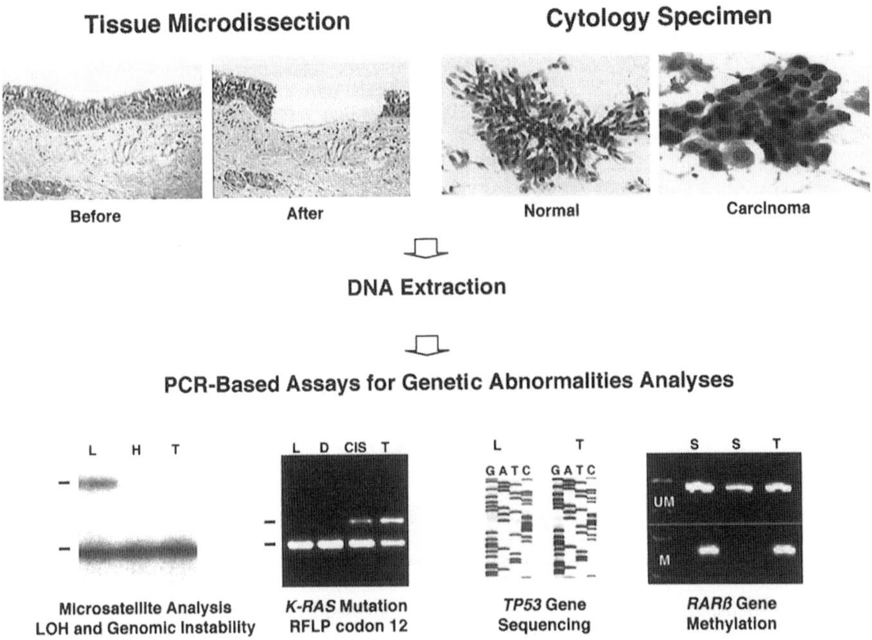

Fig. 2.4. Procedures for investigating molecular lesions in lung cancer. *Upper panel:* tissue microdissection of hyperplastic lesion (left) and examples of cytological specimens (right). *Lower panel:* Molecular analyses of microdissected invasive carcinoma (T), preneoplastic lesions (H, hyperplasia; D, dysplasia; CIS, carcinoma *in situ*) and sputum specimens. Normal lymphocytes (L) are used as a source of constitutional DNA. Typical examples of PCR-based assays for microsatellite (LOH/instability), *KRAS* gene codon 12 mutation, *TP53* mutation sequence, and retinoic acid receptor beta (*RARB*) gene promoter methylation analysis, are shown (UM, unmethylated; M, methylated). Reprinted from II Wistuba *et al. Oncogene* 21, 7298-7306 (2002) by kind permission of Nature Publishing Group

Baksh *et al.*, 2003; Woenckhaus *et al.*, 2003; Shan *et al.*, 2004). LOH may arise through the action of any one of a number of different mutational mechanisms including deletion, recombination and non-disjunction (Figure 2.3). An example of the use of microsatellite markers to detect the loss of alleles at numerous linked loci, the process that constitutes LOH regardless of the precise mutational mechanism involved, is given in Figure 2.4.

Chromosomal gains and losses in NSCLC detected by comparative genomic hybridization

Comparative genomic hybridisation (CGH), a technique which employs competitive fluorescence *in situ* hybridisation (FISH) to detect chromosomal regions that are either amplified or lost (Figure 2.5), has also been used in the chromosomal analysis of lung cancer. This technique is sensitive enough to be able to detect chromosomal deletions as small as 3 megabases (Mb). In a typical experiment, DNA from the normal 'reference' genome might be labelled with a green fluorochrome whereas the tumour ('test') genome could be labelled with a red fluorochrome. The two probes

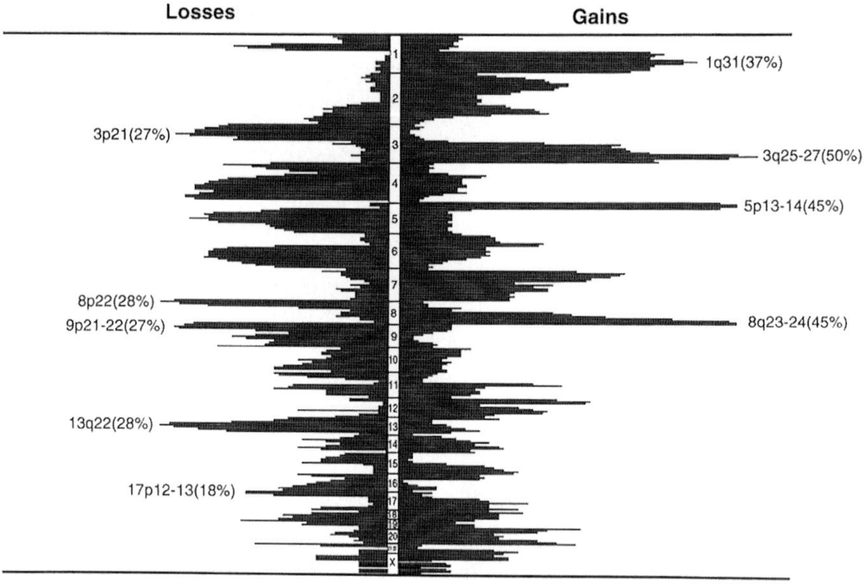

Fig. 2.5. Chromosomal gains and losses in NSCLC as assessed by comparative genomic hybridisation. Genomic imbalances found in a total of 166 NSCLC specimens using comparative genomic hybridisation (CGH). The most frequent sites of gains and losses are indicated with the frequency of affected cases given in parentheses. Reprinted from BF Balsara & JR Testa, *Oncogene* 21, 6877-6883 (2002) by kind permission of Nature Publishing Group

would then be hybridised simultaneously to normal chromosome spreads. Digital image analysis is used to scan the ratio of green to red fluorescence intensity along each chromosome; the relative amount of tumour and reference DNA bound to a given chromosomal DNA region is dependent upon the relative abundance of these sequences in the original DNA samples. If a chromosomal segment has been deleted from the test (tumour) genome, only green fluorescence will be seen on the corresponding segment of the normal chromosomes after hybridisation. If, on the other hand, a duplication of a particular chromosomal segment has occurred, this segment will be indicated by red fluorescence after hybridisation. The use of high-density oligonucleotide microarrays instead of metaphase chromosomes greatly increases CGH resolution (Lindblad-Toh *et al.*, 2000; Kano *et al.*, 2003).

Goeze *et al.* (2002) used CGH to screen 83 lung adenocarcinomas for chromosome imbalances; DNA over-representation was noted at 1q, 8q and 20q whilst deletions were noted at chromosomes 3p, 4q, 6q, 9p, 9q and 13p. However, considerable variability is apparent between CGH studies. Thus, Lindstrom *et al.* (2002) found gains at 11q and 22q in adenocarcinoma, Chujo *et al.* (2002) found gains at 3q, 5p, 8q, 12p and Xq in squamous cell carcinoma, Bjorkqvist *et al.* (1998) found gains at 1q, 3q, 5p, 7p, 8q and 12p in adenocarcinoma and squamous cell carcinoma, Michelland *et al.* (1999) found gains at 1p, 1q, 3q, 5p, 6p, 8q, 12, 17q, 18q, 19, 20 and X, and losses at 2q, 3p, 4, 5q, 8p, 9p, 11, 13q and 17 in NSCLC and neuroendocrine carcinomas. Lu *et al.* (1999) found gains at 1q, 3q, 7p, 8q, 9q, 17q and 20q in NSCLC, Luk *et al.* (2001)

found gains at 1q, 3q, 5p, 8q, 11q, 15q, 19q and 20q in NSCLC, Berrieman *et al.* (2004) found gains on 2q, 3q, 5p, 8q, 11q, 12p and 12q and losses at 3p, 6q, 8p, 9p, 10p, 10q, 17p, 19p and 22q, whilst Pei *et al.* (2001) found gains at 3q, 4q, 7q, 8q and 20p in adenocarcinoma and squamous cell carcinoma. Yokoi *et al.* (2003b) noted losses at 8p, 9p, 13q and 18q and gains of 1q, 2p, 3q, 5p, 6p, 7p, 7q, 8q and Xq in NSCLC whilst gains of 1q, 3q, 5p and 20q and the loss of 5q have been found in various types of NSCLC (Tai *et al.*, 2004). Some consistent similarities between studies are immediately apparent (e.g. gains of 3q, 5p and 8q) but the inherent inter-study variability militates against our being able to draw firm conclusions as to which chromosomes are consistently involved in gains and losses. No significant differences in terms of NSCLC-associated chromosomal gains and losses have been noted between non-smoking women of Caucasian and Chinese origin (Sy *et al.*, 2003).

Finally, CGH detected frequent gains of chromosome 7 in both SCLC and NSCLC (Testa *et al.*, 1994; Yamada *et al.*, 2000), and a gain of 7p in a lymph node metastasis of a NSCLC (Ubagai *et al.*, 2001). Intriguingly, trisomy 7 has been noted in both NSCLC tumour material and surrounding bronchial epithelium of lung cancer patients (Teyssier *et al.*, 1985; Crowell *et al.*, 1996) whilst aneuploidy of chromosome 7 has been reported from NSCLC tissue and associated premalignant lesions (Zojer *et al.*, 2000). It appears unlikely that these findings are attributable solely to the presence of the *EGFR* gene on chromosome 7p12 and so novel oncogenes on chromosome 7, yet to be described, could conceivably be involved in NSCLC.

Loss of heterozygosity in SCLC

In SCLC, cytogenetic studies have identified chromosomes 3p, 5q, 13q and 17p as those chromosomal arms most often deleted in cells derived from the primary tumour (Morstyn *et al.*, 1987; De Fusco *et al.*, 1989; Testa 1996; Testa *et al.*, 1997; Dennis and Stock 1999; Tosolini and Testa 2000; Sattler and Salgia 2003; Figure 2.1) and these are thought to be relatively early events in SCLC tumorigenesis (Endo *et al.*, 1998). In addition, losses of 4q, 5q and 10q are also common (Miura *et al.*, 1992; Petersen *et al.*, 1997; Kim *et al.*, 1998b; Shivapurkar *et al.*, 1999; Cho *et al.*, 2002) whilst loss of heterozygosity (LOH) studies have implicated 8p (Wistuba *et al.*, 1999a), 15q (Kee *et al.*, 2003) and 18q (Takei *et al.*, 1998).

Chromosomal gains and losses in SCLC detected by comparative genomic hybridization

CGH has been used to screen 23 cases of SCLC for chromosomal imbalances (Lui *et al.*, 2001); losses of genetic material were noted at 2q, 3p, 4p, 4q and 13q whilst gains were found at 1p, 2p, 5p, 17q, 19p and 19q. Such findings ought however to be treated with some caution since as with NSCLC, they are not wholly reproducible. Thus, Ashman *et al.* (2002) reported losses of genetic material at 3p, 5q, 10, 16q and 17p and gains at 1p, 1q, 3q and 14q whilst Levin *et al.* (1994b; 1995b) reported gains at 1q, 3q, 5p, 8q and Xq and losses at 3p, 4p, 5q, 8p, 10q, 13q, 16p, 17p and 22q. Ried *et al.* (1994) reported gains at 3q, 5p, 8q and 17q and losses at 3p, 5q, 10q, 13q and 17p. Yamada *et al.* (2000) reported frequent gains of chromosome 8q in SCLC whilst Schwendel *et al.* (1997) reported gains at 3q and 5p and losses at 3p, 4q, 5q, 10q, 13q and 17p (all studies in SCLC tumours).

What do these studies really tell us?

> "Under certain circumstances the delicate mechanism of cell-division may become deranged, and so give rise to various forms of pathological mitoses. The frequency of abnormal mitoses in pathological growths is a most suggestive fact, but it is still wholly undetermined whether the abnormal mode of cell-division is the cause of the disease or the reverse."
>
> E.B. Wilson (1900) *The Cell in Development and Inheritance.* Macmillan, London.

A combination of classical and molecular cytogenetic analysis has demonstrated that chromosomal alterations are very numerous in lung cancer (Tosolini and Testa 2000). Although SCLC and NSCLC appear on the face of it to manifest rather different cytogenetic alterations (Girard *et al.*, 2000), inspection of Table 1 reveals the existence of some chromosomal regions affected in both types of lung cancer. Indeed, Virmani *et al.* (1998) reported allelotyping 54 specific chromosomal regions of which 42 manifested allele loss and 22 of these (52 %) were common between NSCLC and SCLC cell lines. Conversely, different types of NSCLC (viz. adenocarcinomas and squamous cell carcinomas) may differ both with respect to LOH (Ho *et al.*, 2002a) and to gains and losses of genetic material at specific chromosomal loci (Massion *et al.*, 2002; Sy *et al.*, 2004).

Despite these shortcomings, it is widely assumed that the prevalence of LOH represents a rough guide to both the relative importance of a given chromosomal region to lung cancer pathogenesis and the stage of tumorigenesis at which the chromosomal losses are likely to have occurred (Yatabe *et al.*, 2000). Thus, if this assumption is correct, it may be concluded that chromosome arms 3p and 9p are intimately involved in the pathogenesis of NSCLC and occur comparatively early in tumorigenesis (Hung *et al.*, 1995; Kishimoto *et al.*, 1995; Wistuba *et al.*, 1999b; 2000; Zabarovsky *et al.*, 2002; Shan *et al.*, 2004; Senchenko *et al.*, 2004). These losses of heterozygosity appear to occur in chromosomal regions that harbour known tumour suppressor genes (*see* Chapter 3; The Potential Significance of Tumour Suppressor Gene Location in Lung Cancer) whose loss may play a fundamental role in tumour progression (Aoyagi *et al.*, 2001). Some examples of LOH may have prognostic significance and/or could serve as biomarkers for the classification of pathological stage (e.g. Zhou *et al.*, 2000; Bepler *et al.*, 2002).

One interpretational problem with LOH studies is however that cancer cells at an advanced stage often exhibit LOH at a high proportion of those loci tested (e.g. An *et al.*, 2002a). It is therefore by no means clear how much significance to attach to the loss of a given candidate gene in a particular chromosomal region, particularly when multiple candidate genes are screened (Beau-Faller *et al.*, 2003c). Careful study is thus required in order to distinguish tumour type-specific changes from the general background chromosomal instability (CIN) characteristic of most human cancers. CIN is thought to arise as a consequence of a breakdown in cell cycle control (Lengauer *et al.*, 1998; Masuda and Takahashi 2002). The mis-segregation of chromosomes characteristic of CIN may arise as a consequence of defects in the mitotic spindle checkpoint, abnormal centrosome formation, loss of telomere function or failure of cytokinesis (Maser and DePinho 2002; Gisselsson 2003). CIN rarely coexists with microsatellite instability implying that only one type of instability is required to drive tumorigenesis. The structural alteration of chromosomes may also be caused by failure to repair double strand breaks (DSBs) as a result of impaired

DNA damage checkpoints or a faulty DSB repair system. At a molecular level, causes of CIN phenotype acquisition may include *TP53* gene inactivation (Haruki *et al.*, 2001), or mutation in one of a number of other genes involved either in controlling chromosome segregation or the operation of cell cycle damage checkpoints (Cahill *et al.*, 1998; Jallepalli and Lengauer 2001; see below). In NSCLC, CIN represents an independent factor that predicts a poor prognosis (Nakamura *et al.*, 2003b). It however remains unclear to what extent the acquisition of CIN is an early event that then drives tumour progression (Nowak *et al.*, 2002). It is important to realise, however, that

> "CIN does not arise because it accelerates the accumulation of adaptive mutations. Instead, it arises for other reasons, such as environmental factors, and is subsequently finetuned by selection to minimize the time to further cancer progression by means of the inactivation of tumour suppressor genes".
>
> NL Komarova & D Wodarz (2004) The optimal rate of chromosome loss for the inactivation of tumor suppressor genes in cancer. *Proc. Natl. Acad. Sci. USA* 101: 7017-7021.

Clearly, in order to include or exclude the involvement of putative tumour suppressor loci or alternatively to identify more precisely target regions for the identification/positional cloning of candidate tumour suppressor genes, higher resolution studies of the implicated chromosomal region(s) are required. This has usually been achieved by means of (i) determining the *shortest region of overlap* between different deletions, (ii) undertaking searches for (usually rare) homozygous deletions (Todd *et al.*, 1997; Tamura *et al.*, 2002; Senchenko *et al.*, 2004), (iii) performing *loss of heterozygosity* studies using regionally assigned microsatellite markers to search for evidence of allele loss (Lindblad-Toh *et al.*, 2000; Cave-Riant *et al.*, 2002; Wong *et al.*, 2002), (iv) the detection of either more subtle intragenic mutations in specific genes on the cytologically normal counterpart of the deleted chromosome or the loss of expression of this allele through other mechanisms such as DNA methylation, (v) the over-expression *in vitro* of a candidate tumour suppressor gene in lung cancer cells to determine whether or not it induces growth arrest (e.g. *CDKN1B*, Naruse *et al.*, 2000; *WWOX*, Bednarek *et al.*, 2001; *DDX26*, Wieland *et al.*, 2004), (vi) study of the effect of the wild-type candidate tumour suppressor gene in the *in vivo* tumorigenicity assay in nude mice (e.g. *GDF10*; Dai *et al.*, 2004) and finally (vii) the study of the phenotype of transgenic mice deleted for the candidate tumour suppressor gene (e.g. *ROBO1*; Xian *et al.*, 2001).

Practical difficulties with the general approach of searching for LOH are however considerable and include a question mark over the general validity of the 'two-hit' hypothesis (haploinsufficiency may also prove to be important), intra-tumoral heterogeneity, contamination of samples by normal cells, karyotypic complexity (as a consequence of the presence of multiple chromosome copies), the presence of homozygous deletions, functional hemizygosity (due for example to mitotic recombination or chromosomal loss followed by reduplication), gene dosage changes, and artefacts associated with the use of the polymerase chain reaction (PCR) [Tomlinson *et al.*, 2002]. These difficulties notwithstanding, some of the approaches outlined above can nevertheless still result in the identification and characterization of novel candidate tumour suppressor genes (e.g. Kuramochi *et al.*, 2001) or alternatively in the exclusion of genes as candidate tumour suppressors (e.g. Pitterle *et al.*, 1999; Groet *et al.*, 2000). Thus, the 9p21-encoded methylthioadenosine phosphorylase

(*MTAP*) gene was found to be deleted in 19/50 (38%) NSCLC carcinomas, but co-deletion of the closely linked p16 (*CDKN2A*) gene was noted in only half of these cases (Schmid *et al.*, 1998; Mead *et al.*, 1997). This observation led directly to the identification of a novel gene, c86^{fus} (*MTAP*) that lies upstream and in the opposite orientation to the *CDKN2B* gene, and which is frequently deleted in lung cancer (Schmid *et al.*, 2000). Other putative tumour suppressor genes such as *TUSC1* have also been found by microsatellite analysis to be present in a region of chromosome 9p subject to homozygous deletion in both SCLC and NSCLC (Shan *et al.*, 2004). The development of new techniques exploiting a combination of microarray technology and CGH ('microarray-CGH') or 'SNP array hybridization' promises to be extremely useful for genome-wide searches for the DNA copy number abnormalities that are indicative of gene deletion and amplification events in lung cancer (Schwab 1998; Kashiwagi and Uchida 2000; Jänne *et al.*, 2004; Zhao *et al.*, 2004).

Once a candidate tumour suppressor gene has been identified, further functional analysis can be considered e.g. *in vitro* expression of the gene in question in a lung cancer cell line lacking both functional copies. Observation of cell growth suppression or restoration of apoptosis would then lend support to the view that the candidate gene was indeed a tumour suppressor [e.g. *SEMA3B*, Tomizawa *et al.*, 2001; *STK11*, Jimenez *et al.*, 2003; *IGSF4*, Mao *et al.*, 2004]. In the case of the semaphorin 3B (*SEMA3B*) gene, its inducibility by p53 (Ochi *et al.*, 2002), its ability to inhibit lung cancer cell growth and induce apoptosis after reintroduction/re-expression of the wild-type *SEMA3B* gene (Tomizawa *et al.*, 2001), plus its ability to suppress tumour formation in an adenocarcinoma cell line (Tse *et al.*, 2002), together provide compelling evidence that this gene plays a key role in regulating lung cell growth. Finally, the identification of specific domains within a given protein that are critical to the protein's tumour suppressor activity may be accomplished by the introduction of gene constructs containing or lacking particular domains in order to assess whether or not these domains are capable of conferring tumour suppressor activity (e.g. Mao *et al.*, 2003).

The most extensive search for tumour suppressor genes so far conducted has been performed on a 630 kb region of chromosome 3p21.3 which corresponds to a region that displays frequent changes at the earliest stage of lung tumorigenesis (Lerman and Minna 2000). A total of 19 genes were characterized from this region in terms of their structure and expression, and intragenic mutations were sought. Three genes [the E2F-regulated, stress-responsive Blu (*ZMYND10*) gene, *HYAL1* (hyaluronidase 1) and *SEMA3B* (semaphorin 3B)] not only manifested reduced or absent mRNA levels in a Northern analysis of NSCLC and SCLC cell lines, but also exhibited homozygous missense mutations. The sequence analysis of four other genes [*TUSC4* (Npr2L), *TUSC2* (Fus1), *NAT6* (Fus2) and *RBM6*] also revealed missense and nonsense mutations, a frameshift mutation and a ~30 kb deletion (Table 2). This analysis did not however identify a single gene that exhibited a high frequency of mutation in lung cancer. Promoter hypermethylation could not be ruled out since it was not specifically examined during this study. A subsequent study has indicated that the expression of an exogenously introduced wild-type *TUSC2* gene leads to the G_1 arrest and growth inhibition of lung cancer cells, consistent with *TUSC2* being a tumour suppressor gene (Kondo *et al.*, 2001). Finally, Uno *et al.* (2004) have demonstrated that NSCLC cells and tissue frequently display defective myristoylation of the

TUSC2-encoded Fus1 protein: myristoylation of the Fus1 protein is essential for its tumour suppressor activity both *in vitro* and *in vivo*, and its demyristoylation leads to rapid proteasome-dependent degradation. Demyristoylation therefore appears to represent a novel mechanism for tumour suppressor gene inactivation.

Ji *et al.* (2002) studied the effects of the forced expression of three 3p21.3 genes, *TUSC2* (Fus1), *CYB561D2* (101F6) and *HYAL1*, on tumour cell proliferation and apoptosis in human lung cancer cells. The expression of these genes significantly inhibited tumour cell growth, a finding consistent with the presence of multiple contiguous tumour suppressor genes in the 3p21.3 region. Perhaps this is what makes this chromosomal region so important in lung cancer thereby ensuring that 3p consistently crops up in studies of LOH in both SCLC and NSCLC.

The detailed characterization of a 310 kb region of chromosome 11p15.5 (Zhao and Bepler 2001) and a 170 kb region of chromosome 17p13.3 (Konishi *et al.*, 2003), both chromosomal regions whose deletion has been implicated in lung cancer, have also been reported.

CHAPTER 3

Genes Involved in Sporadic Forms of Lung Cancer

> "So we have arrived at.....a genetic paradigm that [provides] a powerful view of cancer. The seemingly countless causes of cancer – tobacco, sunlight, asbestos, chemicals, viruses and many others – all these may work in a single way, by playing on a genetic keyboard, by damaging a few of the genes in our DNA. An enemy has been found and we are beginning to understand its lines of attack".
>
> J. Michael Bishop (1995) Cancer: the rise of the genetic paradigm. *Genes Dev.* 9: 1309-1315.

Some 120 different genes have so far been found to be mutated in human lung tumours or lung tumour-derived cell lines (Table 2). The underlying mutations vary from the subtle (e.g. missense mutations) to the gross (e.g. gene deletions or gene amplification events). Many of these genes represent physiologically plausible candidates for direct involvement in the tumorigenic process and include oncogenes, tumour suppressor genes, apoptosis regulatory genes, telomerase genes and genes involved in DNA repair [reviewed by Fong *et al.* (2002), Osada and Takahashi (2002) and Sekido *et al.* (2003)].

Clearly, however, since multiple gene loci are lost from the chromosomal regions that have been cytogenetically implicated in lung tumorigenesis, it is not always straightforward to identify those genes that are critical for neoplasia to occur and to distinguish them from "bystander loci" that are mutated or ablated as a consequence rather than as a cause of tumorigenesis and which may contribute little or nothing by way of a growth advantage. Even the subtle mutations characterized within gene coding regions may have arisen as a consequence of genome-wide instability and may not necessarily have conferred a growth advantage. Another *caveat* is that it is always possible that some of the mutations described have originated not in the original tumour but rather during subsequent cell culture [*see* Chapter 5; Have Some *TP53* Mutations Occurred During Cell Culture Rather Than in the Tumour?]. In addition, some lesions may have occurred *in vivo* but in response to treatment with anti-tumour drugs e.g. *TOP2A* (Wessel *et al.*, 2002) and *TUBB* (Kelley *et al.*, 2001); such mutations are not included in Table 2.

Table 2 is also very unlikely to be either comprehensive or complete, and new gene loci will need to be continually added to the list as evidence for their involvement in lung cancer pathogenesis emerges. The discussion that follows serves to augment the Table but focuses specifically on the relatively small number of key genes that are either (i) frequently mutated in lung cancer or (ii) already known to be important from studies of other types of tumour or (iii) illustrative of a functional principle that is common to a number of other genes.

Oncogenes

RAS and RAF

The three human *RAS* genes (*KRAS*, *HRAS* and *NRAS*) encode 21 kD proteins which are members of the Ras superfamily of GTPases that play an important role in signal transduction and cell proliferation (Crespo and León 2000). The Ras proteins are localized to the inner surface of the plasma membrane where they serve as a link between the upstream receptor tyrosine kinases EGFR and ErbB2 and downstream serine/threonine kinases. Ras bound to GDP is inactive but can be converted into its active GTP-bound state by guanine nucleotide exchange factors (GEFs) [Figure 3.1]. GTPase-activating proteins (GAPs) return Ras to its inactive state by hydrolysing GTP into GDP. Ras, in its activated state, interacts with its primary effector Raf, a serine/threonine kinase, leading to the activation of MAP kinase kinase (MEK) that in turn activates mitogen-activated protein kinases ERK1 and ERK2. These then translocate to the nucleus and phosphorylate transcription factors such as Ets, myc, Fos and Jun as well as inducing the transcription of cyclin D1 and various growth factors. In practice, there are three functional forms of Raf in humans: C-Raf-1 (also known as Raf-1, and encoded by the *RAF1* gene on chromosome 3p25), A-Raf (*ARAF*; Xp11) and B-Raf (*BRAF*; 7q34).

One way in which lung cancer cells can up-regulate the Ras signal transduction pathway is gene amplification. However, this appears to be a relatively infrequent event in human lung cancer: amplification has so far been noted in $<5\%$ of lung tumours for the *NRAS*, *KRAS* and *RAF1* genes (Table 2).

Studies of a wide variety of cancers have shown that single base-pair substitutions in codons 12, 13 or 61 (in or near the guanine binding domain) of Ras proteins lead to the acquisition of malignant properties. Such activating lesions, which appear to lock the Ras protein into a conformation that promotes cellular proliferation, have also been reported in the *KRAS* gene in at least 20-30 % of NSCLC tumours but are not found in SCLC (Table 2; Mitsudomi *et al.*, 1991a; reviewed by Vachtenheim 1997; Lechner and Fugaro 2000). More specifically, *KRAS* mutations occur almost exclusively in adenocarcinoma (Tsuchiya *et al.*, 1995), the most common lesions being G→T transversions in codon 12 (Rodenhuis and Slebos 1992; Figure 2.4). Intriguingly, the oncogenic activation of K-Ras is frequently associated with over-expression of the mutant allele (Uchiyama *et al.*, 2003).

If highly sensitive PCR-based detection assays are employed, the detected *KRAS* mutation prevalence in NSCLC may be even higher (perhaps up to 50 %) [Mills *et al.*, 1995a; Gao *et al.*, 1997]. By contrast, *HRAS* and *NRAS* missense mutations have been detected only comparatively rarely in lung cancer, in perhaps ~1 % of cases (Table 2; Yuasa *et al.*, 1984; Srivastava *et al.*, 1985; Suzuki *et al.*, 1990; Hajj *et al.*, 1990; Vachtenheim *et al.*, 1995).

An association between rare alleles of a minisatellite variable number repeat polymorphism (VNTR) in the 3' untranslated region of the *HRAS* gene and lung cancer risk has been claimed by a number of groups (Ryberg *et al.*, 1990; Sugimura *et al.*, 1990; Weston and Godbold 1997; Rosell *et al.*, 1999; Lindstedt *et al.*, 1999; Pierce *et al.*, 2000) although in these studies, the type of lung cancer involved has often not been specified. The *HRAS* VNTR is located 1000 bp 3' to the polyadenylation signal of the *HRAS* proto-oncogene. This minisatellite displays a consensus 28 bp repeat

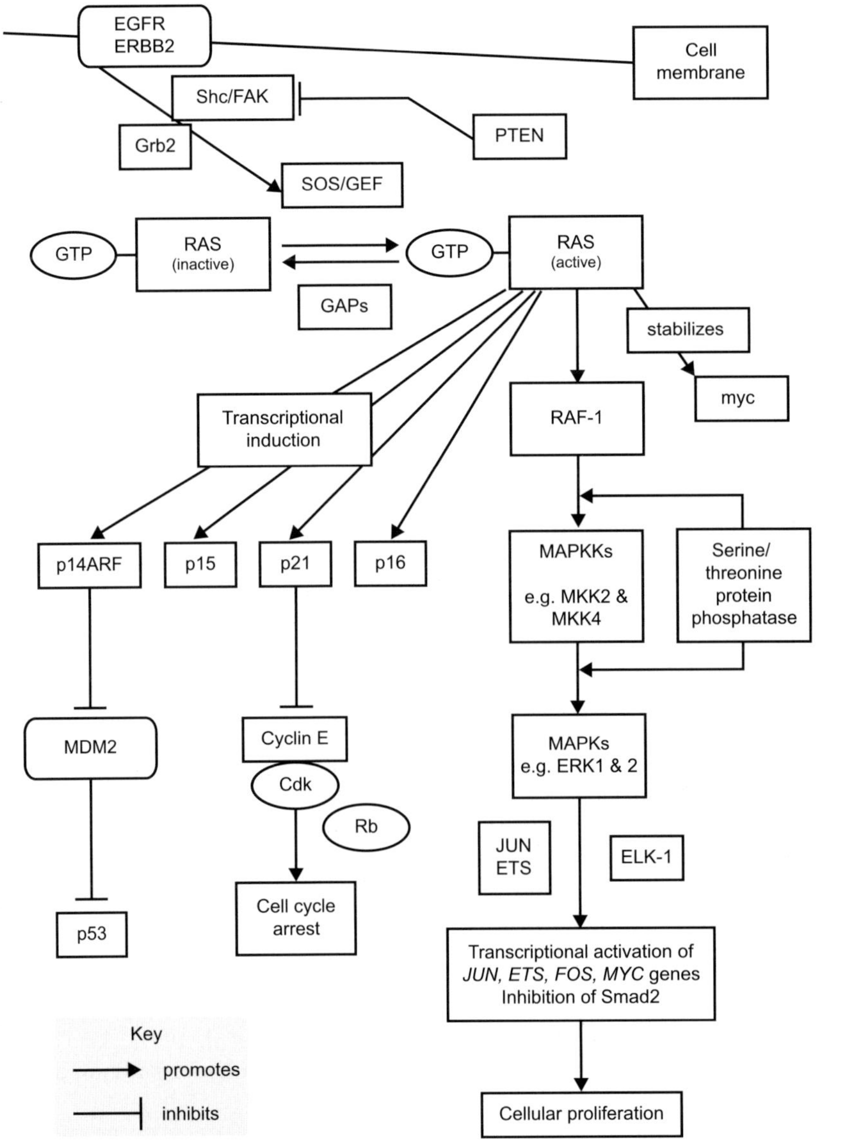

Fig. 3.1. Ras/MAP kinase pathway

unit and occurs as 4 common alleles [varying in size between 1000 and 2500 bp (35-90 repeats) and comprising ~93 % of all alleles], as well as a considerable number of rarer derivative alleles. The *HRAS* minisatellite is known to bind members of the *rel*/NFκB family of transcription factors and the various alleles differ with respect to their enhancer activity over a 5-fold range. It remains unclear, however, whether *HRAS* minisatellite variation constitutes a functional polymorphism in the sense that it exerts a direct effect on the expression of its associated gene, or whether the minisatellite instead represents a 'modifier gene' that can affect the penetrance or expressivity of other loci. Further, some doubt has recently been cast on the conclusions drawn by the original studies. Whilst the mutation rate of the *HRAS* VNTR is much lower than previously estimated, the 'rare alleles' noted in European populations have been found to predominate in Africans (Langdon and Armour 2003). Thus, the original idea of the *HRAS* VNTR being a highly unstable locus at which purifying selection acts against the rare alleles so as to maintain four common alleles in the population may no longer be tenable (Langdon and Armour 2003). Finally, Langdon and Armour (2003) have expressed concern that previous studies reporting an association between rare *HRAS* VNTR alleles and cancer risk may have failed to resolve certain alleles leading to an under-estimation of the rare allele frequency in the population. The proposed association between rare alleles of the *HRAS* VNTR and lung cancer remains therefore to be confirmed or refuted by additional studies.

Ras directly activates PI3K (Rodriguez-Viciana *et al.*, 1996), and *KRAS* mutations appear to enhance the motility of lung adenocarcinoma cells through a pathway involving PI3K/Akt signaling (Okudela *et al.*, 2004). Whether or not this contributes toward non-invasive expansion of lung adenocarcinoma cells is unclear. However, the detection of *KRAS* mutations does appear to be an unfavourable prognostic marker (Slebos *et al.*, 1990; Mitsudomi *et al.*, 1991b; Sugio *et al.*, 1992; Kern *et al.*, 1994; Cho *et al.*, 1997; Fukuyama *et al.*, 1997; Huncharek *et al.*, 1999) although this has not invariably been found (Siegfried *et al.*, 1997a). *KRAS* mutations have also been detected at high frequency in non-malignant lung tissue [e.g. in 29% of non-malignant lung tissue samples as compared to 32% in tumour tissue samples; Nelson *et al.*, 1996] suggesting that *KRAS* mutations may be comparatively early events in tumorigenesis (Cooper *et al.*, 1997). The relative absence of intra-tumour heterogeneity of *KRAS* gene mutations in lung cancer has also been used to argue for *KRAS* mutations being a relatively early event in tumorigenesis (Sagawa *et al.*, 1998). The co-occurrence of *KRAS* and *TP53* mutations has been reported in NSCLC (Mitsudomi *et al.*, 1992) but this appears to be simply the result of the chance coincidence of frequent yet independent and unrelated events.

The majority of *KRAS* mutations, at least in adenocarcinoma, appear to be G→T transversions (Cooper *et al.*, 1997; Siegfried *et al.*, 1997a; Petmitr *et al.*, 2003). Although claims of an association between *KRAS* mutations and tobacco exposure have been made (Slebos *et al.*, 1991; Reynolds *et al.*, 1991; Westra *et al.*, 1993a), Gao *et al.* (1997) reported that the proportion of patients with *KRAS* mutations did not differ between smokers and non-smokers.

Johnson *et al.* (2001) found that somatic activation of the *Kras* oncogene in mice led to an early onset of lung cancer. This was confirmed by Zhang *et al.* (2001) using a strain of *Kras2*-deficient mice. Further, wild-type *Kras2* inhibited colony formation and tumour development by a murine lung tumour cell line containing an activated *Kras2* allele. These data are consistent with the interpretation that *Kras2* is an important tumour suppressor gene in murine lung cancer pathogenesis. In similar vein, Zhang *et al.* (2000c) have shown that the introduction of an anti-*KRAS* ribozyme into human lung cancer cells serves both to inhibit endogenous mutant *KRAS* gene expression and to suppress cellular growth.

MYC

The MYC oncogenes, c-MYC (*MYC*), N-MYC (*MYCN*) and L-MYC (*MYCL1*) encode DNA binding proteins with roles in transcriptional regulation, cell cycle progression and apoptosis (Ryan and Birnie 1996). These proteins possess a transactivation domain at the N-terminus, a nuclear localization signal, and helix-loop-helix and leucine zipper domains at the C-terminus that enable myc to form both homodimers and heterodimers. Myc forms a heterodimer with Max that activates genes involved in growth control (e.g. *CDC25A*) and apoptosis, and represses growth arrest genes (e.g. *GAS1, CDKN1A*). Indeed, Myc may regulate the expression of a wide range of different genes (Patel *et al.*, 2004b).

Myc stimulates the G_1/S transition of the cell cycle by regulating the levels and activity of the cyclins, the cyclin-dependent kinases (cdks) and their inhibitors as well as the Rb-binding transcription factor, E2F (Zajac-Kaye 2001; Figure 3.2). Myc

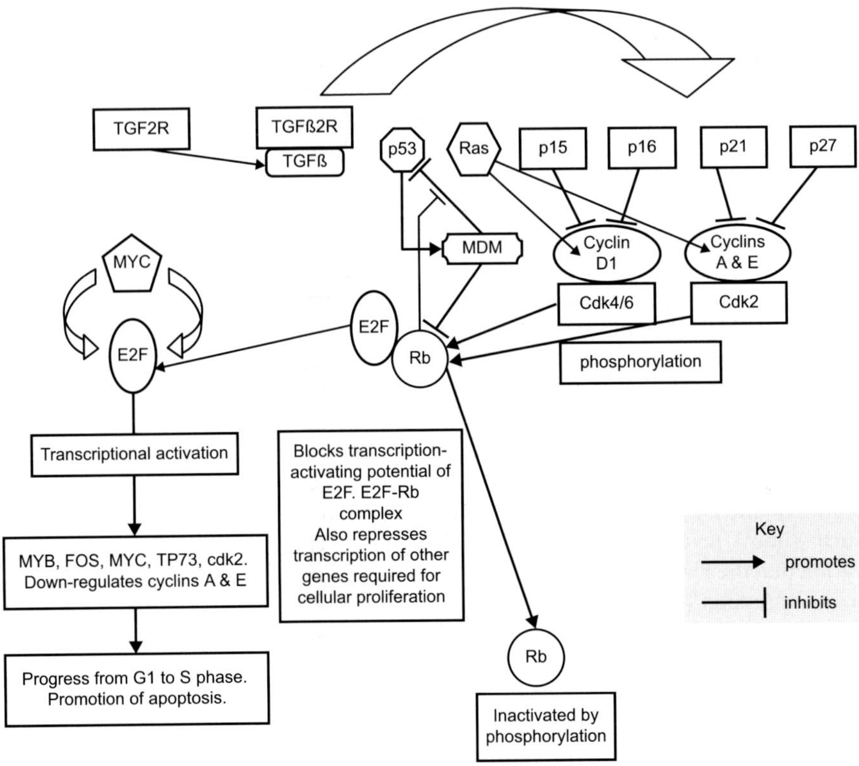

Fig. 3.2. Rb/E2F pathway. Rb is a nuclear phosphoprotein that is tethered to the nuclear matrix. In its hypophosphorylated state during the first part of G1, it binds E2F thereby preventing the transcription factor from activating genes whose products are essential for S phase. Cyclin-cdk complexes phosphorylate distinct sites on Rb thereby releasing E2F and permitting exit from G1 and progress through the cell cycle. Extracellular signals such as p53 or TGFβ stimulate cdk inhibitors p15, p16, p21 and p27 which then bind to and inactivate cyclin/cdk complexes thereby preventing Rb phosphorylation and maintaining the cell in G1 arrest

is also thought to play a role in the induction of apoptosis in the absence of functional p53 (Supino *et al.*, 2001; Nilsson and Cleveland 2003) and acts as a downstream effector of the ErbB2 receptor tyrosine kinase (Hynes and Lane 2001).

In lung cancer, myc usually exerts its transforming potential through gene amplification and consequent over-expression (Little *et al.*, 1983; Prins *et al.*, 1993). The amplification of the *MYCL1* (Nau *et al.*, 1985; Mäkelä *et al.*, 1992) and *MYCN* (Nau *et al.*, 1986; Noguchi *et al.*, 1990) oncogenes in SCLC is only marginally less frequent than *MYC* gene amplification (Johnson *et al.*, 1996; Jänne and Johnson 2000). Indeed, amplification of at least one member of the *MYC* gene family is relatively common in SCLC (25-30 %) but rather less so in NSCLC (< 10 %) [Richardson and Johnson 1993; Johnson *et al.*, 1996; Yokota *et al.*, 1988; Slebos *et al.*, 1989; Jänne and Johnson 2000]. The degree of amplification of these genes usually ranges from between 20 and > 100 copies per cell. The incidence of *MYC* family gene amplification in SCLC has been found to be lower in primary SCLC tumour samples (22 %) than in cell lines (50 %) consistent with it being a relatively late event in lung tumorigenesis (Takahashi *et al.*, 1989a) but also perhaps consistent with it being on occasion a consequence of *in vitro* selection for growth (*see* Chapter 5; Have Some *TP53* Mutations Occurred During Cell Culture Rather Than in the Tumour?). *MYC* family gene amplification has also been found to be more frequent in SCLC cell lines from patients who have been treated with chemotherapeutic agents than in cell lines from untreated patients (Johnson *et al.*, 1996), again consistent with gene amplification being a late event in lung cancer. *MYC* gene amplification is not however an easily interpretable prognostic indicator (Kubokura *et al.*, 2001).

In some cases of SCLC, amplification of the *MYCL1* gene is associated with an intrachromosomal rearrangement that results in the fusion of the *MYCL1* gene to another gene (*RLF*) some 300 kb centromeric to *MYCL1* on the short arm of chromosome 1 (Mäkelä *et al.*, 1995). This serves to generate a novel chimeric transcript with exon 1 of the *RLF* gene being fused to exons 2 and 3 of *MYCL1*. The resulting zinc finger-containing fusion protein may deregulate the tightly controlled expression of the *MYCL1* gene.

In the presence of mitogenic stimuli, deregulation or over-expression of myc induces cellular proliferation by promoting the G_1 to S transition through interaction with E2F and by activating cyclin E/cdk2 complexes (Figure 3.2). In the absence of such stimuli, myc deregulation induces apoptosis. This helps to explain the observation that the majority of cases of SCLC exhibit *TP53* gene mutations as well as *MYC* over-expression (Gazzeri *et al.*, 1994; Bai *et al.*, 1997). Myc activates $p14^{ARF}$ and the p53-dependent pathway of apoptosis and it would appear that inactivation of this $p14^{ARF}$/Mdm2/p53 pathway is required for lung tumorigenesis to occur (Zindy *et al.*, 1998). On the other hand, it would appear as if myc also induces apoptosis in cooperation with Bax employing a mechanism that is p53-independent (Juin *et al.*, 2002; Nilsson and Cleveland 2003). Finally, myc deregulation may also play a role in promoting genomic instability by initiating gene amplification, gene rearrangement and karyotypic instability (Mai and Mushinski 2003).

Elevated levels of expression of *MYC*, *MYCN* or *MYCL1* have been detected in > 80 % of SCLC cell lines and tumours even without gene amplification (Takahashi *et al.*, 1989a). The mechanism underlying this observation is unclear although loss of transcriptional attenuation of the *MYC* and *MYCL1* genes and the antisense expression of the *MYCN* gene have been proposed to be involved (Osada and Takahashi 2002).

Somewhat surprisingly, Barr *et al.* (2000) noted that *MYC* expression suppressed tumour formation by SCLC cells in nude mice. Controversially, these authors proposed that *in vivo*, *MYC* gene expression might serve to suppress tumour formation [mediated they suggested by down-regulation of pro-angiogenic vascular endothelial growth factor (*VEGF*) gene expression] and that the amplification/over-expression of members of the *MYC* gene family in lung cancer cell lines might merely be an artefact of selection for growth *in vitro*. Whilst the latter possibility may have some truth in it, myc actually appears to be capable of inducing VEGF, a finding not inconsistent with the up-regulation of *VEGF* gene expression commonly noted in NSCLC (Table 5). An alternative explanation for myc-mediated tumour suppression could be through a direct apoptotic effect although the precise mechanism involved remains unknown (Juin *et al.*, 2002; Nilsson and Cleveland 2003).

EGFR and ERBB2

Together with Ras, receptor tyrosine kinases are key components of signaling cascades. Important examples are the heregulin (neuregulin) receptor (ErbB2/HER2/neu) encoded by the *ERBB2* gene and the epidermal growth factor receptor encoded by the *EGFR* gene (Gschwind *et al.*, 2004). EGFR regulates epithelial proliferation with EGF, TGFα and amphiregulin as specific ligands. ErbB2 forms heterodimers with other members of the ErbB family and stimulates various signaling pathways including phosphatidylinositol 3-kinase/Akt, JAK-STAT (Figure 3.3; Song *et al.*, 2003) and Ras/Raf/MAPK (Figure 3.1).

The *EGFR* and *ERBB2* genes are both amplified and over-expressed in NSCLC (Tables 2 and 5) whilst the *TGFA* and *AREG* genes, encoding respectively the EGFR ligands TGFα and amphiregulin, have also been reported to be over-expressed in NSCLC (Table 5; Lechner and Fugaro 2000). Over-expression of the *ERBB2* gene is associated with a poor prognosis (Kern *et al.*, 1990).

A different type of mutant EGFR is provided by EGFRvIII, a variant found in a number of different cancers including NSCLC, and caused by the in-frame deletion of exons 2-7 of the *EGFR* gene corresponding to the extracellular ligand-binding domain. It is thought that these frequent somatic rearrangements are brought about by homologous recombination between an *Alu* repeat sequence in intron 7 of the *EGFR* gene and *Alu*-like sequences in intron 1 (Frederick *et al.*, 2000). EGFRvIII displays ligand-independent constitutive autophosphorylation and is inefficiently down-regulated. EGFRvIII constitutively activates PI3K signaling (Moscatello *et al.*, 1998) and promotes the constitutive phosphorylation of extracellular-regulated kinases (ERKs) [Lorimer and Lavictoire 2001]. It appears to enhance the malignant phenotype in SCLC cells (Damstrup *et al.*, 2002). A high incidence of enhanced EGFRvIII expression (16-39 %) has also been noted in NSCLC although this variant is also detectable in normal bronchial epithelium (Garcia de Palazzo *et al.*, 1993; Okamoto *et al.*, 2003). Expression profiling of SCLC cells transfected with the EGFRvIII variant displayed altered mRNAs levels of a number of different genes including *ATF2*, *JUN* and *MYB* (Pedersen *et al.*, 2001).

The genes encoding the two other members of the ErbBB family of receptor-type tyrosine kinases (*ERBB3* and *ERBB4*) are also known to be over-expressed in NSCLC

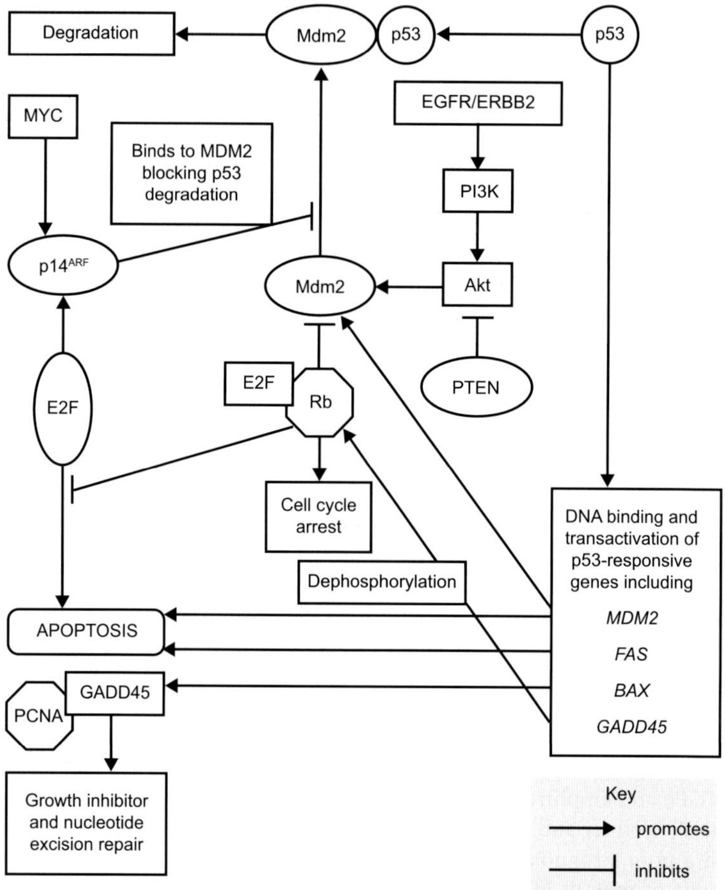

Fig. 3.3. p53 activation and regulation. p53 is responsible for binding to the upstream regions of p53-responsive genes and transactivating them. Mdm2 interacts with p53 thereby inactivating it. E2F up-regulates the expression of $p14^{ARF}$ which binds Mdm2 thereby preventing it from inactivating p53

(Table 5). ErbB3 has been implicated in promoting cell proliferation in lung adenocarcinoma through activation of phosphatidylinositol 3-kinase (PI3K) and Akt (Sithanandam *et al.*, 2003). Phosphorylated Akt promotes cell survival by regulating many of the key effector molecules involved in cell cycle progression, apoptosis and angiogenesis. Further, lung tumour cells display increased activation of the PI3K/Akt pathway, enhanced survival and increased apoptosis in response to inhibition of the pathway (West *et al.*, 2004; Balsara *et al.*, 2004).

MET

The *MET* proto-oncogene has come to prominence by virtue of the germline and somatic mutations that characterize hereditary papillary renal cell carcinoma. The *MET* oncogene encodes a receptor tyrosine kinase that contains extracellular, transmembrane, juxtamembrane and tyrosine kinase domains. MET is the cell surface receptor for hepatocyte growth factor/scatter factor (HGF), a known mitogen for both epithelial cells and many types of tumour cells. MET transduces, via PI3K, HGF-mediated signals that are involved in regulating differentiation, cellular motility, angiogenesis and tissue invasion. Gain-of-function *MET* mutations are known to be associated with increased levels of tyrosine phosphorylation and enhanced kinase activity as compared with the wild-type.

Recently, Ma *et al.* (2003) have also reported missense mutations in SCLC tumours and cell lines. Although the elevated expression/constitutive activation of both MET and HGF has been previously reported in NSCLC (Table 5), Ma *et al.* (2003) showed that the level of *MET* expression did not correlate with the presence of *MET* mutations. Two missense mutations were located in the juxtamembrane domain. When these putative gain-of-function mutations were introduced into an SCLC cell line with minimal endogenous wild-type MET expression, they served to alter cell morphology and adhesion, increased tumorigenicity in *in vitro* assays and increased cellular motility and migration. They were also associated with increased constitutive phosphorylation of several cellular proteins suggesting that the MET/HGF signaling pathway may play an important role in SCLC.

Tumour suppressor genes

TP53

The *TP53* gene, located at 17p13.1, contains 11 exons. It is the most frequently mutated gene in a wide variety of different cancers and is without doubt the best-studied tumour suppressor gene in lung cancer pathogenesis. It encodes a nuclear phosphoprotein, p53, which performs multiple functions in the cell (reviewed by Levine 1997; Hainaut and Hollstein 2000; Woods and Vousden 2001):

(i) p53 is a transcription factor that, once activated by DNA damage and other forms of cellular stress, binds DNA (Figure 3.4) to induce the expression of a large number of genes including *CDKN1A*, *MDM2*, *GADD45A*, *DDB2*, *FAS*, *SFN* (14-3-3σ/stratifin) and *BAX* (Kannan *et al.*, 2000; Hainaut and Hollstein 2000; Amundson *et al.*, 2002; Mirza *et al.*, 2003; Nakamura 2004; Scian *et al.*, 2004; Figure 3.5). Many of these p53-responsive genes contain consensus p53 DNA-binding sites within their regulatory regions (Mirza *et al.*, 2003). p21, encoded by the *CDKN1A* gene, mediates the negative regulation of transcription by p53 (Lohr *et al.*, 2003). p21 also binds proliferating cell nuclear antigen (PCNA) thereby preventing it from acting as a DNA polymerase processivity factor in DNA replication whilst GADD45 competes with p21 for binding to PCNA (Chen *et al.*, 1995) and impedes the function of PCNA in negative growth control (Azam *et al.*, 2001; Figure 3.3). Mdm2 is a negative regulator of both p53 and p21 (Zhang *et al.*, 2004c) whose interaction with p53 is facilitated by YY1 (Sui *et al.*, 2004). p53 is known to mediate arrest of the cell cycle at G_1 and is also involved in regulating the checkpoint at G_2/M. It is the up-regulation of p21 (mediated by LKB1, BRG1, PPARγ and CDX2 among others) and 14-3-3σ (which positively regulates p53) that imposes G_1 and G_2 arrest (Yang *et al.*,

2003). p53 also acts to repress the transcription of various genes including *PCNA*, *RB1* and *NFKB1* (NFκB).

(ii) p53 plays a key role in triggering apoptosis in response to DNA damage, the expression of a cellular oncogene, or the absence of a critical tumour suppressor product such as Rb. p53 controls apoptosis by up-regulating the expression of pro-apoptotic genes such as *BAX* and *FAS* and down-regulating apoptotic inhibitor genes such as *BCL2* and *BIRC5* (survivin).

(iii) Finally, p53 also plays a direct role in DNA replication and repair (Zhou *et al.*, 2001b), acting as a chromatin accessibility factor to bring about global chromatin relaxation (Rubbi and Milner 2003). Three p53-regulated genes, *GADD45A*, *DDB2* and *XPC*, participate in global genomic repair, a sub-pathway of nucleotide excision repair (Amundson *et al.*, 2002). p53 is also thought to play a role in base excision repair (Zurer *et al.*, 2004) and double strand break repair (Kumari *et al.*, 2004).

p53 therefore functions in at least two different ways to maintain genomic integrity: in response to DNA damage, it acts to arrest the cell cycle while the damaged DNA is repaired. However, if that damage is substantial and irreparable, it triggers the cells to undergo apoptosis (Shen and White 2001; Fridman and Lowe 2003; Figure 3.3). When p53 function is impaired, its ability to preserve genomic integrity is inevitably compromised, resulting in a genome-wide increase in the mutation rate both at the chromosomal level and in terms of more subtle lesions at the level of the gene (Morris 2002).

ATM (ataxia-telangiectasia mutated gene), a member of the phosphatidylinositol 3-kinase-related kinase family, plays an important regulatory role in that it phosphorylates p53, thereby abolishing the p53-Mdm2 interaction and allowing p53 to accumulate (Shiloh 2003). ATM also phosphorylates Mdm2 which then interferes with the nucleo-cytoplasmic shuttling of the Mdm2-p53 complex (Maya *et al.*, 2001). Finally, ATM also phosphorylates the tyrosine kinase c-ABL which then contributes to p53 stabilization by phosphorylating Mdm2.

The *TP53* gene is inactivated either by allele loss or more subtle lesions in ~40-60 % NSCLC tumours and ~80-95 % SCLC tumours (Takahashi *et al.*, 1989b; Sameshima *et al.*, 1992; Kishimoto *et al.*, 1992; D'Amico *et al.*, 1992a; Shipman *et al.*, 1996). Such lesions are generally considered to occur at an early stage in lung cancer (Greenblatt *et al.*, 1994) and appear to represent a poor prognostic marker for survival in NSCLC (Mitsudomi *et al.*, 1993; Tomizawa *et al.*, 1999; Niklinska *et al.*, 2000; Huncharek *et al.*, 2000). Conversely, there also appears to be a significant relationship between strong p53 expression and prolonged survival (Tan *et al.*, 2003b). Studies of the relationship between genotype and clinical phenotype have however been more equivocal. Thus, although an association between *TP53* mutation and the degree of differentiation was found for lung adenocarcinoma, none was noted in squamous cell carcinoma of the lung (Zhao *et al.*, 1999). No associations were noted between the presence of *TP53* mutation and histological type or lymph node metastasis (Zhao *et al.*, 1999).

p53 comprises several functional domains. The N-terminus of the protein contains a transactivation domain that is involved in modulating the transcription of target genes. This portion of the molecule also contains a binding site for Mdm2, the main regulator of p53 stability, and a proline-rich region that acts as a protein-binding site and a regulator of apoptosis. At the C-terminus of the molecule lie domains involved in tetramerization, the regulation of DNA-binding activity and nuclear localization. Both the transactivation activity and the pro-apoptotic function of p53 are inhibited by polo-like kinase 1, a marker of cellular proliferation, which binds directly to the tumour suppressor protein (Ando *et al.*, 2004).

The loss or functional inactivation of p53 by mutation [or occasionally by *MDM2* gene amplification; Momand *et al.*, 1998] leads to increased genomic instability which manifests as an increased frequency of gene amplification, aneuploidy, chromosomal rearrangement as well as more subtle lesions some of which may further contribute to the process of tumorigenesis (reviewed by Wang and Harris 1997; Robles *et al.*, 2002). Normally, the centrosome is replicated only once per cell cycle, but cell cycle deregulation consequent to the loss of p53 leads to defective centrosome replication resulting in unequal chromosomal segregation (Fukasawa *et al.*, 1996; Tarapore and Fukasawa 2002). p53 mutation also leads to the loss of the transactivation-independent suppression of homologous recombination, again leading to an increase in genomic instability (Bertrand *et al.*, 2004).

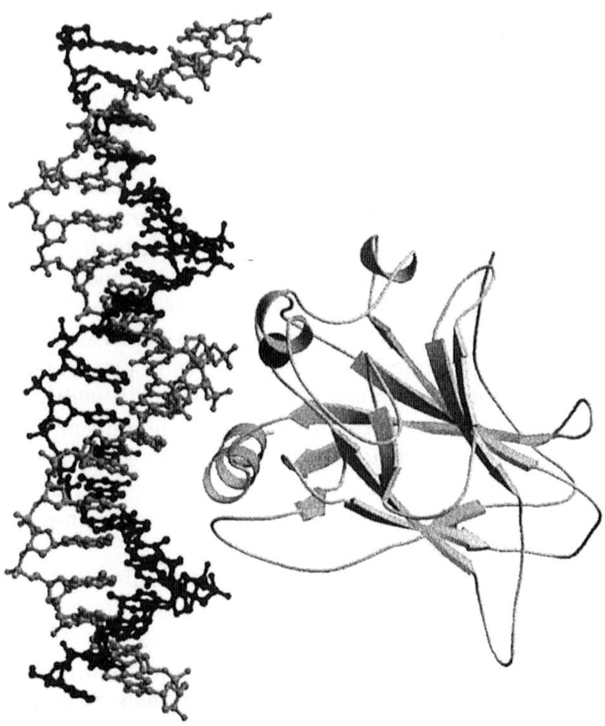

Fig. 3.4. p53 structure. Crystal structure of the core domain of p53 bound to DNA. Reprinted from ACR Martin *et al.* (2002), Integrating mutation data and structural analysis of the *TP53* tumor suppressor protein. *Human Mutation* 19: 149-164. Copyright© John Wiley & Sons Inc. by kind permission of Wiley-Liss, Inc., a subsidiary of John Wiley & Sons, Inc

The majority of characterized somatic *TP53* mutations are single base-pair substitutions (Caron de Fromental and Soussi 1992; Greenblatt *et al.*, 1994; Bennett *et al.*, 1999; Figure 2.4). Of the 2049 subtle lung cancer-associated *TP53* gene mutations listed in version R8 of the IARC *TP53* Mutation Database [http://www.iarc.fr/p53/], 88.4% are single base-pair substitutions, 8.8% are micro-deletions and 1.5% are micro-insertions. The micro-deletions and micro-insertions are readily explicable in terms of 'misalignment mutagenesis' resulting from the repetitive nature of the local DNA sequence environment (Greenblatt *et al.*, 1996). Although much of this mutational spectrum is quite consistent with an endogenous mechanism of mutagenesis (Davidson *et al.*, 2002), the action of exogenous mutagens certainly cannot be excluded. Some 3.9% of mutations listed in the IARC mutation database are 'silent' in that they do not change the amino acid encoded (Hainaut and Hollstein 2000); it is however unclear if some of these silent mutations could affect mRNA splicing efficiency (Takahashi *et al.*, 1990; Sameshima *et al.*, 1990; Holmila *et al.*, 2003) or alternatively whether they are unrelated to any perturbation of p53.

Usually, the majority of mutations in tumour suppressor genes are found to be inactivating (gross deletions, nonsense or frameshift mutations) but *TP53* is an exception to this rule: ~75% of all *TP53* cancer-associated mutations are missense. The probable explanation is that the missense mutations tend to occur disproportionately at 'hotspots' that correspond to specific residues which either lie within the DNA-binding domain of the protein and are involved in making direct contact with DNA, or in residues that are key to protein folding (Hollstein *et al.*, 1996; Skaug *et al.*, 2000; Martin *et al.*, 2002). They are therefore still 'loss-of-function' inactivating mutations even although immunoreactive p53 protein is produced (Top *et al.*, 1995).

Since the activation of transcription of certain p53-responsive genes is critical to the tumour suppressor function of this key transcription factor, missense mutations that interfere with DNA binding will abrogate this role of the protein. Sometimes, this is highly selective. For example, one p53 missense mutant activates the p53-responsive promoter element in the *CDKN1A* gene but not its counterpart in the *BAX* gene (Friedlander *et al.*, 1996). Some p53 missense mutants can derepress the transcription of other genes [e.g. the insulin-like growth factor I receptor (*IGF1R*) gene, thereby promoting mitogenesis (Werner *et al.*, 1996)] whereas other mutants exert a dominant negative effect by reducing the binding of wild-type p53 to the promoters of p53-responsive genes (Willis *et al.*, 2004). Xu and El-Gewely (2003) have reported not only that different p53 mutants generate different downstream expression profiles but also that some genes are responsive to gain-of-function p53 mutants rather than wild-type p53.

Some *TP53* missense mutations induce conformational changes in the p53 protein that serve to stabilize it and prolong its half-life (Bodner *et al.*, 1992; Nishio *et al.*, 1996). How this confers a growth advantage is not altogether clear. Perhaps such gain-of-function p53 mutations (van Oijen and Slootweg 2000) achieve this by up-regulating either a growth inducer such as cyclin G (*CCNG1*; Bates *et al.*, 1996; Smith *et al.*, 1997) or *MYC* (Frazier *et al.*, 1998) or alternatively an anti-apoptotic gene (e.g. *BAG1*, Yang *et al.*, 1999a). This question notwithstanding, the expression of a wild-type *TP53* cDNA in lung cancer cell lines carrying either a homozygous deletion or a missense mutation in their *TP53* genes serves to suppress tumour growth (Takahashi *et al.*, 1992). However, it would appear that p53-mediated growth suppression is due to induction of apoptosis rather than G_1 arrest (Adachi *et al.*, 1996).

The question of whether one or two *TP53* mutations are required to promote tumorigenesis is often poorly articulated. It is, at least in part, a question of the functional consequences of the mutation(s) in question. The classical two-hit model for the mutational inactivation of the *TP53* alleles of this tumour suppressor gene would require two loss-of-function mutations. By contrast, a single dominant negative mutation suffices to produce the same effect, as in the case of the DNA-binding mutants discussed above. Confirmation of the role of dominant negative *TP53* mutations in lung tumorigenesis has come from a transgenic mouse model in which a mutant (His273-bearing) *TP53* gene was expressed under the control of a lung-specific promoter; transgenic mice were found to be significantly more likely to develop lung tumours (Duan *et al.*, 2002). However, haploinsufficiency probably also appears to play a role since gene expression profiling studies have demonstrated that *TP53* gene dosage influences the transcriptional regulation of p53 target genes (Yoon *et al.*, 2002).

Among *TP53* mutations associated with lung cancer, G→T transversions are common (~30 % of the total) and these have been claimed to be attributable to the action of benzo[*a*]pyrene in forming DNA adducts at specific sites (Puiseux *et al.*, 1991; Denissenko *et al.*, 1996; 1997; Kriek *et al.*, 1998; *see* Chapter 5; p53 Mutations, Benzo[*a*]pyrene and Lung Cancer: the Controversy). CpG dinucleotides in the *TP53* gene also represent important somatic mutation 'hotspots' in lung cancer (Pfeifer 2000; Soussi and Béroud 2003). C→T and G→A transitions in these methylated dinucleotides are thought to be due simply to the endogenous process of methylation-mediated deamination of 5-methylcytosine (Rideout *et al.*, 1991; Magewu and Jones 1994; Krawczak *et al.*, 1995; Davidson *et al.*, 2002). Whether DNA methylation also influences benzo[*a*]pyrene adduct formation is however somewhat unclear (Denissenko *et al.*, 1997; *see* Chapter 5; p53 Mutations, Benzo[*a*]pyrene and Lung Cancer: the Controversy). Interestingly, the proportion of single base-pair substitutions that are transitions at CpG dinucleotides is lower in lung cancer (9 % of the total) than in almost all other cancers, presumably due indirectly to the unusually high frequency of G→T transversions (Greenblatt *et al.*, 1994).

The interplay between the short-lived oncogene product Mdm2 and the tumour suppressor gene product p53 is complex but plays an important role in regulating cell growth and apoptosis. Under normal conditions, the *MDM2* gene product, Mdm2, targets p53 for proteasomal degradation but in response to stress, p53 is unaffected by Mdm2 and functions as a transcription factor that induces the transcription not only of the *MDM2* gene but also many other genes involved in both cell death and apoptosis (Vargas *et al.*, 2003; Deb 2003). Mdm2 also regulates p53 by inducing translation of the *TP53* mRNA from two alternative initiation sites to yield two distinct protein products (p53 and p53/47) [Yin *et al.*, 2002]. Translational induction requires Mdm2 to interact directly with the p53 polypeptide. The alternatively translated form of p53, p53/47, lacks both the Mdm2 binding site and the most amino-terminal transcriptional activation domain of p53. Increased expression of p53/47 serves to stabilize p53 in the presence of Mdm2 thereby altering the expression levels of p53-induced gene products. Mdm2 is therefore a key regulator of p53 both by targeting it for degradation and by controlling its synthesis. Some p53 mutants display enhanced transactivating activity in relation to Mdm2 (Ko *et al.*, 2002). In addition, Mdm2 can be stabilized by certain p53 mutations (specifically the 'hotspot' mutations, *see* Chapter 5; p53 Mutations, Benzo[*a*]pyrene and Lung Cancer: the Controversy) potentially allowing Mdm2 to contribute to tumorigenesis through p53-independent mechanisms (Peng *et al.*, 2001). Irrespective of its mode of action, *MDM2* gene expression appears to be a favourable prognostic indicator in NSCLC (Ko *et al.*, 2000).

It should be appreciated that p53 occurs in more than one form (Courtois *et al.*, 2004). These isoforms appear to arise via different mechanisms such as alternative splicing, proteolytic cleavage and, as mentioned above, alternative initiation of translation. Although usually expressed at rather low levels, it is possible that overexpression of one or other of these isoforms could play a role in tumorigenesis.

Under some circumstances, other members of the p53 family, p63 and p73, may to be able to substitute functionally for p53. In common with p53, the various isoforms of p73 induce growth arrest and apoptosis and participate in an autoregulatory feedback loop with Mdm2; Mdm2 is itself transcriptionally activated by p73 and negatively regulates p73 through a mechanism distinct from that involved in p53 inactivation (Zeng *et al.*, 1999). In the presence of mutant p53, p73 retains the ability to transactivate the p21-encoding gene (*CDKN1A*) and suppresses cell growth through the induction of cell cycle arrest and apoptosis (Willis *et al.*, 2003). Similarly, p63 is

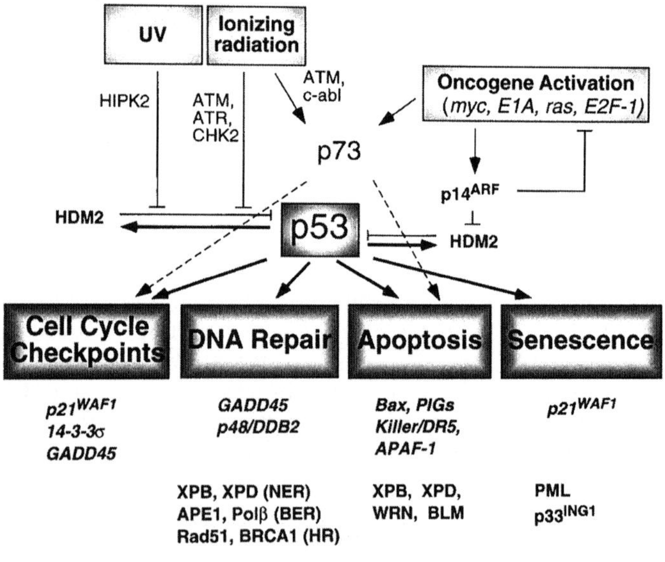

Fig. 3.5. Major pathways of stress-induced p53 stabilization and downstream effects of p53 activation. Genes/proteins involved in each of the signaling pathways controlled by p53 are listed, transcriptional targets on top, functional interactions below. NER: nucleotide excision repair. BER: base excision repair. HR: homologous recombination. The p53 homologue, p73, may perform some of the functions of p53, either redundantly or as a back-up in the case of p53 loss. Reprinted from AI Robles *et al. Oncogene* 21, 6898-6907 (2002) by kind permission of Nature Publishing Group

expressed in at least 6 isoforms that share with p53 an ability to activate transcription and to induce cell cycle arrest and apoptosis (Little and Jochemsen 2002). Our understanding of the role of p63 and p73 in malignancy is however still quite limited (Irwin and Kaelin 2001). There is at least one reported case of a lung cancer cell line which contains a p73 DNA-binding domain mutation in addition to a common gain-of-function p53 mutation (Huqun *et al.*, 2003). Since LOH involving the *TP73* gene occurs fairly frequently in lung cancer, we may surmise that co-inactivation of both *TP53* and *TP73* genes is not an infrequent event.

RB1

The retinoblastoma (*RB1*) gene is an archetypal tumour suppressor gene that is important in regulating the cell cycle during the G_0/G_1 phase (reviewed by Claudio and Giordano 2000; Nevins 2001; Classon and Harlow 2002). The *RB1* gene encodes a nuclear protein (Rb) that, in its hypophosphorylated state, binds to the transcription factor E2F thereby acting as an inhibitor of cell growth (Figure 3.2). When phosphorylated, Rb dissociates from E2F allowing the latter to activate the expression of other genes thereby potentiating cell cycle progression. Rb is a downstream effector of the p53 G_1 arrest pathway and, in its absence, p53 activates the apoptotic pathway. Finally, Rb binds the transcription factor ATF-2 and in so doing up-regulates the expression of TGFβ, leading to growth suppression by inhibiting the interaction between cdk4 and cyclin D1.

The expression of the *RB1* gene is reduced in $>90\%$ of SCLC tumours (Horowitz *et al.*, 1990; Hensel *et al.*, 1990; Shimizu *et al.*, 1994) but in only 14-32% NSCLC tumours (Xu *et al.*, 1991; Reissmann *et al.*, 1993; Shimizu *et al.*, 1994; Tamura *et al.*, 1997; Marchetti *et al.*, 1998a) and is associated with a poorer prognosis (Xu *et al.*, 1994). *RB1* gene inactivation occurs in lung cancer both by deletion and as a result of more subtle lesions (e.g. missense mutations in the functionally critical "pocket" domain) that adversely affect phosphorylation of the protein (reviewed by Kaye and Kubo 2001). In addition, reverse transcriptase-PCR analysis has revealed that ~60% of lung tumour samples exhibit either the absence or a low level of *RB1* mRNA (Gouyer *et al.*, 1998). *RB1* gene promoter hypermethylation does not appear to be the mechanism of inactivation (Gouyer *et al.*, 1998).

Reintroduction of the wild-type *RB1* gene into Rb^- cells leads to growth suppression (Ookawa *et al.*, 1993) whilst experiments employing conditional murine *Rb* mutants have demonstrated that the acute loss of Rb in quiescent cells is sufficient for cell cycle re-entry to occur (Sage *et al.*, 2003). Inactivation of *RB1* leads to rapid growth arrest or apoptosis in $p53^+$ cells but results in cell proliferation in $p53^-$ cells. Loss of Rb function may also lead to genome instability through abrogation of its roles in maintaining chromosome stability, promoting faithful chromosome segregation and in facilitating chromatin remodelling (Zheng and Lee 2002). Many SCLC tumours possess both *TP53* mutations and loss of *RB1* expression, possibly because *TP53* gene mutation allows cells to escape from p53-dependent apoptosis which is itself induced by the inactivation of the Rb pathway. In this context, it is noteworthy that a mouse model for neuroendocrine lung tumours, made by conditional inactivation of both *Tp53* and *Rb1*, manifested a high incidence of lung tumours with morphological and immunohistochemical properties similar to SCLC (Meuwissen *et al.*, 2003).

Loss of Rb function can either occur by *RB1* gene inactivation or by deregulation of other factors responsible for Rb hyperphosphorylation. It would appear as if one component of the p16-cyclin D/cdk4/Rb axis is mutationally inactivated in most types of tumour (Sherr 1996; Betticher *et al.*, 1997b; Tanaka *et al.*, 1998). Whilst expression of the *CDKN2A* and *RB1* genes is inversely correlated (Sakaguchi *et al.*, 1996), the mutational inactivation of these genes appears to be mutually exclusive (Kelley *et al.*, 1995). Since cyclin D inhibits the activity of Rb by stimulating its phosphorylation by cdk4, cyclin D overexpression (Schauer *et al.*, 1994; Betticher *et al.*, 1996) is functionally equivalent to an *RB1* gene lesion. Thus, a given tumour may harbour mutations that lead to inactivation of Rb or p16 or which upregulate cyclin D expression with the net result that Rb and Rb family members are inactivated allowing progression from G_1 into S phase (Kaye 2002). Since elevated cyclin D (*CCND1*) gene expression and reduced *RB1* gene expression have been noted at the resection margins of NSCLC tumours (Betticher *et al.*, 1997a), these changes in expression may be comparatively early events in lung tumorigenesis.

There are two other members of the Rb family: p130 (Rb2) and p107 (Claudio and Giordano 2000). The former is encoded by the *RBL2* gene and is known to be both inactivated (Helin *et al.*, 1997; Claudio *et al.*, 2000) and down-regulated (Baldi *et al.*, 1997; Xue Jun *et al.*, 2003) in lung cancer. The loss of expression of Rb2/p130, a tumour suppressor with functions that are thought to be distinct from those of Rb, is associated with an unfavourable clinical outcome (Caputi *et al.*, 2002).

FHIT

Loss of heterozygosity on the short arm of chromosome 3 is a frequent occurrence in lung cancer. The fragile histidine triad (*FHIT*) gene is a putative tumour suppressor gene located in the FRA3B fragile site region of chromosome 3p14 (Croce *et al.*, 1999). Breakage and rearrangement within the fragile site result in alterations of the structure of the *FHIT* gene and these may be involved in tumorigenesis (Ong *et al.*, 1997; Inoue *et al.*, 1997; Corbin *et al.*, 2002). It has been known for some time that FHIT performs a proapoptotic function (Sard *et al.*, 1999). *FHIT* allele loss or loss of *FHIT* expression has been noted to be frequent (50-70 %) in both SCLC and NSCLC (Sozzi *et al.*, 1996; Fong *et al.*, 1997; Geradts *et al.*, 2000; Veronese *et al.*, 2000; Zöchbauer-Müller *et al.*, 2000; Garinis *et al.*, 2001; Pavelic *et al.*, 2001) whilst intragenic deletions have been found in NSCLC but not in SCLC (Yanagisawa *et al.*, 1996). *FHIT* gene promoter hypermethylation is evident in both SCLC and NSCLC (Table 3). *FHIT* gene hypermethylation was noted in 9 % of the corresponding non-malignant tissues indicating that it can be an early event in lung tumorigenesis (Zöchbauer-Müller *et al.*, 2001a).

Loss of *FHIT* alleles appears to be an indicator of poor prognosis in NSCLC (Burke *et al.*, 1998; Toledo *et al.*, 2004) and is also reportedly associated with p53 over-expression (Lee *et al.*, 20040). The prevalence of *FHIT* allele loss (Sozzi *et al.*, 1997), *FHIT* intragenic deletion (Nelson *et al.*, 1998) and loss of FHIT protein (Sozzi *et al.*, 1998; Pylkkanen *et al.*, 2002) has been claimed to be higher in NSCLC tumours from smokers as compared to non-smokers. Unusual mRNA splice products of this gene have been reported in lung cancer cells with some 80 % of SCLC tumours and 40 % of NSCLC tumours exhibiting abnormal *FHIT* transcripts lacking two or more exons (Sozzi *et al.*, 1996). Although such transcripts may become more prevalent as lung tumorigenesis proceeds (Sato et al., 1999), these transcripts are also present in normal cells (Tokuchi *et al.*, 1999a), suggesting that a role for these transcripts in lung carcinogenesis is unlikely.

Despite an association between *FHIT* gene loss and both increased cellular proliferation and decreased apoptotic potential (Pavelic *et al.*, 2001; Toledo *et al.*, 2004), firm evidence for the direct causal involvement of the *FHIT* gene in lung cancer was until recently lacking, and it was suggested that it might simply represent a marker for 3p deletion. Whilst no consistent effect on cell growth was noted when FHIT$^-$ cancer cells were transfected with the *FHIT* gene in culture, *FHIT* gene replacement did however suppress tumorigenesis in nude mice (Siprashvili *et al.*, 1997). Overexpression of the *FHIT* gene has also been found to inhibit cellular proliferation and to induce apoptosis in adenovirus-*FHIT*-transduced human lung cancer cells (Ji *et al.*, 1999). More convincingly, however, the adenovirus-mediated reintroduction of the *FHIT* gene into FHIT$^-$ lung cancer cell lines served both to suppress tumorgenicity and to induce apoptosis (Roz *et al.*, 2002). Finally, the pathway through which FHIT induces apoptosis and inhibits cell proliferation has recently been shown to involve the tyrosine phosphorylation of the FHIT protein by Src protein kinase (Pekarsky *et al.*, 2004).

PTEN and PIK3CA

PTEN (a protein with homology to protein tyrosine phosphatases and tensin) is a tumour suppressor gene that is frequently mutated in a variety of different tumours. PTEN plays a role not only in integrin-mediated signaling through dephosphorylation of FAK and Shc tyrosine phosphatases (Tamura *et al.*, 1999; Figure 3.1) but also in the regulation of a number of other cellular processes including apoptosis, interactions with the cellular matrix and cell migration (Simpson and Parsons 2001). PTEN is able to inhibit cell growth but not in Rb-deficient cell lines suggesting a requirement for Rb (Paramio *et al.*, 1999). Perhaps the most important function of PTEN is however in regulating both p53 level and activity (Freeman *et al.*, 2003). PTEN also up-regulates β-catenin, leading to transcriptional activation of myc (Figure 3.6). Finally, PTEN negatively regulates phosphatidylinositol 3-kinase/Akt signaling (Vivanco and Sawyers 2002) and may be involved in the induction of anoikis (apoptosis of cells after loss of contact with the extracellular matrix) and inhibition of cell migration (Maehama and Dixon 1999; Kops and Burgering 1999; Yamada and Araki 2001; David 2001; Kandasamy and Srivastava 2002).

The loss of PTEN expression is not uncommon in NSCLC; immunohistochemical analysis has suggested that ~13% of NSCLC tumours have lost the ability to express PTEN (Olaussen *et al.*, 2003). Intragenic *PTEN* lesions have been found in 16-40% of SCLC cell lines and in 8-17% NSCLC cell lines (Kohno *et al.*, 1998; Forgacs *et al.*, 1998; Yokomizo *et al.*, 1998; Okami *et al.*, 1998) whilst promoter hypermethylation serves as a frequent epimutational event in NSCLC (Table 3). *PTEN* gene mutations in lung tumours may lead to constitutive activation of Akt thereby explaining the frequently observed resistance to anoikis in lung cancer (Brognard *et al.*, 2001; Lee *et al.*, 2003a) whilst in other types of cancer, there is some evidence for haploinsufficiency of the *PTEN* gene promoting tumour progression (Kwabi-Addo *et al.*, 2001). Finally, PTEN's role in regulating p53 level and activity may go some way toward explaining why *TP53* and *PTEN* gene lesions are not commonly encountered in the

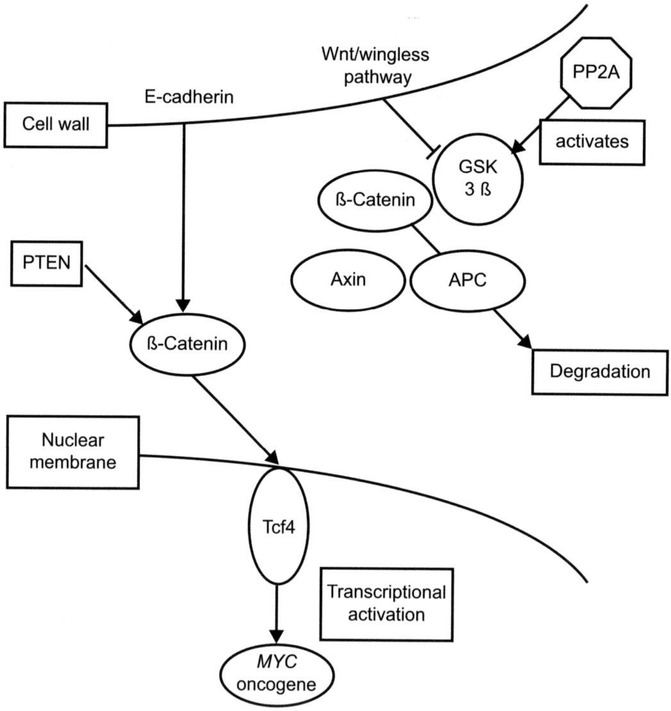

Fig. 3.6. APC-mediated activation of myc via the Wnt pathway. APC is a protein involved in cell adhesion and cytoskeletal function. It regulates β-catenin by forming a complex with axin and glycogen synthase kinase 3 beta (GSK 3β) that binds β-catenin which is then degraded. The availability of free β-catenin is regulated by cadherin-mediated signaling whilst Wnt signaling inhibits GSK 3β thereby stabilizing β-catenin. Free β-catenin crosses into the nucleus to activate the MYConcogene via Tcf4

same lung tumour: the loss of PTEN may decrease the selective pressure on tumours to lose p53 during tumorigenesis (Freeman *et al.*, 2003).

The phosphatidylinositol 3-kinase (PI3K)/Akt signaling pathway is frequently activated in NSCLC (Brognard *et al.*, 2001; Lee *et al.*, 2002c). While PTEN negatively regulates PI3K/Akt signaling, PIK3CA positively regulates it. *PIK3CA* gene amplification leads to increased PIK3CA expression and PI3K activity exerting downstream effects on cellular proliferation, cell adhesion, apoptosis, Ras signaling and cellular transformation. p53 regulates cell survival by inhibiting PIK3CA whilst constitutive activation of PIK3CA induces resistance to p53-dependent apoptosis in PTEN-deficient cells (Singh *et al.*, 2002). Thus, both *PIK3CA* gene amplification and activating mutations of the *PIK3CA* gene (and probably also amplification of the *PIK3CB* gene) could release PI3K from its negative regulation and lead to the abrogation of apoptosis and to the survival of genetically compromised cells. Consistent with their functional equivalence in activating the PI3K/Akt signaling pathway, *PTEN* (loss-of-function) and *PIK3CA* (amplification/activating) gene mutations have been found to be mutually exclusive events, at least in gastric cancer.

CDKN2A (p16) and *CDKN2B* (p15)

p16 (or *p16*^{INK4a}) plays an important role in the cyclin D-Rb pathway of control of cellular proliferation by acting as a cyclin/cyclin-dependent protein kinase (cdk) inhibitor to prevent Rb phosphorylation (Figure 3.2). The observation that lung tumours/cell lines are frequently characterized by either loss of *RB1* or *CDKN2A* gene expression but not both (Otterson *et al.*, 1994; Shapiro *et al.*, 1995a; Rocco and Sidransky 2001), is consistent with the view that the hypophosphorylated form of Rb functions as a growth suppressor. By contrast, mutations in both the *CDKN2A* and *TP53* genes are fairly frequently detected, at least in NSCLC (Sanchez-Cespedes *et al.*, 1999). *CDKN2A* mutations are comparatively rare in SCLC (Otterson *et al.*, 1994). This is probably due to the much higher frequency of *RB1* mutations in SCLC than in NSCLC. Thus, although the cyclin D-Rb pathway is important in both types of cancer, inactivation of the pathway appears to proceed by *CDKN2A* gene inactivation in NSCLC and by *RB1* gene inactivation in SCLC.

The p16 gene (*CDKN2A*) is located on chromosome 9p21, a region that exhibits frequent loss of heterozygosity in NSCLC. Either gross gene deletions or more subtle point mutations of the *CDKN2A* gene are found in ~30 % of NSCLC (de Vos *et al.*, 1995; Pollock *et al.*, 1996; Gazzeri *et al.*, 1998a; Table 2) whilst *de novo* hypermethylation of the 5' CpG island encompassing the promoter of the *CDKN2A* gene inactivates gene expression in a further 20-30 % NSCLC (Merlo *et al.*, 1995; Otterson *et al.*, 1995; Gazzeri *et al.*, 1998a; Belinsky *et al.*, 1998; Table 3). Immunohistochemical analysis of lung tumour tissue is capable of detecting those samples lacking p16 (Okamoto *et al.*, 1994; Taga *et al.*, 1997). Okamoto *et al.* (1995) noted a difference in *CDKN2A* mutation prevalence between metastatic NSCLC and primary tumours, suggesting that *CDKN2A* gene mutation may be a comparatively late event in NSCLC. However, the finding of aberrant methylation of the *CDKN2A* promoter in 19 % of precancerous bronchial lesions (Lamy *et al.*, 2002) argues strongly for this epimutation being an early event.

In contrast to the mutational spectrum of the *TP53* gene in lung cancer where G→T transversions comprise ~40 % of the total, G→T transversions comprise a rather smaller proportion (25 %) of *CDKN2A* gene mutations in NSCLC (Pollock *et al.*, 1996). The double strand breaks responsible for the 9p21 deletions do not occur at specific sites and do not involve specific motifs (Sasaki *et al.*, 2003b). Rather, they occur within a 10 kb region flanking the *CDKN2A* gene. The majority of characterized *CDKN2A* deletion breakpoints have been found to be located in or close to LINE-1 retrotransposon clusters suggesting that these elements could have facilitated the deletions (Florl and Schulz 2003). Sequence microhomologies are evident at the breakpoint junctions, a finding compatible with the rejoining of the broken ends by non-homologous end-joining (Sasaki *et al.*, 2003b; Florl and Schulz 2003).

Considerable interest has been shown in an alternative transcript derived from the *CDKN2A* gene and designated *p14*^{ARF}. *p14*^{ARF} is encoded by a separate exon 1β that lies some 20 kb 5' to exon 1α, plus exons 2 and 3 of the *CDKN2A* gene translated in an alternative reading frame. *p14*^{ARF} is thus quite unrelated structurally to *p16*^{INK4a}. Consistent with its putative tumour suppressor role, *p14*^{ARF} has been shown to induce growth inhibition when transfected into various lung cancer cell lines (Gao *et al.*, 2001). It is believed to function in a feedback loop involving

Mdm2 to regulate p53 stability (Zhang *et al.*, 1998). The p53 and Rb pathways are therefore connected through the action of the two alternative products of the *CDKN2A* gene, *p14*^{ARF}and *p16*^{INK4a}. Deletion of the *CDKN2A* gene would be predicted to disrupt both pathways. The *p14*^{ARF} gene is frequently deleted in both SCLC and NSCLC (Gazzeri *et al.*, 1998b; Nicholson *et al.*, 2001) and hypermethylated in 6-45 % of lung tumours (cf. 31 % for *p16*^{INK4a}) [Esteller *et al.*, 2001a; Mori *et al.*, 2004]. The ratio of *p16*^{INK4a} to *p14*^{ARF} is reported to correlate with the clinicopathological characteristics of the tumour during progression (Moriyama *et al.*, 2002). However, since *p16*^{INK4a} has been reported to be inactivated by either mutation or hypermethylation in some 84 % of NSCLC cell lines whereas *p14*^{ARF} was only inactivated in 55 % of these cell lines, Park *et al.* (2003b) concluded that *p16*^{INK4a} is the primary target of *CDKN2A* gene alterations in NSCLC.

Wang *et al.* (2003c) crossed transgenic mice carrying a dominant-negative p53 mutation with heterozygous *Ink4A/Arf*-deficient mice to explore the possibility that these mutations might synergize in promoting carcinogen-induced lung tumour progression in progeny mice. Doubly heterozygous mice exhibited a > 20-fold increase in lung tumour volume as compared with a 4-fold increase in *Ink4A/Arf*-deficient mice and a 9-fold increase in p53-deficient mice. Moreover, ~80 % of the lung tumours in doubly heterozygous mice were adenocarcinomas as compared to < 10 % of lung tumours in wild-type mice and ~50 % in heterozygous *Ink4A/Arf*- or *p53*-deficient mice. These findings were interpreted as meaning that there was indeed a significant synergistic interaction between these mutations at unlinked loci, at least in mouse. By contrast, in human cells, RNA interference-mediated suppression ('knockdown') of *p14*^{ARF} has been reported to enhance growth in a p53-dependent manner but with negligible tumorigenic effect; whilst suppression of *p16*^{INK4a} failed to influence cellular proliferation, it did synergize with p53 loss to promote cell growth and transformation (Voorhoeve and Agami 2003).

Another member of the p16 family of cdk inhibitors is p15 which is inducible by transforming growth factor β. p15 is encoded by a gene (*CDKN2B*) that is located adjacent to the *CDKN2A* gene on chromosome 9p21 and is commonly co-deleted with the *CDKN2A* gene in NSCLC (Okamoto *et al.*, 1995; Xiao *et al.*, 1995; Washimi *et al.*, 1995). *CDKN2B* gene promoter hypermethylation has also been observed in 11 % of NSCLC tumours and 15 % of neuroendocrine lung tumours (Chassade *et al.*, 2001; Kurakawa *et al.*, 2001).

Apoptosis regulatory genes

Multicellular organisms respond to DNA damage by activating cell cycle checkpoints and DNA repair pathways, and by triggering the process of apoptosis, programmed cell death (Zhou and Elledge 2000; Norbury and Zhivotovsky 2004). Since these processes have coevolved and are inextricably linked, it is not surprising that the biochemical mechanisms that underlie apoptosis overlap to some extent with those that initiate cell cycle arrest and DNA repair.

The apoptotic machinery comprises *sensors* whose role it is to monitor both the extracellular and intracellular environment for abnormalities (e.g. elevated oncogene expression, survival factor insufficiency, hypoxia etc), and *effectors* that can induce apoptosis to bring about cell death (Figure 1.3). The sensors constitute cell surface receptors that bind to and regulate the effectors - either survival or death factors. Thus, survival signals may be transmitted by IGF-1/IGF-2 through IGF-1R and by IL-3 through IL-3R, whilst death signals may be transmitted by Fas ligand binding to the Fas receptor and by tumour necrosis factor alpha (TNFα) binding to TNF receptor 1. Fas, a member of the tumour necrosis factor receptor superfamily, is triggered by interaction with its ligand (FasL) to transmit a death signal to the target cells. This mechanism represents a very important means of bringing about the immune-cytotoxic killing of tumour cells. Death signals induce mitochondria to release cytochrome *c*, which brings about apoptosis through the activation of a 'caspase cascade' that then executes the death programme of destruction of subcellular structures and organelles. The release of cytochrome *c* is regulated by members of the Bcl2 family whose members are either pro-apoptotic (e.g. Bax, Bak) or anti-apoptotic (e.g. Bcl2). Since p53 elicits apoptosis by up-regulating the expression of Bax (Shen and White 2001), the mutational inactivation of the *TP53* gene in 50-60 % of lung cancers results in the loss of a key sensor of DNA damage that is responsible for triggering apoptosis.

Fas expression is greatly reduced in 90 % of SCLC/NSCLC tumours and completely lost in 24 % of SCLC/NSCLC tumours (Viard-Leveugle *et al.*, 2003). Fas ligand is also dramatically reduced in NSCLC but elevated in SCLC tumours (Viard-Leveugle *et al.*, 2003). Lee *et al.* (1999a) reported that some 8 % of NSCLC tumours possessed missense mutations in the Fas (*TNFRSF6*) gene, located on chromosome 10q24, and that most of these were in the "death domain" of Fas known to be involved in the transduction of the apoptotic signal. The human genome also contains another gene (*TNFRSF6B*) that encodes a non-signaling decoy receptor (DcR3) which competes with Fas for FasL binding. Thus, in binding to FasL, DcR3 blocks FasL-dependent natural killer cell activity. Pitti *et al.* (1998) reported that the *TNFRSF6B* gene was amplified and over-expressed in 44 % of lung tumours examined, thereby titrating the death signal away from the Fas receptor. Amplification of the *TNFRSF6B* gene may thus promote tumour cell survival through the inhibition of FasL-mediated apoptosis.

Bcl2 serves to regulate cell death by protecting cells from apoptosis. Normally, Bcl2 is absent from the differentiating cells of the bronchial epithelium. However, 80-90 % of SCLC tumours or cell lines express Bcl2 (Ikeyaki *et al.*, 1994; Jiang *et al.*, 1995; Yan *et al.*, 1996; Kaiser *et al.*, 1996; Kitagawa *et al.*, 1996) whilst the corresponding proportion for NSCLC is between 20 and 50 % (Pezzella *et al.*, 1993; Walker *et al.*, 1995; Table 5). It is thought that the persistence of Bcl2 expression may contribute to the accumulation of oncogenic mutations by suppressing the apoptotic removal of cells following DNA damage thereby potentiating neoplastic cell growth (Sun *et al.*, 2001). It has been claimed that Bcl2 expression is associated with a higher survival probability (Fontanini *et al.*, 1995; Tomita *et al.*, 2003) but this has not been borne out by other studies (Martin *et al.*, 2003). Bcl2 expression is inversely correlated with that of p53 (Fontanini *et al.*, 1995; Kitagawa *et al.*, 1996) presumably because p53 both represses the transcription of the *BCL2* gene and up-regulates the expression of Bax thereby inhibiting Bcl2. Since Bcl2 expression has been shown to occur in pre-cancerous tissue (Ferron *et al.*, 1997; Kalomenidis *et al.*, 2001), it would appear that it is expressed relatively early in lung neoplasia. The significance (if any) of the loss of heterozygosity at the *BCL2* locus noted in 24 % NSCLC tumours (Nagatake *et al.*, 1996) is unclear.

Mutations in various other genes encoding key components of the apoptotic pathway (*FAIM, FADD, CASP3, CASP5, CASP8, CASP10, TNFRSF10A* and *TNFRSF10B*) have been reported in lung cancer. These lesions presumably also serve to suppress apoptosis thereby increasing the potential for transmission of DNA damage (Shivapurkar *et al.*, 2003).

Cell cycle control and DNA damage checkpoint genes

Genes encoding proteins of the cell cycle are commonly mutated in lung cancer (Hall and Peters 1996; Molinari 2000; Sherr 2000; Zhou *et al.*, 2001a). The cell cycle has a cell division phase (M phase) and a long interphase that comprises S phase (DNA synthesis), G_1 phase and G_2 phase (Figure 1.1). Cells normally only enter S phase if they are committed to mitosis whereas non-dividing cells remain quiescent in G_0 phase. Checkpoints serve to regulate the order of events in the cell cycle as well as ensuring that DNA repair is coordinated with cell cycle progression. These checkpoints are regulated by the cyclin/cyclin-dependent kinase (cdk) complexes (that phosphorylate key substrates which then control the various cell cycle transitions), and cdk inhibitors (Figure 1.2). Positive regulators of cell cycle progression include cyclin D1, cyclin E, cdk4, cdc25A and myc. In lung cancer, amplification and/or over-expression of the cyclin A2 (*CCNA2*), cyclin D1 (*CCND1*), cyclin E (*CCNE1*), and *MYC* genes has been observed (Tables 2 and 5). Overexpression of cyclins A2, D and E allows escape from Rb-mediated cell cycle regulation.

Negative regulators of cell cycle transitions include p53, *p14*^{ARF}, Rb, TGFβ, Chk2, p27, Bub1 and the cdk inhibitors p15 and p16. Cdk inhibitors may be divided into two groups, the INK4 (p15, p16) family which bind to and inhibit cdk4 and cdk6-associated activity and the Cip/Kip (p21, p27, p57) family which are able to inhibit cyclin D-dependent kinases as well as cyclins A and E in complex with cdk2 (Figure 3.2). The genes encoding p15 (*CDKN2B*), p16 (*CDKN2A*) and p57 (*CDKN1C*) have all been found to be mutated or inactivated in lung cancer (Tables 2 and 3) but the p21 (*CDKN1A*) and p27 (*CDKN1B*) genes have not (Shimizu *et al.*, 1996).

DNA damage checkpoints act at three different stages in the cell cycle: induction of G_1 arrest, blockage of DNA replication or a G_2 delay, depending on the type of damage experienced and the stage of the cell cycle when the damage is detected. One of the best known proteins involved in the DNA damage response is, of course, p53. Another mitotic checkpoint protein known to be mutated in lung cancer is Chk2 (Table 2). Chk2 (the product of the *CHEK2* gene) is activated upon DNA damage and its phosphorylation of both p53 (effecting its stabilization) and the *BRCA1* gene product is essential for proper DNA repair (Figure 3.7; Deng and Wang 2003). No somatic mutations of the *BRCA1* gene have been reported in lung cancer but the promoter of this gene is hypermethylated in 4 % of NSCLC tumours (Esteller *et al.*, 2001a; Marsit *et al.*, 2004).

The minichromosome maintenance proteins MCM2-7 are a set of related proteins that are essential for replication initiation and elongation in human cells (Tye *et al.*, 1999). MCM function is required for processive DNA replication during S phase; the

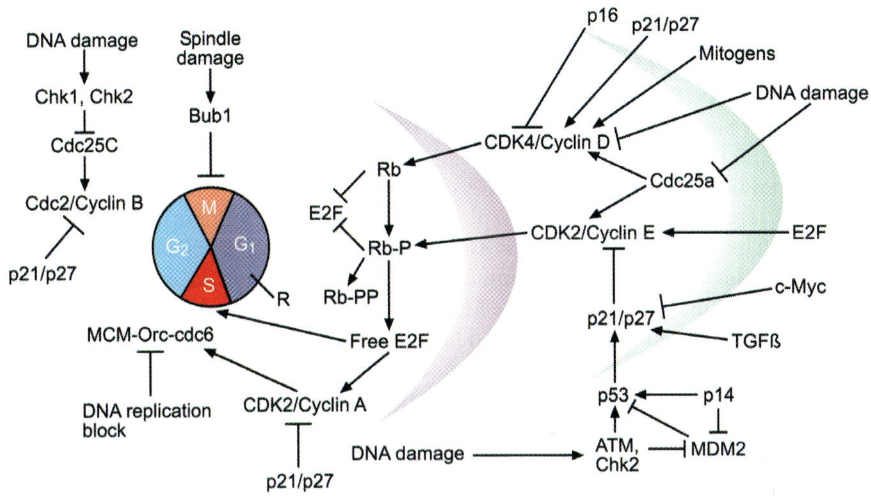

Fig. 3.7. Cell cycle checkpoints and their inactivation in lung cancer. Cell cycle transitions are mediated by cyclin/cyclin-dependent kinase (cdk) complexes that phosphorylate key substrates which allow the various cell cycle transitions to occur in a controlled manner. Cdk complexes are regulated by activating and inhibiting phosphorylation events that can be reversed by the action of the activating cdc25 phosphatases. Cdk complexes are also regulated by the presence/absence of the cyclin subunit, by their subcellular localization and by two families of cdk inhibitors ($p16^{INK4}$ and $p21^{WAF1}$). Cyclin/cdk activity is very low during cellular quiescence (G_0) through inhibition of cyclin D production. Mitogenic stimuli induce cyclin D1 synthesis and transition from G_0 into G_1 phase. Sequential phosphorylation of the Rb protein by cyclinD/cdk4, and later by cyclin E/ cdk2, permits cells to pass through the transition point (R) through release of E2F. Once past the R checkpoint, cells are no longer dependent upon mitogenic signals for subsequent DNA replication and mitosis. Cyclin A/cdk2 functions during S phase in part through regulation of origin recognition complexes (ORC) to allow initiation of DNA replication and to ensure that DNA is replicated only once per cell cycle. E2F function is inhibited as cells progress through S phase via phosphorylation by cyclin A/cdk2. Cyclin/cdk kinases are negatively regulated by the cdk inhibitors, specifically the $p16^{INK4}$ family (p15, p16, p18, p19) and the $p21^{WAF1}$family (p21, p27, p57). p53 stability is regulated negatively by Mdm2 and positively by ATM (which phosphorylates Mdm2) and $p14^{ARF}$ (which sequesters Mdm2). Cell cycle checkpoints can be activated in response to specific stimuli and then inhibit cell cycle transitions. These checkpoints are targets for inactivation in lung cancer through a variety of mechanisms. DNA damage causes cell cycle arrest in G_1 and G_2. In G_1, rapid arrest is initiated through pathways leading to the degradation of cyclin D1 and cdc25A. Cell cycle arrest is maintained in G_1 and G_2 through p53 activation and up-regulation of p21. Different forms of DNA damage activate different checkpoint kinases to trigger cell cycle checkpoints. Thus, ionising radiation activates ATM and Chk2 which phosphorylate p53, leading to its stabilization. In G_2, DNA damage activates Chk1 and Chk2 which phosphorylate cdc25C leading to its cytoplasmic sequestration by 14-3-3 proteins. DNA damage also activates the p53 pathway in G2, which in addition to up-regulating p21, targets cdc2 for cytoplasmic localization through the action of 14-3-3 sigma. Mitotic spindle damage activates Bub1 that inhibits mitosis. Positive regulators of cell cycle transitions targeted for deregulation in cancer include cyclins D and E, cdk4, cdc25A and myc. Negative regulators targeted for inactivation include p53, Rb, TGFβ, $p14^{ARF}$, p27, p16, Bub1, ATM and Chk2. Reprinted from ER McDonald and W El-Deiry (2001) *Annals of Medicine*, 33: 113-122, by kind permission of Taylor & Francis

MCM2-7 complex may well play a role in unwinding the DNA strands prior to the advance of the replicative DNA polymerase. MCM proteins are also thought to affect chromosome dynamics (by interacting with chromatin components and chromatin remodelling enzymes) whilst the MCM2-7 complex may serve as an indirect target of checkpoint signalling mediated by checkpoint kinases ATM and ATR (Cortez *et al.*, 2004). Since MCM proteins act so as both to sense and respond to S phase DNA damage, the loss of MCM function is likely to cause DNA damage and lead to genome instability. However, in the context of lung cancer, decreased MCM2 immunoreactivity appears to be associated with an improved prognosis in NSCLC (Ramnath *et al.*, 2001).

The spindle checkpoint detects both microtubule attachment on the centromeres and their tension, and inhibits metaphase-anaphase transition until any damage is repaired. The spindle checkpoint is essential for the maintenance of genomic integrity (Bharadwaj and Yu 2004) and its components include Mad1, Mad2, Mad3 (BubR1), Bub1, Bub3 and Mps1. The *BUB1* (budding uninhibited by benzimidazole) gene, which is infrequently mutated in lung cancer, encodes a protein kinase that is essential for chromosome segregation and for the control of mitotic progression (Figure 3.7); its inactivation leads to premature mitotic exit. The Mad1 (mitotic arrest deficient) protein, encoded by the *MAD1L1* gene, is the target of the Tax oncoprotein from human T-cell leukaemia virus 1 that presumably destroys a mitotic checkpoint allowing faster viral propagation; the *MAD1L1* gene is also infrequently inactivated in lung cancer.

Mutator (DNA mismatch repair) genes and microsatellite instability

> "Instability is the engine of both tumour progression and tumour heterogeneity, guaranteeing that no two tumours are exactly alike and that no single tumour is composed of genetically identical cells".
>
> C. Lengauer, K.W. Kinzler & B. Vogelstein (1998) *Nature* 396: 643-649.

Studies of the DNA repair capacity of lung cancer patients have sometimes been suggestive of a generalized deficiency in DNA repair (Wei *et al.*, 1996; Zienolddiny *et al.*, 1999; Wei *et al.*, 2000; Rajaee-Behbahani *et al.*, 2001; Shen *et al.*, 2003a). The implication has been that sub-optimal DNA repair capacity might be associated with the risk of developing lung cancer (Spitz *et al.*, 2003). Mechanistically, one possible cause of such a deficiency could be the deletional loss of the genes that encode DNA repair enzymes [e.g. *OGG1*, located within the deletion-prone 3p25-p26 region (Lu *et al.*, 1997; Hardie *et al.*, 2000; Shinmura and Yokota 2001)] or alternatively, reduced levels of their expression (Cheng *et al.*, 2001b).

Another possible cause of genomic instability is a deficiency in DNA mismatch repair such as that noted in hereditary non-polypotic colon cancer (HNPCC). HNPCC is characterized not only by the loss of alleles at numerous loci but also by the gain of additional novel alleles. This *microsatellite instability* (MI) has been extensively investigated in both NSCLC and SCLC. In NSCLC, MI has usually been found in between 19 % and 58 % of tumours (Shridhar *et al.*, 1994; Fong *et al.*, 1995a; Adachi *et al.*, 1995; Ryberg *et al.*, 1995; Hurr *et al.*, 1996; Miozzo *et al.*, 1996; Wieland *et al.*, 1996; Sekine *et al.*, 1997; Caligo *et al.*, 1998; Kim *et al.*, 1998a; Chang *et al.*, 2000; Ahrendt *et al.*, 2000a; Xu *et al.*, 2001c; Petmitr *et al.*, 2002; Woenckhaus *et al.*, 2003). In SCLC, estimates of the prevalence of MI vary more widely: e.g. 16 % (Kawanishi *et al.*, 1997), 45 % (Merlo *et al.*, 1994), 76 % (Chen *et al.*, 1996c) and 100 % (Hurr *et al.*, 1996). One of the problems in interpreting these data is that studies differ dramatically in terms of the identity and number of microsatellite markers employed and also, and perhaps more worryingly, in terms of the definition of microsatellite instability employed. Xinarianos *et al.* (2000a) noted that some 58% of NSCLC tissue samples exhibited reduced expression of MLH1 or MSH2 proteins. The proportion was higher (77%) when only samples manifesting MI were considered (Chang *et al.*, 2000). The finding of MI may be associated with a poor prognosis (Pifarre *et al.*, 1997; Rosell *et al.*, 1997). Interestingly, Suzuki *et al.* (1998a) claimed to have found MI more frequently in NSCLC patients with a family history of malignancy (67%) than in control patients (18%). Somewhat puzzlingly, Hansen *et al.* (2003) found 1 of 21 SCLC cell lines to be deficient in mismatch repair but this cell line did not exhibit MI and appeared to express all the major components of mismatch repair. How and why mismatch repair is dissociated from MI in this SCLC cell line remains unclear.

As yet, no homozygous or compound heterozygous lesions within mismatch repair genes have been found in lung tumour DNA to parallel those reported in HNPCC. However, heterozygous deletions of chromosomal regions 3p21 and 5q11-q13 (which include the mismatch repair genes *MLH1* and *MSH3*) occur in 55% and 42% of NSCLC tumours respectively (Benachenhou *et al.*, 1998). In addition, promoter hypermethylation of the *MLH1* and *MSH2* genes is apparent in a substantial proportion of NSCLC tumours (Table 3) indicating that epimutation at these loci may occur as frequently as LOH. Such findings suggest that some lung tumours probably acquire second hit inactivating mutations (or epimutations) that would compromise mismatch repair thereby serving to increase dramatically the mutation rate in secondary target genes (Duval and Hamelin 2002). Clearly, such mutations might reasonably be expected in those cases of lung cancer that manifest MI. Alternatively, it may be that haploinsufficiency for a given mismatch repair protein may serve to compromise mismatch repair (Ohmiya *et al.*, 2004).

The secondary target genes of the genome-wide somatic hypermutability evident in mismatch repair-deficient cells have not yet been identified in lung cancer. One good candidate target gene would be *TGFBR2* (encoding the transforming growth factor β II receptor) which, by virtue of its possession of a hypermutable polyadenine tract, is frequently mutated in HNPCC. The TGFβ signaling pathway is known to make a significant contribution to lung tumour progression, mainly through cellular proliferation (Park *et al.*, 2002a; Figure 3.8) and a mutation in the *TGFBR2* polyadenine tract has been reported in a case of SCLC with MI (Tani *et al.*, 1997). A further mutation in this sequence has been found in a SCLC tumour but unfortunately no microsatellite instability studies were performed in tandem (Hougaard *et al.*, 1999). *TGFBR2* gene mutations have also been found in 5/7 NSCLC cell lines and 1/21 NSCLC tissues (Kim *et al.*, 2000). Other studies of lung tumour material have however failed to find mutations in the *TGFBR2* polyadenine tract (Abe *et al.*, 1996; Takenoshita *et al.*, 1997; Caligo *et al.*, 1998; Gotoh *et al.*, 1999).

It is interesting to note that MI may also be evident in histologically normal lung tissue distant from the site of NSCLC tumour formation (Park *et al.*, 2000). Since MI has also been noted in lung tissue from individuals with non-malignant conditions (Liloglou *et al.*, 2001), the acquisition of some level of MI may be an early, perhaps predisposing, event in lung cancer tumorigenesis. A correlation has been noted between MI and the presence of *TP53* gene lesions in colorectal cancer cell lines (Cottu

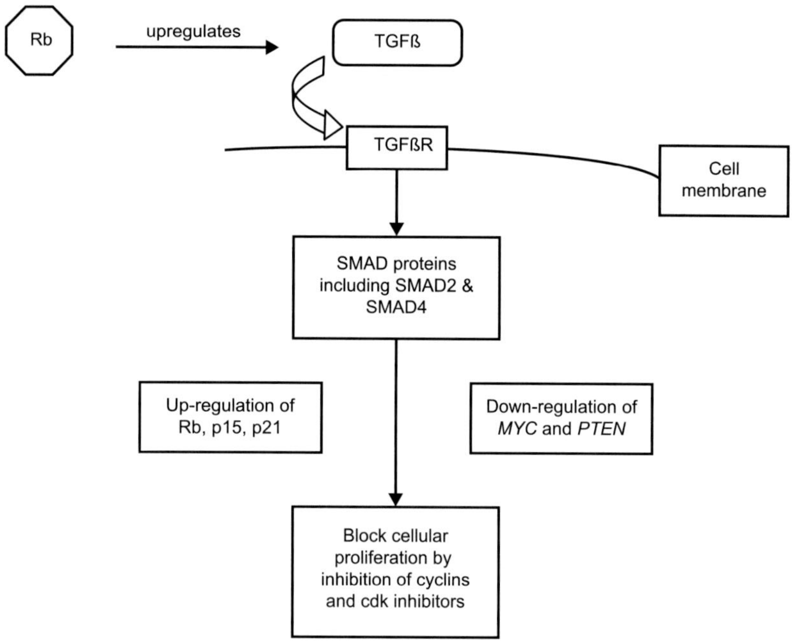

Fig. 3.8. TGFβ signaling pathway

et al., 1996) but it is as yet unclear whether this is a cause-and-effect, and if so, which is the cause and which is the effect.

A different type of microsatellite instability has been observed at certain tetranucleotide repeat markers of type $(AAAG)_n$ in various cancers. This type of instability, termed EMAST (elevated microsatellite alterations at selected tetranucleotide repeats) is not presently understood but is quite distinct from MI as defined above. It does however occur in ~50 % of NSCLC tumours in the absence of conventional MI (Xu *et al.*, 2001a). There is an indication that *TP53* mutations occur more frequently in EMAST(+) tumours than in EMAST(-) tumours (Ahrendt *et al.*, 2000a). It is therefore still somewhat unclear at this stage whether or not MI is a major contributory factor in lung tumorigenesis. It is nevertheless worthy of investigation as a potential marker of use in the early detection of lung cancer.

DNA methylation and lung cancer

> "Since methylation plays an important role in controlling normal cellular development, it follows that aberrations within this mechanism may be implicated in the abnormal gene control which characterizes cancer".
>
> PA Jones (1986) DNA methylation and cancer. *Cancer Res.* 46: 461-466.

DNA methylation is essential for normal mammalian development. It is thought to play a role in both gene regulation and imprinting, may serve as a cue for strand specificity in DNA replication and repair and could conceivably have functioned as a self-defence mechanism to silence transposable elements and proviral DNAs integrated into the genome over evolutionary time. 5-methylcytosine (5 mC) is the most common form of DNA modification in the human genome. Soon after DNA synthesis is complete, target cytosines are modified by a DNA methyltransferase using S-adenosylmethionine as methyl donor. Between 70 % and 90 % of 5 mC in the human genome occurs in CpG dinucleotides, the majority of which appear to be methylated.

Spatially, the distribution of CpG appears to be non-random in the human genome; about 1 % of the genome consists of stretches very rich in CpG which together account for roughly 15 % of all CpG dinucleotides. In contrast to most of the scattered CpG dinucleotides, these *CpG islands* represent unmethylated domains and comprise ~50 % of all non-methylated CpGs in the genome. CpG islands are often located immediately 5' to gene coding regions. In general, gene promoters containing CpG islands are unmethylated regardless of expression whereas promoters lacking CpG islands tend to lose their methylation upon transcription. CpG islands can however acquire a more extensive level of methylation if they are located on the inactive X chromosome or in the vicinity of imprinted genes, or as we discuss below, in tumour tissue (Kopelovich *et al.*, 2003).

It is as yet unclear how tissue-specific methylation patterns are established or how they are altered during tumorigenesis. One possibility is that chromatin alterations stimulate *de novo* methylation, another is that CpG islands associated with the promoters of transcribed genes may be targets for *de novo* methylation which then spreads to the promoters of neighbouring genes (Jones and Baylin 2002). Irrespective of the precise mechanism involved, once established, methylation patterns are heritable and stably reproducible after transmission through the germline. The establishment of cell type-specific methylation patterns in both the soma and the germline begins with global methylation of non-CpG island sequences in the embryo. The final methylation patterns are determined by a specific and highly regulated process of demethylation.

Methylation-mediated inactivation of tumour suppressor genes in lung cancer

The inverse correlation between the level of DNA methylation and gene transcription has been apparent for some time. Not unexpectedly, this correlation is at its strongest in gene promoter regions where methylation of 5-methylcytosine residues is presumed either to reduce the binding affinity of transcription factors (Zhu *et al.*, 2003c) or to allow the binding of transcriptional repressor proteins (Bader *et al.*, 2003). As we have seen, gene promoters often contain CpG islands. Methylation of these CpG islands serves to inactivate gene expression, a process sometimes termed "epimutation". The details of the mechanism are still unclear but it involves the binding of methylcytosine-binding proteins and histone deacetylases to the methylated DNA. In heavily methylated regions, the nucleosomes (composed of histone proteins around which DNA winds) are characteristically tightly compacted and regularly spaced, a configuration that excludes transcriptional activator proteins. Although the methylation of promoter-associated CpG islands in normal somatic tissues is a comparatively rare event, in a variety of different tumours, tumour suppressor genes that possess CpG islands in their promoter regions have often been found to be silenced by *de novo* DNA methylation (e.g. *CDKN2A*, *CDKN2B*, *MLH1*, *RB1*, *VHL* etc; reviewed by Baylin *et al.*, 2001; Garinis *et al.*, 2002; Herman and Baylin

2003; Nephew and Huang 2003; Macaluso *et al.*, 2003; Figure 3.9). Mutations *sensu stricto* are therefore not the only means by which tumour suppressor genes can be inactivated since DNA methylation also represents an important and alternative epigenetic mechanism for silencing these genes. It follows that future molecular genetic studies of cancer genes will have to take into account a gene's methylation status as well as its DNA sequence.

The first hint of an association between DNA methylation and lung cancer came in 1983 with reports of the hypomethylation of the *HRAS* oncogene in lung carcinoma (Feinberg and Vogelstein 1983a; 1983b; Figure 3.9). There was, however, no indication of increased expression of the oncogene accompanying the epimutation and essentially no further reports of oncogene hypomethylation in lung cancer have appeared since. It was not until nine years later that Makos *et al.* (1992) demonstrated the existence of regions of hypermethylation on chromosomes 3p and 17p in lung cancer. Methylation-mediated inactivation of a chromosome 9-encoded tumour suppressor (*CDKN2A*) gene was first reported in lung cancer in 1995 (Shapiro *et al.*, 1995b; Otterson *et al.*, 1995). An increasing number of reports have since appeared detailing evidence for the methylation-mediated inactivation of tumour suppressor genes as a contributory cause of lung cancer (reviewed by Toyooka *et al.*, 2001b and Tsou *et al.*, 2002). Each of these genes possesses a CpG island in their 5' regulatory regions. These regions, which contain the promoter elements of the genes, are normally unmethylated or minimally methylated. However, these regions

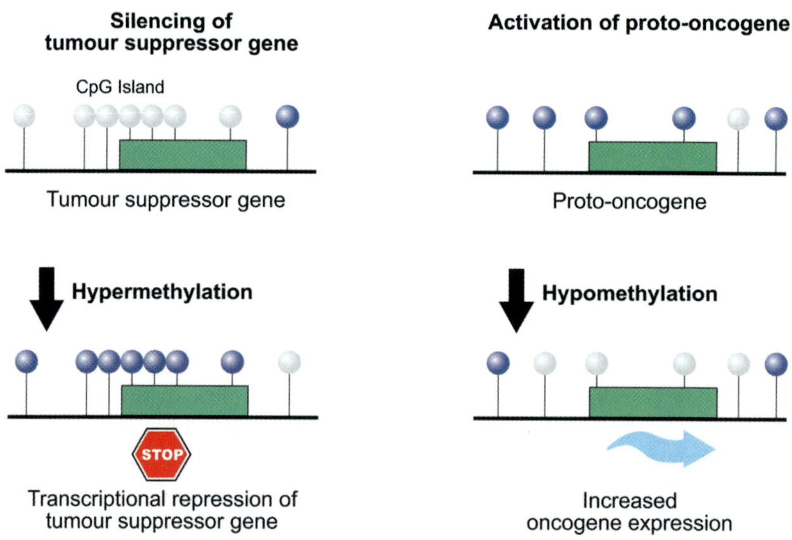

Fig. 3.9. Role of DNA methylation in cancer. Silencing of tumour suppressor genes: aberrant hypermethylation of CpG islands associated with tumour suppressor genes could result in transcriptional repression. Activation of proto-oncogenes: hypomethylation of proto-oncogene promoters might facilitate their increased expression.
Filled circles denote methylated CpG dinucleotides, unfilled circles denote unmethylated CpG dinucleotides. [Reprinted from *Molecular Medicine Today*, vol. 3: PW Laird, Oncogenic mechanisms mediated by DNA methylation, pp223-229, Copyright (1997) by kind permission from Elsevier]

become hypermethylated during tumorigenesis leading to the silencing of the downstream gene (Warnecke and Bestor 2000; Figures 2.4 & 3.9).

Table 3 lists 51 known or putative tumour suppressor genes whose promoters have been shown to become hypermethylated in at least a proportion of cases during lung tumorigenesis. For 22 of these genes (*RASSF1, FHIT, RARB, CDH13, CHEK2, CDKN2A, CDKN2B, CHFR, APC, MLH1, DDX26, GDF10, CASP8, MYO18B, PRKCDBP, IGSF4, ROBO1, ZMYND10, SEMA3B, PTEN, TGFBR2* and *STK11*), conventional inactivating mutations of one sort or another have already been described (Table 2). Of the remainder, at least 22 genes (*DAPK1, TIMP3, MGMT, CDH1, DLC1, CCND2, GSTP1, BRCA1, MSH2, SEMA3B, RBP1, PAX5, RARRES1, PRDM2, FANCF, DKK3, SOCS3, TMEFF2, CAV1, MADH9, RUNX3* and *PYCARD*) represent good candidates for involvement in lung cancer, whether frequent or infrequent (on the basis of their presumed function or because they have already been reported to exhibit either methylation-mediated inactivation, or LOH or subtle intragenic mutations in other types of tumour). In these cases, promoter hypermethylation rather than conventional types of mutation may be the major mechanism responsible for gene inactivation. Thus, aberrant methylation patterns in the 'target of methylation-induced silencing' (Tms1; *PYCARD*) gene have been reported in 41 % of SCLC tumours, 70 % of SCLC cell lines, 40 % of NSCLC tumours and 48 % of NSCLC cell lines (Virmani *et al.*, 2003a). Expression of the *PYCARD* gene serves to inhibit cellular proliferation and induces caspase 9-mediated apoptosis (McConnell and Vertino 2000) and so its loss may confer a survival advantage by allowing cells to escape from apoptosis. Similarly, promoter hypermethylation in lung tumours serves to inactivate the Suppressor of cytokine signaling 3 (*SOCS3*) gene, whose protein product normally inhibits JAK/STAT signalling thereby suppressing cell growth (He *et al.*, 2003). Restoration of SOCS3 expression results in the down-regulation of STAT3, induction of apoptosis and growth suppression (He *et al.*, 2003). The hypermethylation and consequent inactivation of the *DAPK1* and *CASP8* gene promoters would also be predicted to disrupt apoptosis. Promoter hypermethylation in other genes may interfere with a variety of other different cellular mechanisms and pathways including cell cycle control (*CDKN2A, CDKN2B, CHEK2, CHFR* and *SFN*), signal transduction (*MADH9, PTEN, STK11* and *RASSF1A*), protein transport (*CAV1*), DNA repair (*MLH1, MSH2, MGMT, FANCF* and *BRCA1*), tumour cell invasion (*APC*), suppression of metastasis (*CDH1* and *TIMP3*) and response to growth factors (*RARB*).

The genes listed in Table 3 are clearly only some of the loci that are inactivated by promoter hypermethylation during lung tumorigenesis. Indeed, Shiraishi *et al.* (2002a) have isolated and identified some 200 different gene-associated CpG islands (i.e. 1-2 % of the total number in the human genome) that are methylated *de novo* in human lung adenocarcinoma. One such CpG island is associated with the *HOXA5* gene that encodes a homeobox protein that up-regulates the expression of the *TP53* gene (Shiraishi *et al.*, 2002a). This notwithstanding, many CpG islands associated with the clusters of *HOXA* and *HOXD* genes are also methylated in human lung adenocarcinoma (Shiraishi *et al.*, 2002b), a finding which may be interpretable in terms of these two chromosomal regions being especially susceptible to *de novo* methylation.

Aberrant DNA methylation patterns are a common feature of human cancers (Feinberg *et al.*, 2002). However, cancer-associated hypermethylation and hypomethylation may represent quite different processes that involve the programmed activation and deactivation of different sets of genes (Szyf *et al.*, 2004). In lung cancer, the emergence of global genomic hypomethylation during tumorigenesis contrasts sharply with the CpG island hypermethylation manifested by the promoter regions of tumour suppressor genes (Dunn 2003). In squamous cell lung cancer, an association between the extent of global DNA methylation and both the size of the tumour and the stage of the disease (Piyathilake *et al.*, 2001) suggests the possibility of direct involvement of genomic hypomethylation in tumour progression.

Comparison of different stages of NSCLC in terms of the methylation status of various genes has suggested that promoter hypermethylation may be progressive during lung tumorigenesis and that tumour cells possessing methylation-inactivated tumour suppressor genes could acquire a growth advantage as a consequence (Belinsky *et al.*, 1998; Seike *et al.*, 2000; Zöchbauer-Müller *et al.*, 2001). However, promoter hypermethylation in a number of genes including *APC* has also been noted at high frequency in normal lung samples derived from NSCLC patients but not in controls (Brabender *et al.*, 2001; Yanagawa *et al.*, 2003). Although this may reflect neoplastic cell infiltration, it is also possible that the histologically normal tissue has acquired some but not all of the mutations (and/or epimutations) required for lung tumorigenesis in a clonal fashion. Consistent with this postulate, methylation of the *CDKN2A* gene promoter has been noted in preinvasive bronchial lesions (Lamy *et al.*, 2002), whilst methylation of the *RASSF1* gene promoter was detectable in the sputum of two patients some 12-14 months before clinical manifestations of lung cancer became apparent (Honorio *et al.*, 2003). These findings are consistent with the view that aberrant methylation is a comparatively early event in lung tumorigenesis. Intriguingly, Belinsky *et al.* (2004) have reported increased methylation of the *CDKN2A* gene promoter in lung adenocarcinomas from plutonium-exposed Russian workers as compared to lung adenocarcinomas from non-exposed workers.

In a study of 8 different genes in 75 NSCLC tumours, Yanagawa *et al.* (2003) showed that the level of promoter hypermethylation was often nearly as high in non-neoplastic lung tissue as in the corresponding tumour tissue. The disparity was found to be greatest for the *RASSF1* (43% v. 4%) and *RUNX3* (20% v. 3%) genes and so these genes may provide the best diagnostic markers (at least in the context of aberrant methylation patterns) in NSCLC tumorigenesis. By contrast, however, whilst methylation of the gap junction α1 protein (*GJA1*) gene promoter correlated inversely with mRNA and protein expression, a gradient of decreasing *GJA1* gene expression was noted from normal lung tissue to tumour tissue (Chen *et al.*, 2003 g).

Where examined, promoter methylation correlates reasonably well with gene inactivation in lung cancer (Zöchbauer-Müller *et al.*, 2001; Toyooka *et al.*, 2001a). In some cases, gene expression may be restored by treatment with a demethylating agent such as 5-azacytidine, suggesting a good correlation between methylation status and gene expression (Toyooka *et al.*, 2001a; He *et al.*, 2003). Apparently contradictory results have however been reported such as the promoter hypomethylation and concomitant reduced expression of the *CYP2E1* gene in lung cancer (Botto *et al.*, 1994).

A wide variety of genes including those encoding tumour suppressors, DNA repair proteins and proteins with functions related to metastasis and invasion, thus become hypermethylated during lung tumorigenesis. The patterns of DNA methylation at multiple gene loci can actually be used to discriminate between SCLC and NSCLC cells (Virmani *et al.*, 2002; Marchevsky *et al.*, 2004), a finding that may well attest to the functional significance of this phenomenon. These patterns could in principle be used as a classificatory aid and may shed some light on the nature of the underlying differences between SCLC and NSCLC. However, there are various *caveats* when assessing such data: firstly, one has in general to be careful about interpreting studies of DNA methylation in cancer since altered levels of methylation genome-wide are not only a marker of tumorigenesis but also of cells in culture (Smiraglia *et al.*, 2001). Secondly, it is often hard to distinguish cause from consequence (Baylin and Bestor 2002). Although for some genes, a correlation between promoter hypermethylation, gene inactivation and histone acetylation has been noted in lung cancer cells (Momparler 2003; Tani *et al.*, 2004), the temporal order of DNA and protein modification is not always clear and may not be obligatorily related (Bachman *et al.*, 2003). It must therefore be appreciated that even the methylation-mediated inactivation of a known tumour suppressor gene does not automatically imply that this event contributes towards neoplastic transformation – it may simply be a neutral consequence of the multiplicity of genome-wide changes in methylation resulting from the tumorigenic process. This notwithstanding, the potential importance of the epigenetic silencing of tumour suppressor genes in lung cancer is evidenced by the observation that both the inhibition of genome-wide DNA methylation by experimental disruption of the *Dnmt1* methyltransferase gene in a transgenic mouse model and the inhibition of histone deacetylation in wild-type mice serve to reduce the incidence of chemically induced lung cancer (Belinsky *et al.*, 2003). Further, the RNA interference-mediated reduction ('knockdown') in *DNMT1* gene expression in a human NSCLC cell line has been shown to result in a $>80\%$ reduction in promoter methylation as well as in re-expression of the *RASSF1*, *CDKN2A* and *CDH1* genes (Suzuki *et al.*, 2004).

The relationship between gene inactivation by mutation and gene inactivation by promoter hypermethylation is both complex and intriguing. Thus, in NSCLC, a significant correlation between hypermethylation of the *SEMA3B* gene and LOH in the 3p21.3 chromosomal region harbouring this gene has been observed that is suggestive of the operation of a two-hit mechanism, involving both allele loss and epigenetic changes, leading to tumour suppressor gene inactivation (Kuroki *et al.*, 2003). Although mechanistically unrelated, the functional interaction between the two gene inactivating mechanisms, promoter hypermethylation and genomic deletion, appears to be a general phenomenon in tumorigenesis and can in principle be used to identify novel tumour suppressor genes (Esteller *et al.*, 2001b; Zardo *et al.*, 2002). Since epimutations can be functionally equivalent to both LOH and intragenic inactivating lesions, mutational and epimutational changes in different genes may also exert additive or even synergistic effects on tumour progression and patient survival (Sanchez-Cespedes *et al.*, 1999; Kersting *et al.*, 2000; Kim *et al.*, 2003c).

The relationship between mutational inactivation and inactivation by promoter hypermethylation can also be indirect and bi-directional. Thus, promoter methylation-mediated inactivation of the *MGMT* gene encoding the DNA repair protein O6-methylguanine DNA methyltransferase is reportedly associated with a significant increase in the frequency of G→A transitions in CpG dinucleotides in the *TP53* gene (Wolf *et al.*, 2001). If confirmed, this epigenetically-based mutator phenotype (Esteller and Herman 2004) would parallel the predicted consequences of the methylation-mediated inactivation of the *MLH1* and *MSH2* genes: the loss of both alleles of either of these key mismatch repair genes would be expected to lead to the emergence of microsatellite instability and a genome-wide increase in the mutation rate. Conversely, conventional gene mutations can potentially alter non-allelic DNA methylation patterns. Thus, carriers of genotypes containing the 677T allele of the methylenetetrahydrofolate reductase (*MTHFR*) gene appear to exhibit significantly reduced levels of 5 mC in their genomes, and lung tumours in such individuals do not manifest severe global hypomethylation (Paz *et al.*, 2002). Similarly, lung tumours occurring in individuals homozygous for the 2756G allele in the 5-methyltetrahydrofolate-homocysteine methyltransferase (*MTR*) gene have been found to exhibit a significantly lower number of hypermethylated CpG islands in their tumour suppressor genes (Paz *et al.*, 2002). Finally, a nonsense mutation has been reported in the *MBD1* gene, encoding the methyl-CpG binding protein MBD1, in a SCLC cell line (Bader *et al.*, 2003). MBD1 serves to repress transcription from methylated gene promoters (Nakao *et al.*, 2001) and cooperates with *N*-methylpurine-DNA glycosylase (MPG), a key enzyme in base excision repair (Watanabe *et al.*, 2003). Although it remains unclear whether this truncating lesion of MBD1 occurred in the original tumour or whether instead it arose upon cell culture, it could, if consolidated by further studies, provide an example of how a single mutation in a key gene can have genome-wide consequences for gene expression (Prokhortchouk and Hendrich 2002).

Changes in the methylation status of gene promoter regions can thus have both direct and indirect consequences for the progression of lung tumorigenesis. Whilst the methylation-mediated inactivation of certain key genes may directly promote cell proliferation or inhibit apoptosis, the inactivation of DNA repair genes by promoter hypermethylation can have indirect consequences for many other genes by modulating the mutation rate. In this respect, the direct and indirect consequences of epimutation mirror precisely those of conventional gene mutations. Exactly how individual tumour cells acquire their specific *de novo* gene-associated promoter hypermethylation profiles is still unclear. Presumably, every tumour displays a different methylation profile (Paz *et al.*, 2003) and not all the genes inactivated by hypermethylation will contribute to cellular proliferation or to the inhibition of apoptosis (some may even do the opposite). Clonal selection will nevertheless ensure that those cells that obtain the maximum growth and survival advantage from their *de novo* methylation profiles will eventually come to predominate.

We should not forget that there are potential deleterious consequences for the lung cell by increasing the genomic level of DNA methylation. 5-methylcytosine (5 mC) is an 'endogenous mutagen' in that it is subject to high-frequency methylation-mediated deamination, yielding C→T and G→A transitions (Pfeifer 2000). In addition, as we have seen, methylation is likely to contribute indirectly to increased genomic instability by methylation-mediated inactivation of the *MLH1* mismatch repair gene. Finally, 5 mC also appears to enhance adduct formation (and hence mutation) in cells exposed to the exogenous mutagen benzo[*a*]pyrene (Denissenko *et al.*, 1997; Yoon *et al.*, 2001) raising the prospect of the tumorigenic interaction between an epimutation and exposure to a xenobiotic.

Chromatin remodeling

By temporarily altering the nucleosome structure of chromatin, the access of transcription factors to DNA can be modulated. Such 'chromatin remodeling' is brought about by the covalent modification of histones by acetylation, methylation and phosphorylation. Chromatin remodeling has downstream consequences for DNA methylation and gene expression as well as DNA replication, recombination and repair (Huang *et al.*, 2003) and can play an important role in tumorigenesis (Hake *et al.*, 2004).

One family of proteins that contributes to chromatin remodelling complexes is SWI/SNF (Roberts and Orkin 2004). SWI/SNF complexes contain one of two highly homologous ATPases, BRG1 and BRM, that associate with different gene promoters during cellular proliferation and differentiation; thus, BRG1 binds to zinc finger-containing transcription factors whilst BRM interacts with two ankyrin repeat-containing proteins that play a key role in Notch signal transduction (Kadam and Emerson 2003). BRM is encoded by the *SMARCA2* gene on chromosome 9p23-p24 whilst BRG1 is encoded by the *SMARCA4* gene (19p13.3). The *SMARCA4* gene is already known to be mutated in lung cancer (Wong *et al.*, 2000; Table 4). The expression of BRM and BRG1 has been reported to be lost in 10% of NSCLC tumours and ~30% NSCLC cell lines (Reisman *et al.*, 2003) and appears to be an indicator of poor prognosis (Fukuoka *et al.*, 2004). Since BRG1 interacts with a number of other proteins whose genes are also known to be mutated in human lung cancer [viz. BRCA1 (Bochar *et al.*, 2000), Rb (Dunaief *et al.*, 1994), β-catenin (Barker *et al.*, 2001) and LKB1/STK11 (Marignani *et al.*, 2001)], it may be that defects in these genes contribute to lung tumorigenesis by exerting a detrimental effect on chromatin remodeling.

Imprinting and its loss in lung cancer

DNA methylation is also thought to be involved in *imprinting*, defined as the „differential modification of the maternal and paternal contributions to the zygote, resulting in the differential expression of parental alleles during development and in the adult". This parent-of-origin-specific allele silencing appears to be essential for normal mammalian development since parthenogenetic embryos (whether diploid paternal or diploid maternal) do not survive to term: in diploid maternal embryos, fetal development is normal but development of the extra-embryonic membranes is abnormal. In diploid paternal embryos, it is the other way around. Clearly, maternal and paternal chromosomes must differ epigenetically and in such a way that different developmental programmes are followed. Only some 100-200 genes in the human genome are thought to be imprinted and these are often clustered (Verona *et al.*, 2003). Examples include the insulin (*INS*), *H19* and insulin-like growth factor 2 (*IGF2*) genes at 11p15.5 and the Wilms' tumour (*WT1*) gene at 11p13. A common function of imprinted genes is in the control of embryonic growth with paternally expressed genes (e.g. *IGF2*) tending to enhance growth rates and maternally expressed genes (e.g. *H19*) tending to reduce them.

Kondo *et al.* (1996) investigated the potential role of imprinting of the $p57^{KIP2}$ (*CDKN1C*) gene in lung cancer on the grounds that this gene is located on chromosome 11p15 in the vicinity of other imprinted loci (*INS, H19, IGF2*). These workers first confirmed that the *CDKN1C* gene was imprinted (paternal allele normally inactivated, maternal allele normally active) and then went on to demonstrate that there was a bias toward maternal allele loss in lung cancer: maternal alleles were lost in 11/13 (85%) of lung cancer cases bearing 11p15 deletions. Thus, the normal imprinting of the paternal allele allows the second hit to inactivate the maternal allele and hence the tumour suppressor activity of the locus. The loss of imprinting in lung cancer has also been reported for two other 11p15-encoded genes, *IGF2* (Suzuki *et al.*, 1994; Zhan *et al.*, 1995; Kohda *et al.*, 2001) and *H19* (Kondo *et al.*, 1995). It may be therefore that genomic imprinting at 11p15 plays a role in lung tumorigenesis. Loss of imprinting of the chromosome 7q32-located *MEST* gene has been reported in 85% lung adenocarcinomas (Kohda *et al.*, 2001). By comparison, mono-allelic expression was present in all matched normal tissues for the *IGF2* and *MEST* genes, and in both adenocarcinoma and normal tissues for the *H19, SNRPN* (15q12) and *NDN* (15q11-q12) loci. These data suggest that the loss of imprinting of the *IGF2* and *MEST* genes may be involved in the development of lung adenocarcinoma.

The potential significance of tumour suppressor gene location in lung cancer

Deletion mapping and LOH studies have identified nearly 40 distinct regions dispersed on some 21 different chromosomes as being associated with either SCLC or NSCLC tumorigenesis. If many of these regions were indeed to harbour tumour suppressor genes, then it would follow that individual lung tumours are likely to have multiple genes inactivated prior to their becoming clinically manifest (Thiagalingam *et al.*, 2002). The frequency of loss of tumour suppressor gene function is clearly quite variable between loci. It should however be noted that the loss of a particular gene in lung cancer, even the frequent loss, does not automatically imply direct and obligate involvement in the process of tumorigenesis. Close linkage to a second tumour suppressor gene that is the real target for clonal selection could also provide an explanation. Careful mutational analysis of the tumour suppressor gene in question for second (somatic) hits is necessary to provide evidence of direct involvement.

From inspection of Table 2, it is apparent that virtually all the known or putative tumour suppressor genes listed in Table 2 are located in one of the regions known by deletion mapping or LOH studies (*see* Table 1 and Clues to Candidate Genes from Cytogenetic Abnormalities and Loss of Heterozygosity Studies) to be associated with lung tumorigenesis [exceptions include *TP73* (1p36) and *MAD1L1* (7p22)]. Moreover, quite a few of the tumour suppressor loci listed are chromosomally clustered such that 67 of the different putative tumour suppressor genes identified as being involved in lung cancer may be found on 16 different chromosome arms: *TP73, PTPRF* and *MUTYH* on 1p, *LRP1B, CASP8, CASP10, ERCC3, ATF2* and *BUB1* on 2q, *FHIT, DLEC1, MLH1, VHL, ROBO1, BAP, RASSF1, RARB, CTNNB1, OGG1, AR-MET, TGFBR2, TUSC2, NAT6, HYAL1, CTDSPL, RBM6, ZMYND10, SEMA3B* and *TUSC4* on 3p, *EIF4G1* and *FAIM* on 3q, *MSH3* and *APC* on 5q, *MAP2K2* and *PPP1R3A* on 7q, *TNFRSF10A, TNFRSF10B* and *PDGFRL* on 8p, *CDKN2A* and *CDKN2B* on 9p, *DMBT1, TNFRSF6* and *PTEN* on 10q, *SLC22A18, WT1, PTPRJ, PRKCDBP* and *CDKN1C* on 11p, *IGSF4, MEN1, CASP5, FADD* and *PPP2R1B* on 11q, *RB1* and *DDX26* on 13q, *RBL2* and *WWOX* on 16q, *TP53, YWHAE* and *MAP2K4* on 17p, *DPC4, DCC, MBD1* and *MADH2* on 18q, and *SEZ6L, MYO18B* and *CHEK2* on 22q. As already mentioned, these chromosomal locations have previously been implicated in tumorigenesis by virtue of their harbouring cytogenetic abnormalities and exhibiting frequent LOH (*see* Chapter 2; Clues to Candidate Genes from Cytogenetic Abnormalities and Loss of Heterozygosity Studies). On the face of it, this may appear to provide welcome and convincing evidence for the direct involvement of these loci in lung tumorigenesis. However, it should be realized at least for some of the tumour suppressor gene loci putatively involved in lung tumorigenesis that they were initially sought and then positionally cloned on the basis of their chromosomal location (Kohno and Yokota 1999). A clear idea of the total number of tumour suppressor genes likely to be causally involved in the process of lung tumorigenesis will therefore have to await the methodical high resolution LOH scanning of large numbers of lung tumours on almost a genome-wide basis followed by the in-depth mutational and functional characterization of all tumour suppressor loci so identified.

We may also note that there is some correspondence between the locations of epigenetically silenced genes (Table 3) and the sites of frequent deletion/loss of heterozygosity (*see* Table 1 and Chapter 2; Clues to Candidate Genes from Cytogenetic Abnormalities and Loss of Heterozygosity Studies). Thus, the chromosomal regions commonly deleted in NSCLC (3p, 9p, 11q23, 12p13 and 16q24) harbour between them at least 13 tumour suppressor genes that are also known to be inactivated by promoter hypermethylation in NSCLC tumours. For these tumour suppressor genes, it may therefore be that a combination of deletion/loss of heterozygosity and methylation-mediated inactivation serves fairly frequently to bring about the functional loss of both alleles.

In three cases, there is a correspondence between the location of a tumour suppressor gene and the location of a fragile site, defined as a site on a chromosome that is prone to frequent breakage because it contains a highly repetitive simple sequence element. Thus, the *FHIT* gene at 3p14.2 spans the FRA3B fragile site, the *WWOX* gene at 16q23-q24 spans the FRA16D fragile site and the *PARK2* gene at 6q25-q27 spans the FRA6E fragile site. It may be that these tumour suppressor genes are vulnerable to mutational inactivation precisely because of their close proximity to the fragile sites.

The potential significance of oncogene location for chromosomal amplification and gene over-expression in lung cancer

Evidence for the amplification of some 24 different genes has been presented in either SCLC or NSCLC (Table 2). These genes are *NRAS, HRAS, MYC, MYCL1, MYCN, PAX7, RAF1, TP73L, EIF4G1, SKP2, MYB, EGFR, ERBB2, PTK2, CCND1, BIRC2, BIRC3, MDM2, DYRK2, FOXA1, E2F1, TNFRSF6B, TERT* and *TERC*. For nine of these genes (*EGFR, MYC, CCND1, BIRC3, MDM2, DYRK2, FOXA1, ERBB2* and *E2F1*), there is independent evidence for up-regulated gene expression in lung cancer tissue or cell lines (Table 5).

Comparative genomic hybridisation (CGH) has been used to identify 'gains' in a number of different chromosomal regions that may be interpreted as evidence for localized chromosomal duplication/amplification (*see* Chapter 2; Clues to Candidate

Genes from Cytogenetic Abnormalities and Loss of Heterozygosity Studies). The regions include 1p (SCLC), 1q (SCLC and NSCLC), 3q (NSCLC), 4q (NSCLC), 5p (SCLC and NSCLC), 7p (NSCLC), 8q (NSCLC), 11q (NSCLC) and 20q (NSCLC). Let us take the example of the *TERC* gene which encodes the RNA component of telomerase and which is amplified in NSCLC, is located at 3q26. Despite this correspondence, it is not automatically clear that the *TERC* gene is the sole target for selection for a growth advantage within the amplified 3q region because there are likely to be many other genes within this region. Yokoi *et al.* (2003b) examined four candidate genes within the 3q26 region (*TERC, EVI1, SNO, PIK3CA*) for both expression and amplification. Since the *TERC* gene was amplified in a higher proportion of NSCLC cell lines than the others, as well as being up-regulated (by comparison with non-tumour lung tissue) to a greater extent than the others, Yokoi *et al.* (2003b) concluded that the *TERC* gene represented the most probable target for selection within the 3q26 amplicon.

Although subject to the same *caveat*, inspection of Table 2 allows one to propose possible *a posteriori* explanations for most of the above chromosomal regions in terms of cellular selection for the amplified oncogenes they contain. Thus, in lung cancer, amplified oncogenes are present on 1p (*NRAS, MYCL1, PAX7*), 3q (*EIF4G1, TP73L, PIK3CA*), 5p (*SKP2*), 7p (*EGFR*), 8q (*PTK2, MYC*), 11q (*CCND1, BIRC2, BIRC3*) and 20q (*E2F1*). Once again, however, although the spatial coincidence of chromosomal gains and amplified oncogenes is noteworthy, it falls somewhat short of providing direct and definitive evidence for the involvement of these oncogenes in the process of cellular selection.

Another approach to identifying amplified chromosomal regions, *transcriptome mapping*, has been employed by Zhou et al. (2003b). This method attempts to identify chromosomal regions (termed *regions of increased tumour expression* or RITEs) in which clusters of genes exhibit non-random increased expression in tumour samples (10 different types including lung) as compared to normal samples. A total of 61 different RITEs were identified using this approach (1p33, 1p34, 1p36, 1q21, 1q42, 2p12, 2q11, 2q13, 2q21, 2q37, 3q21, 3q23, 3q27, 4p16, 5p15, 5q35, 6p21, 7p22, 7p13, 7p12, 7q11, 7q22, 8q24, 9q22, 9q34, 10p15, 11p15, 11p11, 11q12, 11q13, 12p13, 12p11, 12q13, 12q24, 14q11, 14q32, 15q22, 15q24, 15q25, 15q26, 16p13, 16p12, 16p11, 16q12, 16q22, 16q24, 17p11, 17q11, 17q21, 17q25, 19p13, 19p12, 19q13, 20p13, 20q11, 20q13, 21q22, 22q11, 22q12, Xp11 and Xq28). Some 75% of these regions had already been identified previously by CGH analysis, a proportion that increases one's confidence in the efficacy of the technique of transcriptome mapping. Further examples of amplified oncogenes that map to these regions include *TERT* (5p15), *HRAS* (11p15) and *ERBB2* (17q21). Amplification of both these genes as well as the *TNFRSF6B* Fas ligand decoy receptor gene (20q13) may be selected for on the basis of promoting cellular proliferation and the inhibition of apoptosis respectively. For the remaining 7 genes that are known to be amplified in lung cancer but which do not map to any of the above amplified chromosomal regions, there may be two explanations: either the gene amplification events are comparatively rare in lung cancer, or the limited extent of the amplified region militates against facile detection.

Telomere length and telomerase activity

"Senescence begins and middle age ends, the day your descendants outnumber your friends."
Ogden Nash (1957) *You Can't Get There From Here.*

Telomeres are chromosomal structures that allow the ends of chromosomal DNA to be replicated completely without the loss of bases at the termini. As such, telomeres play a major role in the maintenance of genomic integrity. They are the sites at which the pairing of homologous chromosomes is initiated and contain long arrays (averaging about 10-15 kb) of tandem hexanucleotide repeats, most frequently TTAGGG, that are bound by telomere-binding proteins. Since the DNA replication mechanism is unable to synthesize the end of the chromosome, telomeric DNA shortens by about 100 bp every time the cell divides. Thus, telomere length decreases with age and number of cell divisions, resulting eventually in cellular senescence when the telomeres are no longer able to fulfil their normal protective functions. Chromosomes lacking functional telomeres become highly unstable leading to chromosomal instability, chromosome imbalance and gene amplification via breakage/fusion/bridge cycles (Maser and DePinho 2002; O'Hagan *et al.*, 2002; Desmaze *et al.*, 2003). To counteract this process, germline cells possess a ribonucleoprotein polymerase, *te-*

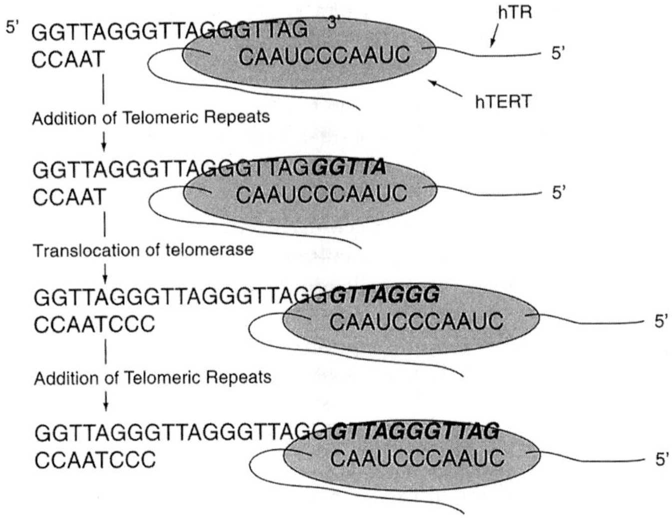

Fig. 3.10. The synthesis of telomeric sequences by telomerase. Telomerase contains RNA-dependent DNA polymerase activity that uses its RNA component, hTR, which is complementary to the telomeric single stranded overhang, as a template in order to synthesize TTAGGG repeats directly on to telomeric ends. The catalytic component of telomerase, hTERT, has reverse transcriptase activity. HTR and hTERT form the core of the telomerase that binds to the telomeric repeats. It functions by adding 6 bases, and then pauses while it repositions the template RNA for the synthesis of the next 6 base repeat. Telomerase is thus processive. The extension of the 3' DNA template end in turn permits additional replication of the 5' end of the lagging strand thereby compensating for the telomere shortening that occurs in its absence. [Reproduced from Fig. 2 (page 486), Chapter 21 by JW Shay, WE Wright & RA Schultz, Role of telomeres and telomerase in aging and cancer, in *Molecular Genetics of Cancer 2nd Ed.*, Ed. JK Cowell, (2001) by kind permission of Taylor & Francis/BIOS, Oxford]

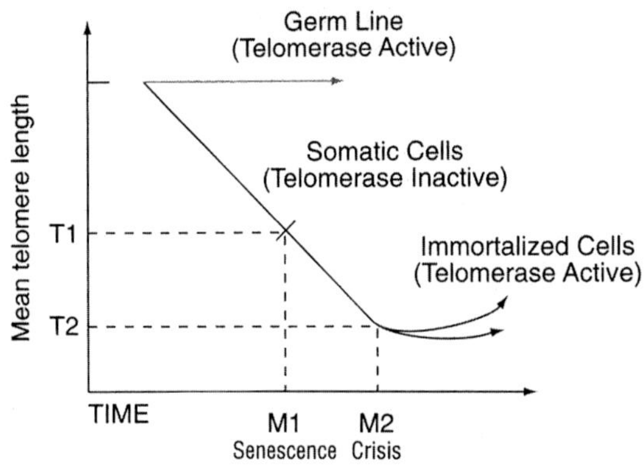

Fig. 3.11. Telomere hypothesis of aging and cancer. In the absence of telomerase, telomeres shorten with each cell division in somatic cells. Cellular senescence begins when there are on average several kilobases of telomeric TTAGGG repeats remaining. It may be that the stimulus for the induction of M_1 is damage signals from a rare telomere that lacks repeats. M_1 requires p53 and Rb. If these molecules are inactivated, the cells continue to divide and the telomeres continue to shorten until the telomeres are so short that further replication is prevented. Human cancer cells often acquire the ability to reactivate/up-regulate telomerase and thus usually have detectable telomerase activity as well as short telomeres. [Reproduced from Fig. 1 (page 484), Chapter 21 by JW Shay, WE Wright & RA Schultz, Role of telomeres and telomerase in aging and cancer, in *Molecular Genetics of Cancer 2nd Ed.*, Ed. JK Cowell, (2001) by kind permission of Taylor & Francis/BIOS, Oxford]

lomerase, which adds hexanucleotide repeats to the 3' end of the telomere thereby maintaining telomere length (Figure 3.10). By contrast, in somatic cells, telomerase (specifically the catalytic subunit of telomerase) is normally down-regulated leading to a reduction in telomere length as the number of cell divisions increases. This 'mitotic clock' effectively counts the number of cell divisions, and limits the replicative capacity of the cell through a process that culminates in cellular senescence and growth arrest (Figure 3.11). Since cancer cells proliferate at a greater rate than their normal counterparts, there is a tendency for them to exhibit reduced telomere length (Ishikawa 1997). Lung cancer is no exception; thus, for example, reduced telomere length has been noted in 23-32% of primary lung tumours (Shirotani *et al.*, 1994; Hiyama *et al.*, 1995b).

Wu *et al.* (2003d) also demonstrated that telomere length is significantly shorter in lymphocytes from lung cancer patients than in control subjects. Although telomere shortening is inversely proportional to age, telomere length in lymphocytes does vary considerably between individuals of the same age. This notwithstanding, the results of Wu *et al.* (2003d) support the view that telomere dysfunction is associated with an increased risk of lung cancer, perhaps by impairing chromosomal stability.

Since allelic loss involving both the *TP53* and *RB1* genes is present in the great majority of lung cancer cases, it may be that the inactivation of these growth regulatory genes serves to hasten telomere shortening by promoting cell division (Hiyama *et al.*, 1995b; Maniwa *et al.*, 2001). Certainly, there is some evidence for a combinatorial effect of telomere dysfunction and p53 deficiency in initiating and/or accelerating tumorigenesis (Chin *et al.*, 1999; Hackett *et al.*, 2001).

In order to avoid senescence and to continue growing, cancer cells often either reactivate telomerase (reviewed by Shay and Bacchetti 1997; Shay *et al.*, 2001a; Figure 3.11) or employ a little understood alternative lengthening of telomeres (ALT) mechanism that serves to maintain telomere length in the ~10-15% of tumour cells that lack telomerase activity but nevertheless harbour long telomeres (Reddel 2003). It is not however clear whether the telomerase present occurs as a result of clonal selection of pre-existing telomerase-positive cells during tumorigenesis or through the induction of the expression of the components of telomerase that are normally tightly repressed (Newbold 2002). Whichever, between 75% and 93% of lung tumours and cell lines have been found to display telomerase activity, which is low to undetectable in normal human lung (Hiyama *et al.*, 1995a; Albanell *et al.*, 1997; Lee *et al.*, 1998a; Wu *et al.*, 1999a; Taga *et al.*, 1999; Marchetti *et al.* 1999; Lonardo and Albanell 2000; Kumaki *et al.*, 2001; Chen *et al.*, 2001b; Yang *et al.*, 2001; Hara *et al.*, 2001; Chen *et al.*, 2002d; Wu *et al.*, 2003c; Hsu *et al.*, 2003). That telomerase reactivation provides a cellular growth advantage is evidenced by the rescue from senescence and increased proliferation of normal human fibroblasts modified by the retroviral-mediated transfer of the catalytic portion of telomerase (Forsythe *et al.*, 2002). Iniesta *et al.* (2004) have reported an association between telomerase expression and loss of heterozygosity on chromosome 3p. Telomerase expression varies between different histopathological classes of lung cancer and may have a prognostic influence (Hara *et al.*, 2001; Kumaki *et al.*, 2001; Gonzalez-Quevedo *et al.*, 2002; Wu *et al.* 2003c; Lantuejoul *et al.* 2004).

Telomerase is a ribonucleoprotein DNA polymerase that possesses a protein component comprising telomerase reverse transcriptase and telomerase-associated protein 1, and an RNA component. Telomerase RNA acts as a template for the addition of telomeric repeat sequences, and is encoded by a gene, *TERC*, located on chromosome 3q26.3. The *TERC* gene is highly expressed in the germline and in tumour cell lines that possess a high level of telomerase activity, but at much lower levels in somatic tissues in which telomerase activity is not detectable. The *TERC* gene has been shown to be amplified in both SCLC and NSCLC tumours (Soder *et al.*, 1997; Sugita *et al.*, 2000; Yokoi *et al.*, 2003b) whilst *TERC* expression is reportedly up-regulated in some human SCLC tumours (Sarvesvaran *et al.*, 1999) as well as in most primary NSCLC tumours and some NSCLC cell lines (Yokoi *et al.*, 2003b). Presumably, gene amplification creates an imbalance between the expression of *TERC* and its regulators (telomerase repressors); this might lead to increased telomerase activity and telomere stabilization.

The prime determinant of telomerase activity, however, is the reverse transcriptase component of telomerase, encoded by the *TERT* gene located at chromosome 5p15. The *TERT* gene is frequently amplified in cell lines and tumours from both SCLC and NSCLC (Zhang *et al.*, 2000a; Saretzki *et al.*, 2002) and *TERT* gene transcription has been shown to correlate with the level of telomerase activity in lung cancer cells (Fujiwara *et al.*, 2004; Lantuejoul *et al.*, 2004). Although there does not appear to be any obligatory association between *TERT* expression/telomerase activity (Fujiwara *et al.*, 2004) and telomere length, the amplification of both the *TERC* and *TERT* genes may nevertheless still contribute to the dysregulation of telomerase in at least a proportion of lung tumours. Consistent with these findings, *TERT* expression has been noted to be almost invariably present in NSCLC cells

(Fujita *et al.*, 2003; Geng *et al.*, 2003) whilst Shibuya *et al.* (2001) have reported progressively increased telomerase activity and elevated *TERT* mRNA expression during the development of squamous cell carcinoma. Yokoi *et al.* (2003b) failed to note any association between the expression levels of *TERC* and *TERT*, but the elevated expression of *TERC* was found to be significantly associated with high telomerase activity. Recent findings suggest that the expression of the *TERT* gene may be modulated by a minisatellite (MNS16A) tandem repeat polymorphism downstream of the gene (Wang *et al.*, 2003f); a reported association between the LL genotype of this MNS16A polymorphism and risk of NSCLC suggests that it could be clinically relevant.

Nakanishi *et al.* (2002) noted expression of *TERC* and *TERT* mRNA in adenomatous hyperplasia of the lung, a putative precursor of adenocarcinoma, suggesting that telomerase activation may be an early step in lung tumorigenesis. An elevated *TERT* expression level in lung tissue prospectively sampled from the area of a future tumour, but prior to clinical diagnosis, has also been claimed to be a marker of incipient squamous cell lung cancer (Snijders *et al.*, 2004). Although both Wang *et al.* (2002b) and Fujita *et al.* (2003) have reported that *TERT* expression was associated with shorter disease-specific survival and shorter disease-free survival, Wu *et al.* (2003c) found that *TERT* expression did not correlate with any clinical parameter.

Several other proteins may play important roles in telomere formation, length control and maintenance e.g. TTAGGG-binding protein (TRF1) and its interacting partners tankyrase 1, TRF1-interacting factor 2 (TIN2), telomeric repeat binding factor 1-interacting protein 1 (PinX1) and protection of telomeres 1 (POT1). TRF1 is encoded by the *TERF1* gene located at 8q13 whilst tankyrase 1 is encoded by the *TNKS* gene, also on chromosome 8. PinX1 is a potent telomerase inhibitor that is encoded by a gene (*TERF1IP*) on 8p23. Depletion of PinX1 has been reported to both increase telomerase activity and to elongate telomeres as well as to increase tumorigenicity in nude mice (Zhou and Lu 2001). The frequent loss of the *TERF1IP* gene (and possibly also the *TNKS* and *TERF1* genes) in lung cancer due to LOH at 8p (and 8q) may thus be significant for the tumorigenic process.

CHAPTER 4

Somatic Mutation in Lung Cancer

From Table 2, we can see that there are now some 120 different genes which are known to harbour somatic mutations in lung cancer, either SCLC or NSCLC. Some of these lesions are likely to be directly causative, others may be merely contributory, whilst still further mutations could be simply consequential to the greatly increased mutation rate that often accompanies tumorigenesis. Mutations that appear to be consequent to chemotherapeutic treatment [e.g. DNA topoisomerase IIα (*TOP2A*; Wessel *et al.*, 2002) and β-tubulin (*TUBB*; Kelley *et al.*, 2001)] have however not been included in Table 2. Also not included are mutations in the mitochondrial genome in both SCLC and NSCLC (Fliss *et al.*, 2000; Sanchez-Cespedes *et al.*, 2001c; Jin *et al.*, 2002; Carew and Huang 2002; Matsuyama *et al.*, 2003; Suzuki *et al.*, 2003b) even although these may correlate with stage progression and prognosis in NSCLC. Can we attempt to make some sense of the known mutational spectrum of lung cancer by grouping the different somatic gene lesions according to functional considerations e.g. common involvement in a particular signaling pathway? This is attempted below by pathway or system, using data from Tables 2, 3 and 5.

Functional consequences of somatic mutation in lung cancer

> "Carcinogenesis can be viewed as a distortion in signal transduction".
> IB Weinstein (1988) The origins of human cancer. *Cancer Res.* 48: 4135-4143.

Ras/MAP kinase pathway

The relatively common activating *RAS* mutations serve to lock Ras into its growth stimulating GTP-bound form (Figure 4.1). It should however be realised that these mutations also promote cell growth through the Rb/E2F pathway. Thus, activated Ras upregulates the synthesis of cyclin D1, thereby promoting the phosphorylation and inactivation of Rb, which in turn leads to the release of E2F (from its Rb-bound form) to transcriptionally activate a variety of genes that are required for cell cycle progression (Figure 4.1). Finally, activated Ras also stabilizes myc that is then able to stimulate cellular proliferation on its own account (Figure 4.1).

Ras has another rather interesting functional role: it induces the expression of $p14^{ARF}$ which inhibits Mdm2, which is in turn required for p53 degradation (Figure 4.1). Ras activation might therefore be predicted to increase the rate of Mdm2-me-

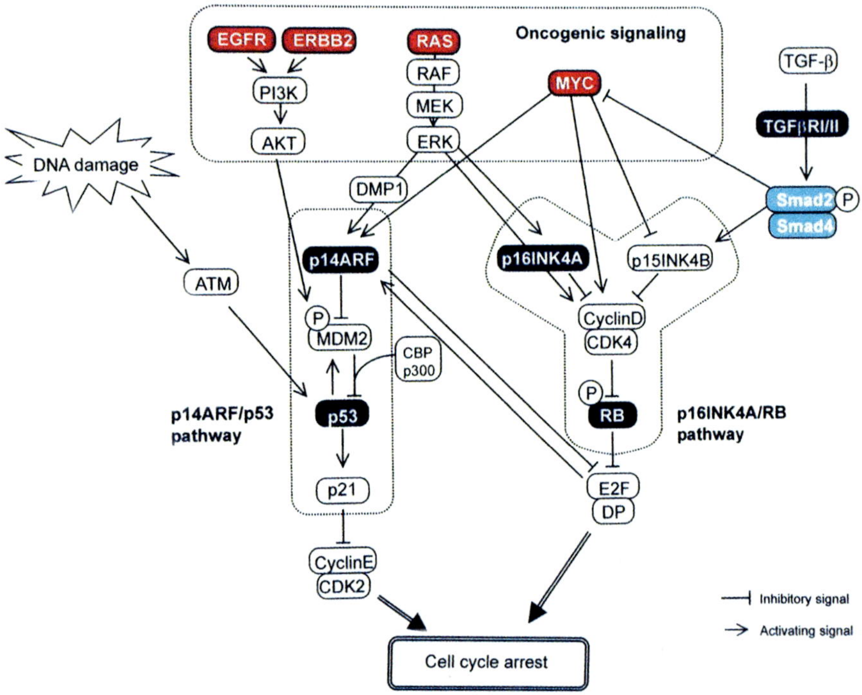

Fig. 4.1. Alterations in the p14^{ARF}/p53 and p16^{INK4A}/Rb pathways in lung cancer. The *p14^{ARF}*/p53 and *p16^{INK4A}*/Rb pathways are functionally linked and play distinct roles as components of the cell cycle checkpoints and growth inhibitory pathways. Signals from various oncogenes and growth factors are also linked to these pathways. Molecules with frequent activating alterations in lung cancer are marked in red whilst proteins encoded by genes that are frequently or infrequently inactivated in lung cancer are denoted by dark blue and light blue respectively. Reproduced from H Osada & T Takahashi, *Oncogene* 21, 7421-7434 (2002) by kind permission of Nature Publishing Group

diated degradation of p53 resulting in the abrogation of normal p53-mediated cell cycle arrest. In this context, *TP53* gene mutation and *MDM2* gene amplification are both likely to have the same effect. However, the relatively frequent co-occurrence of *KRAS* and *TP53* mutations (Mitsudomi *et al.*, 1992; Gao *et al.*, 1997) as well as the co-occurrence of *p14^{ARF}* and *TP53* mutations (Sanchez-Cespedes *et al.*, 1999) suggest that neither *KRAS* nor *p14^{ARF}* gene inactivation is functionally equivalent to abrogation of the p53 pathway by *TP53* gene mutation. A start has been made to identify RAS transformation target genes by using subtractive hybridisation to compare gene expression profiles in transformed and non-transformed cells (Zuber *et al.*, 2000); the data generated from this study have suggested that mutant *HRAS*, *KRAS* and *NRAS* may elicit similar expression profiles implying target genes in common.

Mutations in other genes in the Ras pathway (Figure 4.1) may well have similar effects. Thus amplification of the *RAF1* gene encoding the primary effector of the Ras pathway should increase the transduced signal. So should amplification (or constitutive activation, in the case of EGFRvIII) of the *EGFR* and *ERBB2* genes encoding the transmembrane receptor tyrosine kinases that serve to initiate the signaling process. Similarly, it is not unreasonable to expect that inactivation of the *PTEN* gene, which encodes a protein that inhibits two tyrosine phosphatases required for Ras activation (Figure 3.1), also increases the transduced signal output of the Ras pathway. Missense mutations in the *BRAF* gene, such as the Leu596Val and Val599Glu mutations in NSCLC (Brose *et al.*, 2002), display greatly increased kinase activity (Wan *et al.*, 2004). Such mutations are therefore similar to *RAS* mutations in terms of their ability to induce constitutive ERK activity thereby exerting a profound influence on the G_1/S transition of the cell cycle through dysregulation of a number of cell cycle proteins such as cyclins D and E and p21 (Mercer and Pritchard 2003). Interestingly, it has been noted that *BRAF* and *RAS* mutations are rarely found to be present in the same tumours (Rajagopalan *et al.*, 2002; Kim *et al.*, 2004b), although the cancer types with *BRAF* mutations are very similar to those with *RAS* mutations (Mercer and Pritchard 2003). This argues for their functional equivalence because no additional selective advantage would accrue to the tumour cells by mutation of one of these genes once the other had been inactivated. The increased expression of the *JUN* and *ETS1* genes (Table 5) may also promote lung cell proliferation by enhancing Ras signaling (Figure 3.1; Xiao and Lang 2000).

The Ras effector homologue, RASSF1A, mediates Ras-dependent apoptosis (Khokhlatchev *et al.*, 2002) and so inactivation of the *RASSF1* gene by promoter hypermethylation might be expected to exert an anti-apoptotic effect. Since the RASSF1A protein also serves as a microtubule-binding protein, its loss could affect spindle assembly and chromosome attachment leading to chromosomal instability (Liu *et al.*, 2003b). By interacting with Cdc20, RASSF1A inhibits the anaphase-promoting complex and prevents the degradation of cyclins A and B until the spindle checkpoint becomes fully operational (Máthé 2004). The artificial depletion of RASSF1A by RNA interference leads to cyclin degradation and mitotic progression (Song *et al.*, 2004). The loss of *RASSF1* gene expression in lung tumorigenesis as a consequence of promoter hypermethylation may therefore contribute to cellular proliferation at the same time as increasing chromosomal instability and inhibiting apoptosis.

A missense mutation reported in the *MOS* gene in a NSCLC tumour (Gorgoulis *et al.*, 2001) could in principle be important in promoting lung tumorigenesis. The *MOS* gene encodes a serine/threonine protein kinase that is thought to play a role in cell cycle regulation (Singh and Arlinghaus 1997). The *MOS* gene mutation may adversely affect the MAP kinase pathway since most known activities of Mos are mediated through this pathway.

The reported lung cancer-associated homozygous deletion of the SERK1 (*MAP2K4*) gene (Teng *et al.*, 1997), which encodes mitogen-activated protein kinase kinase 4, may lead to defective activation of JNK and AP-1-dependent transcriptional activity in response to some forms of cellular stress (Cuenda 2000). It is however less obvious why inactivation of the *MAP2K2* and *MAP2K3* genes, both encoding mitogen-activated protein kinase kinases, should lead to a Ras-mediated increase in cellular proliferation.

Rb/E2F pathway

In its hypophosphorylated state during the first part of G_1, Rb binds the transcription factor E2F thereby preventing it from activating genes whose products are essential for S phase. Cyclin-cdk complexes are responsible for phosphorylating Rb thereby releasing E2F and allowing exit from G_1 with subsequent progress through the cell cycle. Cdk inhibitors p15, p16, p21 and p27 inactivate the cyclin/cdk complexes thereby preventing Rb phosphorylation and maintaining the cell in G_1 arrest. The loss of Rb through mutational inactivation (Table 2) or reduced transcription (Table 5) of the *RB1* gene removes regulatory control over E2F resulting in cellular proliferation (Figure 4.1). *MDM2* gene over-expression or amplification may be expected to inhibit Rb directly (Yap *et al.*, 1999) whilst over-expression of the genes encoding cyclins A2 (*CCNA2*), D1 (*CCND1*) and E (*CCNE1*) should all promote Rb phosphorylation leading to release of inhibition of E2F. The loss of the p16 (*CDKN2A*), p15 (*CDKN2B*) and p57 (*CDKN1C*) genes (Table 2) and the observed reduction in expression of p27 (*CDKN1B*) in NSCLC (Table 5) should all contribute to a failure to inactivate the cyclin/cdk complexes that promote Rb phosphorylation, again leading to release of inhibition of E2F.

E2F1 gene over-expression has recently been noted in up to 82% of NSCLCs examined (Gorgoulis *et al.*, 2002) whilst *E2F1* gene amplification has been found to be present in 9% of NSCLC tumours (Gorgoulis *et al.*, 2002). Over-expression of the *E2F1* gene may result from loss of *RB1* gene expression or alternatively, through hyperphosphorylation of Rb (Imai *et al.*, 2004). Over-expression of the *E2F1* gene could in some cases also result from *MYC* gene amplification. Whatever the underlying reasons for the over-expression of the *E2F1* gene, E2F1 is an important transcriptional activator and increased cellular proliferation would be one predicted consequence (Cam and Dynlacht 2003). However, since E2F1 also functions in a DNA damage checkpoint and in p53-dependent apoptosis (Cam and Dynlacht 2003), there may well be other consequences of *E2F1* gene over-expression.

p53 pathway

The loss of the *TP53* gene abrogates both DNA damage-induced cell cycle arrest and p53-mediated apoptosis (Figure 4.2). With p53 non-functional, DNA damage is not repaired and lesions may be propagated to successive cell generations at the same time as apoptosis is compromised. *MDM2* gene amplification and/or over-expression are predicted to lead to excessive inhibition of p53 as would the removal of $p14^{ARF}$ due to inactivation of the *CDKN2A* gene (Yap *et al.*, 1999). Mdm2 over-expression and $p14^{ARF}$ inactivation are considered to be mutually exclusive events in lung cancer tumorigenesis (Eymin *et al.*, 2002). This is because there would be no additional growth advantage conferred by inactivation of one of these factors if the other had already been inactivated thereby compromising the p53 pathway.

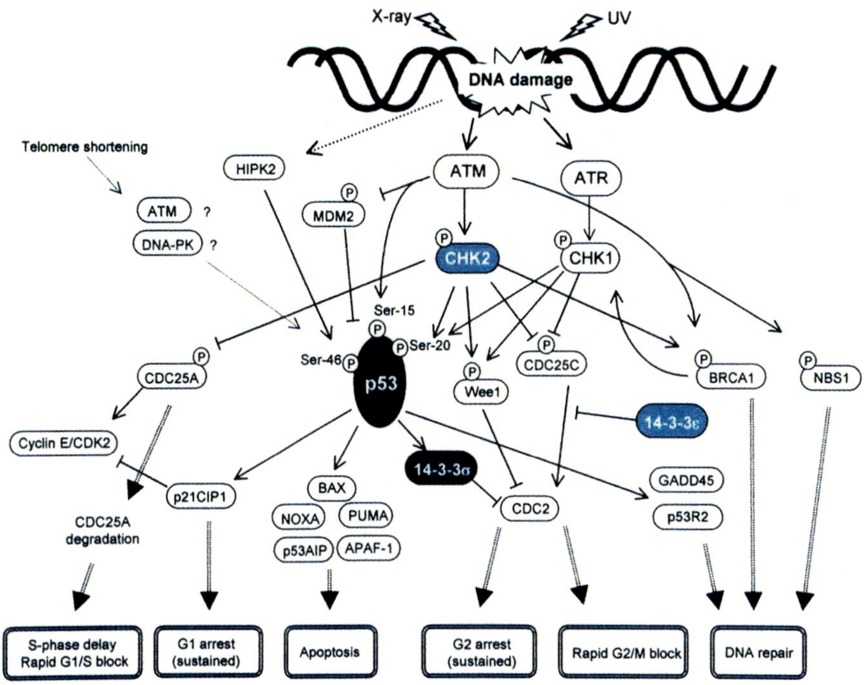

Fig. 4.2. p53 functions and cell cycle checkpoints. p53 plays an important role in the cell cycle checkpoints and is frequently inactivated by mutation in lung cancer. DNA damage activates the ATM and ATR kinases and results in the phosphorylation and consequent activation of p53, leading to the transactivation of p53-responsive genes that encode proteins which function in cell cycle arrest, apoptosis, DNA repair and centrosome duplication. Reproduced from H Osada & T Takahashi, *Oncogene* 21, 7421-7434 (2002) by kind permission of Nature Publishing Group

APC/β-catenin pathway

The Wnt signaling pathway is involved in developmental pattern formation in embryogenesis and leads to tumorigenesis when constitutively activated (Giles *et al.*, 2003; Kikuchi 2003; Nelson and Nusse 2004). Thus, the loss of the *APC* gene through mutation (D'Amico *et al.*, 1992; Sanz-Ortega *et al.*, 1999) removes the negative regulatory control mechanism acting on β-catenin thereby freeing it up to activate the *MYC* oncogene via Tcf4 (Figure 4.3). In addition, if as has been proposed (Jaiswal *et al.*, 1999), there is an inverse relationship between APC and myc levels that may imply mutual co-regulation, one consequence of *MYC* gene amplification in lung cancer could be the down-regulation of the expression of the *APC* gene.

Amplification of the *MYC* gene also promotes cellular proliferation by overcoming Max-mediated transcriptional repression. Thus, in normal development, the growth stimulating action of myc in association with Max can be supplanted by alternative complexes of Max with a group of Mad transcription factors (Mad-Max complexes) that elicit differentiation-inducing signals. Over-expression of myc through *MYC*

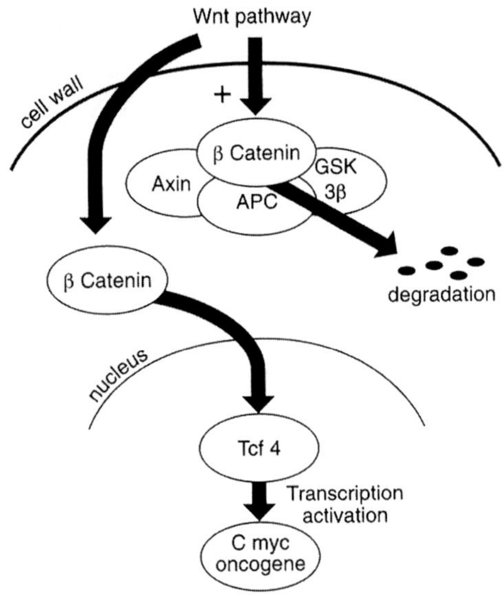

Fig. 4.3. Mechanism of APC activation of the myc oncogene. Reproduced from Fig. 4 (page 238), Chapter 10, Colorectal cancer, by PD Chapman and J Burn (2001) in *Molecular Genetics of Cancer 2nd Ed.*, Ed. JK Cowell, (2001) by kind permission of Taylor & Francis/BIOS, Oxford]

gene amplification can reverse this process shifting the balance back in favour of the formation of myc-Max complexes, thereby hindering differentiation and hence promoting growth.

At least some of the mutations in the β-catenin (*CTNNB1*) gene (Table 2) serve to constitutively activate the Wnt signaling pathway (Giles *et al.*, 2003), with the predicted effect of inhibiting GSK3β and stabilizing β-catenin (Sunaga *et al.*, 2001). Since serine-threonine phosphatase 2A (PP2A) is thought to downregulate β-catenin by dephosphorylating and activating GSK3β (Ruediger *et al.*, 2001), mutational inactivation of the two genes encoding the α and β isoforms of PP2A subunit A (*PPP2R1A* and *PPP2R1B*) may also serve to stabilize β-catenin. Mutations in the LKB1 (*STK11*) gene are not uncommon in NSCLC (Table 4), and since LKB1, a serine-threonine protein kinase, is a regulator of Wnt signaling (Ossipova *et al.*, 2003), such lesions may be responsible for the alterations in Wnt signaling noted in lung cancer.

Impaired expression of the E-cadherin (*CDH1*) gene has been reported in NSCLC (Bremnes *et al.*, 2002b; Fei *et al.*, 2002; Yanagawa *et al.*, 2003) but is negatively associated with tumour differentiation and metastasis, and may be associated with a poor prognosis in NSCLC (Liu *et al.*, 2001; Bremnes *et al.*, 2002b; Hirata *et al.*, 2002; Hazan *et al.*, 2004). Since E-cadherin is known to be induced by the wingless (Wnt) pathway (Ohira *et al.*, 2003), and expression of the *WNT7A* gene is frequently downregulated in lung cancer (Calvo *et al.*, 2000), it may be that the mutational loss of the *WNT7A* gene also plays a role in lung tumorigenesis. Finally, the Wnt pathway is also activated in NSCLC through over-expression of the Dishevelled (*DVL3*) gene, although the upstream events responsible remain unclear (Uematsu *et al.*, 2003).

TGFβ signaling pathway

TGFβ acts as an inhibitor of cellular proliferation by inducing cdk inhibitors and by down-regulating myc. It binds to TGFβII receptor, leading to phosphorylation of the TGFβI receptor, which in turn phosphorylates SMAD proteins thereby inducing their translocation to the nucleus (Wong and Lai 2001; Figure 4.1). TGFβ also promotes the synthesis of the $p15^{INK4B}$ and p21 cell cycle control proteins which block the formation of the cyclin:cdk complexes responsible for Rb phosphorylation. Since the role of the TGFβ signaling pathway is to block cellular proliferation, the described mutations inactivating the TGFβ2 receptor (*TGFBR2*), SMAD2 (*MADH2*) and SMAD4 (*MADH4*) genes (Table 2), and the epigenetic inactivation of the SMAD8 (*MADH9*) gene, would all be predicted to lead to the abrogation of this negative control mechanism (Piek and Roberts 2001; Park *et al.*, 2002a; Cohen 2003; Miyaki and Kuroki 2003). These mutational lesions are however relatively infrequent and cannot fully account for the observation that most lung cancer cell lines have lost their growth inhibitory response to TGFβ. This appears to be caused by loss of *TGFBR2* gene expression due to histone deacetylation (Norgaard *et al.*, 1996; Osada *et al.*, 2001) although the trigger for this change in chromatin modification is unknown. Interestingly, the presence of a mutant p53 has also been shown to diminish growth inhibition by TGFβ in bronchial epithelial cells (Gerwin *et al.*, 1992).

Protein phosphatases

The attention of lung cancer researchers was initially drawn to the protein phosphatases by the frequent inactivation of the *PTEN* gene in both SCLC and NSCLC. The *PTEN* gene encodes a lipid phosphatase that negatively regulates cell survival mediated by the phosphatidylinositol 3-kinase/Akt signaling pathway (Vivanco and Sawyers 2002; Figure 4.4).

A number of other genes encoding protein phosphatases have however also been found to be mutated in human cancer, including those encoding subunits of the serine/threonine phosphatases PP1 and PP2A (Takakura *et al.*, 2001). PP1 also negatively regulates the cell cycle and one of its 10 regulatory subunit genes (*PPP1R3A*) has been found to be inactivated in both SCLC and NSCLC (Table 2). The *PPP1R3A* gene encodes the PP1 regulatory subunit 3 that regulates glycogen metabolism by directing the PP1 catalytic subunit(s) to glycogen particles or to the membranes of the sarcoplasmic reticulum. The PP2A holoenzyme down-regulates the mitogen-activated protein kinase (MAPK) cascade and is involved in DNA replication, cell proliferation and cell fate determination. The *PPP2R1A* and *PPP2R1B* genes, that encode respectively subunits Aα and Aβ of PP2A, are both known to be altered in lung cancer (Table 2). We may surmise that mutational inactivation of PP2A could have similar downstream consequences to the inhibition of the phosphatase. Inhibition of PP2A by okadaic acid is known to result in a significant increase in Jun N-terminal kinase (JNK), leading to increased Jun phosphorylation and AP-1 transcriptional activity (Shanley *et al.*, 2001). Finally, consistent with the hypothesis that PP2A functions as a tumour suppressor, suppression of PP2A expression has been shown to

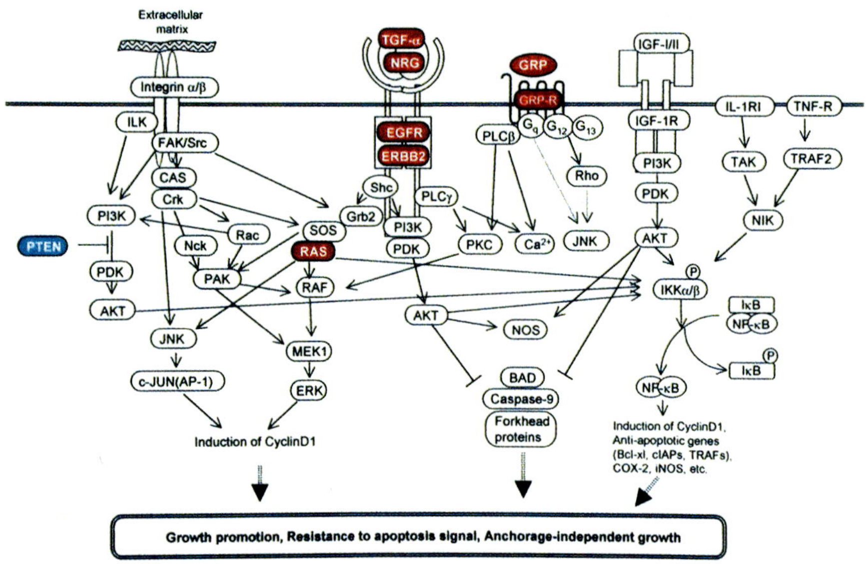

Fig. 4.4. Activation of growth promoting pathways in lung cancer. Cell proliferation is positively regulated by various growth promoting pathways including the MAPK cascades, and the PI3K-Akt, PLC-PKC and NF-κB pathways. Growth factor-mediated pathways may diverge through a tyrosine kinase receptor into the RAS-RAF-MAPK, PI3K-Akt, and PLC-PKC pathways. NF-κB is convergently activated by several distinct stimuli including growth factors, cytokines, and extracellular matrix attachment. The PI3K-Akt signaling pathway phosphorylates and inhibits several apoptosis-inducing genes. Autocrine secretion of the EGF family and/or overexpression of the EGF receptor family are frequently observed in lung cancer. In SCLC, autocrine activation of GRP-mediated signaling is also frequently present, leading to activation of the ERK and JNK cascades. Stimulatory signaling via members of the EGF receptor family and various integrins coordinately promote cell proliferation. Reproduced from H Osada & T Takahashi, *Oncogene* 21, 7421-7434 (2002) by kind permission of Nature Publishing Group

promote tumour formation in nude mice whilst PP2A overexpression serves to reverse the induced tumorigenicity (Chen *et al.*, 2004d).

Two genes, *PTPRF* and *PTPRT*, encoding receptor-like transmembrane protein tyrosine phosphatases have recently also been shown to be mutated in lung cancer (Wang *et al.*, 2004b). These phosphatases are membrane proteins that possess both cytoplasmic and extracellular domains. Expression of wild-type but not mutant forms of PTPRT in human cancer cells inhibited cellular growth, consistent with the protein's postulated role as a tumour suppressor (Wang *et al.*, 2004b).

Retinoic acid-mediated growth inhibition

All-*trans* retinoic acid serves to inhibit cell proliferation, an effect mediated via retinoic acid receptors α, β and γ and EGF activation of the mitogen-activated protein kinase ERK1 (Crowe *et al.*, 2003). Retinoic acid receptor β also induces apoptosis. Lung cancer cells frequently lose expression of the retinoic acid receptor β (*RARB*) gene either through gene rearrangement, LOH or promoter hypermethylation (Tables 2 and 3; Figure 2.4) and as a consequence lose their retinoic acid responsiveness (Geradts *et al.*, 1993). Weber *et al.* (1999) have shown that *RARB* gene transfer into SCLC cells can restore retinoic acid sensitivity to the $RARB^-$ cells, leading to growth inhibition and apoptosis.

DNA repair

Several different, yet overlapping and interwoven, DNA repair processes have been identified in human cells (Tuteja and Tuteja 2001; Moses 2001; Hoeijmakers 2001; Rouse and Jackson 2002) involving the products of at least 133 known DNA repair genes (Wood *et al.*, 2001; http://www.cgal.icnet.uk/DNA_Repair_Genes.html). These will be briefly described. Where genes encoding proteins involved in these pathways have been shown to be compromised in lung cancer, this will be highlighted.

(i) *Base excision repair (BER)*

BER removes modified or damaged bases from DNA and is perhaps the most important mechanism for protecting DNA from oxidative damage whether this occurs as a result of exposure to endogenous reactive oxygen species or the action of exogenous mutagens (Cooke *et al.*, 2003; Slupphaug *et al.*, 2003). Two forms of BER have been documented: a 'short patch' pathway involving the replacement of a single nucleotide and a 'long patch' pathway that fills in gaps of up to 14 bases. Short patch repair is initiated by recognition of the adduct followed by excision of the modified base by DNA glycosylases leaving a single base gap. The single base gap is then filled by apurinic/apyrimidinic endonuclease (*APEX1*), DNA polymerase beta (*POLB*) and DNA ligase I (*LIG1*) or III (*LIG3*) together with the accessory protein XRCC1 (*XRCC1*). XRCC1 stimulates polynucleotide kinase activity at damaged DNA termini and accelerates DNA single-strand break repair. Certain polymorphic variants in the *XRCC1* gene have been reported to be associated with an increased risk of lung cancer (Table 4) as well as an increased risk of single strand break formation (Vodicka *et al.*, 2004). Long patch repair, by contrast, is dependent upon proliferating cell nuclear antigen (PCNA) and DNase IV.

XRCC1 is physically and functionally associated with 8-oxoguanine (or 8-hydroxyguanine) DNA glycosylase (OGG). OGG acts to remove 8-oxoguanine, the major base lesion resulting from exposure to reactive oxygen species (Olinski *et al.*, 2003). Loss of heterozygosity involving the 3p25-located gene encoding this glycosylase, *OGG1*, is apparent in 40-60 % of NSCLC primary tumours (Table 2) whilst missense mutations in the gene have been reported in SCLC (Chevillard *et al.*, 1998). Polymorphic variants in the *OGG1* gene have also been reported to be associated with an increased risk of lung cancer (Table 4). For a detailed description of the putative association between low OGG activity and lung cancer risk (Paz-Elizur et al., 2003), *see* Chapter 6; DNA Repair Activity for Oxidative Damage; the Contribution of OGG1.

MYH is an A/G mismatch-nicking endonuclease that works with OGG to correct A/G and A/C mismatches. Inherited defects in the *MUTYH* gene that encodes human MYH are associated with an increased risk of colorectal cancer as a consequence of base excision repair being compromised (Parker and Eshleman 2003). Moreover, individuals with such *MUTYH* mutations display an increased frequency of somatic G→T transversions in their colorectal tumours. *Myh* null mice have been found to display an increased incidence of lung cancer with a high frequency (75 %) of tumours possessing G→T transversions at the hotspot codon 12 in the *Kras* gene (Xie *et al.*, 2004). Since human lung tumours are known to be characterized by a preponderance of G→T transversions, Al-Tassan *et al.* (2004) screened 276 patients with lung cancer and 106 controls for *MUTYH* gene mutations. Several novel missense mutations were noted as well as a Gly382Asp substitution already reported to be associated with colorectal cancer. However, none of these lesions were over-represented in the cohort of patients by comparison with the controls. Their relevance to a predisposition to lung cancer must therefore remain doubtful.

(ii) *Nucleotide excision repair (NER)*

NER removes oligonucleotide fragments (more than one base) in response to distortion of the DNA structure. One example of DNA damage targeted by NER is the UV-induced pyrimidine dimer but NER also protects DNA against mutations caused by environmental carcinogens (Friedberg 2001). One NER subpathway is *global genomic repair*, which deals with all reparable lesions in the genome (Hanawalt *et al.*, 2003). NER proceeds at a faster rate on the transcribed strand as compared with the non-transcribed strand, hence the term used for the other subpathway, *transcription-coupled repair.*

The first damage-recognizing step of NER involves XPC (*XPC*) and XPA (*XPA*) in association with single stranded DNA-binding protein replication protein (RPA; *RPA1*), ERCC1 (*ERCC1*) and XPE (*DDB2*). Several DNA helicases including XPB (*ERCC3*) and XPD (*ERCC2*) participate in recognition of the lesion and in the removal of the damaged segment by nucleases XPF (*ERCC4*) and XPG (*ERCC5*). XPD (*ERCC2*) is also thought to regulate the activity of the cdk-activating kinase complex and to control mitotic progression (Chen *et al.*, 2003e). The gap filling of the excised patch is performed by DNA polymerases epsilon (*POLE*) and delta (*POLD*) in association with replication factor C (RFC; *RFC1*), proliferating cell nuclear antigen (PCNA; *PCNA*) and RPA. The final step of NER is the ligation of the newly synthesized DNA to the existing DNA by DNA ligase I. Loss of heterozygosity involving the *ERCC3* gene has been reported in ~20 % of lung tumours (Table 2) whilst polymorphic variants in the *XPA*, *ERCC5* and *ERCC2* genes have been found to be associated with an increased risk of lung cancer (Table 4).

(iii) *Double strand break repair*

Double strand breaks can be caused by ionising radiation or reactive oxygen intermediates or can arise when DNA replication forks stall at sites of DNA damage. Double strand break repair acts to repair double strand breaks by employing the proteins of either the homologous recombination apparatus (e.g. RAD51, BRCA1, BRCA2, NBS1; Thompson and Schild 2002) or those of non-homologous end joining [e.g. Ku70P (*G22P1*), Ku80p (*XRCC5*), DNA ligase IV (*LIG4*), ATM (*ATM*), XRCC4 (*XRCC4*) and Artemis (*DCLRE1C*); Lieber *et al.*, 2003].

BRCA1 is thought to play a role in transcription-coupled repair, nucleotide excision repair as well as double strand break repair (Jasin 2002; Hartman and Ford 2002; Deng and Wang 2003; Figure 4.2). It interacts with various proteins that function in DNA repair and recombination, and in cell cycle checkpoints (e.g. RAD50, RAD51, Rb and p53). BRCA1 is phosphorylated by ATM and/or ATR after DNA damage caused by ionising radiation. BRCA1 also interacts with the Fanconi anaemia protein FANCD2, which is similarly phosphorylated by ATM in response to ionising radiation. FANCD2 is activated (monoubiquitinated) in response to DNA crosslinking by a complex of Fanconi anaemia proteins including FANCF (Bogliolo *et al.*, 2002). The *FANCF* gene has been found to be inactivated by promoter hypermethylation in 14% of NSCLC tumours (Marsit *et al.*, 2004). Inactivation of the *BRCA1* gene by promoter hypermethylation has also been reported in NSCLC tumours (Table 3).

(iv) *Mismatch repair*
Mismatch repair corrects errors occurring during DNA replication, repair and recombination that give rise to nucleotides that are incorrectly paired with nucleotides on the opposite DNA strand. The incorrect base(s) usually occur on the newly synthesized strand, as a consequence of the error-prone nature of the DNA replication machinery, and the cell is able to discriminate between the newly replicated and parental DNA strands. Mismatch repair also contributes to genome stability by inducing apoptosis through activation of protein kinases that activate p53 and p73 (Li 1999), and by facilitating the phosphorylation (and hence activation) of checkpoint kinase Chk2 by ATM (Brown *et al.*, 2003). The proteins of the human mismatch repair complexes that are formed at the sites of DNA damage, are encoded by several genes including *MLH1, MSH2, MSH3, MSH6, PMS1, PMS2* and *MUTYH*. Loss of heterozygosity (LOH) for both the *MLH1* and *MSH3* genes has been reported in a considerable proportion of NSCLC tumours (Table 2) whilst promoter hypermethylation has been reported as a cause of inactivation of the *MLH1* and *MSH2* genes in NSCLC (Table 3). A solitary missense mutation in the *MUTYH* gene of uncertain significance has been reported in a lung cancer cell line (Table 2).

A number of other mechanisms of DNA repair exist to deal with specific types of mutation. For example, alkylation of DNA at the O6 position of guanine is a common pro-carcinogenic adduct on account of the preference that O6-alkylguanine has for pairing with thymine during DNA replication. O6-alkylguanine is removed by O6-methylguanine DNA methyltransferase (encoded by the *MGMT* gene) which transfers the offending alkyl group to its own molecule, and in so doing is inactivated (Gerson 2004). The *MGMT* gene is reported to be inactivated by promoter hypermethylation in ~20% of NSCLC tumours (Table 2) whilst a polymorphic variant in the *MGMT* gene has been reported to be associated with an increased risk of lung cancer (Table 4). Loss of MGMT activity results in the persistence of O6-methylguanine adducts that are then misread as adenines by DNA polymerase leading to the occurrence of G:C to A:T transitions (Esteller and Herman 2004; Gerson 2004).

Cell cycle control and DNA damage checkpoint genes

When cells leave G_0 (quiescence), they enter a gap phase (G_1) before S phase (DNA synthesis) proceeds (Figure 1.1). Many signaling pathways exert their influence in G_1. A second gap phase (G_2) then precedes mitosis. The order of events in the cell cycle is regulated by various checkpoints which also ensure that DNA repair is coordinated with cell cycle progression. These checkpoints are regulated by several cyclin/cyclin-dependent kinase (cdk) complexes (that phosphorylate key substrates that then control the various cell cycle transitions), and cdk inhibitors (Figure 1.2). In order to bring about cell cycle arrest, for example after DNA damage, checkpoint control pathways must sense the damage and then transduce the signal to alter the activity of key cell cycle regulators. DNA damage checkpoints act at three different stages in the cell cycle: induction of G_1 arrest, blockage of DNA replication or a G_2 delay, depending on the type of damage experienced and the stage of the cell cycle when the damage is detected (Malumbres and Barbacid 2001).

Mutations in the cell cycle checkpoint genes may thus lead to aberrant cell cycle progression in the presence of DNA damage, leading to genomic instability (Cahill *et al.*, 1998). Indeed, a number of genes encoding proteins of the cell cycle are known to be mutated in lung cancer (Hall and Peters 1996; Molinari 2000; Zhou *et al.*, 2001a). Positive regulators of cell cycle progression include cyclin D1, cyclin E, cdk4, cdc25A and myc. The amplification and/or over-expression of the cyclin A2 (*CCNA2*), cyclin D1 (*CCND1*), cyclin E (*CCNE1*) and *MYC* genes have been noted in lung cancer (Tables 2 and 5). Negative regulators of cell cycle progression include p53, $p14^{ARF}$, Rb, TGFβ, Chk2, p27, Bub1 and the cdk inhibitors p15 and p16. The genes encoding p53 (*TP53*), Rb (*RB1*), p15 (*CDKN2B*), $p14^{ARF}$ and p16 (*CDKN2A*) and p57 (*CDKN1C*) are known to be either mutated or inactivated by promoter hypermethylation in lung cancer (Tables 2 and 3).

Chk2, encoded by the *CHEK2* gene, is a downstream effector of the ATM-dependent DNA damage checkpoint pathway (Castedo *et al.*, 2004; Figure 4.2). This mitotic checkpoint protein is activated upon DNA damage and its phosphorylation of both p53 (effecting its stabilization) and the *BRCA1* gene product is essential for damage-induced arrest at G_1 and G_2 checkpoints and ensuing DNA repair (Figure 4.2). The stabilization of p53 leads to cell cycle arrest in G_1. Chk2 is also involved in mediating p53-independent apoptosis. Missense mutations in the Chk2 (*CHEK2*) gene have been reported in both SCLC and NSCLC (Table 2) and if inactivating, could exert adverse effects on both cell cycle arrest and apoptosis. *CHEK2* gene inactivation may however arise more frequently through promoter hypermethylation (Zhang *et al.*, 2004b).

The spindle assembly checkpoint, which is essential for the maintenance of genomic integrity, modulates the timing of anaphase initiation in mitotic cells that contain improperly aligned chromosomes (Bharadwaj and Yu 2004). It senses both microtubule attachment on centromeres and their tension, and acts so as to inhibit the metaphase-anaphase transition until any damage is repaired (Zhou *et al.*, 2002a). *BUB1* is a member of a family of genes encoding proteins some of which bind to the centromere and all of which are required for a normal mitotic delay in response to spindle disruption (Figure 3.7). The *BUB1* gene encodes a protein kinase required for chromosome segregation and the control of mitotic progression. A missense mutation in the *BUB1* gene has been reported in a lung adenocarcinoma (Gemma *et al.*, 2000a). Were this to turn out to be an inactivating mutation, it would be predicted to cause premature mitotic exit. This type of mitotic checkpoint defect could be one cause of chromosomal instability (CIN), a process which leads to aneuploidy, the acquisition of abnormal chromosome number (Jallepalli and Lengauer 2001; Bharadwaj and Yu 2004).

Another gene whose dysfunction is associated with chromosome instability is the *MAD1L1* gene encoding the mitotic checkpoint protein Mad1 (mitotic arrest deficient). Again, a solitary missense mutation has been reported in the *MAD1L1* gene in an NSCLC tumour (Nomoto *et al.*, 1999). If this were to prove to be an inactivating mutation, then loss of Mad1 function would be expected to result in loss of centromere localization leading to chromosome instability.

The *SKP2* gene encodes S-phase kinase-associated protein 2 (a ubiquitin ligase), an essential component of the cyclin A-cdk2 S-phase kinase. SKP2 is thought to serve as a bridge between SKP1 and the cdk2/cyclin A complex, and is required for the ubiquitin-mediated degradation of p27 thereby positively regulating the G_1-S transition. SKP2 is also involved in down-regulating Rb2/p130 during the G_1-S transition. SKP2 also regulates myc, both at the transcriptional level and by modulating ubiquitinylation-mediated stability (Kim *et al.*, 2003 g). The *SKP2* gene has been reported to be both amplified and over-expressed in ~40 % of SCLC tumours (Yokoi *et al.*, 2002). Since an inverse correlation has been noted between SKP1 and p27 expression (Yokoi *et al.*, 2002), and since the antisense oligonucleotide-mediated down-regulation of *SKP2* gene expression has been found to suppress the growth of SCLC cells (Yokoi *et al.*, 2002), it may well be that *SKP2* gene amplification plays an important role in promoting cell growth in SCLC. Alternatively, since the antisense-mediated down-regulation of the *SKP2* gene induces apoptosis in lung cancer cells (Yokoi *et al.*, 2003a), *SKP2* gene amplification may serve to inhibit apoptosis thereby promoting cell survival.

Another ubiquitin ligase that acts as a mitotic checkpoint protein is Chfr, a protein containing forkhead-associated and RING finger domains. Chfr regulates a prophase delay in cells experiencing microtubule disruption by negatively regulating the activation of Cdc2 kinase at the G_2-M transition (Scolnick and Halazonetis 2000). Thus, cells expressing wild-type Chfr exhibit delay entry into metaphase when centrosome separation is inhibited by mitotic stress. By contrast, tumour cells that have lost Chfr function enter metaphase without delay. Inactivation of the *CHFR* gene [by promoter hypermethylation in 10-20 % of lung tumours (Mizuno *et al.*, 2002; Corn *et al.*, 2003; Table 3) and by mutation in at least 5 % of lung tumours (Mariatos *et al.*, 2003; Table 2)] may therefore lead to the propagation of chromosomal damage by premature passing of the G_2-M damage checkpoint.

Apoptosis signaling pathways

"Amid the miseries of our life on earth, suicide is God's best gift to man"
Pliny the Elder (c.70 AD) *Naturalis Historiae* II.

When apoptosis is inhibited, one of the protective mechanisms safeguarding against tumorigenesis becomes inoperative. In lung cancer, apoptosis has been reported to be frequently compromised, with low apoptotic capacity in cultured lymphocytes being found to be an independent risk factor for lung cancer (OR=2.69, 95% CI 1.18-6.15) after adjustment for age, sex, ethnicity and smoking status (Wang *et al.*, 2003e).

p53 is of course key to the p53-dependent apoptotic pathway and therefore inactivation of the *TP53* gene abrogates this pathway (Figure 4.2). Indeed, *TP53* gene inactivation may be the most frequent method of circumventing apoptosis in lung cancer. This notwithstanding, inactivation of the LKB1 (*STK11*) gene may also influence apoptosis since LKB1 physically associates with p53 and serves to regulate p53-dependent apoptosis (Karuman *et al.*, 2001). Other than *TP53*, there is no single major *TP53*-dependent apoptosis-inducing gene. Rather, the cell response is dependent on the combined effects of the altered expression of numerous *TP53*-responsive genes (Figure 3.3). The reduced expression of Fas and Fas ligand (Viard-Leveugle *et al.*, 2003), the mutational inactivation of the Fas (*TNFRSF6*) gene (Table 2) and the high expression of the *BCL2* gene (Table 5) would also be predicted to inhibit apoptosis (Figure 1.3). Defects in the DR5/TRAIL receptor 2 (*TNFRSF10B*) gene (Table 2) are likely to block the TRAIL-mediated induction of apoptosis. The Fas ligand decoy receptor 3 (*TNFRSF6B*) gene has been noted to be amplified in 44% of lung tumours. This decoy receptor is capable of blocking Fas ligand and therefore *TNFRSF6B* gene amplification may allow tumours to escape FasL-dependent immune attack.

Inactivating mutations have been reported in the genes encoding the pro-apoptotic factors, fas apoptotic inhibitory molecule (*FAIM*), fas-associated via death domain (*FADD*), caspases 3, 5, 8 and 10 (*CASP3*, *CASP5*, *CASP8*, *CASP10*) and DR4/TRAIL receptor 1 (*TNFRSF10A*), in lung tumours or cell lines (Table 2; reviewed by Shivapurkar *et al.*, 2003). These alterations are broadly understandable in that they have been selected on the basis that they confer upon the tumour cells the ability to avoid programmed cell death. The same goes for the reported amplification of the *BIRC2* and *BIRC3* apoptosis inhibitor genes (Table 2), the methylation-mediated inactivation of the caspase 8 (*CASP8*), *PYCARD* ('apoptosis-associated speck-like protein containing a card'/'target of methylation-induced silencing 1') and death-associated protein kinase 1 (*DAPK1*) genes (Inbal *et al.*, 1997; Table 3) and the increased expression of the apoptosis inhibitor 5 (*API5*) [Sasaki *et al.*, 2001b], amphiregulin (*AREG*) [Hurbin *et al.*, 2002], *BIRC3* and *BIRC4* (Hofmann *et al.*, 2002) and survivin (*BIRC5*) [Monzo *et al.*, 1999a] genes observed in NSCLC (Table 5). Cause and effect are as ever difficult to disentangle; thus, since p53 negatively regulates survivin (*BIRC5*) gene expression (Mirza *et al.*, 2002), *BIRC5* gene expression may be, at least in part, up-regulated as an indirect consequence of the *TP53* gene dysfunction that is so common in lung cancer. Since survivin also plays an important role as a molecular sensor of the spindle checkpoint (Bharadwaj and Yu 2004), this protein is precisely positioned at the interface between the regulation of apoptosis and the control of cellular proliferation.

Observations inconsistent with selection for the avoidance of programmed cell death are the over-expression of the pro-apoptotic death receptors DR4/TRAIL receptor 1 (*TNFRSF10A*) and DR5/TRAIL receptor 2 (*TNFRSF10B*) genes in NSCLC tumours (Wu *et al.*, 2000; Spierings *et al.*, 2003), the increased expression of the *TNFSF10* (TRAIL) gene (Table 5), the over-expression of the programmed cell death 6 (*PDCD6*) gene in both SCLC and NSCLC (La Cour *et al.*, 2003) and the loss of expression of the anti-apoptotic *BHLHB2* gene in ~60% of NSCLC tumours (Li *et al.*, 2002; Giatromanolaki *et al.*, 2003). It may be that these changes in gene expression represent part of the pro-apoptotic response mounted by cells in the wake of genomic damage or they may simply be consequential to other genomic changes and therefore not the result of cellular selection. Also inconsistent with selection for the avoidance of programmed cell death is the high level expression of the pro-apoptotic *BAX* gene observed in $>70\%$ of NSCLC tumours (Caputi *et al.*, 1999). It should however be noted that another report has claimed Bax to be expressed in only 26% of lung cancer biopsies (Badillo-Almaraz *et al.*, 2003).

A number of additional genes, whose proteins have been discussed in the context of other signaling pathways and processes, have proapoptotic functions; thus mutational inactivation of these loci would also serve to inhibit apoptosis. These genes include *FHIT* (Sard *et al.*, 1999), *IGF2R* (Gemma *et al.*, 2000b), WW domain-containing oxidoreductase (*WWOX*; Chang *et al.*, 2001), *TP73* and *TP73L* (Jost *et al.*, 1997; Davis and Dowdy 2001; Flores *et al.*, 2002), serine/threonine protein kinase 11 (*STK11*; Karuman *et al.*, 2001) and retinoic acid receptor β (*RARB*). The S phase kinase-associated protein 2 (*SKP2*) gene is both amplified and over-expressed in ~40% of SCLC tumours (Yokoi *et al.*, 2002). Since the antisense-mediated down-regulation of the *SKP2* gene is known to induce apoptosis in lung cancer cells (Yokoi *et al.*, 2003a), *SKP2* gene amplification may serve to inhibit apoptosis thereby promoting cell survival.

Homozygosity for a polymorphic variant (Asn312) in the *ERCC2* gene/protein is reportedly associated with increased apoptotic potential (Seker *et al.*, 2001). The reduced apoptotic potential of the product of the wild-type Asp312 allele could thus be associated with an increased risk of tumorigenesis by promoting the survival of damaged cells (Benhamou and Sarasin 2002). Similarly, the alternative p53 isoforms encoded by the alleles of the Arg/Pro72 polymorphism in the *TP53* gene display markedly different apoptotic potential (Dumont *et al.*, 2003). Finally, the finding that *CASP8*, *CASP10*, *TNFRSF10B*, *TNFRSF6* and *TNFSF6* gene expression correlated with *MYC* gene amplification is intriguing in the light of the fact that *MYC* over-expression is a potent stimulator of apoptosis (Shivapurkar *et al.*, 2002). Presumably, inactivation of the components of the death-inducing signaling complex leads to resistance to apoptotic stimuli that is essential to cellular survival.

Autocrine and paracrine growth factors

> "A substance which has the power of promoting growth has been obtained from very young mice, and termed embryo extract. [This] extract has extremely powerful growth-promoting properties. Furthermore, this extract behaves upon tissues *in vitro* in the same manner as extract of a malignant tumour. It is possible, therefore, that the discovery of this tissue extract may have important results with regard to the discovery of the causation and treatment of cancer".
> G.R. de Beer (1924) *Growth*. Edward Arnold, London.

Autocrine and paracrine signaling may play an important role in stimulating tumour cell growth. Indeed, a variety of growth factors and growth factor receptors have now been implicated in the pathogenesis of both SCLC and NSCLC (reviewed by Kalemkerian 2000). In SCLC, there is now a body of evidence supporting the autocrine and paracrine involvement of specific neuropeptides such as gastrin, gastrin-releasing peptide, neuromedin B, neurotensin, insulin-like growth factor I, stem cell factor (c-kit ligand), vasoactive intestinal peptide, cholecystokinin and arginine-vasopressin (Siegfried *et al.*, 1999; Kalemkerian 2000; Moody *et al.*, 2003; Coulson *et al.*, 2003). All of these neuropeptides play a role in signaling by binding to seven-pass transmembrane-spanning receptors coupled to G-proteins, and may represent useful markers of SCLC (Stieber *et al.*, 1999; Yegen 2003). Similarly, a number of growth factor/ growth factor receptor pairs (e.g. epidermal growth factor, TGFα, PDGF, insulin-like growth factors I and II, and vasoactive intestinal polypeptide) have been implicated in NSCLC proliferation (Kalemkerian 2000).

Angiogenesis

> "The normal blood vessels of the organs in which the tumor is developing are disturbed by chaotic growth, there is a dilatation and spiralling of the affected vessels, marked capillary budding and new vessel formation, particularly at the advancing border".
> E. Goldman (1907) *Lancet* ii: 1236-1240.

> "However slowly neoplastic cells divide, they must of course obtain additional food if they are to multiply; and however benign the growing tumors, their demands can be peremptory".
> Peyton Rous (1966) Nobel Lecture presentation speech

Angiogenesis is the process by which new capillaries are formed from pre-existing vasculature. It is normally controlled and modulated via a delicate balance between pro-angiogenic and anti-angiogenic factors. However, when this balance is lost during tumorigenesis, inappropriate vessel growth ensues (Tonini *et al.*, 2003). Indeed, neoplastic lung cells are known to release a variety of pro-angiogenic factors that induce the formation of new vasculature, without which the tumours would be unable to grow (Hanahan and Folkman 1996; Ohta and Watanabe 2000; Carmeliet 2003; Onn and Herbst 2003). In general, high vascularity is associated with tumour progression and the higher the microvessel density, the poorer the prognosis (Ushijima *et al.*, 2001). Probably the best characterized inducer of angiogenesis in lung cancer is vascular endothelial growth factor (VEGF; Mattern *et al.*, 1996; Fontanini *et al.*, 1997b; Ohta *et al.*, 1997; Tsao *et al.*, 1997; Decaussin *et al.*, 1999; Masuya *et al.*, 2001; Yuan *et al.*, 2000; 2002; O'Byrne *et al.*, 2003; Stefanou *et al.*, 2004a). A strong correlation has been noted between the presence of a *TP53* gene mutation (Niklinska *et al.*,

2001b), p53 nuclear accumulation (Fontanini *et al.*, 1997b) and VEGF expression. In the same vein, over-expression of the ErbB2 receptor has been shown to result in the induction of basal level VEGF expression (Yen *et al.*, 2000). This may be of considerable significance in the light of the amplification of the *ERBB2* gene noted in 3-5% of NSCLC tumours (Table 2).

Other factors thought to be involved in promoting lung cancer angiogenesis include neuropilins 1 and 2 which act as receptors for VEGF and the semaphorins (Kawakami *et al.*, 2002), angiopoietin 1 (Tanaka *et al.*, 2002), interleukin 8 (Masuya *et al.*, 2001; Yuan *et al.*, 2002; Chen *et al.*, 2003b), cyclooxygenase-2 (Kim *et al.*, 2003e), interleukin 10 (Hatanaka *et al.*, 201), endothelin-1 and its receptor (Ahmed *et al.*, 2000), N-cadherin (Nakashima *et al.*, 2003), platelet-derived endothelial cell growth factor (Koukourakis *et al.*, 1997) and basic fibroblast growth factor (Takanami *et al.*, 1996; Volm *et al.*, 1997; Ito *et al.*, 2002b). These pro-angiogenic molecules represent potential targets for anti-tumour vasculature therapy (Yano *et al.*, 2003). Anti-angiogenic molecules include angiopoietin 2 (Tanaka *et al.*, 2002) and thrombospondins 1 and 2 (Fontanini *et al.*, 1999; Hawighorst *et al.*, 2001).

Folkman and Kalluri (2004) have proposed that when the delicate angiogenic balance between pro-angiogenic growth factors and their inhibitors is disrupted during tumorigenesis, this disruption is strongly influenced by the genetic make-up of the individual cancer cell as well as its microenvironment within the tumour. This could explain not only why cancer can progress at different rates in different individuals, but also why some individuals enter the lethal phase of cancer whereas others do not, despite carrying tumour cells within their organs (NB. *in situ* tumours are commonly found in autopsied trauma victims). If this intriguing postulate turns out to be correct, then an important goal of future clinical research will be to find means to impede or prevent the development of disease in those individuals whose genetic constitution and/or tumour genetic make-up favours the progression of potentially harmless *in situ* tumours into lethal cancers. Folkman and Kalluri (2004) therefore foresee a day dawning when cancer, including lung cancer, can be effectively treated as a chronic yet wholly manageable disease.

Evasion of host immunity

> "Tumours are subjected to.....potential attack by numerous arms of the immune system. It is difficult to think of a more fertile breeding ground for a mutator phenotype".
>
> C. Lengauer, K.W. Kinzler & B. Vogelstein (1998) *Nature* 396: 643-649.

Different mechanisms are employed by cancer cells to evade host immune surveillance. In lung cancer, class I major histocompatibility complex antigen expression is down-regulated (Redondo *et al.*, 1991). Mutations of the β2-microglobulin (*B2M*) gene on chromosome 15q21-q22 are occasionally found in lung cancer (Chen *et al.*, 1996a) and these lesions serve to down-regulate MHC class I expression. The transporter associated with antigen presentation, TAP-1, is also known to be mutated or down-regulated in lung cancer (Korkolopoulou *et al.*, 1996; Chen *et al.*, 1996b: Singal *et al.*, 1996) leading to the loss of antigen-presenting ability.

Lung cancer cell lines have been shown immunohistochemically and by reverse transcription-PCR to express FasL (Niehans *et al.*, 1997). Further, co-culture of a lung carcinoma cell line with a Fas-sensitive T-cell line led to the induction of apoptosis of the T-cells (Niehans *et al.*, 1997) suggesting that the peripheral deletion of tumour-reactive T-cell clones might be a possible mechanism to evade host immunity.

Metastasis

> "Attempts at encapsulation by the animal are frustrated by the tumour invading the tissues bordering on it. Such a tumour is termed malignant and constitutes the most redoubtable form of cancer. Not only is the abandonment by the cells of their former functions deleterious, but their invasion of other tissues, and the mechanical disturbances produced by the presence of a large mass of tissue in undesirable positions, is dangerous to life".
>
> G.R de Beer (1924) *Growth*. Edward Arnold, London.

The high mortality rate associated with lung cancer is associated with the metastatic spread of lung tumour cells to the blood and lymph vessels and thence to other tissues via the circulation. The molecular mechanisms that underlie metastatic progression have yet to be elucidated but involve growth factors, chemokines, cell-cell adhesion molecules and extracellular proteases (Bogenrieder and Herlyn 2003).

Early detection of metastases

Since metastases kill a large majority of lung cancer patients, their early detection is a priority. One way to achieve this is through the identification of tumour-specific lesions in distant metastases, potentiated by the fact that mutational patterns in the primary tumour are often conserved in metastases. For example, the *TP53* gene mutations characterized in primary lung tumours have been subsequently noted in brain metastases (Schlegel *et al.*, 1992). The detection of *KRAS* or *TP53* mutations in lymph nodes can also serve to confirm occult metastases (Ahrendt *et al.*, 2002) and to distinguish multiple primary lung carcinomas from metastases of tumours from other sites (Kandioler *et al.*, 1996; Lau *et al.*, 1997; Matsuzoe *et al.*, 1999a; Shimizu *et al.*, 2000a; van Rens *et al.*, 2002; van der Sijp *et al.*, 2002).

The early detection of micrometastases in individuals with detectable levels of circulating lung tumour cells can in principle be potentiated by means of a reverse transcription-polymerase chain reaction (RT-PCR) assay. RT-PCR has been successfully used to detect the extremely low levels of expression of some lung tumour markers [e.g. carcinoembryonic antigen (Kurusu *et al.*, 1999), mucin 1 (Salerno *et al.*, 1998), the palate, lung and nasal epithelium carcinoma-associated gene LUNX (*PLUNC*) (Iwao *et al.*, 2001), chromogranin A (Begueret *et al.*, 2002)] in blood.

Detection of metastases of non-pulmonary origin

Metastases of non-pulmonary origin are also fairly common in the lung and can be difficult to distinguish from lung adenocarcinomas (Flint and Lloyd 1992). Since the clinical course of such disease is different from lung adenocarcinoma, it is highly desirable to be able to discern the origin of the tumour prior to the initiation of treatment. Molecular markers of other types of cancer can be used to identify the origin of metastases found in the lungs. For example, *CDX2* homeobox gene expression is a marker of colorectal adenocarcinoma (Barbareschi *et al.*, 2003).

In their microarray-based study of lung tumours, Bhattacharjee *et al.* (2001) presented evidence for a hierarchical cluster of samples that expressed known markers of colorectal carcinoma including galectin 4 (*LGALS4*), carcinoembryonic antigen-related cell adhesion molecule 1 (*CEACAM1*), cadherin 17 (*CDH17*), *CDX1* and *CDX2*. On this basis, these authors concluded that this cluster probably represented metastatic adenocarcinomas from the colon. Bhattacharjee *et al.* (2001) also noted another cluster of lung adenocarcinomas that exhibited a non-lung 'signature': the expression of breast-associated markers, oestrogen receptor (*ESR1*) and mammaglobin 1 (*SCGB2A2*) were consistent with the interpretation of metastasis from the breast.

The use of molecular biological markers can, at least in principle, also be used to predict the sites of metastases (D'Amico *et al.*, 2001). Indeed, recent work on breast cancer has shown that sub-populations of tumour cells can display a tissue-specific expression profile that appears to predict the site of metastasis (Kang *et al.*, 2003). Similar organ-specific expression profiles have been reported for human SCLC metastases in mice (Kakiuchi *et al.*, 2003) but it remains unclear whether these profiles are already present in the parental cells.

Identification of genes involved in the metastasis of lung tumours

A number of candidate genes have been identified that may play a role in regulating the metastatic potential of neoplastic lung cells (Yokota *et al.*, 2003). In some cases, the over-expression of specific genes has been correlated with metastasis e.g. *HOXD3*. The over-expression of *HOXD3* in human lung cancer cells induces the coordinate expression of a variety of other genes (including those encoding matrix metalloproteinase 2, and integrins α3 and β3) thought to play a role in metastasis (Hamada *et al.*, 2001; Lim and Jablons 2003) and enhances cellular motility and tissue invasiveness through both TGFβ-dependent and -independent pathways (Miyazaki *et al.*, 2002).

Other examples of genes whose over-expression has been correlated with lung cancer metastasis include TGFβ (*TGFB*; Saji *et al.*, 2003), *ERBB2* (Yu *et al.*, 1994), *RAD17* (Sasaki *et al.*, 2001e), stromelysin 3 (*MMP11*; Delebecq *et al.*, 2000), galectin 3 (*LGALS3*; Yoshimura *et al.*, 2003), tumour-associated antigen L6 (*TM4SF5*; Kao *et al.*, 2003), chemokine receptor 7 (*CCR7*; Takanami 2003), thymosin β4 (*TMSB4X*; Ji *et al.*, 2003), eukaryotic translation initiation factor 4a isoform 1 (*EIF4A1*; Ji *et al.*, 2003), the chromosome 11q13-located metastasis-associated lung adenocarcinoma transcript 1 (*MALAT1*) gene that encodes a non-coding RNA (Ji *et al.*, 2003) and caveolin-1 (*CAV1*; Ho *et al.*, 2002b). By contrast, the reduced expression of several other genes has also been correlated with lung cancer metastasis e.g. E-cadherin (*CDH1*; Bremnes *et al.*, 2002a), mucins 2 and 6 (*MUC2*, *MUC6*; Nishiumi *et al.*, 2003), collapsing response mediator protein-1 (*CRMP1*; Shih *et al.*, 2001), retinoblastoma-like protein 2 (*RBL2*; Baldi *et al.*, 1997), kangai 1 (*KAI2*; Higashiyama *et al.*, 1998) and CD9 antigen (*CD9*; Funakoshi *et al.*, 2003).

The use of cDNA microarrays in large scale transcriptional profiling promises to greatly speed up the process of identification of further genes involved in the metastasis of lung tumours (Gemma *et al.*, 2001; Koshikawa *et al.*, 2002; Kikuchi *et al.*, 2003; de Lange *et al.*, 2003). Perhaps the most important study to date has been that of Ramaswamy *et al.* (2003) which explored the differences between primary tumours and metastases by comparing the gene expression profiles of adenocarcinoma metastases of multiple tumour types (including lung) to those of unmatched primary adenocarcinomas. A 17-gene 'expression signature' was found to distinguish primary from metastatic adenocarcinoma. This gene expression signature contained 8 up-regulated genes [small nuclear ribonucleoprotein F (*SNRPF*), elongation initiation factor 4E-like 3 (*EIF4EL3*), heterogeneous nuclear ribonucleoprotein A/B (*HNRPAB*), deoxyhypusine synthase (*DHPS*), securin (*PTTG1*), type 1 collagens α1 and α2 (*COL1A1* and *COL1A2*) and lamin B1 (*LMNB1*)] and 9 down-regulated genes [γ2 actin (*ACTG2*), myosin light chain kinase (*MYLK*), myosin heavy chain 11 (*MYH11*), calponin 1 (*CNN1*), MHC class II, DPβ1 (*HLA-DPB1*), Runt-related transcription factor 1 (*RUNX1*), metallothionein 3 (*MT3*), nuclear hormone receptor TR3 (*NR4A1*) and RNA binding motif 5 (*RBM5*)]. Importantly, in the light of the previous studies of specific genes involved in metastasis, none of the above genes represented individual markers of metastasis; only the signature as a whole was found to contain predictive information.

Although the role in metastasis of the changes in gene expression noted by Ramaswamy *et al.* (2003) is still unclear, their findings are consistent with a model in which the metastatic potential of a given tumour is encoded in the bulk of that tumour rather than emerging from rare cells with a metastatic phenotype. It may be therefore that the metastatic ability of the cells in a given tumour is pre-programmed either by their oncogenic mutations or by the genetic background of variation that serves to modify metastatic efficiency (Hunter 2004). This notwithstanding, since the metastatic signature identified by Ramaswamy *et al.* (2003) appears to be tissue-independent, it may have the potential to be developed as a diagnostic test that could be used to predict clinical outcome using gene-expression profiles of primary tumours at the time of diagnosis.

Other miscellaneous genes

> "So he went home, very angry indeed and horribly scratchy; and from that day to this every rhinoceros has great folds in his skin and a very bad temper, all on account of the cake-crumbs inside".
>
> Rudyard Kipling (1902) *Just So Stories*

A considerable number of putative tumour suppressor genes have been identified by virtue of (i) their chromosomal location corresponding to regions exhibiting loss of heterozygosity and (ii) the characterization of more subtle intragenic lesions. However, in many cases virtually nothing is known of their precise function. A variety of other genes have been found to be mutated in lung cancer but it is not always easy to put forward a hypothesis to explain why such mutations could, for example, confer a growth advantage or permit escape from apoptosis. Some of the reported mutations have been characterized in NSCLC or SCLC cell lines and it remains quite possible that the reported lesions actually occurred during cell culture and were not present in the original lung tumour. Some of the following examples of genes known to be mutated in lung cancer (see Table 2 for details) may be consequential rather than causal and the accompanying tentative *post hoc* explanation may therefore represent no more than a 'Just So' story.

(i) *PAX7 gene amplification*
PAX genes are frequently expressed in cancer cells and their expression appears to be a requirement for cell growth and survival (Muratovska et al., 2003). PAX7 is known to play a role in migration during embryogenesis and it is possible that *PAX7* gene amplification has played a role in metastasis by virtue of these migratory properties.

(ii) *MYB gene deletions and amplification*
The *MYB* gene has been noted to be up-regulated in haematopoietic tumours. Myb is a transcription factor that activates various genes (e.g. *PTGS2, BCL2, MYC*), genes that play key roles in cellular proliferation, angiogenesis, apoptosis and metastasis.

(iii) *RET gene mutations and HGF over-expression*
RET is a receptor tyrosine kinase, gain-of-function mutations in which confer uncontrolled proliferation. Loss-of-function *RET* mutations have however also been reported in lung cancer (Futami *et al.*, 2003). Over-expression of the RET ligand, hepatocyte growth factor/scatter factor (Rygaard *et al.*, 1993; Olivero *et al.*, 1996; Harvey *et al.*, 1996), which is a potent mitogen for both normal and neoplastic lung epithelium (Singh-Kaw *et al.*, 1995), has also been reported in NSCLC where it is associated with a poor prognosis (Siegfried *et al.*, 1997b).

(iv) *IGF2R LOH and missense mutation*
Over-expression of insulin-like growth factors and their receptors may play a role in promoting cell cycle progression and in inhibiting apoptosis (Quinn et al., 1996; Moschos and Mantzoros 2002; LeRoith and Roberts 2003; Coleman and Grimberg 2003; Figure 4.4). Transgenic over-expression of IGF-II induces spontaneous lung tumours in mice (Moorehead et al., 2003). It is however unclear how LOH involving IGF2R could promote lung tumorigenesis.

(v) *YWHAE gene deletion*
The 14-3-3 family of proteins, which mediate signal transduction by binding to phosphoserine-containing proteins, are frequently dysregulated in human cancers (Mackintosh 2004). This family of proteins also plays a role in cell cycle regulation and the control of apoptosis by their participation in directing nucleocytoplasmic transport (Brunet et al., 2002; Masters et al., 2002). The *YWHAE* gene encodes 14-3-3ε, a protein that plays a key role in neuronal development, and which interacts with Raf1, the Ras-binding protein Rin1, tuberin, and the cell division cycle proteins Cdc25A and Cdc25B. Its loss by deletion in lung cancer may promote cellular transformation by abrogating the protein's role either in cell cycle arrest or senescence.

(vi) *MEN1 LOH and micro-mutations*
The function of menin remains unclear although it is known to bind to JunB, NFκB, SMAD3, GFAP and probably p53 (Poisson *et al.*, 2003).

(vii) *WT1 LOH*
WT1 (Wilms' tumour 1), a known tumour suppressor, encodes a zinc finger transcription factor that activates a variety of growth factor genes and genes implicated in cellular differentiation, particularly in the context of the development of the genitourinary system.

(viii) *EIF4G1 gene amplification*
The deregulation of translational initiation is increasingly recognised as being of critical importance during tumorigenesis (Watkins and Norbury 2002; Pandolfi 2004; Rosenwald 2004). Amplification and/or over-expression of the eukaryotic translation initiation factor 4γ1 (*EIF4G1*) gene could lead to increased production of the translation initiation complex eIF4F which may facilitate the accelerated growth and division of neoplastic cells.

(ix) *NOTCH3-associated translocation*
Dang *et al.* (2000) reported a novel balanced t(15;19) translocation occurring in an aggressive metastatic lung carcinoma. The breakpoint of the translocation was mapped to ~50 kb upstream of the *NOTCH3* gene (19p13.1-p13.2) which was found to be highly expressed in the cell line derived from the adenocarcinoma. Intriguingly, these authors reported a number of other *NOTCH3*-expressing NSCLC cell lines that also manifested translocations involving chromosome 19. Notch3 is a transmembrane receptor that plays an important role in the Notch signaling pathway of cellular differentiation (Radtke and Raj 2003); constitutive Notch signaling is known to be involved in leukaemogenesis and has been shown to induce cell cycle arrest in SCLC cells (Sriuranpong *et al.*, 2001). More recently, Dang *et al.* (2003) have shown that ectopic constitutive expression of *NOTCH3* in a transgenic mouse model leads to altered lung morphology and delayed development, consistent with a role for *NOTCH3* in lung tumorigenesis through inhibition of terminal differentiation.

(x) *MST1R LOH and missense mutation*
The macrophage stimulating 1 receptor (*MST1R*) gene encodes the tyrosine kinase RON, a member of the MET proto-oncogene family. Since this gene is located at chromosome 3p21.3, it is frequently lost in both SCLC and NSCLC. However, in lung cancer, only a solitary missense mutation of uncertain significance has been reported (Angeloni *et al.*, 2000). In experimental model systems, constitutive activation of the *MST1R* gene can promote transformation, invasive growth and metastasis (Santoro *et al.*, 1996; Willett *et al.*, 1998; Peace *et al.*, 2001) whilst over-expression in a transgenic mouse model is associated with the formation of lung tumours (Chen *et al.*, 2002f). Although there is some evidence from the study of NSCLC cell lines for a paracrine role for MST1R/RON in lung cancer (Willett *et al.*, 1998), this remains far from clear.

(xi) *PARK2 exon 2 deletions*
Parkin is involved in protein degradation as a ubiquitin protein ligase. Exon 2 deletions have been noted in lung adenocarcinoma cell lines (Cesari *et al.*, 2003) and *PARK2* gene expression is frequently reduced in lung tumour tissue (Picchio *et al.*, 2004). Restoration of *PARK2* gene expression by transfection of wild-type *PARK2* cDNA into the *PARK2*-deficient lung adenocarcinoma cell line did not lead to cell cycle changes or to any change in cellular proliferation. However, ectopic *PARK2* expression did serve to reduce *in vivo* tumorigenicity in nude mice, suggesting that *PARK2* might have a tumour suppressor role in lung cells.

Order and timing of mutations and changes in gene expression in lung cancer

> "One of the beauties of genetics as a science is the clear cause and effect relationships that can be inferred, relating genotype to phenotype. The genetics of cancer forces us to re-examine our simple notions of causality, such as those embodied in Koch's postulates; how does one come to grips with words like 'necessary' and 'sufficient' when more than one mutation is required to produce a phenotype and when that phenotype can be produced by different mutant genes in various combinations?"
> B. Vogelstein and KW Kinzler (1993) The multstep nature of cancer. *Trends Genet.* 9: 138-141.

In 1990, Fearon and Vogelstein put forward a multistep model for colorectal cancer. They proposed that these cancers arise by the sequential acquisition of a number of different mutations (possibly between three and seven) in distinct growth regulatory genes. Thus, it was proposed that tumours would grow by a process of clonal evolution driven by mutation, each new mutation conferring an additional growth advantage to a cell and its subsequent progeny. With characteristic prescience, the authors then noted that "the relevance of the colorectal model to other tumor types, such as small cell carcinoma of the lung. . ...is beginning to gather support".

One of the most detailed studies of mutational order in human neoplasia is that of Barrett *et al.* (1999), performed on a premalignant condition known as Barrett oesophagus which predisposes to oesophageal adenocarcinoma. These authors prospectively studied oesophageal biopsies for LOH at 5q, 9p, 13q, 17p and 18q and screened samples for subtle lesions in the *TP53* and *CDKN2A* genes. They concluded that there was no obligate involvement of a given gene or chromosomal region and no obligate order of the different mutational events observed. Indeed, there were found to be multiple genetic routes to cancer in Barrett oesophagus, a situation that the authors described as 'a pattern of events probably shared by most human neoplasms'. According to Hanahan and Weinberg (2000), these routes necessarily involve the acquisition of different capabilities (viz. self sufficiency in growth signals, insensitivity to growth inhibitory signals, evasion of apoptosis, limitless replicative potential, sustained angiogenesis, tissue invasion and metastasis) but these capabilities may be acquired in different orders (Figure 4.5; *See* Chapter 1; Cancer, Signaling and Acquired Capabilities). Further, certain capabilities may be acquired simultaneously. Thus, for example, loss of function of p53 can facilitate angiogenesis and promote resistance to apoptosis at the same time as promoting genomic instability. By contrast, other capabilities may only be acquired cooperatively through the occurrence of two or more distinct genetic changes.

As we have seen, a large number of different genetic changes occur in both SCLC and NSCLC during lung cancer tumorigenesis from hyperplasia and metaplasia through primary tumour formation to metastasis (Greenberg *et al.*, 2002). These changes include the occurrence of activating and inactivating mutations, gene amplification, the over- or under-expression of certain genes, chromosomal aneuploidy, telomerase expression and promoter hypermethylation. Additional genes may also play a role later on in lung neoplasia by promoting angiogenesis, evasion of host immunity or metastasis.

Evidence for mutational changes in specific genes or chromosomal regions being early events has come indirectly from simple prevalence data or from the analysis of

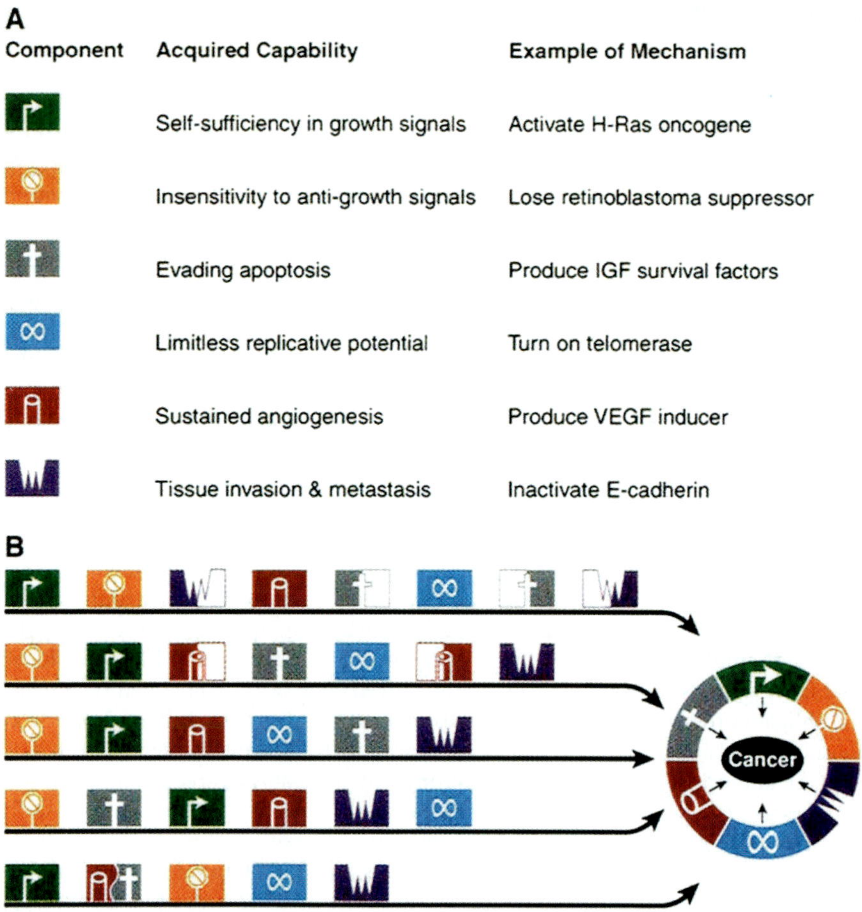

Fig. 4.5. Parallel pathways of tumorigenesis. According to Hanahan and Weinberg (2000), virtually all cancers acquire the same six 'hallmark capabilities' (A) although both the mechanism and the chronological order (B) of acquisition will vary. [Reprinted from *Cell* 100: D Hanahan and RA Weinberg, The hallmarks of cancer, pp57-70, Copyright (2000) by kind permission from Elsevier]

a series of a particular type of tumour identified as being at different stages. Such evidence may however also come from the observation of LOH or inactivating gene mutations in putative precancerous lesions such as adenomatous hyperplasia of the lung (reviewed by Niklinski *et al.*, 2001; Mori *et al.*, 2001; Gabrielson 2003). Thus, LOH at 3p (Sozzi *et al.*, 1995; Hung *et al.*, 1995; Kohno *et al.*, 1999b; Wistuba *et al.*, 1999b; Wistuba *et al.*, 2000; Nishisaka *et al.*, 2000; Sikkink *et al.*, 2003), 9p (Kishimoto *et al.*, 1995; Kohno *et al.*, 1999b), 9q and 16p (Takamochi *et al.*, 2001) and 17p (Sozzi *et al.*, 1992; Kitaguchi *et al.*, 1998) seem to be relatively early events on the grounds that they are often evident in preneoplastic lesions. Whilst chromosome 3p deletions are thought to precede *TP53* gene mutations (Chung *et al.*, 1995), it is still unclear how early mutations involving the 9p-located *CDKN2A* gene actually occur

(Okamoto *et al.*, 1995; Niklinska *et al.*, 2001a). LOH at chromosomal regions 3p21, 5q21 and 9p21 has been noted in normal bronchial cells adjacent to NSCLC tissue (Sanz-Ortega *et al.*, 2001; Caballero *et al.*, 2001) consistent with their being early events in NSCLC. LOH involving the *APC* gene at 5q21 (Sanz-Ortega *et al.*, 1999) and the short arm of chromosome 8 (Wistuba *et al.*, 1999a) have also been claimed to be relatively early events in NSCLC.

Subtle lesions in the *TP53* gene (17p) have been noted in atypical alveolar hyperplasia (Slebos *et al.*, 1998), bronchial dysplasia (Sundaresan *et al.*, 1992; Sozzi *et al.*, 1992; Wistuba *et al.*, 1999b) and in atypical epithelial lesions in patients with idiopathic pulmonary fibrosis (Kawasaki *et al.*, 2001). In one case, two dysplasia samples, taken 9 months apart from the same tumour-free patient, yielded one and two *TP53* mutations respectively (Chung *et al.*, 1996). Franklin *et al.* (1997) reported that a single G→T transversion in codon 245 of the *TP53* gene, identified in a smoking individual with widespread dysplastic changes in the respiratory epithelium but with no overt lung carcinoma, was detectable in bronchial epithelium from 7/10 locations in both lungs. This wide distribution of a single mutation suggests that a single progenitor bronchial epithelial clone may expand to populate swathes of the bronchial mucosa. Consistent with this view, *TP53* mutations have also been found in 9 % of apparently tumour-free surgical margins in patients with NSCLC (cf. 30 % of primary tumours) who had undergone complete pulmonary resection (Jassem *et al.*, 2004).

Although *KRAS* mutations were initially claimed to occur as late events in NSCLC (Sugio *et al.*, 1994), they are probably early events in adenocarcinoma (Sagawa *et al.*, 1998) and have been noted in atypical alveolar hyperplasia (Westra *et al.*, 1996; Cooper *et al.*, 1997). Inactivation of the *FHIT* gene (3p14) through either LOH or subtle mutation has been reported in idiopathic pulmonary fibrosis (Uematsu *et al.*, 2001) and in "preinvasive bronchial lesions" (Fong *et al.*, 1997; Sozzi *et al.*, 1998).

Bcl2 expression has been shown to occur in pre-malignant bronchial lesions (Ferron *et al.*, 1997; Kalomenidis *et al.*, 2001), a finding that suggests that it is expressed at a relatively early stage in lung neoplasia. Similarly, Akt is expressed in a high proportion of bronchial dysplasia samples suggesting that Akt activation may be an early event in lung tumorigenesis (Tsao *et al.*, 2003). Lantuejoul *et al.* (2003) also reported a steady increase in VEGF levels from low grade to high grade dysplasia whilst levels of neuropilins 1 and 2 (VEGF receptors) were found to increase from lung dysplasia to microinvasive lung carcinoma.

Telomerase expression has been noted in 26 % of cells in normal bronchial epithelium from lung carcinoma patients and in 23 % of peripheral lung samples, while it was found in a higher proportion of cells from abnormal bronchial epithelium in these patients viz. hyperplasia, 71 %; metaplasia, 80 %; dysplasia, 82 %; carcinoma *in situ* (100 %) consistent with telomerase expression being both an early and a steadily accumulating event in lung tumorigenesis (Yashina *et al.*, 1997).

Brambilla *et al.* (1999) reported loss of p16 (*CDKN2A*) expression in 12 % of moderate dysplasia and in 30 % of carcinoma *in situ* whilst cyclin D1 (*CCND1*) overexpression was noted in 6 % of hyperplasia and metaplasia, 17 % of mild dysplasia and 46 % of moderate dysplasia. Lonardo *et al.* (1999) have also reported over-expression of cyclins D1 (*CCND1*) and E (*CCNE1*) in preneoplasia. Taken together, these studies suggest that Rb function may be compromised at an early stage in lung tumorigenesis, in preinvasive bronchial lesions, by alterations in the expression levels of the proteins that regulate Rb phosphorylation and function in G_1 arrest.

Available data on the order and timing of LOH, mutational lesions and changes in gene expression are however often contradictory. Further, it is far from clear that the various types of 'precancerous lesion', held in the cited studies to provide evidence of the early occurrence of genetic changes, will invariably and inevitably go on to become neoplastic. Moreover, it should be remembered that 'pre-malignant' dysplastic lung tissue is likely to constitute a clonal patchwork of cell types manifesting multiple molecular abnormalities (Franklin *et al.*, 1997; Park *et al.*, 1999; Sikkink *et al.*, 2003). The theory of *field cancerization* proposes that dysplastic lung tissue has an increased risk for the development of pre-malignant lesions due to a range of multiple pre-existing genetic abnormalities.

Although our knowledge of the relative timing of such events is relatively crude at this stage, Figure 4.6 represents a rough pictorial summary of our current understanding of the process at the molecular genetic level. This schema can however only be tentative since the data are still sketchy, open to different interpretations, and may even be downright contradictory. Moreover, mutational order may well differ between different types of lung cancer; thus, LOH on chromosomes 3p and 9p appear

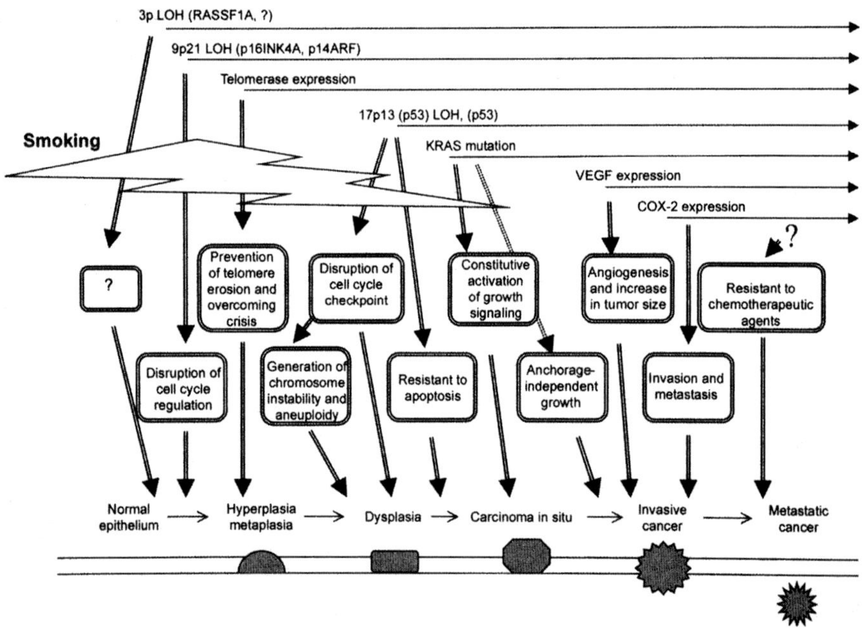

Fig. 4.6. Multi-step progression of lung carcinogenesis. Exposure to carcinogens (e.g. in cigarette smoke) is thought to trigger a series of events that give rise to the accumulation of genetic and epigenetic alterations that together convert normal bronchial epithelium into hyperplastic, metaplastic and dysplastic lesions, carcinoma *in situ*, and finally overt squamous cell carcinoma. Adenocarcinoma may also develop in a similar multi-step fashion via 'pre-malignant lesions' such as atypical adenomatous hyperplasia. The putative biological consequences of the various genetic and epigenetic alterations are indicated. Reproduced from H Osada & T Takahashi, *Oncogene* 21, 7421-7434 (2002) by kind permission of Nature Publishing Group

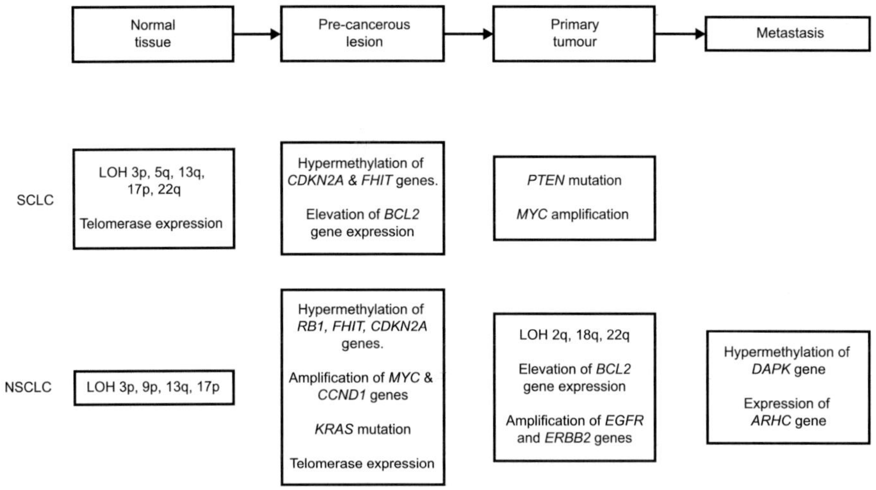

Fig. 4.7. Molecular genetic model for lung cancer progression

to be early events in squamous cell carcinoma but late events in adenocarcinoma (Kohno *et al.*, 1999b). The Barrett oesophagus example cited above should also ensure that we bear in mind that there may not be any obligate involvement of a given gene or chromosomal region and that mutational events may not invariably occur in the same order. What is clear is that SCLC and NSCLC appear to differ with respect not only to the nature of the mutational events but also to the relative timing of these events (Figure 4.7). Although we are at the stage of being able to recognize these differences, we are as yet no nearer to being able to use such mutational data in a prognostic context (Brechot *et al.*, 2002).

Detection of mutations or aberrant methylation in sputum or plasma/serum from lung cancer patients

"The small [lung] tumors. . ..will probably never be diagnosticated, unless pathognomonic cells in the sputum direct attention to the possible existence of tumor in the respiratory system".
Isaac Adler (1912) *Primary Malignant Growths of the Lungs and Bronchi*

Detection of mutations or aberrant methylation in sputum

DNA analysis from cells derived from sputum or bronchial lavage is a very promising approach to the early (preclinical) diagnosis of lung cancer. Thus, the detection of mutations in the *KRAS* and *TP53* genes (Takeda *et al.*, 1993; Mills *et al.*, 1995b; Lehman *et al.*, 1996; Scott *et al.*, 1997; Ahrendt *et al.*, 1999; Behn *et al.*, 2000; Kersting *et al.*, 2000; Ferretti *et al.*, 2000; Dai *et al.*, 2000; Keohavong *et al.*, 2004), promoter hypermethylation in the *CDKN2A*, *DAPK1*, *RASSF1*, *CDH1*, *GSTP1*, *RARB*, *CDH13*, *APC*, *MLH1*, *MSH2* and *MGMT* genes (Ahrendt *et al.*, 1999; Kersting *et al.*, 2000; Palmisano *et al.*, 2000; Belinsky *et al.*, 1998; 2002; Soria *et al.*, 2002a; Chan *et al.*, 2002; Ng *et al.*, 2002; Wang *et al.*, 2003b; Zöchbauer-Müller *et al.*, 2003; Topaloglu *et al.*, 2004; Grote *et al.*, 2004), *HNRPA2B1* gene over-expression (Tockman *et al.*, 1997; Fielding *et al.*, 1999), expression of the gastrin-releasing peptide (*GRP*) gene (Lacroix *et al.*, 2001; Saito *et al.*, 2003), LOH in various regions including 1p34-p36, 3p21-p23, 9p21-p24, 13q12-q13 and 17p13 (Demopoulos *et al.*, 2002; Arvanitis *et al.*, 2003), telomerase activity (Yahata *et al.*, 1998; Arai *et al.*, 1998; Xinarianos *et al.*, 2000b; Sen *et al.*, 2001; Dikmen *et al.*, 2003) and microsatellite instability (Miozzo *et al.*, 1996; Ahrendt *et al.*, 1999; Arvanitis *et al.*, 2003) have all been achieved using sputum samples or bronchial lavage. More recently, Schmidt *et al.* (2004) have demonstrated that microsatellite alterations may be detected using cell-free bronchial lavage supernatants.

Some mutations (or epimutations) have been detectable in sputum or bronchial lavage samples taken prior to clinical diagnosis, either prospectively or retrospectively (Mao *et al.*, 1994; Mills *et al.*, 1995b; Chen *et al.*, 2000; Tockman and Mulshine 2000; Honorio *et al.*, 2003). Further, as many as 12 % of controls (actually defined as individuals without lung cancer but with one of a number of non-oncological conditions including bronchitis, asthma and pneumonia) have been shown to possess *KRAS* gene mutations in their sputum (Ronai *et al.*, 1996). Without additional work, the prognostic value that can be attached to such findings remains unclear. In principle, early detection provides the possibility of surgical or chemotherapeutic intervention at the earliest possible stage thereby optimising the chances of a successful treatment. There is however some way to go before such tests are deemed to be reliable owing to problems with sensitivity and specificity (Tockman 2000). For instance, as far as mutation detection is concerned, the concordance between sputum and matched tumour may be as high as 92 % (Chen *et al.*, 2002e) but can sometimes be rather lower e.g. 60-70 % (Nakajima *et al.*, 2001; Zhang *et al.*, 2003b).

For promoter methylation analysis, a similar situation pertains; thus, for the *MLH1* gene, there is a 72 % concordance between aberrant methylation in sputum samples and aberrant methylation in the matched tumours (Wang *et al.*, 2003b), and a 93 % concordance between *MSH2* gene promoter methylation and loss of *MSH2* gene expression (Wang *et al.*, 2003b).

Screening sputum samples for lung tumour marker expression is also of potential importance. One way to do this is by reverse transcript PCR (RT-PCR), taking advantage of the fact that we can isolate tumour-specific mRNA from the cell-free bronchial lavage supernatant of patients with lung cancer (Engel *et al.*, 2004). Thus, for example, an RT-PCR assay for *MAGEA1* expression in sputum has yielded positive results in 2.1 % of non-cancer controls and 54 % of lung cancer patients, as compared to 83 % of the cancer tissues (Jheon *et al.*, 2004).

Detection of mutations or aberrant methylation in plasma/serum

The discovery of tumour-derived DNA in the plasma of cancer patients has opened up new possibilities in cancer detection and monitoring. Indeed, tumour-derived genetic alterations have been detected in DNA derived from the plasma or serum of lung cancer patients (Sozzi *et al.*, 2003). The use of such a minimally invasive procedure to identify tumour-specific lesions to be used as surrogate tumour markers has great potential not only for diagnosis and prognosis, but also to monitor the course of the disease (Ziegler *et al.*, 2002; Wong and Lo 2003) and the response to therapy (Kimura *et al.*, 2004). Thus, tumour cell-derived *TP53* gene mutations

(Silva *et al.*, 1999; Gonzalez *et al.*, 2000; Andriani *et al.*, 2004), LOH in various chromosomal regions (Sanchez-Cespedes *et al.*, 1998; Allan *et al.*, 2001) and microsatellite alterations (Sozzi *et al.*, 1999; Gonzalez *et al.*, 2000; Sozzi *et al.*, 2001; Beau-Faller *et al.*, 2003a; 2003b; Andriani *et al.*, 2004; Khan *et al.*, 2004) have all been detected in plasma. Specificity and sensitivity values appear encouraging (Sozzi *et al.*, 2001) and the tests can in principle be used to detect genetic alterations either in advance of a clinical diagnosis (Allan *et al.*, 2001) or during the course of the disease to monitor progression (Gonzalez *et al.*, 2000). Ultimately, non-invasive tests for lung cancer-associated mutations may become available. For example, Gessner *et al.* (2004a) claim to have demonstrated the feasibility of detecting *TP53* gene lesions in exhaled breath condensate taken from NSCLC patients.

Promoter hypermethylation may also be detected in plasma, with Oshita *et al.* (2003) claiming that a methylation-specific PCR assay can detect as few as one tumour cell per 1000 normal cells. Esteller *et al.* (1999) demonstrated that aberrant methylation of at least one of the *CDKN2A*, *DAPK1*, *GSTP1* and *MGMT* genes was detectable in 68 % of NSCLC tumours; of the tumours displaying aberrant methylation, some 73 % also exhibited abnormal methylation in the matched serum samples. Similarly, the majority (55-88 %) of patients with a methylated *CDKN2A* gene promoter in lung tumour tissue also exhibited hypermethylation of this promoter in plasma samples (An *et al.*, 2002b; Bearzatto *et al.*, 2002). However, since only 60-80 % of lung cancer patients display aberrant *CDKN2A* gene promoter methylation in their tumours, the utility of this test in a diagnostic context would be limited without employing further markers.

Finally, the reverse transcription-polymerase chain reaction (RT-PCR) assay may be used to detect the extremely low levels of expression of some lung tumour markers [e.g. carcinoembryonic antigen (Kurusu *et al.*, 1999; Castaldo *et al.*, 2003), Lunx (*PLUNC*; Mitas *et al.*, 2003), mucin 1 (*MUC1*; Salerno *et al.*, 1998), PGP9.5 (*UCHL1*; Fleischhacker *et al.*, 2001) and hnRNP-B1 (*HNRPA2B1*; Fleischhacker *et al.*, 2001] in blood that may allow the early detection of micrometastases in individuals with detectable levels of circulating lung tumour cells.

Early diagnosis and identification of prognostic factors

> "Present-day medicine treats these cases purely symptomatically with the sole object of relief, and interest attaching to an accurate diagnosis is mainly theoretical and scientific. It is not to be wondered at that the physician takes little interest in types of diseases that offer not the slightest hope of therapeutic success [but] this is a practical illustration of how unwise it is to attempt to set limits to the progress of science".
>
> Isaac Adler (1912) *Primary Malignant Growths of the Lungs and Bronchi*

Much emphasis has been placed upon the early molecular diagnosis of disease so that clinical intervention is feasible before it is too late (Petty and Miller 2000). The hope therefore is that molecular and cytological techniques will permit the identification of reliable early clonal markers of lung tumorigenesis that exhibit high sensitivity and specificity (Gazdar *et al.*, 2000). The enormous potential of molecular genetic markers in this regard is evidenced by results from the Johns Hopkins Lung Project showing that *TP53* gene mutations can be detected retrospectively in sputum samples taken from patients at least one year prior to a clinical diagnosis of lung cancer being made (Mao *et al.*, 1994).

A considerable number of potential prognostic factors have been described both for SCLC (Feld *et al.*, 2000) and NSCLC (Lau *et al.*, 2000b). Indeed, molecular genetic analyses of lung cancer have yielded a multitude of different prognostic indicators, and a wide range of different properties and characteristics have, in recent years, been reported to predict independently poor outcome. These include:

- Chromosomal instability (CIN; Nakamura *et al.*, 2003b).
- Microsatellite instability (Pifarre *et al.*, 1997; Rosell *et al.*, 1997) and a high degree of genomic damage (de Juan *et al.*, 1998).
- Telomerase activity (Hara *et al.*, 2001; Kumaki *et al.*, 2001; Gonzalez-Quevedo *et al.*, 2002; Wu *et al.* 2003c) and expression of the telomerase reverse transcriptase catalytic subunit (*TERT*) gene (Wang *et al.*, 2002b).
- Loss of heterozygosity on chromosomes 3p (Mitsudomi *et al.*, 1996; Burke *et al.*, 1998; Iniesta *et al.*, 2004) and 11p15 (Bepler *et al.*, 2002).
- *MYC* gene amplification (Noguchi *et al.*, 1990).
- *MDM2* gene amplification (Dworakowska *et al.*, 2004).
- Increased expression of N-myc (*MYCN*) (Funa *et al.*, 1987).
- Expression of cyclins B1 (*CCNB1*; Arinaga *et al.*, 2003; Yoshida *et al.*, 2004), D1 (*CCND1*; Jin *et al.*, 2001; Ikehara *et al.*, 2003) and E (*CCNE1*; Hayashi *et al.*, 2001a).
- Expression of cyclin-dependent kinase inhibitor 1B (*CDKN1B*; Ishihara *et al.*, 1999) but also (inconsistently) abnormal expression of *CDKN1B* (Hirabayashi *et al.*, 2002) and reduced expression of *CDKN1B* (Tsukamoto *et al.*, 2001).
- Loss of expression of the *RB1* (Xu *et al.*, 1994), *RBL2* (Caputi *et al.*, 2002), *PDCD4* (Chen *et al.*, 2003c) and *IGSF4* (TSLC1) [Uchino *et al.*, 2003] genes.
- Loss of expression of the *CDKN2A*, *BAX*, and *WEE1* genes (Gessner *et al.*, 2002; Gonzalez-Quevedo *et al.*, 2002; Esposito *et al.*, 2004; Yoshida *et al.*, 2004).
- Hypermethylation of the *RASSF1* (Kim *et al.*, 2003b; 2003c; Dammann *et al.*, 2003), *FHIT* (Maruyama *et al.*, 2004), *CDKN2A* (Fu *et al.*, 2003), *FANCF* (Marsit *et al.*, 2004) and insulin-like growth factor-binding protein-3 (*IGFBP3*; Chang *et al.*, 2002) genes.
- Loss of *FHIT* expression (Burke *et al.*, 1998; Toledo *et al.*, 2004).
- Loss of expression of heterogeneous nuclear ribonucleoprotein A2/B1 (*HNRPA2B1*; Wu *et al.*, 2003b) and interleukin 10 (*IL10*; Soria *et al.*, 2003b).
- Low expression of UDP-*N*-acetyl-alpha-D-galactosamine:polypeptide *N*-acetyl galactosaminyltransferase 3 (*GALNT3*; Dosaka-Akita *et al.*, 2002; Gu *et al.*, 2004), *N*-acetylglucosaminyltransferase V (*MGAT5*; Dosaka-Akita *et al.*, 2004), Diablo homologue (*DIABLO*; Sekimura *et al.*, 2004), *KIT* (Rohr *et al.*, 2004), *PTEN* (Bepler *et al.*, 2004), ribonucleotide reductase M1 (*RRM1*; Bepler *et al.*, 2004), connective tissue growth factor (*CTGF*; Chang *et al.*, 2004a), and motility related protein-1 (*CD9*; Higashiyama *et al.*, 1995b).
- Elevated expression of cell division cycle protein cdc25B (*CDC25B*; Sasaki *et al.*, 2001d), the p53 inhibitor, polo-like kinase (*PLK1*; Wolf *et al.*, 1997), matrix metalloproteinase 9 (*MMP9*; Simi *et al.*, 2004), collagen XVIII (Chang *et al.*, 2004b), thymidylate synthase (*TYMS*; Nakagawa *et al.*, 2002), intercellular adhesion molecule-1 (*ICAM1*; Shin *et al.*, 2004) and osteopontin (*SPP1*; Schneider *et al.*, 2004a).

- Reduced expression of α3-integrin (*ITGA3*; Adachi *et al.*, 1998), osteonectin (*SPARC*; Schneider *et al.*, 2004a), peroxisome proliferator-activated receptor-γ (*PPARG*; Sasaki *et al.*, 2002c), E-cadherin (*CDH1*; Liu *et al.*, 2001) and the *let-7* non-coding microRNA (Takamizawa *et al.*, 2004).
- Over-expression of β1-integrin (*ITGB1*; Oshita *et al.*, 2002), hepatocyte growth factor/scatter factor (*HGF*; Siegfried *et al.*, 1997b), α-actinin-4 (*ACTN4*; Yamagata *et al.*, 2003), α-1,3-fucosyltransferase type IV and VII genes (*FUT4* & *FUT7*; Ogawa *et al.*, 1996), retinoic acid receptor β (*RARB*; Khuri *et al.*, 2000), dihydrodiol dehydrogenase (*AKR1C2*; Hsu *et al.*, 2001), caspase 3 (*CASP3*; Takata *et al.*, 2001), prothymosin-α (*PTMA*; Sasaki *et al.*, 2001f) and glucose transporters Glut1 and Glut3 (*SLC2A1* & *SLC2A3*; Younes *et al.*, 1997).
- Expression of vascular endothelial growth factors A (*VEGF*; Nakashima *et al.*, 2004; Kaya *et al.*, 2004) and C (*VEGFC*; Li *et al.*, 2003b), hypoxia-inducible factor 1α (*HIF1A*; Swinson *et al.*, 2004), neuron-specific enolase (*ENO2*; Ferrigno *et al.*, 2003), caveolin-1 (*CAV1*; Yoo *et al.*, 2003), Y-box binding protein YB1 (*NSEP1*; Gessner *et al.*, 2004b), SPARC/osteonectin (*SPARC*; Koukourakis *et al.*, 2003), CD151 antigen (*CD151*; Tokuhara *et al.*, 2001a); apoptosis inhibitor AAC-11 (*API5*; Sasaki *et al.*, 2001b), survivin (*BIRC5*; Escuin and Rosell 1999), Ets-1 transcription factor (*ETS1*; Takanami *et al.*, 2001b), receptor-binding cancer antigen RCAS1 (*EBAG9*; Oizumi *et al.*, 2002) and metallothioneins (Joseph *et al.*, 2001; Cherian *et al.*, 2003).
- Co-expression of neuropilins 1 and 2 (*NRP1* & *NRP2*; Kawakami *et al.*, 2002).
- Over-expression of ERBB2 (Tateishi *et al.*, 1991; Kern *et al.*, 1994; Hsieh *et al.*, 1998; Nemunaitis *et al.*, 1998; Korrapati *et al.*, 2001; Selvaggi *et al.*, 2002; Meert *et al.*, 2003)
- Over-expression of epidermal growth factor receptor and amplification of the *EGFR* gene (Hirsch *et al.*, 2003; Selvaggi *et al.*, 2004). Over-expression of EGFR and ERBB2 together (Onn *et al.*, 2004). Phosphorylation but not over-expression of EGFR (Kanematsu *et al.*, 2003).
- The presence of the *GSTM1* null allele (Sweeney *et al.*, 2003).
- The presence of *KRAS* mutations (Slebos *et al.*, 1990; Silini *et al.*, 1994; Fukuyama *et al.*, 1997; Nemunaitis *et al.*, 1998; De Gregorio *et al.*, 1998; Huncharek *et al.*, 1999).
- The presence of *TP53* mutations (Horio *et al.*, 1993; Mitsudomi *et al.*, 1993; de Anta *et al.*, 1997; Huang *et al.*, 1998; Tomizawa *et al.*, 1999; Mitsudomi *et al.*, 2000; Huncharek *et al.*, 2000; Ahrendt *et al.*, 2003).
- The accumulation/over-expression of p53 (Quinlan *et al.*, 1992; Morkve *et al.*, 1993; Ishida *et al.*, 1997; D'Amico *et al.*, 1999; Pollan *et al.*, 2003).
- Co-expression of p53 and ERBB2 (Saad *et al.*, 2004).
- High expression of both p53 and integrin β1 (Oshita *et al.*, 2004).
- The presence of D-loop region mitochondrial DNA mutations (Matsuyama *et al.*, 2003).

It can thus be seen that a considerable number of prognostic factors have been proposed, usually in NSCLC. In a prognostic context, it scarcely matters whether these genetic markers are causative of lung tumorigenesis, whether they contribute to the development of the disease or alternatively, whether they are merely consequential.

What matters is whether possession of the marker provides predictive information as to the proliferative potential of the tumour cells, the relative aggressiveness of the tumour, the likelihood of metastasis, the response to chemo- or radio-therapy, and as an ultimate arbiter, patient survival. In practice, however, many of these putative prognostic markers have been insufficiently researched for one to be able to assess their probable utility in terms of patient management (reviewed by Brundage *et al.*, 2002).

The number of identified genes whose expression is either up- or down-regulated during the process of lung tumorigenesis is likely to expand enormously as the powerful new microarray techniques of gene expression profiling become more widely available (*see* Chapter 7; Studies of the Expression of Multiple Genes by Microarrays and Similar Techniques). Some of these genes may prove to be useful biomarkers for early detection (Brambila *et al.*, 2003) or for classification of pathological stage (Furmaga *et al.*, 2003; Beau-Faller *et al*, 2003a). The preclinical use of a combination of molecular epidemiological risk assessment based upon environmental exposure variables, assessment of genetic predisposition and the identification of early markers of lung tumorigenesis promises to be a powerful tool in potentiating early detection (Wild *et al.*, 2002; Field and Youngson 2002; Solan and Werner-Wasik 2003). Other markers may have predictive or prognostic value in the context of survival or the likelihood of developing metastatic disease (Jang *et al.*, 2001; Kijima *et al.*, 2003) or could prove useful in the optimisation of the chemotherapeutic treatment regimen (Rosell *et al.*, 2003; Solan and Werner-Wasik 2003).

Molecular genetics of chemotherapy and chemoresistance

> "The poisons are our principal medicines, which kill the disease and save the life."
> RW Emerson (1860) *The Conduct of Life.*

Considerable effort is currently being expended to target lung cancer cells more specifically by focussing upon the particular sensitivities of these cells that are conferred by those selfsame genetic or epigenetic changes that accompanied the tumorigenic process. Indeed, data on inherited polymorphic variants of DNA repair enzymes (e.g. *ERCC1*, *XPD*; Ryu *et al.*, 2004) as well as tumour-specific information on mutation, epimutation and gene expression promise to improve greatly our prediction of response to chemotherapy on an individual basis (Rosell *et al.*, 2004a).

Tailor-made chemotherapeutic regimens for the individual tumour may be still some way off, but considerable progress has already been made in terms of being able to match specific lung tumours to the most efficacious chemotherapeutic agent. The best and perhaps even defining example to date is gefitinib, employed against lung tumours from a sub-group of NSCLC patients who possess activating mutations in their epidermal growth factor receptor (*EGFR*) genes. Gefitinib is a small molecule tyrosine kinase inhibitor that acts intracellularly to block downstream signal transduction from the EGFR (Averbuch 2003). Such blockade can be very effective, even in advanced NSCLC tumours, as a means to reduce cellular proliferation, and to inhibit survival signals, angiogenesis and metastasis. The majority of

NSCLC patients however exhibit no response to gefitinib, whilst ~10% of patients display a rapid and often dramatic clinical response. Lynch *et al.* (2004) reported that 8/9 patients with gefitinib-responsive NSCLC possessed heterozygous activating mutations (missense or in-frame micro-deletions, some recurrent) in the tyrosine kinase domain of their *EGFR* genes. *In vitro*, these EGFR mutants exhibited increased tyrosine kinase activity in response to epidermal growth factor and increased sensitivity to inhibition by gefitinib. These results were published simultaneously with those of Paez *et al.* (2004) who reported a particularly high frequency of EGFR activating mutations in their Japanese patients. The mutant EGFR molecules are thought to selectively activate Akt and STAT signaling pathways that promote cell survival by inhibiting apoptosis (Sordella *et al.*, 2004). A sub-group of NSCLC patients therefore possess EGFR mutations that lead to increased growth factor signaling and cellular proliferation but which confer sensitivity to the tyrosine kinase inhibitor gefitinib. This is the first example in the sphere of lung cancer treatment where mutational information derived from the tumour has been shown to be important in identifying the most appropriate chemotherapeutic regimen for the patient concerned.

As with lung tumorigenesis itself, resistance to chemotherapy is multifactorial. The underlying mechanisms are however poorly understood (Rosell *et al.*, 2001). Mutations in a number of different genes nevertheless appear to confer resistance to antitumour drugs e.g. topoisomerase I (*TOP1*; Tsurutani *et al.*, 2002), topoisomerase IIα (*TOP2A*; Binaschi *et al.*, 1992; Kubo *et al.*, 1996; Dingemans *et al.*, 2001; Mirski *et al.*, 2000), β-tubulin (*TUBB*; Monzo *et al.*, 1999b) and p53 (*TP53*; Blandino *et al.*, 1999; Vogt *et al.*, 2002). Some doubt has however been expressed as to the authenticity of the β-tubulin mutations on account of possible confusion with the very frequently encountered *TUBB* pseudogene sequences (de Castro *et al.*, 2003).

The overexpression of multidrug resistance-associated proteins (MRP) has been noted in cell lines derived from lung tumours that have become resistant to chemotherapeutic agents (Oguri *et al.*, 2000; Savaraj *et al.*, 2003). This may however be an indirect consequence of *TP53* gene mutation since p53 is a negative regulator of MRP gene expression (Wang and Beck 1998; Oshika *et al.*, 1998b). The over-expression of the cisplatin-inducible gene (*ATF4*) encoding activating transcription factor 4 has also been reported to correlate with cisplatin sensitivity and could prove useful in a predictive context (Tanabe *et al.*, 2003). Finally, an increase in the expression of the ribonucleotide reductase large subunit 1 (*RRM1*) gene has been shown to be associated with gemcitabine resistance in NSCLC cell lines (Davidson *et al.*, 2004; Rosell *et al.*, 2004b).

The molecular basis of chemoresistance can be investigated by techniques such as comparative genomic hybridization (CGH). For instance, CGH has been used to screen a human lung adenocarcinoma cell line, exposed to varying concentrations of a topoisomerase inhibitor, for genomic changes: the amplification of chromosome 11q23-qter was detected that encompassed the myeloid/ lymphoid leukaemia (*MLL*) gene within this region, as well as the loss of chromosome 17 and deletions of 2p and 2q (Struski *et al.*, 2001).

A number of different molecular markers, including aberrant expression of p53 and the down-regulation of Bcl2, have been claimed to be predictive of response to both chemotherapy (Rusch *et al.*, 1995; Kawasaki *et al.*, 1998; Dingemans *et al.*, 1999;

Nakanishi *et al.*, 1999; Kandioler-Eckersberger *et al.*, 1999; Rodriguez-Salas *et al.*, 2001; Rosell *et al.*, 2002; Gregore *et al.*, 2003; Kumar Biswas *et al.*, 2004) and radiotherapy (Langendijk *et al.*, 1995; Hayakawa *et al.*, 1998; Matsuzoe *et al.*, 1999b; Hwang *et al.*, 2001) in lung cancer. Similarly, inherited polymorphic variants in the XPD (*ERCC2*) and *XRCC1* genes have been claimed to be useful prognostic markers for NSCLC patients undergoing platinum chemotherapy (Gurubhagavatula *et al.*, 2004). For the future, transcriptional profiling promises to improve our understanding not only of the mechanisms of resistance to anti-tumour agents but also the prediction of drug resistance (Volm *et al.*, 2002d; Sarries *et al.*, 2002) and ultimately the chance of survival after chemotherapy (Ohira *et al.*, 2002; Whiteside *et al.*, 2004; Ikehara *et al.*, 2004). It is anticipated that microarrays will also play an important role in predicting the chemosensitivity of specific tumours (Staunton *et al.*, 2001).

CHAPTER 5

Genetic Approaches to Studying the Association Between Smoking and Lung Cancer

"A custom loathsome to the eye, hateful to the nose, harmful to the brain, dangerous to the lungs, and in the black stinking fume thereof, nearest resembling the horrible Stygian smoke of the pit that is bottomless".

King James I & VI (1604) *A Counter Blaste to Tobacco*

Tobacco smoke contains a considerable number of putative mutagens and carcinogens (Hecht 1999) and epidemiological studies have provided very strong evidence for an association between cigarette smoking and lung cancer. Thus, smokers are estimated to have a ~10-fold higher risk of dying from lung cancer than non-smokers and to display a lifetime cumulative probability of dying from lung cancer of between 8% (females) and 24% (males), not conditional on surviving other causes of death (Thun *et al.*, 2002). However, although only 5-10% of all lung cancers occur in individuals without a prior history of smoking, only 5-10% of smokers get lung cancer. I do not regard it as my remit here (nor am I qualified) to stray into either epidemiological or toxicological research in order to interpret the myriad published studies or to explore the undoubtedly highly complex relationship between cause and effect. It is nevertheless pertinent to review what is currently known about the genetic changes hitherto characterized in the lung cells of smokers and smoking lung cancer patients, and to compare them to those in non-smokers and non-smoking lung cancer patients. One of the difficulties in making sense of this field is that many studies have chosen to focus on one group or another rather than adopting a traditional case-control study approach. This notwithstanding, the investigation of the nature, location and timing of smoking-related genetic changes constitutes a vitally important area of inquiry. If properly conducted, it holds out the promise of eventually placing the epidemiological data upon a sound mechanistic footing, firmly based upon precepts of molecular and cellular biology. This then can be used to good effect in prevention, early detection and therapy (Hu *et al.*, 2002).

Those seeking to elucidate the biological mechanisms underlying the epidemiological association between lung cancer and smoking have, by and large, envisaged a schema similar to that presented by Hecht (2002; 2003):

(i) Continual cigarette smoking leads to chronic exposure to carcinogens.
(ii) The carcinogens are then either metabolically deactivated and excreted harmlessly, or metabolically activated to intermediates that have the potential to react with DNA.
(iii) The intermediates react with DNA forming covalently bound products known as *DNA adducts*.

(iv) The DNA adducts may then be repaired by cellular DNA repair enzymes. However, if they are not, they may persist during DNA replication leading to the introduction of a permanent mutation.
(v) Although cells with damaged DNA can be removed by apoptosis, if mutations have occurred in key genes, cellular proliferation may ensue and apoptosis may be abrogated leading to tumour development.

At present, we are still a very considerable way from being able to confirm and explain this series of linked postulates. Evidence for a causal connection between cigarette smoking and genetic damage leading to lung tumorigenesis has nevertheless come from a variety of different sources.

(i) *In vitro* studies of DNA adduct formation and mutagenesis consequent to exposure to chemicals present in tobacco smoke. This topic deserves a volume in its own right but is only briefly addressed here in the context of the *TP53* gene and benzo[*a*]pyrene in *p53 mutations, benzo[a]pyrene and lung cancer.*
(ii) Cytogenetic changes specifically associated with smoking.
(iii) Case-referent studies exploring associations between the possession of a given allele of a specific gene with risk of lung cancer in both smokers and non-smokers.
(iv) Epigenetic changes (e.g. methylation) specifically associated with smoking.
(v) Differences in mutation frequency or in the nature of the mutational spectrum between smokers and non-smokers.
(vi) Differences in lung/lung cell gene expression between smokers and non-smokers.

Some of these different types of study will now be explored in more detail.

Genetic changes associated with smoking

> "If the tobacco be the poison that its enemies declare it to be, it is eminently slow in its action, for every workhouse, lunatic asylum, and charitable institution has its grey-haired votaries to the pipe."
> Editorial, *The Lancet*, June 1st 1872: The tobacco controversy, p770.

Loss of heterozygosity and smoking

> "In men who die of lung cancer, all of whom were smokers, the remainder of the epithelial lining shows changes that may be considered as preliminary stages to the development of lung cancer, including early invasion. There are few such changes in nonsmokers, but they increase rapidly with the amount of cigarette smoking".
> O Auerbach, AP Stout, EC Hammond, L Garfinkel (1964) The role of smoking in the development of lung cancer. *Proc. 5th Natl. Cancer Conf.* 497-501.

A variety of studies have claimed an association between cigarette smoking and loss of heterozygosity (LOH) at a number of chromosomal loci (Ishikawa *et al.*, 2002). Thus, LOH at chromosomes 3p21 (Hirao *et al.*, 2001; Sanchez-Cespedes *et al.*, 2001a; Zienolddiny *et al.*, 2001; Ho *et al.*, 2002a), 3p14 (Mao *et al.*, 1997; Marchetti *et al.*, 1998b; Nelson *et al.*, 1998; Tseng *et al.*, 1999b; Zienolddiny *et al.*, 2001), 5q11-q13

(Zienolddiny *et al.*, 2001), 6q (Sanchez-Cespedes *et al.*, 2001a), 9p21 (Sanchez-Cespedes *et al.*, 2001b; Zienolddiny *et al.*, 2001), 13q14 (Ho *et al.*, 2002a), 16p (Sanchez-Cespedes *et al.*, 2001a), 17p13 (Sanchez-Cespedes *et al.*, 2001a; Ho *et al.*, 2002a) and 19p (Sanchez-Cespedes *et al.*, 2001a) has been reported to occur more frequently in smokers as compared to non-smokers. Yoshino *et al.* (2003) have even claimed that the extent of LOH on different chromosomes is associated with the degree of smoking.

LOH at 3p14, the location of the fragile site FRA3B, may also occur more frequently in current smokers than in former smokers (Mao *et al.*, 1997; Tseng *et al.*, 1999b; Stein *et al.*, 2002). Whilst LOH at the FHIT locus at chromosome 3p14 has been found to occur more frequently in smokers as compared to non-smokers (Sozzi *et al.*, 1997; Marchetti *et al.*, 1998b), this has not been found in all studies (Pylkkanen *et al.*, 2002). Some reports are difficult to interpret and even harder to compare and contrast owing to small sample size (e.g. Powell *et al.*, 2003b), differences in lung cancer type or the definition of smoking status employed.

To avoid some of these difficulties, Wong *et al.* (2002) used 84 microsatellite markers in a search for LOH in 42 adenocarcinomas from non-smokers and 29 adenocarcinomas from smokers. The adenocarcinomas of non-smokers exhibited frequent LOH on 2q, 6p, 10p, 13q, 16q, 17q, 19p, 19q, 20p and 20q whilst LOH on 1q, 2p, 3p, 3q, 7q, 8p, 9p, 9q, 10q, 11q, 13q, 14q, 17p, 18q and 21q was evident in both smokers and non-smokers. These authors concluded that tumorigenesis in the lungs of smokers and non-smokers is associated with an overlapping series of chromosomal abnormalities that could provide evidence for the existence of distinct genetic pathways to lung cancer in the two groups. It should be noted, however, that microsatellite instability has also been claimed to be higher among smokers than among non-smokers (Petmitr *et al.*, 2002) and this could represent a potential source of confusion. It remains to be seen whether the above listed spatially localized differences in LOH claimed between smokers and non-smokers will be confirmed by further studies.

Since there appears to be an association between possession of a *TP53* mutation and genome-wide losses of heterozygosity (Pellegrini *et al.*, 1999; Zienolddiny *et al.*, 2001), it could be that *TP53* mutation is responsible for a generalized, albeit secondary, increase in genome instability. For further discussion of this topic, the interested reader is referred to several recent review articles for further details on the molecular epidemiology of smoking and lung cancer (Alavanja 2002; Hu *et al.*, 2002; Shields 2002; Wistuba *et al.*, 2002).

Aberrant DNA methylation in lung cancers of smokers

Hypermethylation of the *CDKN2A* gene promoter has been claimed to be associated with smoking status in NSCLC (Kim *et al.*, 2001a) and may even occur prior to any clinical evidence of neoplasia becoming apparent (Kersting *et al.*, 2000). However, *CDKN2A* promoter methylation has been found to occur more frequently in former smokers than in current smokers (Jarmalaite *et al.*, 2003). Further, methylation of the *CYP1A1* gene promoter in lung tissue has been reported in 33 % of heavy smokers, 71 % of light smokers and 98 % of non-smokers, defined as ex-smokers as well as never-smokers (Anttila *et al.*, 2003); this suggests that the loss of *CYP1A1* promoter methylation may be involved in the induction of CYP1A1 expression. Similarly, Pulling *et al.* (2003) have reported that more lung adenocarcinomas from nonsmokers than smokers exhibit *MGMT* promoter methylation.

A number of other reports have appeared of aberrant promoter methylation, involving a variety of different genes, in the bronchial epithelium of smokers and former smokers, smokers with cancer and cancer-free smokers (Kim *et al.*, 2001b; Soria *et al.*, 2002a; Belinsky *et al.*, 2002; Yanagawa *et al.*, 2002; Honorio *et al.*, 2003; Toyooka *et al.*, 2003b; Zöchbauer-Müller *et al.*, 2003; Kim *et al.*, 2003b; Kaplan *et al.*, 2003; Marsit *et al.*, 2004; Kim *et al.*, 2004c; Kim *et al.*, 2004d; Chang *et al.*, 2004e; reviewed by Zöchbauer-Müller *et al.*, 2002). Quite what the results of these studies are trying to tell us is as yet unclear. Thus, extreme caution should be exercised in interpretating these studies if only because methylation profiles differ markedly between men and women (Chan *et al.*, 2002), between different races (Toyooka *et al.*, 2003b), between different types of lung tumour or tumour cells (Toyooka *et al.*, 2001b; Virmani *et al.*, 2002; Honorio *et al.*, 2003), and finally even between individuals on the basis of their possession of specific polymorphic methylenetetrahydrofolate reductase (*MTHFR*) and 5-methyltetrahydrofolate-homocysteine methyltransferase (methionine synthase; *MTR*) gene variants (Paz *et al.*, 2002; Heijmans *et al.*, 2003). It is clear that for any study on the effect of smoking on DNA methylation to yield credible results, it must be very carefully controlled.

Pulling *et al.* (2004) have reported early and frequent promoter hypermethylation and transcriptional inactivation of the death-associated protein kinase (*Dapk*) gene in a murine model of lung cancer. Methylation-mediated inactivation of the gene was found to occur in mice exposed to both mainstream cigarette smoke and the tobacco carcinogens 4-(methylnitrosamino)-1-(3-pyridyl)-1-butanone (NNK) and vinyl carbamate. How these findings relate to the relatively frequent methylation-mediated hypermethylation of the *DAPK1* gene promoter observed in NSCLC is unclear, particularly since no correlation has been noted between smoking behaviour and *DAPK1* gene hypermethylation (Soria *et al.*, 2002a).

Telomerase activity and smoking

Frequent (69%) expression of telomerase reverse transcriptase (*TERT*) has been reported in the biopsied tissue of heavy smokers (Soria *et al.*, 2003a). Further, Xinarianos *et al.* (1999) have reported a correlation between telomerase activity in NSCLC tissues and current smoking status. However, only a non-significant trend was noted between telomerase activity and smoking exposure.

Gene expression studies of normal lung and lung tumour tissue from smokers and non-smokers

Although the over-expression of a number of different genes has been claimed to be associated with smoking, these findings should be viewed within the context of the very considerable number of genes known to be over-expressed in lung cancer tissue (*see* Chapter 7; Studies of the Expression of Individual Genes). This notwithstanding, evidence has been presented for the up-regulation of the *CYP1A1* (McLemore *et al.*, 1990), *CYP1B1* (Spivack *et al.*, 2003), *NQO1* (Spivack *et al.*, 2003) and *ALOX15* (arachidonate 15-lipoxygenase) [Zhu *et al.*, 2002] genes in the normal lung tissue of smokers as compared to non-smokers, and of the D40/AF15q14 (*CASC5*) gene in lung tumour tissue taken from smokers as compared to non-smokers (Takimoto *et al.*, 2002). Similarly, the expression of the macrophage migration inhibitory factor (*MIF*) gene in NSCLC tissue of heavy smokers was found to be significantly higher than in tumour tissue from non-smokers (Tomiyasu *et al.*, 2002). In an *in vitro* assay, expression of the human *CDKN1A* gene in alveolar epithelial cells has been reported to be up-regulated by exposure to tobacco smoke condensate (Marwick *et al.*, 2002). Similarly, xanthine oxidase (*Xdh*) gene expression is increased in cultured rat pulmonary microvascular endothelial cells exposed to tobacco smoke condensate (Kayyali *et al.*, 2003). In a study of NSCLC tumours, Esposito *et al.* (1997b) found that immunohistochemically detectable p53 expression was altered in $>40\,\%$ of lung tumours from smoking patients while none of the lung tumour samples from non-smoking patients exhibited detectable p53 protein.

Shriver *et al.* (2000) analysed the expression of the gastrin-releasing peptide receptor (*GRPR*) gene in lung tissue and cultured airway cells from 78 individuals. The activation of GRPR in human airways is associated with a proliferative response of bronchial cells to gastrin-releasing peptide and is thought to be part of an autocrine loop for growth. *GRPR* gene expression was detected in significantly more female non-smokers than male non-smokers, and more female short term smokers than male short term smokers. Female smokers also exhibited GRPR expression at a lower level of smoke exposure than male smokers. Since the *GRPR* gene is X-linked (Xp21.2-p22.3), it could be that the presence of two copies of the gene in women renders them more susceptible to any carcinogenic effects of tobacco smoke.

The large-scale transcriptional profiling of lung cancer tissue from smokers and non-smokers is still in its infancy. However, microarray expression analysis has already been used to study gene expression in adenocarcinomas. Indeed, various groups claim to have successfully used expression profiling to identify genes whose expression discriminates between the lung adenocarcinomas of smokers and non-smokers (Miura *et al.*, 2002; Powell *et al.*, 2003a; *see* Chapter 7; Use of Microarrays to Study Changes in Gene Expression in Smoking-Related Lung Tumours). Microarray analysis has also been used to compare the gene expression patterns of the airway epithelium of smokers and non-smokers in the absence of neoplasia (*see* Chapter 7; Use of Microarrays to Study Changes in Bronchial Epithelial Cell Gene Expression Consequent to Smoking). Thus, Kaplan *et al.* (2003) noted that smokers exhibited a greatly increased level of expression of the imprinted gene, *H19*. Although loss of imprinting of the *H19* gene was not detected in association with this increase in gene expression, the authors blithely predicted that loss of imprinting would eventually manifest itself "as the burden of smoking increases and as the epithelium undergoes transition from normal to neoplastic". Finally, microarray analysis has purportedly provided evidence for the specific up-regulation of antioxidant-related genes (genes of the catalase/superoxide dismutase family or genes involved in glutathione metabolism or redox balance) in smokers without obvious lung disease as compared to non-smokers (Hackett *et al.*, 2003).

Differences in mutation frequency/mutational spectra between smokers and non-smokers

Tobacco smoke contains many chemicals that can interact with DNA to form DNA adducts (Phillips 2002). In a study of normal bronchial epithelial cells from non-cancerous tissue derived from 22 patients with lung cancer, Rojas *et al.* (2004) found a high level of benzo[*a*]pyrene 7,8-diol-9,10-epoxide-*N*(2)-deoxyguanosine [BPDE-N(2)-dG]. Four-fold inter-individual differences in the DNA adduct level were noted in smokers (BPDE-N(2)-dG adducts/10^8 nucleotides, range = 36-175, mean = 84.7 $\pm$ 38.4) as compared to a 3-fold inter-individual difference in non-smokers (range = 19.7-62.4, mean = 37.6 $\pm$ 22.2). If unrepaired, such adducts could give rise to G$\rightarrow$T transversions. Although the mean adduct level is significantly higher in smokers, it should be noted that the ranges exhibited by the smoking and non-smoking groups are overlapping to some extent.

Various studies have reported a higher *TP53* mutation prevalence in the lung tumours of smokers (55-58 %) than in those of non-smokers (10-26 %) [Husgafvel-Pursiainen and Kannio 1996; Takagi *et al.*, 1998; Gealy *et al.*, 1999; Ahrendt *et al.*, 2000b; Husgafvel-Pursiainen *et al.*, 2000]. More specifically, a number of studies have reported a significantly higher frequency of G$\rightarrow$T transversions in the *TP53* genes of smokers as compared to never-smokers (30 % v. 10-15 %; Hainaut and Pfeifer 2001; Pfeifer *et al.*, 2002; Toyooka *et al.*, 2003a). Toyooka *et al.* (2003a) have however pointed out that this could be largely attributable to the particularly high frequency of such mutations in female smokers.

Although various claims have been made for an association between the incidence of *KRAS* mutations in lung tissue and tobacco smoke exposure (Slebos *et al.*, 1991; Reynolds *et al.*, 1991; Westra *et al.*, 1993a; Gealy *et al.*, 1999; Noda *et al.*, 2001; Ahrendt *et al.*, 2001), the sizes of the non-smoker samples were sometimes quite small. Both Gao *et al.* (1997) and Pulling *et al.* (2003) reported that the proportion of patients with *KRAS* mutations did not differ between smokers and non-smokers. Moreover, Vahakangas *et al.* (2001) have reported that the incidence of codon 12 or 13 *KRAS* mutations was no different between long-term ex-smokers and non-smokers. *KRAS* mutations also appear to be frequent in histologically normal cells taken from outside lung adenocarcinomas (Keohavong *et al.*, 2001) and from atypical alveolar hyperplasia, a potential precursor lesion from which adenocarcinomas may arise (Westra *et al.*, 1996; Slebos *et al.*, 1998). Finally, G$\rightarrow$T transversions have been noted in the *KRAS* genes of lung tumours removed from non-smokers exposed to polycyclic aromatic hydrocarbon-rich coal combustion emissions (DeMarini *et al.*, 2001).

Codons 12 and 14 of the *KRAS* gene have been shown to be hotspots for both BPDE- and 4-(methylnitrosamino)-1-(3-pyridyl)-1-butanone (NNK)-induced adduct formation whilst the other mutation hotspots at codons 13 and 61 were found to exhibit little or no adduct formation (Feng *et al.*, 2002b; Tretyakova *et al.*, 2002; Ziegel *et al.*, 2003). At the same time, BPDE adducts formed at codon 12 were found to be repaired with approximately half the efficiency of those at codon 14 (Feng *et al.*, 2002b) suggesting that the basis for the mutational hotspot at codon 12 of the *KRAS* gene might be a combination of preferential DNA damage and poor repair. Additionally, Hu *et al.* (2003) have provided evidence for an effect of cytosine methylation on DNA adduct formation at codon 14 although not at codon 12. The interpretation of these data is however subject to many of the same problems as for the comparable series of *TP53* data discussed in some detail below. Thus, for example, there is considerable evidence that reactive oxygen species, whether of endogenous or exogenous origin, are involved in the mutation of the *RAS* hotspot codons (Kamiya *et al.*, 1992; Kamiya *et al.*, 1995; Le Page *et al.*, 1995). In conclusion, direct evidence to support the contention that cigarette smoking leads to an increase in the frequency of nuclear point mutations in bronchial epithelial cells is still lacking.

Finally, significantly higher levels of mitochondrial DNA damage were found in a comparison of bronchoalveolar lavage tissues from smokers and non-smokers (Ballinger *et al.*, 1996). Contradictory results were however reported by Coller *et al.* (1998) who performed a comparative study of bronchial epithelial cell mitochondrial DNA mutation in smokers and non-smokers including several pairs of smoking and non-smoking twins. The authors claimed their assay to be very sensitive with mutations being detected at a frequency as low as 10^{-6}. The mutational spectra were not however found to differ significantly and it was suggested that endogenous factors were more likely to be responsible for the observed mitochondrial DNA mutations than exogenous factors. Mitochondrial genome instability is a virtually ubiquitous phenomenon in human cancer and has been considered to be caused by a combination of damage by reactive oxygen species, slipped-strand mispairing and defective DNA repair (Bianchi *et al.*, 2001).

p53 mutations, benzo[*a*]pyrene and lung cancer: the controversy

> "Aus dem Krebserzeugenden Tabakteer ist durch Destillation ein Produkt mit den spektrographischen Eigenschaften und der Fluorescenz des 1:2-Benzpyrens gewonnen worden. Dieses Produkt ist stark krebserzeugend; es verursacht die Entstehung von bösartigen Tumoren, die sich durch ausbreitendes, eindringendes und zerstörendes Wachstum wie Carcinome entwickeln. Auf Grund dieser Versuchsergebnisse glauben wir die hohe krebserzeugende Wirkung des totalen, aus zahlreichen Substanzen zusammengesetzten Tabakteeres dem hier untersuchten Stoff – dem Tabakbenzpyren – zuschreiben zu dürfen".
> [Through distillation from the cancer-producing tobacco tar, a product with the spectrographic properties and the fluorescence of benzopyrene has been extracted. This product is highly cancer-producing; it causes the formation of malignant tumours which develop like carcinoma through expanding, penetrating and destructive growth. On the strength of these test results, we believe that we can attribute the highly cancerous effect of the total tobacco tar, which is composed of many different substances, to the substance analysed in our tests - tobacco benzopyrene.]
>
> A.H.Roffo (1940) Krebserzeugendes Benzpyren, gewonnen aus Tabakteer. *Z. Krebsforsch.* 49: 588-597.

BPDE [(±)-anti-7β,8α-dihydroxy-9α,10α-epoxy-7,8,9,10-tetrahydro-benzo[*a*]pyrene], a metabolite of benzo[*a*]pyrene, is one of the most potent mutagens and carcinogens known. BPDE binds to DNA to form adducts at the N2 position of guanine (Friedberg *et al.*, 1995), and when such adducts remain unrepaired, their misreading by DNA polymerase during DNA replication gives rise predominantly (~70 %) to G→T transversions. Since G→T transversions in the *TP53* gene are frequent (~35 %) in lung tumours at the same time as being comparatively rare in other cancers, drawing a causative link between agent and mutational signature has had a certain intuitive appeal [Bennett *et al.*, 1999; Hecht 1999; Hainault and Pfeifer 2001]. Intriguingly, it has been known for some time that there is significant inter-individual variation in terms of the binding of benzo[*a*]pyrene to DNA (Harris *et al.*, 1976). A functional link between benzo[*a*]pyrene and the *TP53* gene is also worthy of note: benzo[*a*]pyrene transcriptionally activates the *TP53* gene thereby recruiting p53 to the cellular response to the genotoxic effects of the mutagen (Ramet *et al.*, 1995; Pei *et al.*, 1999).

The BPDE-induced mutagenesis model

The suggestion of "a direct link between a defined cigarette smoke carcinogen (benzo[*a*]pyrene) and human cancer mutations" was first made by Denissenko *et al.* (1996). These authors proposed that exposure to benzo[*a*]pyrene represented a direct causative agent for human p53 (*TP53*) gene mutation *in vivo* and that, as a constituent of tobacco smoke, benzo[*a*]pyrene was *ipse facto* directly responsible for the increased lung cancer risk associated with tobacco consumption. Using data from the International Association for Research on Cancer (IARC) *TP53* Mutation Database (http://www.iarc.fr/p53), G→T substitutions in the *TP53* gene in the lung cancers of smokers were found to occur disproportionately at the hypermutable 'hotspot codons' 157, 248 and 273 which were also noted experimentally to be sites of DNA adducts formed by BPDE (Denissenko *et al.*, 1996; Figure 5.1) and subsequently also by other polycyclic aromatic hydrocarbons (Smith *et al.*, 2000). Denissenko *et al.* (1996) claimed that one of these hotspots (codon 157) was smoking-associated and not found in other types of cancer. Although the paper by Denissenko *et al.*

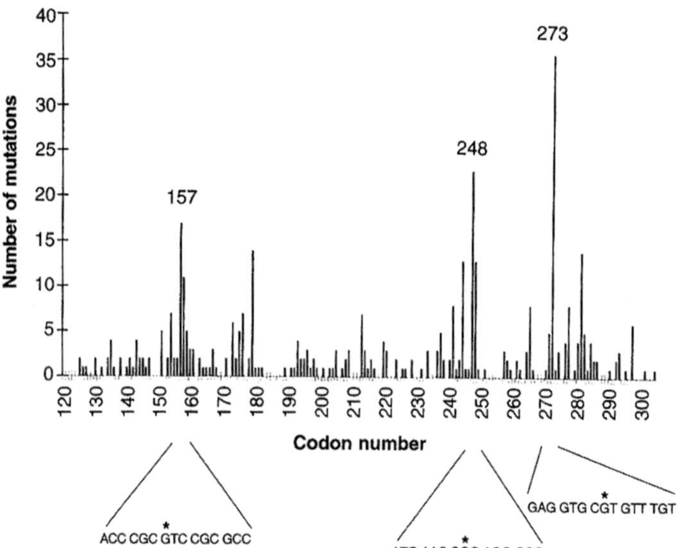

Fig. 5.1. Frequency of TP53mutations in lung cancer by codon position. The sequences surrounding the mutational hotspot codons 157, 248 and 273 are indicated; the mutated G nucleotides within these sequences are marked with asterisks. Reprinted with permission, from MF Denissenko *et al.* (1996) *Science* 274: 430-432. Copyright [1996] AAAS

(1996) clearly represented a landmark publication in terms of the claims made, both the experimental approach adopted and the conclusions drawn have been criticised in several different respects by Krawczak and Cooper (1998) among others. These criticisms relate not only to the shortcomings and limitations of the study but also to a certain lack of clarity and consistency in the conclusions as presented:

- *TP53* G→T substitutions at hotspot codons 157, 248 and 273 have been frequently reported from tumours other than lung, in tissues that are likely to be inaccessible to benzo[*a*]pyrene or its metabolite BPDE. The 'smoking-associated' hotspot at codon 157 cannot therefore be regarded as smoking-specific.
- An *in vitro* study of mutational spectra associated with exposure to various polycyclic aromatic hydrocarbons (PAHs) has shown that mutational hotspots do not invariably correspond to sites of DNA adduct formation and that the DNA sequence context may play a key role in PAH-induced mutagenesis (Bigger *et al.*, 2000).
- Adduct formation at the *TP53* hotspot codons may not be a property unique to benzo[*a*]pyrene (Vähäkangas 2003). Other chemical substances can also form adducts at codons 175, 248 and 273 e.g. *N*-hydroxy-4-acetylaminobiphenyl (Feng *et al.*, 2002a), a potential carcinogen which has also been shown to be associated with the induction of G→T substitutions in cultured cells (Besaratinia *et al.*, 2002). Since the methylating agent *N*-methyl-*N*-nitro-*N*-nitrosoguanidine (MNNG) has also been shown to induce G→T transversions in specific hotspots in the *HPRT* gene via O6-methylguanine adduct formation (Tomita-Mitchell *et al.*, 2003), it would be of considerable interest to establish the mutation profile of MNNG for the *TP53* gene.
- BPDE mutagenesis can induce nucleotide substitutions other than G→T transversions. Thus, not only are BPDE adducts capable of giving rise to G→A and G→C substitutions as well as G→T substitutions (Jelinsky *et al.*, 1995) but substitutions in neighbouring 'non-targeted' bases may also occur (Kramata *et al.*, 2003). BPDE mutagenicity is known to be highly dependent upon both the local DNA sequence environment and adduct stereochemistry (Ponten *et al.*, 2001). Further, oxidative damage by reactive oxygen species generated by endogenous cellular processes is known to result in G→T transversions and G→A transitions (Wang *et al.*, 1998c).
- Control data (i.e. mutation profiles from non-smoking lung cancer patients) were lacking from the study of Denissenko *et al.* (1996). Indeed, for reasons that remain unclear, data from non-smokers were actively excluded. Such data are essential to the validity of any such analysis and could have been used to demonstrate that the putative mutational signature of benzo[*a*]pyrene was indeed smoker-specific and did not alternatively reflect either other environmental sources of exposure to the carcinogen or another carcinogen with a similar mutational signature.
- The term 'hotspot' is perhaps potentially misleading here since only a relatively small proportion (currently 5.7 %) of characterized lung cancer-associated *TP53* gene mutations are actually G→T substitutions in the three hotspot codons, 157, 248 and 273 (IARC Database, Update R7, September 2002 release containing 2003 *TP53* lung cancer-associated mutations). This implies that the BPDE-induced mutagenesis model is at best unlikely to be able to account for more than a small minority of cases of lung cancer.

- ~60 % of human lung cancers contain one or more mutations in the *TP53* gene but these are neither necessary nor sufficient for tumour formation.

In response to some of these criticisms, Denissenko and colleagues presented some mutation data from non-smoking lung cancer patients (Pfeifer *et al.*, 1998). However, the only noticeable differences regarding the distribution of mutations along the *TP53* gene occurred at amino acid residues known to represent mutation hotspots in other types of cancer (Hollstein *et al.*, 1996) and involved tissues that were unlikely to have been directly accessible to benzo[*a*]pyrene. Since the remaining mutations did not allow any effective discrimination between smokers and non-smokers, the question as to whether benzo[*a*]pyrene exposure plays a role in this regard, remained unclear.

Endogenous *versus* exogenous causes of mutation

> "I will consider. . ...the Damoclean possibility that certain normal endogenous cellular processes are inherently mutagenic and that this intrinsic mutagenesis is a significant factor in the etiology and pathogenesis of some human cancers".
>
> LA Loeb (1989) Endogenous carcinogenesis: molecular biology into the 21st century. *Cancer Res.* 49: 5489-5496.

Pfeifer (2000) noted that "a disproportionately high number of mutations in the *TP53* gene are found at methylated [CpG] dinucleotides", a finding also reported by Yoon *et al.* (2001), Soussi and Béroud (2003) and Hussain and Harris (2000). Indeed, the hotspot codons 157, 248 and 278 studied by Denissenko *et al.* (1996) are CG-containing codons and the CG→TG and CG→CA transitions at these locations (but not the G→T transversions) are compatible with a model of methylation-mediated deamination of 5-methylcytosine, a common endogenous mechanism of mutation in the human genome.

Prior to the publication of the paper by Denissenko *et al.* (1996), Krawczak *et al.* (1995) had conducted a meta-analysis of the somatic cancer-associated *TP53* mutational spectrum observed in a variety of tissues and demonstrated that, upon the exclusion of ~12 % of tissue- and site-specific single base-pair substitutions, the somatic spectrum closely resembled the overall germline mutational spectrum of other genes. This similarity suggested that the majority of somatic *TP53* mutations were of endogenous origin and selected for during the process of cellular transformation (Krawczak *et al.*, 1995). The report of Denissenko *et al.* (1997) which demonstrated that methylation of CpG dinucleotides in the *TP53* gene increases susceptibility to BPDE-induced adduct formation *in vitro*, did nothing to invalidate this argument. Rather, it highlighted the difficulties inherent in distinguishing between exogenous and endogenous causes of mutation *in vivo*.*

* Endogenous 5-methylcytosine at CpG dincleotides within *TP53* codons 154, 157, and 248 also appears to exert a protective effect on neighbouring guanine residues with respect to N7- and O6-methylation by the carcinogen 4-(methylnitrosamino)-1-(3-pyridyl)-1-butanone (NNK) [Ziegel *et al.*, 2004].

Patterns of mutagenesis are however also known to be critically dependent on the local DNA sequence environment. A recent analysis of DNA folding has served to emphasize the point that the same missense mutations may have both exogenous and endogenous causes (Wright *et al.*, 2002). This study reported that the majority of the hypermutable bases in the *TP53* gene are located immediately adjacent to the stems of stable hairpin loop structures. The explanation put forward to account for this finding was that these stem-loop structures, whose formation is promoted by the presence of inverted repeat sequences, contain unpaired or mispaired bases that (i) are vulnerable to mutation and (ii) block transcription (or DNA replication?) which serves to increase the time of exposure to the mutagenic influence. This influence could be either endogenous (e.g. oxygen radicals) or exogenous (e.g. mutagens such as benzo[*a*]pyrene).

The above *caveats* notwithstanding, Lewis and Parry (2004) have used BPDE-induced mutation data generated in an *in vitro* bacterial supF tRNA gene system to develop an algorithm that attempts to predict the BPDE-associated mutational spectrum of the *TP53* gene. This algorithm incorporates a number of different parameters including target sequence specificity and DNA curvature. The predicted *TP53* mutational spectrum displayed a strong resemblance to that derived from human lung cancer data, even including the mutation hotspots at codons 157, 248 and 273 (although intriguingly not the codon 249 hotspot) that correspond to the sites of BPDE adduct formation. This suggests that the target site specificity of BPDE may be heavily influenced by the physicochemical properties of the DNA sequence of the gene in question.

A re-examination of the BPDE-induced mutagenesis model

In the year 2000, Rodin and Rodin (2000), from the same research centre as Denissenko and colleagues, took issue with what they saw as the 'almost consensus opinion that the major carcinogenic risk of tobacco smoke is in its direct mutagenic action on DNA of cancer-related genes' They revisited the question of the claimed association between BPDE-induced adduct formation and lung cancer by performing a detailed meta-analysis of the somatic lung cancer-associated *TP53* mutational spectrum. This detailed and highly complex molecular epidemiological re-analysis of a similar dataset to that used by Denissenko *et al.* (1996) may be summarized as follows:

- Despite a difference in the relative mutation rate of the transcribed DNA strand as compared to the non-transcribed strand (a phenomenon recently confirmed by Green *et al.*, 2003), the intra-strand mutational spectra of smoking and non-smoking cases were similar for both strands (Figure 5.2). Moreover, the distribution along the *TP53* gene of G→T substitutions, a hallmark of BPDE-induced mutagenesis, was found to be indistinguishable, not only between smokers and non-smokers, but also between lung cancers in smokers and cancers in smoke-inaccessible tissues (Figure 5.3). The absence of differences was deemed to be incompatible with the BPDE-induced mutagenesis model and it was proposed that the major determinant of the lung cancer-associated spectrum of *TP53* mutations was likely to be selection rather than genotoxicity.

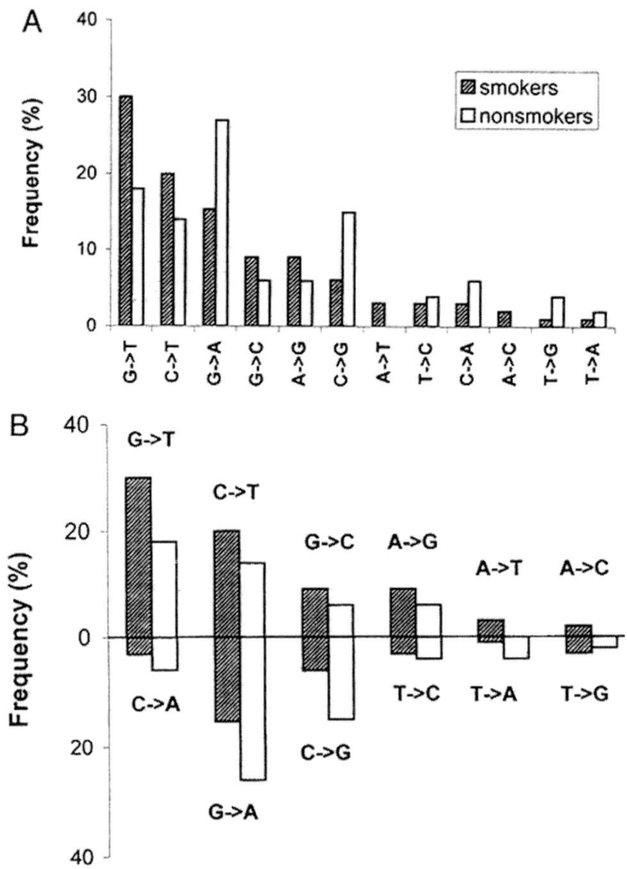

Fig. 5.2 A,B. *TP53* mutational patterns in smokers and non-smokers. *TP53* mutational patterns of lung adenocarcinomas from smokers (99 mutations) and non-smokers (50 mutations). Since the DNA strand with original lesions is identifiable only for (G→T, C→A), (G→C, C→G) and (C→T, G→A) pairs of complementary substitutions, only these three pairs were used to test the BPDE hypothesis against the strand asymmetric repair alternative. **A** In a standard (strand non-specific) representation, the patterns of smokers and non-smokers show a difference, but one that was statistically insignificant. **B** The same patterns as in A but organized as pairs of complementary transitions and transversions; the upper part may represent a nontranscribed strand, the lower part, a transcribed strand.
Reproduced with kind permission, from SN Rodin & AS Rodin (2000) Human lung cancer and p53. *Proc. Natl. Acad. Sci. USA* 97: 12244-12249. Copyright (2000) National Academy of Sciences, USA

- Since G→T substitutions at codon 157 have also been found in smoke-inaccessible tissues, such mutations cannot invariably and unequivocally be attributed to the action of benzo[*a*]pyrene.
- Silent *TP53* mutations were studied in both smokers and non-smokers as a way of distinguishing the imprint of mutability from that of selection. The BPDE-induced mutagenesis model would predict that lung cancer tissue from smokers should contain many more silent G→T substitutions than lung cancers from non-smo-

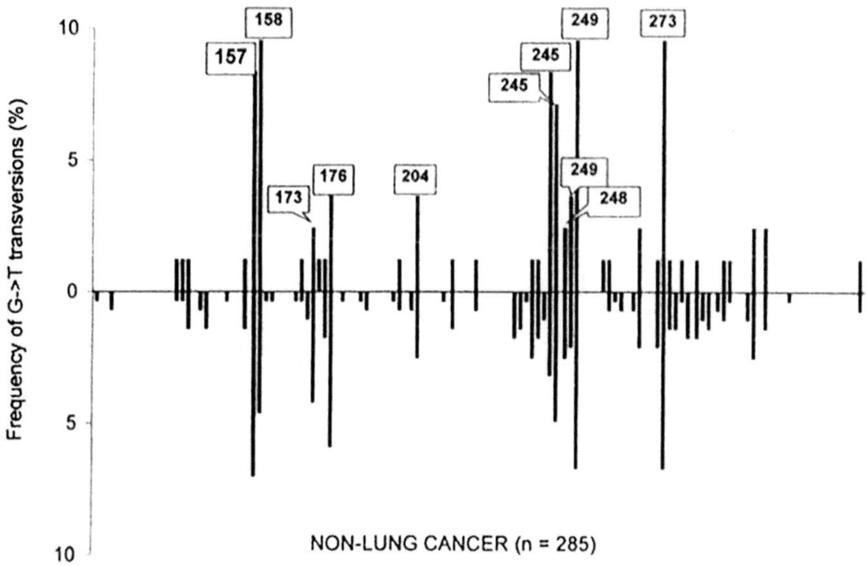

Fig. 5.3. Comparison of *TP53* mutational spectra from lung cancers with non-lung cancers. Comparison of the *TP53* G→T transversion spectrum of lung tumours from ever-smokers with tumours in non-lung tissues least accessible to smoke. Hotspot and "warmspot"" codons are indicated. Codon 248, one of the strongest BPDE adduct formation hotspots, appears not to be a G→T transversion hotspot in either lung cancer or non-lung cancer mutational spectra. Reproduced with kind permission, from SN Rodin & AS Rodin (2000) Human lung cancer and p53. *Proc. Natl. Acad. Sci. USA* 97: 12244-12249. Copyright (2000) National Academy of Sciences, USA

kers and more frequently in codons 248, 267 and 282 (all CGG) because the 3' Gs in these codons all exhibit strong BPDE adduct signals. Examination of the data in the IARC *TP53* Mutation Database revealed that although lung cancers manifested a higher proportion of silent G→T substitutions than non-lung cancers (21 % vs. 2.7 %), the G→T substitutions in lung cancer patients occurred at codons which had not been reported as BPDE adduct targets.

- Rodin and Rodin (2000) argued that a higher primary mutation rate would be incapable of explaining why 63 % of smoking lung cancer patients have *TP53* gene mutations as opposed to 31 % of non-smokers. [A similar disparity between smokers and non-smokers has been noted by other authors (Husgafvel-Pursianen *et al.*, 2000; Vahakangas *et al.*, 2001; Gealy *et al.*, 1999)]. A higher mutation rate would only serve to increase the *absolute* number of mutations, not their *relative* prevalence. Changes of relative prevalence are however potentially explicable by invoking a higher selective pressure in smokers, operating upon stress-responsive genes such as *TP53*.
- The predisposing influence of tobacco smoke to lung cancer was postulated to be exerted via the induction of cell stress rather than through an increased mutation rate. More specifically, Rodin and Rodin (2000) put forward a model whereby cigarette smoking could aggravate oxidative DNA damage and non-genotoxic forms of stress that might then overload the endogenous processes of strand-

asymmetric repair thereby promoting the selection of tumorigenic stem cells carrying mutations in stress-responsive genes such as *TP53*. This alternative to the BPDE-induced mutagenesis model would still be compatible with the finding that a majority of reported *TP53* base substitutions involve either specific amino acids of known structural or functional importance or those that are conserved evolutionarily between the p53 molecules of different species (Skaug *et al.*, 2000; Martin *et al.*, 2002; Greenblatt *et al.*, 2003).

Hainaut and Pfeiffer (2001) and Pfeifer *et al.* (2002) have responded by challenging virtually all aspects of Rodin and Rodin's (2000) paper, but their criticisms have been almost exclusively fuelled by doubts about the validity of the data employed by the latter authors. Whilst Rodin and Rodin (2000) based their analyses upon the April 1999 release of the IARC *TP53* Mutation Database (http://www.iarc.fr/p53/ ; Hernandez-Boussard *et al.*, 1999a; Olivier *et al.*, 2002), Hainaut and Pfeiffer (2001) employed the April 2000 release. Hainaut and Pfeiffer (2001) also excluded data that they considered unreliable due to either technical problems (e.g. the sequencing of cloned PCR products) or occupational exposure to known mutagens (e.g. uranium and coal miners; DeMarini *et al.* 2001). As a result, the emerging mutation profiles and their interpretation differed substantially from those provided by Rodin and Rodin (2000).

Pfeifer *et al.* (2002) have sought to continue the debate by stressing the importance of the observation of a higher prevalence of *TP53* G→T substitutions in the lung cancers of smokers (30%) as compared to the lung cancers of non-smokers (12%) [Figure 5.4]. Recently, Rodin and Rodin (2002) revisited this issue and made it clear that they do not consider this point to be a bone of contention. Instead, however, they essentially confirmed their earlier (2000) findings of a lung-specific strand-asymmetric DNA repair bias (i.e. the inhibition of repair specifically involving the non-transcribed strand in smokers) and reiterated their contention that smoking may exert its effects indirectly through preferential repair rather than through a direct genotoxic effect of benzo[*a*]pyrene. Pfeifer and Hainaut (2003) promptly responded by saying that although it might be 'a theoretical possibility', they did not think that this postulate was supported by the available evidence.

Rodin and Rodin (2002) hinted however at an alternative explanation: oxidative stress from components of cigarette smoke could lead to the generation of 8-oxoguanine (8-oxo-dG). 8-oxo-dG is known to occur at increased levels both in lung tumour tissue (Jaruga *et al.*, 1994) and in peripheral lung tissue in lung cancer patients (Inoue *et al.*, 1998). It is removed by base excision repair involving the OGG protein encoded by the *OGG1* gene. This gene is located on chromosome 3p26 within one of the regions frequently deleted in lung cancer, and its loss may serve to compromise the excision repair process. Thus, it could be that 8-oxo-dG, rather than benzo[*a*]pyrene or another PAH, gives rise to the excess of G→T substitutions observed in lung cancer. Whether 8-oxo-dG is repaired with a strand bias or not is however contentious (Le Page *et al.*, 2000; Thorslund *et al.*, 2002). This notwithstanding, it would appear as if OGG, although capable of removing 8-oxo-dG-induced lesions in human cells, is incapable of acting so as to reduce G→T substitutions induced by BPDE (Yamane *et al.*, 2003). Finally, subtle mutations in the *OGG1* gene are rare in lung cancer and it is unclear whether one hit (a haploinsufficiency

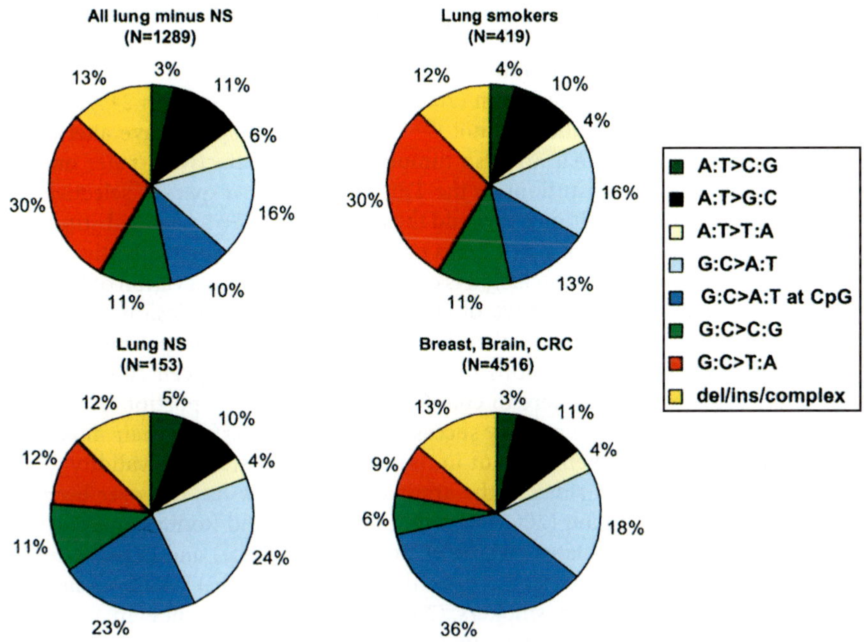

Fig. 5.4. *TP53* mutational spectra in lung cancers and non-lung cancers. *TP53* mutational spectra in lung cancers from smokers and non-smokers (NS), and non-lung cancers, breast, brain and colorectal (CRC). Data are from the January 2002 update of the IARC database (see text) but exclude data derived from cell lines and metastases. The total numbers (N) of mutations are indicated. Del/ins/complex denotes deletions, insertions and complex mutations. Reproduced from GP Pfeifer *et al. Oncogene* 21, 7435-7451 (2002) by kind permission of Nature Publishing Group

model) would be sufficient to compromise the efficiency of the excision repair process.

In their latest restatement of their position, Rodin and Rodin (2004a) have made it clear that they believe there to be lung-specific (but not smoke-specific) causes of *TP53* G→T transversion in addition to polycyclic aromatic hydrocarbons and that "a direct mutagenic action is not the only smoke-associated cause of the prevalence of this class of p53 mutations in lung cancer". The debate continues apace (Rodin and Rodin 2004b; Pfeifer and Hainaut 2004).

Have some *TP53* mutations occurred during cell culture rather than in the tumour?

Rodin and Rodin (2002) have also claimed that cultured cells derived from lung cancers (but not from other types of cancer) exhibit a significant additional excess of G→T substitutions in the *TP53* gene when compared to *TP53* mutations in the parental primary tumour. This is most intriguing since these cells once in culture are no longer going to be exposed to any constituent of tobacco smoke. There would seem to be two possible explanations for a cell culture-associated increase in the proportion of G→T substitutions relative to other types of mutation: (i) selection for these lesions through conferment of a growth advantage mediated by the cell culture conditions or (ii) *de novo* mutation in culture with a bias toward the generation of G→T substitutions. Rodin and Rodin (2002) pointed out that codons 245, 248 and 272, together with codons that are not G→T mutation hotspots, have alternative amino acid substitutions that are also tumorigenic. This effectively rules out the possibility that G→T substitutions in the *TP53* gene confer a specific selective advantage under culture conditions. It would thus appear that *TP53* G→T substitutions arise more often *in vitro* than other types of mutation, quite independently of benzo[*a*]pyrene exposure. This conclusion is consistent with the observation of Wistuba *et al.* (1999c) that as many as 40% of characterized *TP53* gene lesions may have arisen during cell culture and have therefore not been present in the parental tumour material. If those *TP53* gene lesions that have occurred during cell culture rather than in the original tumour were to constitute a significant proportion of the mutations in the IARC database used by successive authors to perform their molecular epidemiological studies, then it is not unreasonable to question the validity of the reported findings on both sides of the argument. This view has however been rejected by Pfeifer and Hainaut (2003) who repeated Rodin and Rodin's analysis (but once again using a slightly later version of the IARC database) and failed to find a statistically significant difference between the frequency of G→T transversions in primary lung tumours from smokers (30.0%) and the frequency in cell lines (36.1%). Further, Pfeifer and Hainaut (2003) made the not unreasonable point that comparisons between primary tumours and cell lines are meaningless so long as paired specimens originating from the same source are not used. They thus felt confident in restating their long-held view that the "abundance and sequence specificity of G→T substitutions [in *TP53*] in lung tumors is best explained by a direct mutagenic action of polycyclic aromatic hydrocarbon compounds present in cigarette smoke".

Other problems for the BPDE-induced mutagenesis model

Recent findings of Toyooka *et al.* (2003a) have also created difficulties for the BPDE-induced mutagenesis model. Although significant differences were noted in the frequencies of G→T substitutions between smokers (30%) and never-smokers (15%) for all lung cancer cases, this difference was found to be mainly due to data from female smokers rather than male smokers (Figure 5.5). This finding is not without precedent since a higher frequency of G→T substitutions in females than in males was first reported by Kure *et al.* (1996). Toyooka *et al.* (2003a) speculated that this might be explicable in terms of a "higher susceptibility to tobacco carcinogens in women". This view is consistent with the finding of higher BPDE adduct levels in female lung cancer patients than in their male counterparts (Kure *et al.*, 1996; Cheng *et al.*, 2001a), and could help to account for a possible gender difference in the risk of developing lung cancer (Haugen 2002). The findings of Toyooka *et al.* (2003a) could also be explicable in terms of an unintentional bias in the selection of experimental subjects. What is clear is that studies of lung cancer that do not carefully match cases and controls by gender lay themselves open to the possibility of bias.

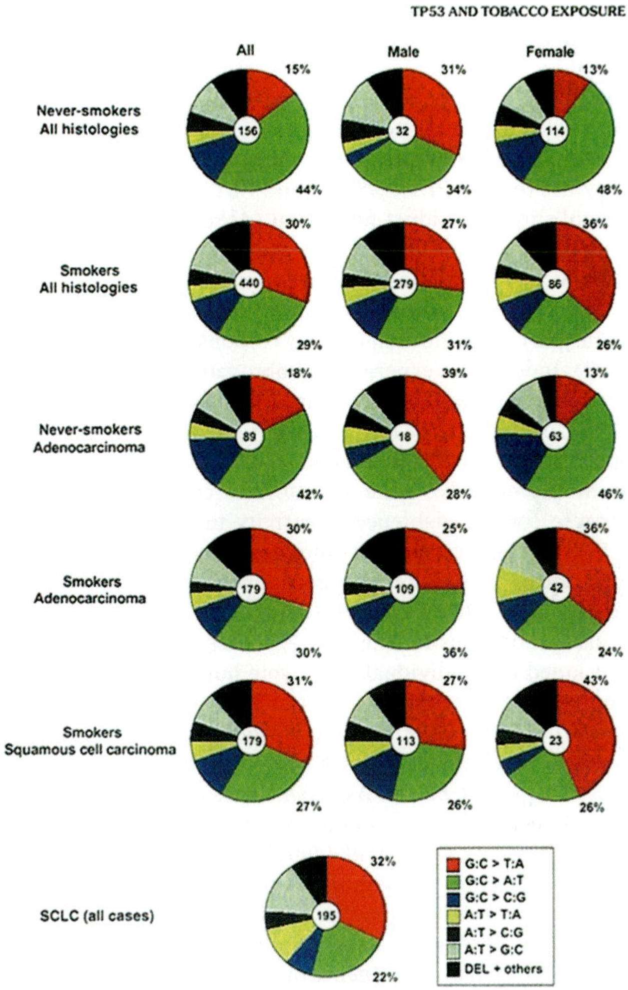

Fig. 5.5. *TP53* mutational spectra in lung cancer. *TP53* mutational spectra in different types of lung cancer from smokers and never-smokers by gender. The numbers in the centre of each circle specify the numbers of cancer cases analysed. Reproduced from KO Toyooka *et al.* (2003) The *TP53* gene, tobacco exposure and lung cancer. *Human Mutation* 21: 229-239. Copyright© John Wiley & Sons Inc. by kind permission of Wiley-Liss, Inc., a subsidiary of John Wiley & Sons, Inc

Another interpretational problem could arise if individuals with lung cancer were to be found to be especially sensitive to BPDE-induced damage. Li *et al.* (2001a) performed a case-control study with 221 newly diagnosed cases of lung cancer and 229 healthy controls matched for age, sex, ethnicity and smoking status. Using a ^{32}P-postlabeling technique to measure BPDE-induced DNA adducts in blood lymphocytes, the patients were found to exhibit a significantly higher level of DNA adduct formation than the controls. This observed association between the level of *in vitro* BPDE-induced DNA adducts and lung cancer risk is consistent with the view that some individuals are particularly sensitive to BPDE-induced DNA damage, that these individuals may have a suboptimal ability to remove BPDE-DNA adducts and that as a consequence, these individuals may be at increased risk of lung cancer after exposure to benzo[*a*]pyrene. This vulnerability may be due to possession of particular combinations of polymorphic alleles of specific xenobiotic metabolising enzymes (e.g. *CYP1A1*, *GSTM1* and *GSTP1*) that serve to influence the likelihood of benzo[*a*]pyrene adduct formation (Alexandrov *et al.*, 2002; Perera *et al.*, 2002; *see* Chapter 6, Evidence for Genetic Susceptibility to Lung Cancer Derived From Polymorphism-Disease Association Studies).

The situation is also complicated by the fact that expression of the p53 protein forms part of the normal cellular response to BPDE-induced damage (Ramet *et al.*, 1995). Indeed, the experimental abrogation of p53 activity in transgenic mice has been shown to lead to an increase in the frequency of lung tumours in mice exposed to benzo[*a*]pyrene by intratracheal instillation (Tchou-Wong *et al.*, 2002). The loss of p53 activity is of course a common feature of lung tumorigenesis where it is often an early event; *TP53* gene mutation (or conceivably polymorphism) may therefore serve to confer heightened sensitivity to the mutagenic effects of benzo[*a*]pyrene.

Polymorphic variants in other genes may also exert an influence. Thus, smoking lung cancer patients heterozygous for the *ERCC2* Asn312 allele have been claimed to be less likely (OR=0.43, 95% CI 0.20-0.89) to possess *TP53* mutations than individuals homozygous for the wild-type Asp312 allele (Gao *et al.*, 2003b). The authors concluded that individuals who were homozygous for the wild-type allele and who smoked might be at greater risk of *TP53* gene mutation. Were this conclusion to be borne out by further studies, it would represent another potential confounding factor in case-referent studies where cases differed from controls in terms of their possession of the *ERCC2* Asp/Asn alleles at codon 312. It would appear increasingly likely that genetic variation at multiple loci is involved in determining *TP53* gene mutability. Thus, Casse *et al.* (2003) reported that A→G transitions in the *TP53* gene were found significantly more frequently in individuals with NSCLC who were homozygous or heterozygous for the Gln399 *XRCC1* allele than in individuals homozygous for the Arg399 allele. Finally, an increased prevalence of G→T and C→A transversions in the *TP53* genes of lung cancer patients was reported to be associated with the Gln allele of the Gln/Glu185 *NBS1* (Nijmegen breakage syndrome) gene polymorphism (Medina *et al.*, 2003). Any of these polymorphic variants could in principle serve as a confounding factor in disease-polymorphism association studies that either have low sample sizes or which have employed poorly matched controls.

Hashimoto *et al.* (2000) have reported that the proportion of G→T substitutions in the *TP53* gene appears to differ between distinct subtypes of lung adenocarcinoma. These authors concluded that whereas one adenocarcinoma subtype might be associated with smoking, the other might manifest a preponderance of mutations originating through endogenous mechanisms. In similar vein, the proportion of G→T substitutions differs between adenocarcinoma (the most frequent type of lung cancer in women and never-smokers) and squamous cell carcinoma (Gao *et al.*, 2003a). If such findings are confirmed by further work, it could indicate that different types of lung cancer are characterized by different mutational mechanisms and hence different mutational spectra. It would then follow that the uncritical use of database material for comparisons between smokers and non-smokers without careful cross-matching by lung cancer type could be a hazardous exercise. Once again, this highlights the critical importance not only of the quality of the *TP53* mutation data employed in the above-mentioned studies, but also how those data are partitioned and used.

The key importance of the quality of the IARC database

Another source of experimental bias was highlighted by the finding of a large number of discrepancies in the epidemiological classification of lung cancer cases between different IARC database releases (Paschke 2000; 2001). Thus, the January 1998 version of the database contained information on smoking habit for 379 cases, whereas the updated July 1998 version, with new 146 cases added, had the same information logged for only 221 entries. Paschke (2000) pointed out that the changes in data quality and sample size would have had serious implications for the scientific analysis of the *TP53* mutational spectra. Indeed, he opined, "the current version of the database does not support the view that there is a difference between smokers and non-smokers in terms of mutational hotspots". However, Hainaut and Pfeiffer (2001) and Hainaut *et al.* (2001) responded by stating that periodic adjustment of the database was necessary, and by rejecting Paschke's criticism as invalid on the grounds of the inappropriateness of the dataset he had employed. The strength of feeling apparent in this ongoing controversy is indicative of the importance of the quality, accuracy and reliability of the data in the IARC dataset. Different groups have access to the same data. Each group tends to use different updates of the dataset for their analyses, probably excludes a different set of entries for essentially very sound but perhaps different reasons, and then proceeds to test the dataset in different ways. It is indeed a salutary lesson for scientists neutrally observing this controversy from afar, that the same dataset can be used by different and highly reputable researchers to justify very different conclusions. It is therefore pertinent to mention a number of issues pertaining to the quality of the entries in the IARC database which, if left unconsidered, might well lead to confusion and potentially to misinterpretation of results:

- By June 2003 (R8 release), the IARC database contained some 18,585 somatic mutation entries including some 2049 pertaining to lung cancer. However, the lung cancer entries included mutations found in pre-cancerous tissues, metastases and cell lines derived from such tissues, so that some of the mutations in the database may not have been directly involved in lung tumorigenesis or could even have arisen during cell culture (Wistuba *et al.*, 1999c; Rodin and Rodin 2002). The latter would have come to attention as a consequence of their conferring a growth advantage upon cultured cells rather than upon tumour cells *in vivo*. If present in the IARC database, such mutations could potentially lead to serious misinterpretation of data.
- Similarly, some *TP53* mutations confer chemoresistance upon lung tumour cells and may therefore have arisen during treatment (reviewed by Toyooka *et al.*, 2003a). Such lesions are not associated with the process of tumorigenesis *per*

se but rather confer a growth advantage upon cells in an environment made artificial by treatment of patients with antitumour drugs.

- The majority of mutation studies performed on the *TP53* gene have opted to screen only exons 5-9 inclusive since this is where most of the pathological mutations are believed, on empirical grounds, to occur (Hainaut and Hollstein 2000; Hakkarainen *et al.*, 2004). This is however clearly a self-fulfilling prophecy. Current mutational spectra are therefore likely to be skewed by the under-ascertainment of mutational lesions that lie outwith the 'hotspot region'. Indeed, Casey *et al.* (1996) performed a rather more complete screen of the *TP53* gene (exons 2-11 inclusive) and detected a total of 63 mutations in 137 NSCLC tumours; significantly, some 17% of the mutations found occurred outwith the hotspot region.
- The quality of the data on smoking status in the IARC database is generally poor (currently confined to smoker, non-smoker and ex-smoker categories). Rodin and Rodin (2000) have pointed out that there are a disproportionately small number of ex-smokers in the IARC database as compared to the general population suggesting possible bias of ascertainment. Further, the quantity of data available for analysis (344 "ever smokers" and 75 "never-smokers") in the IARC database release originally used by Rodin and Rodin (2000) was also poor, with a serious paucity of unequivocal never-smokers. Finally, it has been estimated that some 30-50% of 'non-smokers' have been exposed to environmental tobacco smoke for at least 1 hour per day for at least one year (Veglia *et al.*, 2003), a finding which must cast some doubt upon the reliability of the 'non-smoker' category (Brownson *et al.*, 2002).*
- Since the IARC database comprises published material, it is of course also open to various sorts of reporting bias (Hernandez-Boussard *et al.*, 1999b). In addition, the mutational status of only one *TP53* allele per tumour sample has usually been established.
- Some *TP53* alleles [in between 0.5% (Soussi and Béroud 2003) and 5% (Rodin *et al.*, 2002) of tumours] display multiple mutations some of which are likely to be merely neutral bystanders that have 'hitchhiked' a lift by virtue of their close linkage with the *bona fide* pathological lesion (Rodin *et al.*, 1998; Tseng *et al.*, 1999a; Soussi and Béroud 2003; Maley *et al.*, 2004). Indeed, Rodin *et al.* (2002) have proposed that the initial strongly tumorigenic mutation (the 'driver' mutation) could, by one of a number of different mechanisms, promote the increased mutability of the same *TP53* allele. These mutations then find their way into the IARC database regardless of their consequences for the structure and function of the p53 protein.
- The database does not include data on either the presence of potentially functional polymorphisms on the *TP53* alleles (e.g. Wang *et al.*, 1999b) or the ethno-geographic origin of the clinical material. Both may be important since, for example,

* It is pertinent to note that Sir Richard Doll gave this matter some thought more than 50 years ago: "All reports agree that cancer of the lung occurs among 'non-smokers'. The definition of a 'non-smoker' has, however, varied and has never been so strict as to exclude persons who have smoked only one cigarette, one cigar or one pipe of tobacco. Such a definition would have little interest in England where its use would probably result in no men at all being classified as non-smokers. It would not in any case, delimit a class of persons who had never been exposed to tobacco smoke, since persons who do not themselves smoke, breathe air containing smoke produced by others"; R. Doll (1953) Mortality from lung cancer among non-smokers. *Br. J. Cancer* 7: 303-312.

the p53 proteins corresponding to the alternative alleles of the Arg/Pro72 polymorphism manifest markedly different apoptotic potential (Dumont *et al.*, 2003) and associations between specific *TP53* polymorphic alleles or haplotypes and lung cancer susceptibility have been claimed (Biros *et al.*, 2001a; 2001b; Papadakis *et al.*, 2002; Wu *et al.*, 2002; Hiraki *et al.*, 2003). Moreover, significant differences in *TP53* alleles and haplotype frequencies occur between different ethnic groups (Wu *et al.*, 2002). This could easily lead to the misinterpretation of results derived from database-reliant studies if case and control groups are compared that have not been very carefully matched in terms of their ethno-geographic origin. Similarly, failure to take into account the potential influence of functional polymorphisms at other loci (e.g. the Pro/Ser187 alleles of the *NQO1* polymorphism, the *NAT2* slow acetylator genotype, and the Gln399 *XRCC1* variant; *see* Table 4 and Chapter 6; Interpreting the Role of Xenobiotic Metabolizing Enzyme Polymorphisms in Lung Cancer) could also yield misleading conclusions.

- The curators of a very similar *TP53* (UMD-p53) mutation database (http://www.umd.necker.fr; Béroud and Soussi 2003) freely admit that between 2% and 5% of their entries are likely to be incorrect as a consequence of either laboratory or reporting errors (Béroud and Soussi 2003). We may assume that the IARC database is very likely to suffer from a similar problem.
- The above points are not meant in any way as criticism of the IARC database curators themselves - they have done a highly commendable job under extremely difficult circumstances. Rather, they are intended to highlight the very wide range in quality of the original literature reports that constitute the source material for the database.

In conclusion, the BPDE-induced mutagenesis hypothesis has produced a very lively controversy that shows as yet no sign of dying down. It is, however, in the opinion of the author, by no means clear at this juncture that BPDE is necessarily responsible either for the mutations observed at the hotspot codons in the *TP53* gene or for the lung tumorigenesis in the patients in whom these lesions were originally characterized. A final assessment of the validity of the genetic epidemiological arguments for or against the BPDE-induced mutagenesis hypothesis requires the creation of a sufficiently large, informative and easily accessible *TP53* mutation database of excellent quality.

Putting the p53/BPDE-induced mutagenesis controversy in its proper context

It should be noted that it is, of course, quite unrealistic to expect lung tumorigenesis across the board to be explicable in terms of mutation in a single gene even if that gene encodes a protein product as important as p53. Even though *TP53* mutations are very common in SCLC and fairly common in NSCLC, lung cancer can and does occur in the absence of such lesions. Conversely, *TP53* mutations (including mutations in the hotspots mentioned above) have been shown to occur in non-tumorous tissue taken from peripheral lung samples from lung cancer patients (Hussain *et al.*, 2001) as well as in atypical alveolar hyperplasia (Slebos *et al.*, 1998) and atypical epithelial lesions from patients with idiopathic pulmonary fibrosis (Kawasaki *et al.*,

2001). A mutant *TP53* gene is therefore neither necessary nor sufficient for lung tumorigenesis to occur.

The genetics of nicotine addiction

"He that taketh tobacco saith he cannot leave it, it doth bewitch him."
King James I & VI (1604) *A Counter Blaste to Tobacco*

"The tobacco habit is commonly regarded, even by those who are devoted to the weed, as useless, filthy, and expensive; and I have met with few persons who did not regret have formed it."
John Hinds (1882) *The Use of Tobacco*. Cumberland Presbyterian Publishing House, Nashville, TN.

Nicotine is thought to exert its cognitive and addictive effects by interacting with neuronal nicotinic acetylcholine receptors in the brain. Nicotine has been reported to have a number of biological effects: increasing oxidative stress, sensitizing cells to genotoxic/xenobiotic stress, activating the transcription factor NFκB and the Akt signaling pathway, inducing or suppressing apoptosis, and stimulating angiogenesis (Crowley-Weber *et al.*, 2003; Heeschen *et al.*, 2003; Jain 2001; West *et al.*, 2003; Zhu *et al.*, 2003b; Mai *et al.*, 2003; Jin *et al.*, 2004). NFκB is itself involved in a variety of acute and chronic inflammatory diseases (Wright and Christman 2003) and may play a role in lung tumorigenesis (Ballaz and Mulshine 2003). Thus, nicotine could conceivably play both a direct and an indirect role in promoting lung tumorigenesis (Minna 2003). It is, however, as a consequence of its putative addictive properties that considerable attention has been paid to its role in determining inter-individual differences in smoking behaviour.

Evidence for a genetic influence on smoking behaviour (in terms of both initiation and persistence) has come from twin studies (Carmelli *et al.*, 1992; Heath and Martin 1993; Heath *et al.*, 1995) and from studies of familial clustering of smoking history (Cheng *et al.*, 2000). Attempts to attribute these effects to specific genetic loci have however met with more limited success (Lerman and Niaura 2002; Lerman and Berretini 2003). A search for susceptibility genes for nicotine dependence has claimed to have pinpointed regions on chromosomes 2, 4, 10, 16, 17 and 18 as potentially harbouring gene loci that influence this behaviour (Straub *et al.*, 1999). A significant association has also been noted between a polymorphism in the chromosome 5p15-encoded dopamine transporter gene, *SLC6A3* and smoking status (Sabol *et al.*, 1999; Lerman *et al.*, 1999; Ling *et al.*, 2004; Erblich *et al.*, 2004); in the initial study, individuals with allele 9 of the *SLC6A3* gene were reported to be significantly less likely to be smokers, especially if they also possessed the A2 allele of the dopamine receptor D2 (*DRD2*; 11q23) gene. However, these early results have not always been successfully reproduced or replicated by other studies (Spitz *et al.*, 1998; Singleton *et al.*, 1998; Jorm *et al.*, 2000; Vandenbergh *et al.*, 2002). Smoking behaviour may also be influenced by a possible interaction between the 5-HTTLPR functional polymorphism in the promoter of the 17q-encoded serotonin transporter (*SLC6A4*) gene and neuroticism (Hu *et al.*, 2000; Lerman *et al.*, 2000). Polymorphic variants of the tyrosine hydroxylase (*TH*; 11p15; Olsson *et al.*, 2004) and nicotine acetylcholine receptor α4 subunit (*CHRNA4*; 20q13; Feng *et al.*, 2004) gene have also been reported to be associated with nicotine addiction/tobacco dependence. However, no association was noted between polymorphic variants in the *CYP1A1*, *GSTM1*, *GSTP1*, *GSTT1* and *NAT2* genes and tobacco consumption (Smits *et al.*, 2004).

Another gene that has been considered as a candidate for influencing smoking behaviour is *CYP2D6* (22q13). The *CYP2D6* gene encodes polypeptide 6 of subfamily IIA of cytochrome P450, an enzyme responsible both for nicotine inactivation and the activation of tobacco-related pro-carcinogens such as the nitrosamines. Some 3-10 % of individuals in the Caucasian population are deficient in CYP2D6 metabolism as a consequence of their possession of two *CYP2D6* null alleles whilst 1-2 % of Caucasians exhibit ultrarapid metabolism by virtue of *CYP2D6* gene amplification. Among smokers, the frequency of individuals with impaired nicotine metabolism (heterozygotes and homozygotes for the *CYP2D6* null allele) has been reported to be lower than in the control group (Pianezza *et al.*, 1998). Indeed, Saarikoski *et al.* (2000) demonstrated that *CYP2D6* genotypes associated with increased metabolic capacity occur at a significantly elevated frequency in smokers. The *CYP2D6* locus displays significant variability both in the Caucasian population and between different ethnogeographic groups. It remains to be seen what influence this variation has on nicotine metabolism, dependence and smoking behaviour (Xu *et al.*, 2002b). The interested reader is also referred to several recent reviews on the genetics of nicotine dependence/tobacco addiction (Walton *et al.*, 2001; Tyndale 2003; Batra *et al.*, 2003).

CHAPTER 6

Evidence for Genetic Susceptibility to Lung Cancer Derived from Polymorphism-disease Association Studies

"If. . .. as seems probable, the stimulus to cancer growth is an inoculable something,. . ..it does not follow that the consequence of stimulus is not determined by an inheritable factor".
C.B. Davenport (1912) *Heredity in Relation to Eugenics*

"Susceptibility to cancer does not mean that cancer will occur. It means that preliminary cellular events have occurred, so that a randomly occurring carcinogenic stimulus, or a spontaneous mutation that would cause no apparent damage in some individuals, would trigger the transformation to neoplasia in others".
D.G.Miller (1980) On the nature of susceptibility to cancer. *Cancer* 46: 1307-1318.

Polymorphisms and polymorphism-disease association studies

The term *polymorphism* is commonly defined as a "mendelian trait that exists in the population in at least two phenotypes, neither of which occurs at a frequency of less than 1 %". Most polymorphisms are expected to be neutral with respect to fitness. However, those polymorphisms that occur either within gene coding or promoter regions may affect either the structure or function of the gene product or the expression of the gene and may therefore have the potential to be of phenotypic or even pathological significance. It should be noted in passing that not all polymorphisms are of the single nucleotide (SNP) variety. Indeed, some gene-associated polymorphisms in the human genome involve triplet repeat copy number, gene deletion, gene duplication, intragenic duplication, micro-insertion, inversion, gene fusion and gene copy number (Cooper 1999).

The mechanisms by which polymorphisms are maintained in human populations are likely to be varied. Under a strictly neutralist model, no selection on the alleles of a polymorphic locus would be assumed and the frequency of an allele would therefore increase simply by *genetic drift* (the change of allele frequency due to random sampling). Such "transient polymorphisms" could remain at a low frequency in the population before eventually being lost or might instead increase in frequency under the influence of either genetic drift or positive selection until one allele reaches fixation. Most known polymorphisms are probably of this type. However, if the alternative alleles are not entirely neutral with respect to fitness, the DNA polymorphisms may be maintained by selection pressure, possibly overdominant selection ("balanced polymorphisms").

Association studies employing polymorphisms located within or in close proximity to potential candidate genes can be a powerful approach to the epidemiological analysis of complex disorders including cancer (Tabor *et al.*, 2002; Ahsan and Rundle 2003). Indeed, most mutational information related to complex or multifactorial disorders is of association type and has emerged mainly from case-control studies. One of the problems inherent in interpreting and comparing the results of disease association studies, however, is that even studies involving the same gene will often vary with respect to the polymorphism(s) employed, the population examined, the sample size, the exact definition of disease phenotype, the selection of controls, and the statistical methodology adopted (Hirschhorn *et al.*, 2002; Cardon and Palmer 2003; Ioannidis *et al.*, 2003).

In practice, the relationship between a polymorphism and a disease susceptibility should be interpreted in terms of *linkage disequilibrium*. Linkage disequilibrium is said to be present when certain alleles at one locus occur with certain alleles of another locus on the same chromosome at frequencies greater than can be attributed to chance alone. For our purposes, it can be considered to be due to a mutation occurring in a gene a number of generations ago. This mutant gene has then increased in frequency within the population, affected individuals of succeeding generations inheriting not only the mutant gene but also the particular alleles of neighbouring polymorphisms. Gradually, the relationship between the marker alleles and the mutant gene will decay due to recombination. For this reason, marker/disease associations from different populations are extremely difficult to interpret; although both the polymorphic allele frequency and linkage phase can be established in any one population, either or both can differ dramatically between populations.

- Inter-individual susceptibility to the development of lung cancer may be conferred by genetic polymorphisms in a number of different types of gene (Wu *et al.*, 2004b). In principle, alternative polymorphic alleles could exert their influence through their differential effects on smoking behaviour and nicotine addiction, nicotine metabolism, mutagen/carcinogen metabolism and detoxification, DNA repair, apoptosis, cell cycle control, signal transduction etc (Spivack *et al.*, 1997; Shields 2002; Wiencke 2002; Kiyohara *et al.*, 2002; Table 4). Some of these polymorphic alleles have been proposed to influence the degree of DNA adduct formation (and hence DNA damage) but the pathways through which they exert their influence on lung cancer risk are probably much more complex than at first thought.

The forerunners of current molecular epidemiological studies of lung cancer include the early search for HLA associations (Ford *et al.*, 1981) and the finding of a preponderance of extensive debrisoquine metabolizers (due to CYP2D6 polymorphism) in lung cancer patients (Ayesh *et al.*, 1984). There are now, however, hundreds of papers describing case-control studies of genetic predisposition to lung cancer (Table 4).

Polymorphism-disease association studies in lung cancer

"Even at birth the whole individual is destined to die, and perhaps his organic disposition may already contain the indication of what he is to die from".
Sigmund Freud (1924) *The Dissolution of the Oedipus Complex*

Introduction and overview

As we have seen, familial aggregation of lung cancer is comparatively rare and only one potential 'locus of major effect' has so far been mooted (Bailey-Wilson *et al. 2004*) that could confer increased susceptibility to the disease. On the other hand, a range of studies have consistently pointed to a heritable contribution (albeit relatively small) to the variance of lung cancer susceptibility. We may therefore envisage a polygenic model of inherited predisposition to lung cancer in which a large number of alleles each confer a relatively small genotypic risk (Ponder 2001; Balmain *et al.*, 2003; Kiyohara *et al.*, 2004; Rebbeck *et al.*, 2004; Loktionov 2004; Wu *et al.*, 2004b). Thus, individuals possessing only a few such alleles would be at reduced risk of lung cancer whilst those with many could be at greatly increased risk. Alleles conferring relative risks of <2.0 will rarely give rise to multiple case families and hence will be very difficult or even impossible to identify by genetic linkage studies (Risch and Merikangas 1996). The quest for low penetrance alleles that confer an increased risk of lung cancer has therefore relied almost exclusively on case-control studies, comparing the frequency of the variants between cancer cases and age-, sex- and race-matched controls.

Since only 5-10% of smokers develop lung cancer, considerable interest has been shown in attempts to identify those genetic factors that might predispose to the disease by increasing an individual's susceptibility to the detrimental effects of environmental carcinogens (Wu *et al.*, 2004b). The physiological role of xenobiotic-metabolizing enzymes is to detoxify toxic chemicals but, in performing their task, these enzymes may also activate procarcinogens. Since these enzymes are often polymorphic, with alternative alleles differing with respect to the level of functional enzyme produced, it follows that different genotypes may be associated with the differential activation of procarcinogens. Possession of a particular allele may thus determine an individual's exposure to a given carcinogen and hence influence their cancer risk (reviewed by Smith *et al.*, 1995; Taningher *et al.*, 1999; Nair and Bartsch 2001; Kiyohara *et al.*, 2002; Gemignani *et al.*, 2002). Similar arguments apply to polymorphic variants of DNA repair genes that confer differential repair capacity on their bearers and hence also influence lung cancer risk. To date, polymorphic variants in nearly 50 genes have been claimed to be associated with either an elevated or a reduced risk of lung cancer (Table 4). These variants will now be reviewed together with the genes/proteins with which they are associated, and the possible mechanistic pathways through which they exert their influence on lung cancer risk.

Cytochrome P450 enzymes

The cytochrome P450 enzymes comprise a super-family of heme-containing electron transport molecules that are involved in the oxidative metabolism of a wide range of substrates including steroids, drugs and xenobiotic compounds. They catalyse the insertion of an oxygen atom into substrate thereby converting procarcinogens into their DNA reactive electrophilic forms. The cytochrome P450 family of proteins is encoded by at least 60 different CYP P450 genes and is extremely polymorphic. Polymorphic variants in a total of 8 different CYP P450 genes have been reported to be associated with an altered risk of lung cancer (viz. *CYP1A1*, *CYP1B1*, *CYP2A6*, *CYP2A13*, *CYP2D6*, *CYP2E1*, *CYP3A4* and *CYP3A5*; Table 4). These genes and their associated variants will now be briefly described.

CYP1A1: CYP1A1 is associated with aryl hydrocarbon hydroxylase activity and catalyses the first step in the metabolism of polycyclic aromatic hydrocarbons (PAHs). CYP1A1 is inducible by its xenobiotic substrates; the substrate (e.g. PAH) first binds to the intracellular aryl hydrocarbon receptor leading to dimerization of the receptor to a translocator protein (ARNT) that is required for its translocation to the nucleus (see below). In the nucleus, the ligand-receptor complex then *trans*-activates the *CYP1A1* gene by binding to its cognate recognition site on its promoter.

A number of different polymorphisms have been described in the human *CYP1A1* gene although none of them appear to be responsible for the uncommon phenotype of poor inducibility of CYP1A1 (Anttila *et al.*, 2000). An *Msp*I polymorphism 250 bp 3' to the polyadenylation site is the best known, and has been the most frequently utilized, in lung cancer association studies. The m2 allele (associated with the presence of the *Msp*I site) yields a CYP1A1 product that displays increased activity by comparison with the wild-type, and has been reported in a range of studies to be associated with an elevated risk of lung cancer (Table 4). A second *CYP1A1* polymorphism involves the replacement of Ile by Val at amino acid residue 462 within the heme-binding region of the protein. Whether or not this variant alters the kinetic properties or biological activity of CYP1A1 is the subject of some controversy (Crofts *et al.*, 1994; Persson *et al.*, 1997). This notwithstanding, it has been claimed that this exon 7 mutation is associated with increased CYP1A1 inducibility (Crofts *et al.*, 1994). Carriership of the Val462 allele has been associated with an increased risk of lung cancer (Table 4).

CYP1B1: A second member of the CYP1 family, CYP1B1, activates a wide range of PAHs and aromatic and heterocyclic amines; it is expressed in the lung and like CYP1A1 is PAH-inducible. Watanabe *et al.* (2000) have reported an association between the possession of alternative alleles of the Ala/Ser119 polymorphism and the risk of squamous cell lung carcinoma.

CYP2A6: CYP2A6 is the major liver-expressed CYP P450 enzyme involved in nicotine metabolism; nicotine is mainly metabolised to cotinine. CYP2A6 is also thought to be involved in the metabolic activation of tobacco-specific nitrosamines, particularly those with relatively long alkyl chains (Kamataki et al., 1999). A number of amino acid substitution and deletion polymorphisms have been reported from the *CYP2A6* gene in Caucasian populations. However, the frequencies of these polymorphic variants vary considerably between populations; thus, the deletion alleles are at their most frequent in the Asian population (15-20 %; Raunio *et al.*, 2001). Although it has been suggested that the important role of CYP2A6 in nicotine metabolism might translate into a relationship between *CYP2A6* genotype and smoking behaviour, no such influence has been unequivocally demonstrated (Oscarson 2001). It may be, however, that individuals possessing homozygous *CYP2A6* gene deletions (poor metabolizers) are unable to activate tobacco-derived nitrosamines and consequently could display a reduced risk for smoking-related lung cancer (Miyamoto *et al.*, 1999; Kamataki *et al.*, 2002). Studies performed in Caucasian populations have yielded more equivocal results than in the Japanese population, possibly because the spectrum of polymorphic variation is quite different.

CYP2A13: CYP2A13 is predominantly expressed in the respiratory tract and plays an important role in the metabolic activation of tobacco-specific nitrosamines such as 4-methylnitroso-1-(3-pyridyl)-1-butanone (NNK). An Arg257Cys CYP2A13 variant has been detected at frequencies of 1.9 %, 14.4 % and 7.7 % in white, black and Asian populations respectively (Zhang *et al.*, 2002). Functional analysis has shown that the Cys257 variant is some 37-56 % less active than the wild-type Arg257 form of the CYP2A13 protein towards all substrates tested (Zhang *et al.*, 2002). Heterozygosity and homozygosity for the Cys257 variant has been found to be associated in a Chinese population with a significantly reduced risk of lung adenocarcinoma (OR=0.41, 95 % CI 0.23-0.71) [Wang *et al.*, 2003h]. On this basis, it might seem reasonable to surmise that individuals possessing significantly reduced levels of CYP2A13 may manifest a reduced sensitivity to xenobiotic toxicity resulting from the CYP2A13-mediated metabolic activation of procarcinogens in the respiratory tract. However, a separate study, on a European population, has reported that a null variant (Arg101-Term) of CYP2A13 was associated with a dramatically elevated risk of SCLC (OR=9.9, 95 % CI 1.9-52.2) [Cauffiez *et al.*, 2004]. If this finding is replicated, then it may be necessary to explain the positive association with lung cancer by postulating the reduced metabolism of ingested carcinogens by individuals bearing the null variant.

CYP2C19: CYP2C19 plays a key role in the metabolism of a variety of drugs including the anticonvulsant, mephenytoin. A number of different CYP2C19 polymorphic variants have been described that are responsible for inherited differences in the ability to metabolise mephenytoin (Wedlund 2000). The poor metabolizer phenotype occurs with a frequency of 13-23 % in Oriental populations but only in 2-5 % of Caucasians. Shi and Chen (2004) have reported that the poor metabolizer phenotype is over-represented in Chinese lung cancer patients as compared with matched controls (OR=3.2, 95 % CI 1.5-6.9).

CYP2D6: Also known as debrisoquine/sparteine hydroxylase, CYP2D6 is a constitutive enzyme that catalyses the hydroxylation or demethylation of more than 20 % of drugs metabolised in the human liver (Cascorbi 2003). Some 7-10 % of Caucasians possess inactivating mutations of the *CYP2D6* gene and consequently lack CYP2D6 activity. Conversely, 1-3 % of Caucasians possess gene duplications. As a consequence of this polymorphic variation, there are a range of different phenotypes (poor, intermediate, extensive and ultra-rapid metabolizers) in Caucasian populations that correspond to different genotypic combinations at the *CYP2D6* locus. Different *CYP2D6* genotypes are also associated with different adduct levels (Kato *et al.*, 1995). This notwithstanding, apart from the study of Bouchardy *et al.* (1996) which reported an increased risk of lung cancer in association with elevated CYP2D6 activity, the majority of published studies have not been supportive of a link between CYP2D6 polymorphism and lung cancer susceptibility.

CYP2E1: CYP2E1 is involved in the metabolic activation of nitrosoamines, particularly those with relatively short alkyl chains (Kamataki *et al.*, 1999). A variety of different polymorphisms in the *CYP2E1* gene have been documented leading to inter-individual and inter-population differences in CYP2E1 expression. Different *CYP2E1* genotypes are also associated with different adduct levels (Kato *et al.*, 1995). Several studies have reported an association between *CYP2E1* genotype and lung cancer risk but these studies are inconsistent in terms of which genotype is claimed to confer susceptibility to the disease (Table 4).

CYP3A4: CYP3A4 is the most abundantly expressed cytochrome P450 enzyme in the liver and is of great importance for drug metabolism. Expression of CYP3A4 manifests considerable inter-individual variation, presumably due to polymorphic variation at the *CYP3A4* locus. A solitary study has so far implicated variation in the *CYP3A4* gene as a risk factor in SCLC (Dally *et al.*, 2003).

CYP3A5: As with CYP3A4, CYP3A5 plays a role in drug metabolism. However, unusually, CYP3A5 expression is confined to only a small minority of Caucasians (Kuehl *et al.*, 2001; Hustert *et al.*, 2001) although it is expressed in a rather higher proportion (~60 %) of blacks. One of a number of different polymorphic variants (CYP3A5*1) was found to be under-represented in lung cancer patients relative to normal controls (Yeh *et al.*, 2003) suggesting that genetic variation at the *CYP3A5* locus could play a role in determining lung cancer susceptibility.

Aryl hydrocarbon receptor (AHR) and AHR nuclear translocator

The aryl hydrocarbon receptor (AHR) is a ligand-activated transcription factor that dimerizes with the AHR nuclear translocation (ARNT) protein. The AHR/ARNT heterodimer then translocates to the nucleus where it *trans*-activates a battery of genes (including *CYP1A1*, *CYP1A2* and *CYP1B1* by binding to xenobiotic response elements (XREs) in their promoter regions (Ma *et al.*, 2001). The potential importance of the AHR/ARNT complex in promoting the conversion of procarcinogens to carcinogens by activating CYP1A1 is evidenced by the loss of benzo[*a*]pyrene carcinogenicity in mice lacking AHR (Shimizu *et al.*, 2000b). Intriguingly, AHR also appears to have another cellular role, mediated by its interaction with Rb, in cell cycle control (Elferink *et al.*, 2003). Activated AHR is thought to act as an environmental sensor to provide a cell cycle checkpoint that arrests cell cycle progression, in cells exposed to adverse environmental stimulation, prior to the onset of DNA replication.

The *AHR* and *ARNT* genes are located on chromosomes 7p15 and 1q21 respectively. Various polymorphisms have been described in the *AHR* gene and there is some evidence that these may alter the ability of the AHR/ARNT complex to induce *CYP1A1* gene expression (Cauchi *et al.*, 2001; Wong *et al.*, 2001; Harper *et al.*, 2002). Several polymorphisms have also been described in the *ARNT* gene although no correlation was noted with CYP1A2 activity (Scheel *et al.*, 2002). Whilst it is possible that some of these variants may increase or decrease susceptibility to the carcinogenic effects of PAHs, no evidence has so far been presented for an association between possession of a given polymorphic variant and the risk of lung cancer.

Glutathione S-transferases

The glutathione *S*-transferases (GSTs) constitute a super-family of enzymes that catalyse the conjugation of a variety of electrophilic compounds (including potential carcinogens) with reduced glutathione thereby protecting the cell against the toxic effects of endogenous electrophilic substrates such as the by-products of reactive oxygen species as well as exogenously-derived xenobiotics including benzo[*a*]pyrene epoxide. The GST superfamily comprises a wide variety of isozymes (α, μ, π, ω, θ and ξ) encoded in the human genome by at least 15 genes on 6 chromosomes. Polymorphic variants of each isozyme may confer differences in cancer susceptibility as well as differences in resistance to anti-tumour drugs (Strange *et al.*, 2001; Townsend and Tew 2003). Polymorphic variants in a total of four GST genes (*GSTM1*, *GSTM4*, *GSTP1* and *GSTT1*) have so far been reported to be associated with an increased risk of lung cancer (Reszka and Wasowicz 2001; Table 4). These variants include the null alleles of the *GSTM1* and *GSTT1* genes, resulting from frequent (30-60 %) gene deletions in the general population. Deficiency of these GST enzymes may therefore decrease the efficiency of detoxification of procarcinogens, thereby increasing both the formation of aromatic/hydrophilic DNA adducts and hence the consequent risk of tumorigenesis, including lung cancer.

N-acetyltransferases

N-acetyltransferases transfer an acetyl group to the N atoms of aromatic amines (in what is usually a deactivation process) or to the O atoms of hydroxylated arylamine metabolites (usually activation). In human, there are two genes, *NAT1* and *NAT2*, that encode functional *N*-acetyltransferases with overlapping specificities. Both enzymes display a wide range of polymorphism with both rapid and slow acetylators (Hein 2002). Slow acetylation is thought to be due to enzyme molecules that are poorly expressed, relatively unstable or which display reduced catalytic activity. Polymorphic variants of both NAT1 and NAT2 may exert an influence over the metabolic activation of potentially carcinogenic arylamines. Since it is the slow acetylator genotypes of both NAT1 and NAT2 that appear to be associated with an elevated risk of lung cancer, we may infer that it is the *N*-acetylation detoxification step that serves as the critical arbiter of risk in the context of lung tumorigenesis.

Cytosine DNA methyltransferase 3β

Cytosine DNA methyltransferase 3β (DNMT3B) is essential for *de novo* methylation, imprinting and mammalian development (Okano *et al.*, 1999; Kaneda *et al.*, 2004). In tumour cells, and together with other DNA methyltransferases, DNMT3B potentiates gene silencing by promoter hypermethylation (Rhee *et al.*, 2002) and is deemed to be essential for cancer cell survival (Beaulieu *et al.*, 2002). Since it is also frequently over-expressed in a range of different tumour types including lung cancer, it is not altogether surprising to find that a promoter variant of the *DNMT3B* gene is associated with increased risk of lung cancer (Shen *et al.*, 2002). However, Sato *et al.* (2002) failed to observe any correlation between hypermethylation of the *CDKN2A* and *RASSF1* gene promoters and expression of any of the DNMT genes including *DNMT3B*.

Methylenetetrahydrofolate reductase

Methylenetetrahydrofolate reductase (MTHFR) catalyses the conversion of 5,10-methylenetetrahydrofolate to 5-methyltetrahydrofolate, a co-substrate for homocysteine remethylation to methionine. Intriguingly, and based upon data derived from studies of other cancers, dietary folate may interact with *MTHFR* genetic variation to modify cancer risk (Huang 2002).

A common polymorphic variant in the *MTHFR* gene (677C→T) that results in the substitution of Val for Ala at residue 222 of the MTHFR protein, is associated with reduced enzyme activity and increased thermolability (Frosst *et al.*, 1995). Reduced MTHFR activity is predicted to impair folate metabolism resulting in high homocysteine levels (although the effect of the 677C→T polymorphism will be dependent upon dietary folate). Since this variant may affect *S*-adenosylmethionine addition, it may also have the potential to influence DNA methylation. It is therefore interesting to note that an association has been found between aberrant DNA methylation and possession of the 677T *MTHFR* allele in both normal tissue and primary lung tumours (Paz *et al.*, 2002). However, an alternative mechanism to explain the association between the MTHFR variant and lung cancer risk could lie with the effect of the variant on mutation frequency; indeed, the frequency of C→T transitions in the methylatable CpG dinucleotide has been known for some time to be sensitive to the concentration of the methyl donor (Shen *et al.*, 1992).

The role of the *MTHFR* C677T polymorphism in conferring lung cancer risk remains somewhat contentious (see Table 4). This notwithstanding, on the basis that lung cancer patients carrying the 677T allele may experience more rapid progression of the disease, it has been suggested that these individuals might benefit from dietary folate (Alberola *et al.*, 2004).

Microsomal epoxide hydrolase

Microsomal epoxide hydrolase serves to catalyse the hydrolysis of reactive and toxic epoxides generated during the oxidative metabolism of PAHs. Although basically a detoxification reaction, some of the trans-dihydrodiols generated as a result of the hydrolysis are substrates for additional metabolic changes to highly toxic, mutagenic and potentially carcinogenic polycyclic hydrocarbon diol epoxides (e.g. 7,8-dihydrodiol of benzo[*a*]pyrene). Since epoxide hydrolase is involved in both the detoxification and the activation of procarcinogens, polymorphism in the epoxide hydrolase (*EPHX1*) gene has been explored as a risk factor in lung carcinogenesis. A number of studies (Table 4) appear to show a decreased risk of lung cancer in association with the His113 variant and an increased risk in association with the Arg139 variant. Since these variants appear to differ from the wild-type in terms of both immunoreactive protein and enzymatic activity (Hassett *et al.*, 1994), both would appear to be polymorphisms with functional effect.

Transforming growth factor β1 receptor

Transforming growth factor β (TGFβ) displays potent anti-mitogenic and pro-apoptotic effects that are mediated in part via SMAD signaling proteins. Once activated, TGFβ binds directly or indirectly to the TGFβ2 receptor leading to the phosphorylation of TGFβ1 receptor. The TGFβ1 receptor then exerts its influence on growth by phosphorylating specific SMAD proteins and inducing their translocation to the nucleus where they modulate the expression of target genes (Attisano and Wrana 2002). Zhang *et al.* (2003a) have reported an association between a polymorphic variant in the *TGFBR1* gene and the risk of NSCLC suggesting that an alteration of TGFβ1 receptor-mediated SMAD signaling could serve to modulate lung cancer risk.

NAD(P)H:quinone oxidoreductase

NAD(P)H:quinone oxidoreductase (NQO1) is a key flavin-containing detoxification enzyme that plays an important role in protecting cells against oxidative stress as well as the actions of mutagens and potential carcinogens (Ross and Siegel 2004). NQO1 exerts its protective effect by preventing the generation of reactive oxygen species and by reducing environmental carcinogens such as nitroaromatic compounds and heterocyclic amines. A number of studies have suggested that a Ser187 variant of NQO1, which displays no detectable NQO1 activity (and which has a prevalence of ~20 % in the Caucasian population), is associated with a reduced risk of lung cancer (Table 4). It is at present unclear how this null variant of NQO1 exerts its influence although it is quite possible that the effect is mediated through the reduced activation of procarcinogens.

Surfactant protein B

Surfactant protein B is a 79 amino acid hydrophobic protein that plays an important role in surfactant function and lung homeostasis. Deficiency of this protein is associated with pulmonary alveolar proteinosis and neonatal respiratory distress syndrome. Since surfactant protein B-deficient mice are susceptible to hyperoxic lung injury (Tokieda *et al.*, 1999), it may be that reduced levels of this protein could render the lung vulnerable to injury and hence could be associated with an increased risk of tumorigenesis. In this regard, it is interesting that Seifart *et al.* (2002) have observed an association between specific alleles of a polymorphic marker in the surfactant protein B (*SFTPB*) gene and risk of lung cancer.

Myeloperoxidase

Myeloperoxidase (MPO) is a dimeric iron-containing heme protein, found in neutrophil granulocytes and in the lysosomes of monocytes, that is involved in the destruction of invading bacteria, viruses and fungal cells as well as malignant and nonmalignant nucleated cells. Intranuclear MPO may also play a role in protecting DNA against damage resulting from oxygen radicals and individuals with MPO deficiency are prone to a high incidence of malignant tumours. A number of reports have explored the relationship between a polymorphism in the promoter of the *MPO* gene and the risk of lung cancer (Table 4). The G→A transition at nucleotide position −463 reduces binding of the transcription factor Sp1 to the *MPO* gene promoter thereby lowering *MPO* gene expression. The majority of the reports suggest that possession of the A allele is associated with a decreased risk of lung cancer.

Van Schooten *et al.* (2004) have shown that *MPO* −463 AA/AG genotypes are associated with reduced MPO activity in bronchoalveolar lavage fluid as compared to the GG genotype. Further, these authors studied AA, AG and GG individuals, matched for smoking status, age and gender, and noted that AA individuals displayed lower DNA adduct levels (median 0.62 adducts/10^8 nucleotides) than AG individuals (median 1.51 adducts/10^8 nucleotides) or GG individuals (median 3.26 adducts/10^8 nucleotides). Van Schooten *et al.* (2004) also claimed to have detected a genotype-dependent correlation between the amount of inhaled tar and the measured DNA adduct level. These findings are certainly compatible with the molecular epidemiological data which suggest that the A allele is associated with a decreased risk of lung cancer.

Cyclooxygenase 2

The cyclooxygenases play a role in the production of prostaglandins from arachidonic acid, and prostaglandins are known to promote tumour growth, angiogenesis and metastasis. Cyclooxygenase 2 (COX-2) is frequently over-expressed in lung cancer and is an indicator of poor prognosis in NSCLC. COX-2 may contribute to lung tumorigenesis by stimulating angiogenesis, modulating xenobiotic metabolism, and inhibiting immune surveillance and apoptosis, whilst COX-2 inhibitors have been successfully used to suppress the growth and metastasis of NSCLC tumours. Campa *et al.* (2004) found that carriers of the C allele of a 3' UTR polymorphism in the COX-2 (*PTGS2*) gene had a significantly increased risk of lung cancer. It remains unclear whether this is a direct effect of the polymorphic allele (mediated perhaps via an influence on mRNA stability) or whether the polymorphism may instead be in linkage disequilibrium with an additional variant that is the actual functional polymorphism in this case.

Glutathione peroxidase

Glutathione peroxidase is an antioxidant enzyme with a key role in host defence against toxic reactive oxygen species generated as by-products of cellular metabolism. However, reactive oxygen species are also thought to serve as subcellular messengers which play a role in the regulation of genes that are involved in defence against oxidative stress. Glutathione peroxidase is decreased under conditions of oxidative stress, and the subsequent increase in the level of peroxide may act as a second messenger to regulate the expression of anti-apoptotic genes (Miyamoto *et al.*, 2003b). It follows that cancer risk may be influenced by either the depletion or elevation of cellular glutathione peroxidase. The report of an association between a polymorphism in the glutathione peroxidase (*GPX1*) gene and lung cancer (Table 4) is therefore of considerable interest.

Receptor for advanced glycation end products

The receptor for advanced glycation end products (RAGE) is a member of the immunoglobulin superfamily and serves as a specific cell surface interaction site for *advanced glycation end products*, the products of non-enzymatic glycation/oxidation of proteins/lipids. Reactive glycation products are present in aqueous extracts of tobacco as well as in tobacco smoke in a form that can react rapidly with proteins to form advanced glycation end products (Cerami *et al.*, 1997). An interaction between RAGE and a ligand triggers cell signaling pathways involving p21, MAP kinases and NFκB, thereby promoting both the initiation and the propagation of the inflammatory response. Schenk *et al.* (2001) have reported the disproportionate occurrence of homozygotes for a promoter variant (A-388) in the RAGE (*AGER*) gene in NSCLC patients.

Sulphotransferase SULT1A1

The sulphotransferases are a family of enzymes involved in the detoxification and bioactivation of both endogenous and exogenous compounds. One of these enzymes, SULT1A1, is polymorphic; the replacement of the wild-type Arg213 residue by His results in a reduction in both enzyme activity and thermostability. Both heterozygosity and homozygosity for the His variant are associated with an increased risk of lung cancer (Liang *et al.*, 2004).

Matrix metalloproteinases

Matrix metalloproteinases (MMPs) are zinc-dependent endopeptidases that are capable of cleaving extracellular matrix components (e.g. collagen, laminin, fibronectin) as well as non-matrix substrates such as growth factors and cell surface receptors. MMPs are involved in multiple steps of tumour metastasis including tissue invasion, migration, extravasation, angiogenesis and tumour growth (Yoon *et al.*, 2003). In human, more than 20 MMP genes have been characterized. A number of promoter polymorphisms have been identified that influence the expression of the MMP genes and which may be associated with cancer predisposition (Ye et al., 2000). Promoter polymorphic variants in the *MMP1* and *MMP2* genes have been reported to be associated with an increased risk of lung cancer (Zhu *et al.*, 2001; Yu *et al.*, 2002).

Neutrophil elastase

Neutrophil elastase is a powerful serine protease that is capable of degrading most proteins of the extracellular matrix. It has been proposed that an imbalance between neutrophil elastase and α1-antitrypsin can result in inflammation and tissue damage (Kawabata *et al.*, 2002) and may promote cancer progression (Sun and Yang 2004). Taniguchi *et al.* (2002) reported an association between polymorphic variants in the promoter of the neutrophil elastase (*ELA2*) gene and lung cancer; the higher risk alleles were shown to be associated with an increased level of expression in reporter gene experiments. The association with lung cancer may therefore be due to an altered proteinase-antiproteinase balance in the lung that could be detrimental to the maintenance of homeostasis.

Heat shock protein HSC70

The constitutively expressed heat shock protein HSC70 acts as a 'chaperonin' that mediates correct folding of newly translated polypeptides and stabilizes nascent proteins to prevent their aggregation. HSC70 also plays a role in the repair of damaged proteins and the removal of proteins that have become irremediably compromised. More specifically, it is known to interact with mutant forms of p53 that have adopted aberrant non-native conformations (Merrick *et al.*, 1996). Rusin *et al.* (2004) reported that a polymorphic variant in the HSC70 (*HSPA8*) gene, associated with a 20 % reduction in *HSPA8* gene expression, was also associated with a decreased risk of lung cancer. It is as yet unclear how and why this variant might reduce the risk of lung tumorigenesis.

Interleukin 1-beta

Interleukin 1-beta is an important mediator of inflammation and tissue damage and is required for tissue invasion and angiogenesis (Voronov *et al.*, 2003). Two interleukin 1-beta (*IL1B*) gene promoter polymorphisms have been previously reported to be associated with risk of gastric cancer. More recently, Zienolddiny *et al.* (2004) reported that the T allele of a –31 C/T polymorphism and the C allele of a –511 C/T polymorphism were over-represented in lung cancer cases. These polymorphic alleles are in linkage disequilibrium with each other in the Caucasian population. The T allele of the –31 polymorphism lies in the constitutive TATA box of the *IL1B* gene promoter and appears to affect adversely the expression of the gene.

Interpreting the role of xenobiotic metabolizing enzyme polymorphisms in lung cancer

"If your experiment needs statistics, you ought to have done a better experiment".
[Lord] Ernest Rutherford

Those studies that have claimed a positive association between possession of a particular metabolic polymorphism genotype and an increased or decreased risk of lung cancer are summarized in Table 4. As we have seen, in some cases, the basis for a reduced risk is claimed to be the reduced metabolic activation of tobacco smoke procarcinogens (e.g. *CYP1A1, NQO1*). In other cases, the reduced metabolism of tobacco smoke carcinogens is said to be the risk factor (e.g. *NAT1* and *NAT2, GSTM1* and *GSTT1*). Evidence for the operation of such mechanisms is still rather sparse but the intermediate steps that could be involved are illustrated by the association claimed between certain polymorphic alleles at *CYP1A1*, *GSTM1* and *GSTP1* loci and benzo[*a*]pyrene adduct formation (Alexandrov *et al.*, 2002; Perera *et al.*, 2002). Alternatively, the influence of the polymorphism may be more indirect. Thus, Gilliland *et al.* (2002) have claimed that individuals with at least one G104 *GSTP1* allele or who lack the wild-type C609 *NQO1* allele, exhibit an increased likelihood of *MGMT* and/or *CDKN2A* gene promoter hypermethylation.

Interpreting the results of polymorphism-DNA association studies is not always straightforward. A representative selection of results of studies from the best-understood xenobiotic metabolizing enzyme systems will now be presented in order to illustrate some of the problems and pitfalls inherent in the interpretation of these studies.

Glutathione S-transferases: Glutathione *S*-transferase P1 (GSTP1) is involved in detoxifying polycyclic aromatic hydrocarbons (PAHs) such as benzo[*a*]pyrene diol epoxides (BPDE) so as to prevent BPDE adduct formation (Fields *et al.*, 1998). A *GSTP1* gene polymorphic variant (Val105) has been reported to confer lower catalytic efficiency towards PAHs than the wild-type Ile105 (Zimniak *et al.*, 1994; Watson *et al.*, 1998; Sundberg *et al.*, 2002). Consistent with these findings, several studies have reported an increased risk of lung cancer associated with possession of the Val105 *GSTP1* allele (Ryberg *et al.*, 1997; Perera *et al.*, 2002; Stücker *et al.*, 2002). In apparent contradiction of these results, however, the Val105 variant has also been variously reported to confer greater protection against BPDE adduct formation [Hu *et al.*, 1999], to be associated with lower DNA adduct levels (Grzybowska *et al.*, 2000) and not to be associated with any difference in adduct level (Ozawa *et al.*, 1999). One wholly reasonable explanation for these apparently contradictory findings could be that GSTP1 catalytic efficiency differs between the *in vitro* and *in vivo* situations. However, shortcomings in experimental design could also have led to erroneous conclusions being drawn. Indeed, the importance of using carefully matched controls in polymorphism-disease association studies of lung cancer is underscored by the quite different frequencies of homozygosity for the *GSTP1* Val105 variant in different ethnic groups: 26% (African-Americans), 6% (Caucasians) and 10% (Latinos) [Weiserbs *et al.*, 2003].

With GSTM1, another glutathione *S*-transferase that catalyses the conjugation of PAHs to glutathione, a very frequent *GSTM1* gene deletion polymorphism (allele frequency = ~50%) has been associated with decreased glutathione *S*-transferase activity (Nakajima *et al.*, 1995) and an increased propensity to form PAH-DNA adducts (Kato *et al.*, 1995; Rojas *et al.*, 2000). Although this is consistent with the idea that individuals lacking the *GSTM1* gene should exhibit increased levels of carcinogen activation and a decreased ability to detoxify the carcinogens, only a slightly increased risk of lung cancer has been noted in association with the null allele (Brockmoller *et al.*, 1998; Belogubova *et al.*, 2004). Whilst Nazar-Stewart *et al.* (1993) found that smokers possessing two copies of the *GSTM1* gene have about a third of the risk of lung cancer of smokers with the *GSTM1* gene deletion, a number of less dramatic albeit still suggestive associations have been noted by other workers (e.g. Ruano-Ravina *et al.*, 2003). However, a meta-analysis of 43 published case-control studies has failed to provide any evidence for an increased risk of lung cancer among carriers of the *GSTM1* null allele (Benhamou *et al.*, 2002). Nor was any evidence obtained for an interaction between *GSTM1* genotype and either smoking status or cumulative smoke exposure in modulating lung cancer risk (Benhamou *et al.*, 2002). With many of the published studies on glutathione *S*-transferase gene polymorphisms, interactions with other non-allelic polymorphic variants may also occur [e.g. between the non-null *GSTM1* allele and *GSTP1* variants (Kihara *et al.*, 1999; Butkiewicz *et al.*, 2000; Mohr *et al.*, 2003)], thereby further complicating the interpretation of study findings. Once again, the frequency of homozygosity for the *GSTM1* null allele varies between different ethnic groups: 27% (African-Americans), 50% (Caucasians) and 45% (Latinos) [Weiserbs *et al.*, 2003] indicating the critical importance of using carefully matched controls.

Cytochrome P450 enzymes: CYP1A1 encodes the cytochrome P450 enzyme primarily responsible for the metabolic activation of carcinogenic polycyclic aromatic hydrocarbons. Benzo[*a*]pyrene, for example, is metabolised by cytochrome P450 enzymes to epoxides which are then converted to diol epoxides by epoxide hydrolases. These diol epoxides are capable of reacting with DNA. Whilst the expression of CYP1A1 has been found to correlate with aromatic/hydrophobic DNA adduct levels in the non-tumour lung tissue of currently smoking lung cancer patients (Mollerup *et al.*, 1999), the possession of certain *CYP1A1* alleles has been reported to be associated with an increased risk of lung cancer (see Table 4). *CYP1A1* polymorphic variants that modulate CYP1A1 levels have also been associated with higher adduct levels in newborn blood samples (Whyatt *et al.*, 1998; 2001) but inconsistently in lung cancer patients (Schoket *et al.*, 1998; Patorelli *et al.*, 1998). Ratnasinghe *et al.* (2001b) have however reported no detectable association between *CYP1A1* genotype and lung cancer risk.

An Ile/Val polymorphism at residue 462 within the heme-binding region of the CYP1A1 molecule has reportedly been associated with an increased risk of lung cancer in Japanese (Hayashi *et al.*, 1991) but *in vitro* studies have failed to demonstrate any obvious difference in the kinetic properties of the two forms of the protein (Zhang *et al.*, 1996; Persson *et al.*, 1997). Homozygosity for the Val462 allele has been variously reported to be associated with increased lung cancer risk in a Brazilian cohort (Hamada *et al.*, 1995), of borderline significance in a German cohort (Drakoulis *et al.*, 1994) but not associated with an increased risk of lung cancer in a Finnish cohort (Hirvonen *et al.*, 1992). Negative or inconsistent findings such as these tend to argue against the functional importance of *CYP1A1* polymorphisms and their importance as risk factors in lung tumorigenesis. A recent meta-analysis of reported studies nevertheless concluded that homozygosity for the Val462 allele may confer a slightly increased risk of lung cancer (OR=1.54; Le Marchand *et al.*, 2003). Interactions with polymorphic variants at other loci may however occur as in the case of the *CYP1A1* Val462 variant and the *GSTM1* null genotype (Hayashi *et al.*, 1992; Hung *et al.*, 2003). The basis for the putative interaction between *CYP1A1* and *GSTM1* alleles has been proposed to lie with modulation of BPDE adduct levels (Rojas *et al.*, 1998; 2000; Alexandrov *et al.*, 2002) although this does not always appear to be in evidence (Cheng *et al.*, 2001a).

N-acetyltransferases: N-acetyltransferases are detoxifying enzymes that activate aromatic and heterocyclic amines to electrophilic intermediates that can potentially act as carcinogens. Polymorphisms in the human *NAT1* and *NAT2* genes are associated with variable acetylation capacity but have only inconsistently been associated with lung cancer risk (Bouchardy *et al.*, 1998; Hein *et al.*, 2002; Table 4). Specifically, the slow acetylator genotypes/phenotypes are often found to occur disproportionately in lung cancer patients as compared to their fast acetylator counterparts. The highest risk appears to be associated with a combination of the *GSTM1* null allele and the slow acetylator *NAT2* genotype (Nyberg *et al.*, 1998; Hou *et al.*, 2001). This allele combination is also associated with a high aromatic DNA adduct level (Hou *et al.*, 2001). Intriguingly, in those adenocarcinoma patients under the age of 65 who possessed the *NAT2* slow acetylator genotype, Oyama *et al.* (1997a) found an excess of *TP53* gene mutations suggesting that the impaired metabolism of a carcinogen could be associated with an increased risk of *TP53* gene mutation.

In addition to the specific interpretational difficulties mentioned above, there are a number of general problems that serve to impede assessment of the validity of reported disease association studies, not the least being the fact that they cannot always be replicated. The possibility of confounding factors (e.g. population stratification, life style, nutritional status; Mooney *et al.*, 1997; Au *et al.*, 2001) is omnipresent whilst the use of poorly selected and inappropriate controls is not infrequent. Since metabolic gene polymorphic variants can differ quite widely in frequency between different human populations (Garte *et al.*, 2001), the selection of appropriate controls is absolutely vital. Although many of the genes being studied in the context of a potential lung cancer association display a number of different polymorphisms, risk is rarely if ever stratified according to haplotype (i.e. specific combinations of alleles at the same locus). Inevitably, valuable information is lost as a consequence. Another *caveat* is reporting bias; there will be a tendency only to report significant associations with some negative findings not being published at all.

Many (probably most) of the reported disease association studies reported have failed to distinguish between different types of lung cancer – thus, even SCLC and NSCLC are frequently lumped together as "lung cancer". This is somewhat naïve because different types of lung cancer are associated with mutations at different loci, and specific carcinogens with their own distinct mutational spectra might reasonably be expected to differ in terms of their mutational target genes. One example of this phenomenon may be the positive association noted between the *CYP2E* c2/c2 polymorphism and risk of squamous cell lung carcinoma, a finding that is not at all evident in lung adenocarcinoma (Oyama *et al.*, 1997b). Similarly, the significant odds ratio reported by Stücker *et al.* (2002) for an association between a *GSTP1* variant and lung cancer was largely attributable to cases of SCLC.

Interactions between polymorphic variants at different loci

As we have seen, interactions between polymorphic alleles of non-allelic loci appear to be quite frequent. The rationale for this is exemplified by appreciating that a given procarcinogen is often metabolised in a two-step process: firstly, CYP1A1 catalyzes the insertion of an oxygen atom into a substrate [e.g. a polycyclic aromatic hydrocarbon (PAH)] in a phase I reaction that converts the procarcinogen into its DNA-reactive form (e.g. PAH diol epoxide). Secondly, such activated carcinogens are then detoxified by glutathione *S*-transferase (GSTM1 among them). It follows that the combination of an over-efficient CYP1A1 with a deficient or defective GSTM1 could greatly increase lung cancer risk after PAH exposure. Consistent with this postulate, a number of groups have reported increased lung cancer risk in association with possession of the m2/m2 or Val/Val *CYP1A1* genotype combined with the null *GSTM1* allele (Nakachi *et al.*, 1993; Alexandrie *et al.*, 1994; Kihara *et al.*, 1995; Garcia-Closas *et al.*, 1997; Chen *et al.*, 2001c). Other combinations of genotypes that have been claimed to confer susceptibility to lung cancer include *CYP1A1* and *EPHX1* (Lin *et al.*, 2000; To-Figueras *et al.*, 2001), *GSTT1* and *GSTM1* (Kelsey *et al.*, 1997) and *CYP1A1* and *CYP2D6* (Sobti *et al.*, 2003).

An increasing number of reports are tending to look at multiple polymorphisms at different loci simultaneously (e.g. Miller *et al.*, 2002b; Cajas-Salazar *et al.*, 2003b; Wang *et al.*, 2004a). For example, el-Zein *et al.* (1997a) observed that "a combination of several versions of 'unfavourable' metabolizing genes (viz. *CYP2D6*, *CYP2E1*, *GSTM1* and *GSTT1*) is strongly associated with lung cancer". However, studies such as this one, which seek combinatorial effects of polymorphisms at multiple loci, may unless they both articulate and test a specific well-formulated hypothesis, be at risk of failing to allow for multiple testing in their significance assessment. Indeed, some associations may be found only in a given age or ethnic group (Garte 1998) or in one sex but not the other (Wu *et al.*, 1997). It may be that in some cases such findings simply allow one to conclude that if one looks hard enough for an association, one will find one sooner or later. It should therefore be no surprise that significant associations are sometimes found contradictorily with alternative alleles in different studies which have focussed on different types of lung cancer or different ethnic groups (e.g. Wu *et al.*, 1997; Oyama *et al.*, 1997b). In one study, an increased risk of squamous cell carcinoma was even claimed to have been found in subjects "having at least one m1 allele or having at least one m2 allele" (Song *et al.*, 2001)! Not surprisingly, "disconfirmation" of previously published positive findings is a common occurrence in this research area (e.g. Houlston 2000; Loriot *et al.*, 2001; Xu *et al.*, 2002a).

Interactions between polymorphic alleles and common tumorigenic lesions may also occur. For example, p53 degradation relies upon two alternative pathways that either depend on Mdm2 and ubiquitin or are independent of both (Asher *et al.*, 2002). The latter pathway is regulated by NAD(P)H quinone oxidoreductase 1, encoded by the *NQO1* gene. Asher *et al.* (2003) have shown that several p53 hotspot mutant proteins (viz. Arg175His, Arg248His, Arg273His) are resistant to ubiquitin-independent degradation although they remained sensitive to Mdm2-ubiquitin-mediated degradation. This resistance to degradation with one pathway appears to be due to increased binding of the p53 variant to NQO1 and implies that these arginine residues may be essential for the process of ubiquitin-independent degradation. This finding may go some way toward explaining why it is the wild-type Pro187 *NQO1* allele that is associated with an increased risk of lung cancer rather than the Ser187 null allele (Wiencke *et al.*, 1997; Chen *et al.*, 1999; Lewis *et al.*, 2001; Xu *et al.*, 2001b; Hamajima *et al.*, 2002; Sunaga *et al.*, 2002; Table 4). The functional (Pro187) allele of NQO1 could thus be capable of stabilizing the hotspot mutant p53 proteins in lung cancer cells thereby explaining the apparent 'protective' effect of the alternative Ser187 null allele.

Interactions between xenobiotic metabolizing enzyme gene polymorphisms, lung cancer risk and smoking behaviour

In some cases, [e.g. *NQO1* (Xu *et al.*, 2001b); *NAT2* (Zhou *et al.*, 2002d); *GSTP1* (Miller *et al.*, 2003c; 2003d) and *GSTM1* (Pinarbasi *et al.*, 2003; Alexandrie *et al.*, 2004], the risk of lung cancer associated with a given polymorphic allele has been claimed to be related to smoking behaviour. In their assessment of lung cancer risk, Jourenkova-Mironova *et al.* (1998) reported significant interactions between 'pack years' of smoking and the combined *GSTM3* and *GSTP1* genotype, or the combined *GSTM3* and *GSTM1* genotype. Similarly, the OR associated with homozygosity for the *MMP2* promoter −1306C allele has been found to be 2.2 overall but 5.6 for light smokers and 10.2 for heavy smokers (Yu *et al.*, 2002) whilst the increased risk of lung cancer for specific *SULT1A1* genotypes was found to be confined to smokers and was claimed to be related to 'cumulative smoking dose' (in pack-years) [Liang *et al.*, 2004]. Homozygosity for the *GSTT1* null allele has been claimed to be a 'possible risk factor' in light smokers whereas in heavy smokers, it was apparently associated with a reduced risk of lung cancer (Alexandrie *et al.*, 2004). Further studies have however failed to note any modifying effect of polymorphisms on lung cancer risk in relation to cumulative smoking dose (e.g. *GSTM1, GSTT1, GSTP1*; Schneider *et al.*, 2004b).

Other examples of interactive effects with smoking exposure have been reported (Wu *et al.*, 2004b). Thus, Bouchardy *et al.* (1996) found the *CYP2D6* genotype to be a risk/protective factor only in heavy smokers. Whilst increasing levels of smoking increased lung cancer risk in smokers with the highest CYP2D6 activity level, it was those who smoked heavily and who exhibited the highest CYP2D6 levels who were at the greatest risk of lung cancer, perhaps through increased metabolic activation of polycyclic aromatic hydrocarbons. Concordantly, Saarikoski *et al.* (2000) reported that the prevalence of CYP2D6 ultarapid metabolizers (due to *CYP2D6* gene amplification) was four-fold higher in heavy smokers than in never-smokers. These data are consistent with the view that elevated levels of CYP2D6 are associated with a higher probability of being a heavy smoker.

Another example of a possible interactive effect is provided by the CYP3A4*1B allele that appears to modify the lung cancer risk in smokers in a gender-specific way – thus, an increased cancer risk pertains for women (OR=3.04, 95% CI 0.94-9.90) but not for men (OR=1.00, 95% CI 0.56-1.81) [Dally *et al.*, 2003]. Finally, a reduced risk of lung adenocarcinoma associated with a specific *CYP2A13* genotype (encoding a CYP2A13 isoform known to exhibit a reduced efficiency in activating the tobacco-derived carcinogen NNK) was confined to light smokers rather than heavy smokers (Wang *et al.*, 2003h). These examples therefore go some way to supporting the contention that cigarette smoking can in some cases modify the association between a given xenobiotic metabolizing enzyme gene polymorphism and lung cancer risk. Whether or not these findings are going to be reproducible in future studies is at present unclear.

Interpreting the role of DNA repair enzyme polymorphisms in lung cancer

> "DNA repair appears pivotal to the maintenance of genome integrity, and genetic alterations in repair capacity, due to single nucleotide polymorphisms or mutation, may account for inter-individual differences in cancer susceptibility".
>
> MD Evans & MS Cooke (2004) *BioEssays* 26: 533-542.

Some 133 different human genes are now known to encode proteins that are directly involved in DNA repair (Ronen and Glickman 2001; Wood *et al.*, 2001). A number of DNA polymorphism-disease association studies have reported that certain polymorphic alleles of genes encoding DNA repair enzymes (viz. *XRCC1*, *ERCC2*, *ERCC5*, *XPA*, *OGG1*, *MGMT*) may be associated with an increased or decreased risk of lung cancer (e.g. Sugimura *et al.*, 1999; Lunn *et al.*, 2000; Auckley *et al.*, 2001; Butkiewicz *et al.*, 2001; Divine *et al.*, 2001; David-Beabes and London 2001; Park *et al.*, 2002b; Ratnasinghe *et al.*, 2001a; Spitz *et al.*, 2001; Goode *et al.*, 2002; Table 4). Counter-intuitively, some of these reports have involved polymorphic alleles predicted to compromise DNA repair activity yet which are said to be associated with a 'protective effect' (i.e. decreased risk of lung cancer) [David-Beabes and London 2001]. One explanation may be that under certain circumstances, reduced DNA repair capacity may target damaged cells for apoptosis and will therefore ultimately tend to be protective for lung cancer.

Evidence for a direct effect of the polymorphism on DNA repair protein function is however often lacking. Thus, the Ser326 polymorphism in the *OGG1* gene, which is claimed to be associated with increased lung cancer risk (Sugimura *et al.*, 1999; Le Marchand *et al.*, 2002), has not been found to be associated with any reduction in 8-

oxoguanine DNA glycosylase activity in human lymphocytes (Janssen *et al.*, 2001). Higher aromatic/hydrophobic adduct levels have however been associated with possession of specific *XRCC1*, *XRCC3* and *ERCC2* alleles (Matullo *et al.*, 2001; Palli *et al.*, 2001; Pastorelli *et al.*, 2002; Hou *et al.*, 2002; Matullo *et al.*, 2003). In the *XRCC1* gene, the Gln399 polymorphic variant, which is thought to modulate the genotoxicity of the tobacco-specific nitrosamine NNK (Abdel-Rahman and el-Zein 2000), is reportedly associated with a higher frequency of sister chromatid exchange (Lei *et al.*, 2002), an increased level of single strand breaks (Vodicka *et al.*, 2004), increased susceptibility to vinyl chloride-induced DNA adduct formation (Li *et al.*, 2003e) and a higher A→G transition frequency in the *TP53* gene (Casse *et al.*, 2003) whereas the Trp194 variant is associated with increased G2 cell cycle delay (Hu *et al.*, 2001). Finally, an elevated frequency of chromosomal aberrations has been noted in association with either the Asn312 or the Lys751 allele of the *ERCC2* gene whilst single strand breaks have been found to be more frequent in individuals with the *ERCC2* Lys751 allele and the *ERCC5* His1104 allele (Vodicka *et al.*, 2004; Harms *et al.*, 2004; Affatato *et al.*, 2004).

As with polymorphisms in genes encoding xenobiotic metabolising enzymes, the risk of lung cancer associated with a given polymorphic allele has sometimes been claimed to be related to smoking behaviour. Thus, the odds ratio (OR) for homozygosity for the *ERCC2* Asp312 allele in association with risk of lung cancer was reportedly 5.3 overall but not significantly different from unity in either never-smokers or heavy smokers (Butkiewicz *et al.*, 2001). The ORs associated with homozygosity for the *ERCC2* Asn312 and Gln751 alleles were reportedly 1.5 and 1.1 respectively but 2.6 for non-smokers and 0.7 for heavy smokers (Zhou *et al.*, 2002b). Similarly, the OR associated with homozygosity for the *XRCC1* Gln399 allele was 1.3 overall but 2.4 for non-smokers and 0.5 for heavy smokers (Zhou *et al.*, 2003a). Another such example is provided by a glutathione peroxidase (*GPX1*) variant that appears to be a risk factor for lung cancer in old smokers (OR=3.3) but 'protective' (OR=0.6) in young never-smokers (Yang *et al.*, 2004b). Although no significant association was found between alleles of the *XRCC3* Thr241Met polymorphism and lung cancer risk, a significantly increased risk of lung cancer has been noted among heavy smokers with the Met241 genotype (Wang *et al.*, 2003 g). Finally, among Japanese lung cancer cases, the gene-environment interaction between current smoking and possession of three or more *APEX1* Glu148 or *XRCC1* Gln399 alleles was shown to be statistically significant (OR=2.4, 95% CI 1.00-9.2) suggesting that the combinatorial possession of these polymorphic alleles might modify the risk of lung cancer attributable to cigarette exposure (Ito *et al.*, 2004a). Once again, it remains to be seen whether or not the results of these studies are confirmed by replication.

Somewhat vague statements by authors such as a "decreased risk for the homozygous variant genotype among heavier smokers" (David-Beabes and London 2001) or a "genotype not associated with lung cancer risk in never-smokers or heavy smokers" (Butkiewicz *et al.*, 2001) or "associated with increased risk of lung carcinoma in certain subgroups [but] very heavily dependent on the degree of smoking" (Wang *et al.*, 2003 g) can make the interpretation of some studies somewhat tricky and inter-study comparisons very difficult. Although such findings/conclusions may indicate a highly complex relationship between the polymorphic genotype and smoking exposure, they could equally well just reflect the authors' determination to rescue a research project by repeated phenotype partitioning and re-testing until a reportable significance level is attained. In their recent paper on genotype frequencies, cumulative smoking dose and lung cancer risk, Alexandrie *et al.* (2004) carefully added the rider: 'Due to the multiple comparisons made, we cannot exclude the possibility that some of these associations may represent chance findings'. Most authors, however, still seem relatively unaware of the dangers of multiple testing (Krawczak *et al.*, 2001) and the need for study replication (Ioannidis *et al.*, 2001; Cooper *et al.*, 2002).

From the above examples, it should be apparent that there are a variety of recurring problems, both practical and interpretational, with polymorphism-disease association studies. The utility of this molecular epidemiological approach ought therefore to be seen as limited. For this reason, it is recommended that it should be used simply as an initial test in order to ascertain whether a particular gene/protein is worthy of further functional investigation rather than as a substitute for functional studies (for a detailed discussion of the relative advantages and disadvantages of the polymorphism-disease association approach, see Ahsan and Rundle 2003). This notwithstanding, it may be that future studies of cancer susceptibility will still benefit from the use of microarrays to screen individual blood and tumour samples for the presence of multiple (but carefully selected) polymorphic alleles of genes encoding xenobiotic metabolizing enzymes or DNA repair proteins (Landi *et al.*, 2003).

DNA repair activity for oxidative damage; the contribution of OGG1

A number of studies have suggested that the DNA repair capacity of lung cancer patients may be compromised (Wei *et al.*, 1996; Zienolddiny *et al.*, 1999; Rajaee-Behbahani *et al.*, 2001; Shen *et al.*, 2003a), a finding which if verified, could imply that sub-optimal DNA repair capacity might increase the risk of developing lung cancer (Spitz *et al.*, 2003). One of the several different DNA repair processes, base excision repair, removes single modified or damaged bases from DNA and is regarded as the most important mechanism for protecting DNA from oxidative damage irrespective of whether this occurs as a result of exposure to endogenous reactive oxygen species or the action of exogenous mutagens (Cooke *et al.*, 2003). The repair process is initiated by recognition of the DNA adduct followed by excision of the modified base by DNA glycosylases leaving a single base gap. This gap is then filled by apurinic/apyrimidinic endonuclease (APE), DNA polymerase beta and DNA ligase I together with the accessory protein XRCC1 (Marsin *et al.*, 2003).

One of the glycosylases, 8-oxoguanine (or 8-hydroxyguanine) DNA *N*-glycosylase (OGG), acts to remove 8-oxoguanine, the major base lesion resulting from exposure to reactive oxygen species, ionising radiation, and also tobacco smoke (Olinski *et al.*, 2003; Fortini *et al.*, 2003). If not removed, 8-oxoguanine frequently mispairs with adenine during DNA replication giving rise to G→T transversions, a major endogenous source of mutation in human cells. Rodin and Rodin (2002) have suggested that oxidative stress, resulting from exposure to certain components of cigarette smoke, could lead to the generation of 8-oxoguanine. Thus, it may be that in a situation in which OGG activity (and hence 8-oxoguanine removal) is compromised, 8-oxoguanine (rather than a polycyclic aromatic hydrocarbon of exogenous origin, such as benzo[*a*]pyrene) could give rise to some of the G→T substitutions that are observed in excess in lung cancer. This view is not inconsistent with the observation that 8-oxo-dG occurs at increased levels both in lung tumours (Jaruga *et al.*, 1994) and in peripheral lung tissue from lung cancer patients (Inoue *et al.*, 1998). OGG, although capable of removing 8-oxoguanine-induced lesions in human cells, is however incapable of removing G→T substitutions induced by benzo[*a*]pyrene; these substitutions are instead repaired by nucleotide excision repair (Yamane *et al.*, 2003).

The gene encoding 8-oxoguanine DNA *N*-glycosylase, *OGG1*, is located on chromosome 3p25 within a region prone to frequent deletion in lung cancer. Hence, considerable interest has been expressed in the idea that the loss of 8-oxoguanine *N*-DNA glycosylase activity could play a key role in lung tumorigenesis by compromising base excision repair. Loss of heterozygosity of the 3p25 region is apparent in some 40-60% of non-small cell lung cancer (NSCLC) primary tumours (Lu *et al.*, 1997; Kohno *et al.*, 1998b; Chevillard *et al.*, 1998; Wikman *et al.*, 2000; Hardie *et al.*, 2000; Shinmura and Yokota 2001). Homozygous null *Ogg1* mice exhibit a 3-fold increased accumulation of 8-oxoguanine and manifest a substantial increase in spontaneous mutation frequency (Minowa *et al.*, 2000). Such mice also display an increased incidence of lung cancer with a high frequency (75%) of tumours possessing G→T transversions at the hotspot codon 12 in the *Kras* gene (Xie *et al.*, 2004). However, as far as humans rendered somatically hemizygous for the *OGG1* gene by 3p25 deletion are concerned, it remains unclear whether haploinsufficiency compromises the efficiency of the excision repair process.

Various polymorphic variants of the *OGG1* gene have been reported including Arg/Gln at amino acid residue 46, Ala/Ser at residue 85, Arg/His at residue 154, and Ser/Cys at residue 326 (Sugimura *et al.*, 1999; Wikman *et al.*, 2000; Ito *et al.*, 2002a; Le Marchand *et al.*, 2002). The latter polymorphism is relatively frequent in human populations with the Cys variant occurring in 22% of Caucasians and 42% of Japanese (Le Marchand *et al.*, 2002). The 326Cys allele has been claimed to be associated with an increased risk of various types of NSCLC (Sugimura *et al.*, 1999; Ito *et al.*, 2002a; Le Marchand *et al.*, 2002). However, comparison of the results of the different studies indicates that there is as yet no clear and consistent association between NSCLC and a specific genotype. This may in part be because different ethnic groups (i.e. Caucasians and Japanese) have been used in these studies. Further, the association may be limited to certain types of lung cancer (viz. squamous cell lung cancer and adenocarcinoma) and the relative proportion of these different forms of NSCLC may well differ between studies.

The evidence for a direct effect of the polymorphic OGG variants is still somewhat equivocal. Thus, neither allele of the Ser/Cys326 polymorphism was found to be associated with any reduction in 8-oxoguanine DNA *N*-glycosylase activity in either human lymphocytes (Janssen *et al.*, 2001) or colorectal tumour tissue (Park *et al.*, 2001). In addition, initial *in vitro* studies of the purified Ser326 and Cys326 forms of the OGG protein indicated that both forms excise 8-oxoguanine at very similar rates (Dherin *et al.*, 1999). More recently, however, it has been claimed that the Ser326 form of OGG is more efficient in human lung cells at suppressing 8-oxoguanine-induced mutations than the Cys326 form of the protein (Yamane *et al.*, 2004). Similarly, Audebert *et al.* (2000a; 2000b) performed biochemical and/or kinetic analyses of the Gln46 and His154 forms of OGG showing both to be defective in terms of their catalytic capacity. Park *et al.* (2001), however, reported that the Arg/His154 variant was not associated with any reduction in 8-oxoguanine DNA *N*-glycosylase activity in colorectal cancer tissue.

The potential role of 8-oxoguanine DNA *N*-glycosylase in the repair of oxidative damage has been highlighted by a recent report, a molecular epidemiological case-referent study of 68 non-small cell lung cancer (NSCLC) cases and an identical number of age- and sex-matched controls (Paz-Elizur *et al.*, 2003). The study involved only non-smokers or current smokers; former smokers were excluded. OGG activity was measured in both peripheral blood cells of patients and controls and from the non-tumour lung tissue of patients. The assay specifically developed for this purpose measures the ability of OGG to remove an 8-oxoguanine residue from a radio-labelled synthetic DNA substrate; the authors claimed it to yield results that were both accurate and reproducible. The mean OGG activity was found to be significantly lower in blood cells from patients (5.8 U/μg protein, 95% CI=5.5 - 6.2 U/μg protein) than controls (7.1 U/μg protein, 95% CI=6.8 - 7.3 U/μg protein) [$p < 0.001$]. Among both controls and patients, no statistically significant differences were noted between men and women or between smokers and non-smokers. However, a small statistically significant difference was observed between controls (but not patients) < 60 years of age and those > 60 years. This is consistent with the previously reported age-associated decrease in OGG activity measured in human lymphocytes (Chen *et al.*, 2003d). The distribution of OGG activity values also differed between patients and controls with some 41% of patients, but only 4% of controls, exhibiting OGG activity values ≤ 5.5.

Paz-Elizur *et al.* (2003) tested a total of 8 individuals over the space of 3 years and found that OGG activity values remained relatively stable during this time period (average coefficient of variation = 7%, 95% CI=4-10%). This is encouraging particularly in the light of previous findings suggesting quite wide inter-individual and intra-individual ranges, albeit for *OGG1* mRNA levels measured in peripheral blood cells (Hanaoka *et al.*, 2000). Clearly, Paz-Elizur *et al.* (2003) could not obtain lung samples from the controls. However, lung and blood samples were obtained from 7 NSCLC patients. Linear regression analysis revealed a good fit between OGG values measured in lung and blood from the same individuals ($R^2 = 0.86$, $p = 0.003$). This shows that OGG activity values taken from peripheral blood cells are indicative of OGG activity in the lung and can therefore serve as a reliable surrogate marker for lung OGG activity. In order to rule out the possibility that the OGG values measured in the patients could have been influenced by blood-borne factors emanating from the tumour, the authors looked to see if there was a correlation between OGG activity and the time post-surgery at which the blood sample was taken. No such correlation was found. This is consistent with previous findings that the human *OGG1* gene is constitutively expressed and not up- or down-regulated by exposure to either pro-oxidant or anti-oxidant treatments (Mistry and Herbert 2003).

Three different statistical approaches were then adopted by Paz-Elizur *et al.* (2003) to explore the apparent association between NSCLC and low OGG activity. First, when OGG activity was treated as a continuous variable, adjusted for age and smoking status, the odds ratio (OR) for lung cancer associated with a 1 Unit decrease in OGG activity was 1.9 (95% CI=1.3–2.8, $p < 0.001$). Second, when OGG activity values were partitioned above and below the median, the OR for lung cancer associated with low activity was 5.2 (95% CI=1.9–14.1). Finally, when OGG activity values were divided by tertiles of OGG activity in the control subjects, and using the tertile with the highest OGG activity as the referent group, the OR for lung cancer for the lowest tertile was 4.8 (95% CI=1.5–15.9). Thus, the association held up regardless of whether the entire groups were compared or whether they were partitioned.

Paz-Elizur *et al.* (2003) found there to be no significant interaction between smoking status and reduced OGG activity implying that being a smoker and exhibiting a low OGG activity level are independent risk factors for NSCLC. This is important because it meant that the possibility that the OGG activity assay might be confounded in some way by smoking could effectively be excluded. In addition, it meant that the combined effect of low OGG activity and smoking could legitimately be assessed. Not surprisingly, smoking on its own was found to be strongly associated with NSCLC (OR=18, 95% CI=6.0–53). Making the assumption that both patient and control groups are representative of the population at large, the Odds Ratio may be used to approximate the Relative Risk. The Relative Risks of NSCLC for smokers with OGG activity values of 6.0 or 4.0 were respectively estimated to be 34-fold or 124-fold higher than for non-smokers with an OGG activity level of 7.0. By contrast, the Relative Risks of NSCLC for non-smokers with these OGG activity values were estimated to be only 1.9-fold or 7.0-fold higher than those for non-smokers with an OGG activity level of 7.0.

In their well-designed and carefully controlled study. Paz-Elizur *et al.* (2003) have therefore provided some convincing evidence for OGG activity level being an independent risk factor for NSCLC. Clearly, for the results of any epidemiological study to be credible, they must be compatible with a biologically plausible hypothesis. In this case, the working hypothesis proposed by Paz-Elizur *et al.* (2003) was as follows: lower OGG levels result in a reduced ability to repair oxidative DNA damage. In smokers, inhaled tobacco smoke increases oxidative DNA damage. Therefore, in smokers who manifest a reduced OGG activity level (and who hence have compromised base excision repair potential), mutations may be expected to accumulate at a faster rate than in either smokers with higher OGG activity or in non-smokers with a low OGG activity level. Further studies on this topic will undoubtedly be reported in the coming months and years. However, as the authors themselves state, what are now needed are prospective epidemiological studies. Nevertheless, if these initial findings are borne out by further work, the strong inverse relationship between OGG activity level and risk of NSCLC will provide a powerful argument for the proponents of an interventionist strategy to bring about smoking cessation in individuals specifically targeted by virtue of their low OGG activity levels.

Although the design of the Paz-Elizur study was impressive, its main weakness must be the relatively low number of cases and controls. Despite the strong association noted between OGG activity level and risk of NSCLC, the study could nevertheless be open to criticism on the grounds of sampling bias. There is also an issue regarding the selection of controls. Although one assumes that the patients and controls are matched with respect to racial origin, this is nowhere stated. Whilst the patients were carefully selected individuals with operable NSCLC who had not received prior chemotherapy or radiation therapy, the controls were employees, retired employees and relatives of employees of the Sheba Medical Center and Weizmann Institute. Population stratification cannot entirely be excluded here. Finally, it may be noted that a prior history of cancer was an exclusion criterion for controls selected for this study. If compromised DNA repair due to low OGG activity were indeed a risk factor for other types of cancer as well as lung cancer, then the active removal of individuals with a prior history of cancer (unclear how many) from the control group could have served to remove those with the lowest OGG levels, thereby unintentionally introducing some bias. The strength of the association observed should, however, be sufficient for the potential shortcomings in the design of this study not to have affected either the results or the conclusions drawn.

The strong association observed between OGG activity level and risk of NSCLC contrasts somewhat with the results of the rather equivocal *OGG1* polymorphism-disease association studies reported to date (Table 4). There may be several reasons for this. The polymorphism-disease association studies were not nearly so well performed as the study of Paz-Elizur *et al.* It may also be that the association studies have not so far been performed with optimal genetic markers. Thus, the majority of the polymorphism-disease association studies performed to date have involved the Ser/Cys326 *OGG1* polymorphism despite the fact that no evidence has so far been provided for any difference in activity between the Ser326 and Cys326 forms of OGG (Dherin *et al.*, 1999; Janssen *et al.*, 2001; Park *et al.*, 2001). Future work should therefore focus on those *OGG1* polymorphisms where the alternative allelic forms encode proteins that differ in terms of their biological activity. It should however be appreciated that there may also be inherited variants in other unlinked genes that play a role in influencing *OGG1* gene expression and hence OGG protein levels.

Paz-Elizur *et al.* (2003) made no mention of the potential interaction between low OGG activity level, inherited *OGG1* variants, and the somatic inactivation of the *OGG1* gene that is a consequence of the frequent deletion of the 3p25 region in NSCLC. It may be that those incipient lung tumours that experience somatic inactivation of one of their *OGG1* alleles, but which also occur in individuals with low OGG levels (irrespective of whether or not this is genetically determined), may experience a markedly increased mutation frequency that greatly increases the probability of mutation at other gene loci, thereby potentially accelerating the process of tumorigenesis.

Clearly, however, inter-individual differences in OGG activity levels are not going to be the only factor in predisposing smokers (and to a lesser extent non-smokers) to NSCLC. Other DNA repair proteins also exhibit inter-individual differences in activity levels. Some of the genes encoding these proteins have already been shown to possess polymorphic alleles (e.g. XRCC1, ERCC2, ERCC5, XPA, MGMT) that encode protein isoforms that differ in terms of their repair activity (Ruttan and Glickman 2002; Mohrenweiser et al., 2002; de Boer 2002). Other such variants are predicted to be associated with compromised DNA repair (Xi et al., 2004). It is likely that an inherited predisposition to NSCLC will eventually be found to be dissectible into contributions from genetic variants in a number of different DNA repair protein genes. Different combinations and permutations of such variants could account for DNA repair capacity being a continuous variable in human populations (Wood et al., 2001; Goode et al., 2002). We must also remember that inherited allelic variants in nearly 50 different genes have so far been found to display a positive association with an increased risk of lung cancer (Table 4). Some of these genes encode xenobiotic metabolising enzymes, polymorphic variants of which may play a role in lung cancer susceptibility by modulating the level of harmful metabolites of exogenously derived mutagens or alternatively the level of mutagenic metabolites (Wu *et al.*, 2004b). The likely complexity of the genetics probably means that for the time being, analysis of genetic susceptibility to NSCLC will continue to be performed on an individual gene or protein basis. The OGG activity assay described by Paz-Elizur et al. (2003) would appear to provide a robust test for activity level but could ultimately be replaced by DNA markers if a straightforward and reproducible relationship is found between OGG activity level and OGG1 genotype.

TP53 gene polymorphisms and lung cancer risk

A number of polymorphisms have been noted within the coding region of the human *TP53* gene (viz. codons 21, 36, 47, 72 and 213; *see* Hainaut and Hollstein (2000) and http://www.iarc.fr/p53/Polymorphism.html). Two of these, the rare C→T transition in codon 47 and the common G→C transversion in codon 72, alter the amino acid sequence of the p53 protein. Alleles of the latter Arg/Pro72 polymorphism have been shown to manifest markedly different apoptotic potential (Dumont *et al.*, 2003; Schneider-Stock *et al.*, 2004). In addition, the Pro72 allele is associated with the reduced expression of p53's downstream target gene, *CDKN1A* (Su *et al.*, 2003a). Intriguingly, in patients with advanced lung cancer who are constitutionally heterozygous for the Arg/Pro72 polymorphism and who manifest LOH at the *TP53* locus in their lung tumour tissue, the Arg allele is preferentially retained (Papadakis *et al.*, 2002). If confirmed by additional studies, this would be consistent with the view either that the Arg72 allele promotes growth and/or inhibits apoptosis or that the Pro72 allele inhibits growth and/or promotes apoptosis.

Associations between specific *TP53* polymorphic alleles/haplotypes and lung cancer susceptibility have been claimed in various studies to a greater or lesser extent (Kawajiri *et al.*, 1993; Jin *et al.*, 1995; Murata *et al.*, 1996; 1998; Pierce *et al.*, 2000; Biros *et al.*, 2001a; 2001b; Papadakis *et al.*, 2002; Wu *et al.*, 2002; Hiraki *et al.*, 2003). However, significant differences in alleles/haplotype frequencies occur between different ethnic groups (Wu *et al.*, 2002). This could easily lead to misinterpretation of results derived from database-dependent studies (*see* Chapter 5; p53 Mutations, Benzo[*a*]pyrene and Lung Cancer: the Controversy) if case and control population groups are compared that are not very carefully matched in terms of their ethno-geographic origin (as proposed by Paschke 2000). Thus, for example, the report of a significant difference in *TP53* allele frequencies between SCLC and NSCLC cell lines (Ueda *et al.*, 2003) may owe more to ethno-geographic differences between the individuals from whom the cells were derived, than to any differential influence of the *TP53* polymorphism on lung tumorigenesis. No difference in allele distribution was noted in a study that compared centenarians and younger controls once gender and ethno-geographic origin had been taken into account (Bonafè *et al.*, 1999). Finally, a recent meta-analysis of *TP53* polymorphism-disease association studies failed to provide any evidence for a role for *TP53* polymorphism in susceptibility to lung cancer (Matakidou *et al.*, 2003).

CHAPTER 7

Gene Expression Studies in Lung Cancer

"Eventually, the techniques of nucleic acid chemistry should allow us to itemize all the differences in nucleotide sequence and gene expression that distinguish a cancer cell from its normal counterpart, and perhaps at that point, the steps involved in carcinogenesis will cease to be in doubt".
J. Cairns (1981) The origin of human cancers. *Nature* 289: 353-357.

No cancer can be understood simply in terms of the gene mutations within the cells of the tumour, however well characterized these lesions may be. Since each mutation, whether gross or subtle, is likely to result in multiple and perhaps numerous changes to the expression levels of other genes and their encoded proteins, the first step towards understanding the molecular basis of tumorigenesis lies in the determination of the gene/protein expression profiles of the cell type in question and how they have changed during tumorigenesis. Only then can the similarities as well as the differences between different tumours be properly compared and contrasted. The dissection of the molecular heterogeneity that exists both between and within individual tumours is essential in order to improve the accuracy of both diagnostic and prognostic predictions and to assess the likely response to therapy.

Studies of the expression of individual genes

A large number of studies have focussed on the induced expression, under-expression or over-expression of individual genes in lung cancer. Arguments have usually been presented for the special significance of these genes to some aspect of lung tumorigenesis and/or to their potential prognostic importance (Caputi *et al.*, 2002; Hommura *et al.*, 2002; Shoji *et al.*, 2002). We must however bear in mind that in excess of 13,500 genes are expressed in the human lung and that some 70% of these are also expressed in other tissues (Brentani *et al.*, 2003). We do not yet know the proportion of these genes whose expression is altered in lung cancer nor have we yet identified the genes whose expression becomes deregulated first and which therefore represent the best candidates for key mediators of the tumorigenic process.

Perhaps the first report of the altered expression of a protein in lung cancer was that of Gahmberg *et al.* (1979) who demonstrated that aryl hydrocarbon hydroxylase (now known as the cytochrome P450 family member, CYP1A1) activity was induced in lung cancer. This finding was soon confirmed by Kouri *et al.* (1982) who sensibly added the *caveat*: "whether the higher AH levels are the cause or the result of the primary lung cancer remains to be determined".

Studies of the expression of individual genes have now identified a considerable number of loci (> 270) which are either up- or down-regulated in lung tumours and/or cell lines (Table 5). The majority of genes (~70 %) whose expression has been examined in lung cancer tumour tissue or cell lines have been shown to be up-regulated. The up-regulation of some of these genes, for example those encoding *CCNA2, CCNE1, WNT7A, ERBB2, E2F1, MOS, TP73, JUN, MDM2*, may contribute directly to cellular proliferation but the likely effect of other examples of elevated expression (e.g. *RB1, TP53*) is sometimes rather less clear. In some cases (e.g. Sasaki *et al.*, 2001e; 2003a), no clear difference in the expression of a particular gene was noted between tumour tissue and normal lung tissue but the gene in question was still considered to be of potential importance on the grounds of a correlation between gene expression and either tumour progression or patient survival. In the case of ~5 % of the genes listed in Table 5, contradictory findings have been reported from different gene expression studies (denoted by ↑↓); the most likely explanation is that the expression of the genes concerned differs between lung tumour types and/or sub-types. It should be appreciated, however, that studies of the expression of individual genes are inherently prone to bias since the choice of gene may be somewhat arbitrary and/or is likely to reflect the preconceived views of the individual researchers involved with respect to its potential role in lung tumorigenesis.

The underlying cause of the change in gene expression may not however always be obvious. Some cases may reflect subtle indirect or downstream effects on expression as a consequence of the process of tumorigenesis. In other cases, however, observed changes in gene expression may simply be a direct consequence of gross rearrangements such as loss of heterozygosity, chromosomal aneuploidy or gene amplification (Masayesva *et al.*, 2004), even if this is not explicitly stated (or indeed considered!) in the articles in question (see the discussion below of the article by Fujii *et al.*, 2002). From inspection of Table 5, it can be seen that the expression of genes located on chromosomal arms 3p, 8p and 10q is reduced rather more often than it is increased. Since these regions are often lost in lung cancer (*see* Chapter 2; Clues to Candidate Genes from Cytogenetic Abnormalities and Loss of Heterozygosity Studies), some of the observed reductions in gene expression may be due to chromosomal loss rather than locus-specific *cis*-acting effects on gene expression. Similarly, the expression of genes on chromosomal arms 1q, 3q, 5q, 7p, 11p, 11q, 12q and 14q is increased rather more often than it is decreased (Table 5). Since several of these regions are known to be frequently amplified in lung cancer, some of the observed increases in gene expression may be due to chromosomal gains rather than locus-specific effects on gene expression. The interested reader should note that similar results have been reported by assessing the chromosomal location of differentially expressed genes by 'transcriptome mapping' (*see* Chapter 7; Gene Expression Profiling Using Approaches Other Than Microarrays).

Also difficult to interpret are cases of isoform switching as a consequence of the differential expression of alternatively spliced gene products in lung tumours or cell lines. Examples include the neuron-restrictive silencer factor (*REST*; Coulson *et al.*, 2000), actinin-4 (*ACTN4*; Honda *et al.*, 2004), cell-cell adhesion molecule (*CEACAM1*; Wang *et al.*, 2000a), poly(rC) binding protein (*PCBP4*; Pio *et al.*, 2004), vascular endothelial growth factor (*VEGF*; Cheung *et al.*, 1998), neural cell adhesion molecule (*NCAM1*; Moolenaar *et al.*, 1992), CD44 antigen (*CD44*; Miyoshi *et al.*,

1997), cytokeratin 8 (*KRT8*; Tojo *et al.*, 2003), RNA-binding motif protein 6 (*RBM6*; Timmer *et al.*, 1999), polymerase beta (*POLB*; Bhattacharyya *et al.*, 1999), selenium-binding protein 1 (*SELENBP1*; Chen *et al.*, 2004b), telomerase reverse transcriptase (*TERT*; Fujiwara *et al.*, 2004) and heterogeneous nuclear ribonucleoprotein A2/B1 (*HNRPA2B1*; Sueoka *et al.*, 1999). In addition, Junker *et al.* (2003) have found that a panel of SCLC cell lines differed with respect to their ability to splice out a retained intron in the 5' UTR of *HYAL1* mRNA. Such studies can really only be properly interpreted in the context of lung cancer when the molecular basis of alternative splicing has been properly elucidated, when the population of alternatively spliced variants has been adequately described (Hui *et al.*, 2004) and when the biological functions of the different protein isoforms in any given case are fully understood.

Some HOX genes that are normally expressed in pulmonary embryogenesis, are re-expressed in lung cancer tissue (Lechner *et al.*, 2002). However, these same cells also express HOX genes that are not normally expressed in lung, so the suggestion that lung tumour cells recapitulate the gene expression patterns of cells in the early stages of lung development (*see* Chapter 7; Lung Tumour Cell Ontogeny May be Determined by Gene Expression Pathways that Recapitulate Lung Development) is only partially correct (Lechner *et al.*, 2002).

The expression of some genes appears to be highly variable even within the same lung cancer type. One example of this is thyroid transcription factor 1 (TTF1), a homeodomain-containing transcription factor that plays a key role in lung development, cell growth and differentiation. In normal lung tissue, TTF1 expression is confined to bronchial and alveolar epithelial cells (Fabbro *et al.*, 1996). In order to determine the prevalence of TTF1 expression, and to assess its potential prognostic utility, Tan *et al.* (2003c) examined 126 NSCLC tumours for TTF1 expression; 51 % expressed TTF1, 49 % did not. However, TTF1 expression did correlate with tumour type (68 % of adenocarcinomas were TTF1-positive as compared to 21 % of squamous cell carcinomas) and the loss of TTF1 expression was associated with a poor prognosis. By contrast, Stenhouse *et al.* (2004) found that 75 % of adenocarcinomas displayed TTF1 positivity as compared to 8 % of other types of lung cancer. TTF1 expression may also prove useful in distinguishing between primary lung cancers and metastatic lesions (Chhieng *et al.*, 2001; Roh and Hong 2002).

The studies listed in Table 5 are divided roughly equally between those that measured protein expression by immunohistochemical methods and those that measured gene expression at the transcriptional (mRNA) level. Only ~18 % of the studies listed measured the documented change in expression at both mRNA and protein levels. Studies that assessed expression only at the mRNA level (~40 % of the total) are open to criticism on the grounds that they would miss any effects of post-transcriptional processing. Post-transcriptional changes in mRNA stability do occur, as noted for the *MYC* and *FOS* genes whose mRNA transcripts have been found to be inappropriately stabilized in lung cancer cell lines (Bernasconi *et al.*, 2000). The corollary to this example is that provided by the promyelocytic leukaemia (PML) protein which was found by immunohistochemical staining to be virtually absent from SCLC tumours despite being expressed at the mRNA level (Zhang *et al.*, 2000b). It should be appreciated that if post-transcriptional regulation were to prove to be the norm rather than the exception (and bearing in mind that mRNA translation often becomes deregulated during tumorigenesis; Holland *et al.*, 2004; Pandolfi 2004; Rosenwald 2004), or if many RNA transcripts were not to encode protein products (Sémon and Duret 2004), then measured mRNA levels would provide a very inadequate and inappropriate measure of the levels of the encoded protein products in lung cancer cells and tissues (Chen *et al.*, 2002c).

The differential expression of the two alleles of a given gene in the same individual can also play a role in the inter-individual variation of expression levels of specific gene products (Yan *et al.*, 2002a; 2002b; Lo *et al.*, 2003) and in some cases this appears to be heritable (Cheung *et al.*, 2003). In the context of lung cancer, it is pertinent to note that the two alleles of the O6-methylguanine DNA-methyltransferase (*MGMT*) gene are frequently differentially expressed in normal lung tissue (Heighway *et al.*, 2003). This is consistent with there being a genetic component to the inter-individual variation of MGMT levels and may be one cause of inter-individual differences in DNA repair capacity.

It has been claimed that SCLC cells can be distinguished from NSCLC cells by virtue of the 1000-fold higher level of expression of the achaete-scute homologue 1 (*ASCL1*) gene in SCLC (Westerman *et al.*, 2002). Similarly, expression of the surfactant protein B (*SFTPB*) gene is detectable in 60% of lung adenocarcinomas but not in squamous cell or large cell carcinoma (Khoor *et al.*, 1997). Finally, it has been claimed that chromogranin A and histidine decarboxylase, putative markers of neuroendocrine differentiation, may be useful in distinguishing between SCLC and non-neuroendocrine lung carcinoma (Begueret *et al.*, 2002; Matsuki *et al.*, 2003). If confirmed, these findings could prove to be important both because they will contribute to our understanding of the molecular and cellular differences between these different forms of lung cancer and because they may be used as diagnostic markers.

Studies of the expression of multiple genes by microarrays and similar techniques

> "Genomics and gene expression experiments are sometimes described as 'fishing expeditions'. Our view is that there is nothing wrong with a fishing expedition if what you are after is 'fish', such as new genes involved in a pathway, potential drug targets or expression markers that can be used in a predictive or diagnostic fashion".
>
> D.J. Lockhart & E.A. Winzeler (2000) Genomics, gene expression and DNA arrays. *Nature* 405: 827-836.

Introduction

Studies such as those outlined above invariably employ fairly well-characterized genes. However, it may be that new insights into underlying tumorigenic mechanisms must await the discovery of novel genes with new functions; a more global and less restrictive approach is therefore required. One of the most powerful approaches available to characterize lung cancer cells at the level of gene expression is through the use of microarrays since these "DNA chips" provide the means to determine patterns of gene expression on a genome-wide basis (Ladanyi and Gerald 2003). Typically, such microarrays consist of thousands of oligonucleotides or cDNA sequences, deposited robotically onto a solid support. Hybridization with mRNA derived from the cells or tissue of interest then permits the quantitative assessment of the level of gene expression at thousands of loci simultaneously (Granjeaud *et al.*, 1999; Lockhart and Winzeler 2000). This technology should potentiate the development of a new molecular taxonomy based on tumour type-specific expression profiles or 'signatures'; differences in gene expression identified between normal and neoplastic cells may then serve as tumour-specific markers of use in diagnosis, prognosis and treatment (Scherf *et al.*, 2000; Goodwin *et al.*, 2001; Mao 2002; Yanagisawa *et al.*, 2003b; Borczuk *et al.*, 2004; Fukuoka *et al.*, 2004). Microarrays can also be used to provide answers to a host of entirely new biological questions by studying the coordinate expression of genes, identifying the functions of unknown genes whose expression is either up- or down-regulated during tumorigenesis, and by dissecting gene expression pathways and networks.

The main objectives of microarray studies are generally either *class comparison*, *class discovery* or *class prediction* (Miller *et al.*, 2002c). Class comparisons assess whether expression profiles differ between classes e.g. between two subtypes of lung cancer. Class discovery attempts to identify subclusters or structure among tissue specimens or among genes e.g. to identify previously unrecognised subtypes of lung cancer. Class prediction attempts to predict a clinical phenotype from expression profiling data e.g. prognostic predictions or likelihood of adverse drug response. Two general statistical approaches to tumour classification using microarray data from tumour-control tissue comparisons have been routinely employed (Churchill 2002; Slonim 2002): *Supervised analysis* involves a search for genes whose expression patterns correlate with an 'external' parameter such as survival, presence of metastasis, or response to therapy. This approach is most effective for class comparison or class prediction. The alternative approach is *unsupervised analysis* in which no external parameter is used as a guide and the expression data are used to search for patterns without any *a priori* expectation of the number or type of sub-groups that are likely to emerge; the most commonly used form of unsupervised analysis is *hierarchical cluster analysis*. Unsupervised analysis is most suitable for class discovery. Since each method of analysis has its pros and cons, the significance of the findings of any given study is ideally assessed by means of more than one type of data analysis.

A considerable number of expression profiling studies have been performed with the aim of classifying and sub-classifying tumours (Miller *et al.*, 2002c). Several principles appear to have emerged:

(i) Cell lineage has a key role in determining the expression profile. Thus, array profiles from a primary and a metastatic tumour from the same patient tend to be more similar than profiles generated for primary tumours from different individuals.

(ii) Array profiles may define distinct prognostic subgroups and these subgroups are frequently associated with cell lineages. However, it is usually not clear if individual components of the composite expression 'signature' are causative or merely consequential.

(iii) Information from the composite expression of a group of genes, functionally related by virtue of their belonging to the same biological pathway, is usually more important than that derived from an individual gene.

(iv) Although pathways ultimately define the expression profiles, key events in the tumorigenic process (e.g. gene inactivation or amplification) induce downstream changes in specific pathways that are associated with distinguishable expression profiles.

Microarray studies of tumour suppressor gene-induced gene expression in lung tumour cells

> "Complexity is essential and the analysis of complex array becomes the most critical issue in expression genomics. The power of microarray data is not in viewing the technology as a collection of individual Northern blots, but in generating a composite image of the expression profile of a cell."
>
> LD Miller *et al.* (2002) Optimal gene expression analysis by microarrays. *Cancer Cell* 2: 353-361.

Using a low density cDNA array, Song *et al.* (1999) were the first to study the expression profile in human lung cancer cells of 30 genes (involved in the control of the cell cycle checkpoint and/or apoptosis) in response to the ectopic expression of p53. Kannan *et al.* (2000) greatly extended this work by performing microarray analysis on a p53-inducible human lung cancer cell line. Among those genes whose expression was found to be induced by p53 were cell cycle inhibitors and apoptosis regulatory proteins. Further work on this model system was described in Kannan *et al.* (2001a). Inhibition of protein synthesis was used to distinguish between primary and secondary target genes regulated by p53, the former being $<20\%$ as abundant as the latter. In addition to cell cycle [p21 (*CDKN1A*), cyclin E (*CCNE1*), *TGFB*] and apoptosis-related [Fas (*TNFRSF6*), *BAK1*] genes, the primary targets of p53 induction included genes involved in cell adhesion [thymosin (*TMSB4X*)], signaling (*HRAS*), transcription (*ATF3*) and DNA repair (*BTG2, DDB2*). p53 was shown to up-regulate proapoptotic and cell cycle inhibitors and to down-regulate anti-apoptotic and cell cycle genes. Kannan *et al.* (2001b) then used a more extensive oligonucleotide microarray system to demonstrate that p53-regulated genes may be subdivided into early- and late-induced categories. The emerging identification of the primary target genes of p53 represents an important step toward dissecting its tumour suppressor function as well as determining the role of its mutant forms in driving lung tumorigenesis.

A similar study profiling changes in gene expression in a lung cancer cell line as a consequence of *PTEN* gene over-expression suggested that *PTEN* over-expression may inhibit lung cancer invasion by down-regulating the expression of a number of other genes (Hong *et al.*, 2000). Using microarrays, Jimenez *et al.* (2003) also examined the consequences at the transcriptional level of the ectopic over-expression of the putative tumour suppressor gene *STK11* in lung adenocarcinoma cells. These authors noted the deregulation of 100 genes involved in cell proliferation, apoptosis and cell adhesion. Since the expression of several p53-responsive genes was modified in these cells, it is possible that growth suppression in adenocarcinoma cells over-expressing *STK11* may be mediated by p53. Finally, Agathanggelou *et al.* (2003b) used microarrays to identify the different genes that are differentially regulated by *RASSF1*, a putative tumour suppressor gene that is subject to epigenetic inactivation in NSCLC. The use of microarrays for the identification of 'target genes' whose expression is up- or down-regulated by the over-expression of a specific gene known to be mutated and/or inactivated in lung cancer, is likely to increase in the future (e.g. Russo *et al.*, 2003a).

Microarray studies of gene expression in lung tumorigenesis

In principle, microarray analysis promises to allow one to relate the clinicohistopathological differences between SCLC and NSCLC to expression levels of individual genes (Manda *et al.*, 2000; Petersen and Petersen 2001), thereby eventually allowing molecular differentiation between different types of NSCLC (Kikuchi *et al.*, 2003; Petty *et al.*, 2004) or even between different subclasses of a given tumour type e.g. adenocarcinoma (Bhattacharjee *et al.*, 2001; Garber *et al.*, 2001, see Figure 7.1) and high-grade neuroendocrine tumours (Jones *et al.*, 2004). Thus, microarray analysis has been used to identify 17 genes over-expressed in NSCLC (squamous cell carcinoma) relative to normal tissues (Wang *et al.*, 2000b) including connexin 26 (*GJB2*), plakophilin 1 (*PKP1*) and a number of keratin-related genes. Also over-expressed was the ataxia-telangiectasia (*ATM*) gene that encodes a protein which activates the p53-mediated cell cycle checkpoint through phosphorylation of p53.

Garber *et al.* (2001) examined the gene expression profiles of lung adenocarcinoma using cDNA microarrays and were able to use these profiles to (i) subclassify adenocarcinomas into various sub-groups that corresponded broadly with the degree of tumour differentiation (Figure 7.1) and (ii) to identify three subtypes that predicted favourable, intermediate and poor clinical outcomes. Thus, pulmonary surfactant A1 was expressed in groups 1 and 2 but was poorly expressed in group 3. Expression of surfactant proteins A1, B and C correlated strongly with the expression of thyroid transcription factor 1 which is known to be involved in the regulation of surfactant gene expression and is a marker of primary lung adenocarcinoma. Many of the group 3 adenocarcinomas were metastatic and so it is of interest that group 3 adenocarcinoma shared with large cell lung cancer the increased expression of genes involved in tissue remodelling [e.g. urokinase receptor (*PLAUR*) and cathepsin L (*CTSL*)] and angiogenesis [vascular endothelial growth factor C (*VEGFC*) and peroxisome proliferator-activated receptor γ (*PPARG*)].

Despite differences in sample acquisition, analytical methods and analysis platforms, Bhattacharjee *et al.* (2001), in a study of 203 lung tumours and samples of normal lung, independently observed two of the three adenocarcinoma sub-groups noted by Garber *et al.* (2001). The type II pneumocyte markers, thyroid transcription factor (*TITF1*) and surfactant proteins B, C and D (*SFTPB, SFTPC, SFTPD*) and a number of other markers including cytochrome b5 (*CYB5*), selenium-binding protein 1 (*SELENBP1*), cathepsin H (*CTSH*) and mucin 1 (*MUC1*) were found to exhibit a certain parallelism in terms of their expression in the different adenocarcinoma sub-groups between the two studies. Further, squamous cell carcinomas, which are characterized by keratinization, were found in both studies to exhibit high-level expression of keratin genes, S100 calcium-binding protein A2 (*S100A2*), the keratinocyte-specific protein, stratifin (*SFN*) and p63 (*TP73L*). In addition, markers of proliferation such as proliferating cell nuclear antigen (*PCNA*) and thymidylate synthase (*TYMS*; Nakagawa *et al.*, 2004) were strongly expressed in SCLC, the most rapidly dividing of lung tumours. Finally, both Bhattacharjee *et al.* (2001) and Garber *et al.* (2001) observed that genes associated with detoxification and antioxidant properties

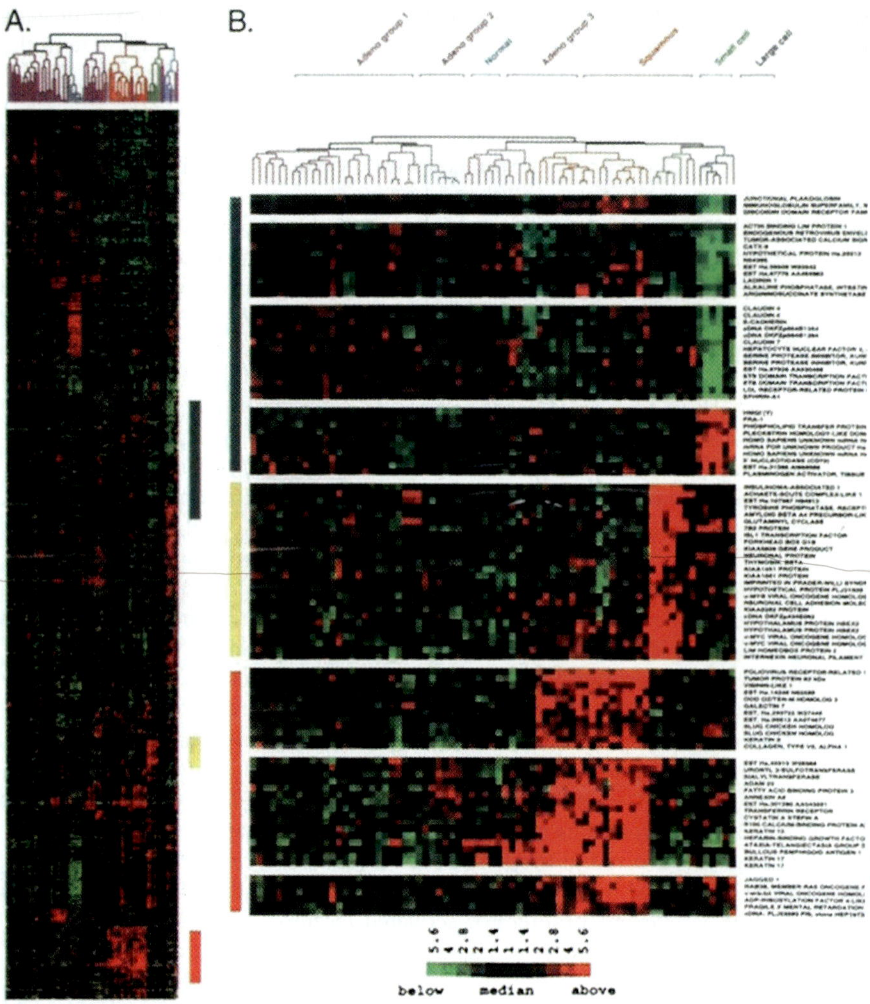

Fig. 7.1 A,B. Gene expression profiling of lung cancer by cDNA microarray analysis. Squamous, small cell and large cell lung tumours express a unique set of genes. **A** Hierarchical clustering sorted 918 cDNA clones and 73 lung tissues based upon similarity in gene expression. Gene clusters relevant to lung tumour types were extracted from the larger cluster of 918 clones in the regions indicated by the coloured bars and expanded on the right to include gene names. A row in the cluster indicates expression of a specific gene across all 73 lung tissues. A column indicates the tissue in which the gene is expressed. Red, green and black squares indicate that expression of the gene is greater than, less than, or equal to the median level of expression across all 73 lung tissues, respectively. Grey represents missing or poor quality data. **B** (Top) Gene clusters relevant to large cell tumours (blue bar).
(Middle) Gene clusters relevant to small cell tumours (yellow bar).
(Bottom) Gene clusters relevant to squamous lung tumours (red bar). The scale bar reflects the ± fold increase (red) or decrease (green) for any given gene relative to the median level of expression across all samples. Reproduced with kind permission, from ME Garber *et al.* (2001) Diversity of gene expression in adenocarcinoma of the lung. *Proc. Natl. Acad. Sci. USA* 98: 13784-13789. Copyright (2001) National Academy of Sciences, USA

(glutathione peroxidase, glutathione *S*-transferases, carboxylesterase and aldo-keto reductases) were highly expressed both in lung squamous cell carcinomas and in a sub-type of lung adenocarcinoma. Broadly speaking, molecular classification has thus not only tended to confirm conventional clinicohistopathological classificatory schemes, but it also promises to extend them.

Meyerson *et al.* (2004) cross-compared the expression profiling studies of Bhattacharjee *et al.* (2001), Garber *et al.* (2001) and Sugita et al. (2002) in order to identify genes that were over-expressed in more than one study. For SCLC, the following genes were highlighted by these means: achaete-scute homologue 1 (*ASCL1*), fork-head box G1B (*FOXG1B*), insulinoma-associated 1 protein (*INSM1*), islet-1 transcription factor (*ISL1*), tripartite motif-containing 9 (*TRIM*) and an imperfectly identified thymosin β gene. For squamous cell lung carcinoma, a different set of genes were highlighted including ataxia-telangiectasia group D-associated protein (*ATM*), bullous pemphigoid antigen 1 (*BPAG1*), collagen VII α1 (*COL7A1*), galectin 7 (*LGALS7*), keratins 5, 6 and, 17 (*KRT5, KRT6, KRT17*), S100 calcium-binding protein A2 (*S100A2*) and tumour protein p63 (*TP73L*).

In the same vein, cDNA microarray profiling has been used to classify high-grade neuroendocrine carcinoma into two groups independent of large-cell neuroendocrine carcinoma and small cell lung carcinoma (Jones *et al.*, 2004). Since one of the newly identified sub-groups manifested a significantly better clinical outcome than the other, it can be seen how expression profiling can lead to advances not only in diagnosis and classification, but also in terms of potential treatment.

Microarray-based studies have contributed toward a description of the similarities between the gene expression networks exhibited by different types of lung cancer (Hellmann *et al.*, 2001) as well as the differences between them (Anbazhagan *et al.*, 1999; Manda *et al.*, 2000; Turney *et al.*, 2004). Thus, Virtanen *et al.* (2002) noted overlaps in gene expression profiles between some adenocarcinoma and SCLC cell lines but it was unclear whether this was because adenocarcinoma cell lines differentiate toward SCLC pathology or because clonal expansion of SCLC subcomponents of adenocarcinomas takes place. Kettunen *et al.* (2004) found that 39 % of the 25 most up-regulated and down-regulated genes were common to both squamous cell lung carcinoma and adenocarcinoma.

One of the most comprehensive microarray studies of human lung cancer performed to date has been that of Yamagata *et al.* (2003). These authors compared the gene expression profiles of normal lung, primary NSCLC tumours of different types, and metastatic lung tumours and were able to identify groups of genes that were helpful in differentiating between the different tumour groups. Thus, keratin 5 (*KRT5*) and bullous pemphigoid antigen 1 (*BPAG1*) were highly expressed in squamous cell carcinoma whilst folate receptor (*FOLR1*) and mucin 1 (*MUC1*) were highly expressed in adenocarcinoma. Interestingly, the large cell carcinomas proved to be difficult to cluster into one grouping on the basis of expression profiling. Indeed, large cell carcinomas often appear to be more closely related to either adenocarcinomas or squamous cell carcinomas. Studies such as these promise to provide the means not only to classify lung tumours into biologically meaningful groupings and to identify novel genes associated with these distinctions, but also to allow us to predict both the biological and clinical behaviour of specific lung tumours.

McDoniels-Silvers *et al.* (2002) claimed to be able to discern certain trends in terms of the nature of the gene products up- and down-regulated during lung cancer tumorigenesis. Thus, there was said to be a tendency for the expression of oncogenes and genes encoding apoptosis regulatory proteins, and enzymes of DNA synthesis, repair and recombination to be reduced, whereas the expression of genes encoding stress response proteins, modulators, effectors, intracellular transducers, receptors and cell adhesion molecules was considered to be increased. Similarly, Nacht *et al.* (2001) claimed that genes encoding proteins with roles in detoxification and with antioxidant properties were up-regulated in SCLC whilst genes encoding small airway-associated proteins were up-regulated in adenocarcinoma. Finally, in apparent contrast to the situation adduced from the studies of single gene expression (*see* Chapter 7; Studies of the Expression of Individual Genes), Nakamura *et al.* (2003c) claimed that 9.4% of genes were up-regulated and 17.4% of genes were down-regulated in Stage IA NSCLC tumour specimens studied by cDNA microarray.

The use of microarrays to determine the downstream consequences of the inactivation of a given gene for gene expression is potentially very important in the molecular genetic analysis of lung tumorigenesis. Using cDNA microarrays, Fernandez *et al.* (2004) have identified a number of genes that are up-regulated in lung adenocarcinomas carrying mutations in the putative tumour suppressor gene, *STK11*. Some of these genes encode proteins that are involved in signal transduction (the PI3K/PTEN pathway), control of transcription and ubiquitinization.

One general question still to be resolved is the extent to which tumours derived from quite different tissues exhibit similarities in their patterns of gene expression. Chung *et al.* (2002) presented the results of a hierarchical clustering analysis of breast and lung tumour expression data. These authors noted that breast and lung tumours exhibited a set of genes whose expression was common to both tissues (a 'proliferation cluster'). This gene set contains genes that are involved in regulating the cell cycle as well as genes that encode proteins involved in DNA replication and chromosome architecture. This proliferation 'signature' is evident in a wide variety of different tumours and appears to correlate, at least *in vitro*, with cellular growth rates. If as would appear likely, the proliferation signature also reflects the *in vivo* growth rate of the tumour cells in question, it could come to be used as a prognostic indicator.

Microarrays can be used to identify genes whose expression can be used to discriminate between normal and neoplastic lung tissue (Hofmann *et al.*, 2004). Microarrays have also been used in pilot attempts to identify genes that might play a specific role in lung tumorigenesis by, for example, contributing to decreased apoptotic ability (Singhal *et al.*, 2003b), increased tissue invasiveness (Chen *et al.*, 2001a; Lader *et al.*, 2004) or angiogenesis (Tanaka *et al.*, 2002), loss of differentiation (Creighton *et al.*, 2003a), cell-cell interactions (Fromigué *et al.*, 2003), clinical outcome (Beer *et al.*, 2002; Wigle *et al.*, 2002), tumour recurrence (Gordon *et al.*, 2003), the development of chemo- and radio-resistance (Henness *et al.*, 2004) and metastatic potential (Gemma *et al.*, 2001; Kikuchi *et al.*, 2003; Peng *et al.*, 2004; Hoang *et al.*, 2004; *see* Chapter 4; Metastasis). Microarrays may also play an important role in predicting the chemosensitivity of specific tumours (Staunton *et al.*, 2001), in assessing the consequences for lung gene expression of pre-operative chemotherapy (Ohira *et al.*, 2002; Whiteside *et al.*, 2004) and in identifying potential prognostic indicators for use in lung cancer patients who have been given chemotherapy (Ikehara *et al.*, 2004) (*see* Chapter 4; Molecular Genetics of Chemotherapy and Chemoresistance). The potential utility of microarray analysis even extends to the identification of genes whose expression may underlie the anti-invasive and/or anti-metastatic properties of putatively anti-tumorigenic compounds (Chen *et al.*, 2004a). Finally, expression profiling by microarray analysis has even proven successful in using the very small lung cancer tissue samples derived from fine needle aspiration biopsies (Lim *et al.*, 2003).

Ultimately, microarray-based expression profiling must be integrated with other analytical techniques that can be used to describe the state of the genome and proteome of a given lung tumour sample or cell line. Thus, the future analysis of lung tumour biopsies should employ comparative genomic hybridization to assess chromosomal gains and losses, microsatellite markers to assess the extent of loss of heterozygosity, mutational screening and immunohistochemical detection of protein expression to identify inactivated genes, and methylation analysis to identify genes inactivated by epigenetic changes. Only then, placed within the context of the totality of the structural changes in the genome, can the changes in gene expression observed during lung tumorigenesis be rationalized in terms of their causes and consequences.

Use of microarrays to study changes in gene expression in smoking-related lung tumours

The expression of specific genes in the lung tumour tissue of smokers and non-smokers has been reviewed in Chapter 5. The advent of microarray technology has however allowed researchers to widen their search for genes whose expression might be dysregulated in smokers or which might be used to distinguish between the lung tumours of smokers and non-smokers. In one of the first studies of its kind, Miura *et al.* (2002) attempted to use microarray expression analysis to differentiate between adenocarcinomas from smokers and non-smokers. The expression of some 28 genes was claimed to be significantly higher in smokers than in non-smokers including Ras-associated protein RAB4 (*RAB4A*), ribosomal protein L22 (*RPL22*), destrin (*DSTN*), a TATA-binding protein-binding protein, α-tubulin (*TUBA1*), glioma tumour suppressor candidate region gene 2 (*GLTSCR2*) and the 18 kDa signal peptidase complex. Conversely, the expression of some 17 genes was claimed to be significantly lower in smokers than in non-smokers including thymidine kinase 2 (*TK2*), UDP-Gal:βGlcNAc β1,3-galactosyltransferase 4 (*B3GALT4*), α2 integrin (*ITGA2*), neural cell adhesion molecule 1 (*NCAM1*), cytochrome P450 IID 7a (*CYP2D*) and carnitine/ acylcarnitine translocase (*SLC25A20*). We should however be careful not to read too much into the results of this single study since the sample size was low (which would have tended to over-emphasize stochastic inter-individual differences) and the definition of smoking status was potentially problematic.

Use of microarrays to study changes in bronchial epithelial cell gene expression consequent to smoking

Microarrays also form the methodological backbone of the emerging discipline of toxicogenomics, the measurement of genome-wide changes in gene expression as a consequence of exposure to toxic substances (Shioda 2004). As yet, however, few studies have focussed on the consequences for bronchial cell gene expression of exposure to tobacco smoke. Yoneda *et al.* (2001; 2003) have begun to explore the expression of smoke-inducible genes in human bronchial epithelial cells. Meanwhile, Hackett *et al.* (2003) assessed the level of expression of 44 antioxidant-related genes in bronchoscopy-derived airway epithelium of smokers and non-smokers; significant up-regulation of expression was noted for 16/44 of these genes in cells derived from smokers.

By far the most comprehensive expression profiling study to date on the effects of cigarette smoke on the bronchial epithelium has been performed by Spira *et al.* (2004). These authors recruited 34 smokers, 18 former smokers and 23 never-smokers, and obtained bronchial airway epithelial cells from them by fibreoptic bronchoscopy. RNA was then extracted, labelled and hybridised to Affymetrix GeneChips containing ~22,500 different human transcripts. Samples taken from the left and right bronchi in the same individual proved to be highly reproducible (R^2=0.92), as were samples from the same individual taken 3 months apart (R^2=0.85). A total of 7,119 genes were found to be expressed at a measurable level in the bronchial epithelial cells of the majority of never-smokers whilst some 2,382 genes were expressed in all 23 never-smokers. As far as the latter set of genes (corresponding to the 'normal airway transcriptome') is concerned, genes associated with oxidant stress, ion and electron transport, chaperone activity, vesicular transport, ribosomal structure and binding functions were found to be over-represented. Multiple linear regression analysis indicated that only a small portion of the variation between subjects could be attributed to age, gender or race. When the expression profiles of current smokers and never-smokers were compared under highly stringent conditions, a total of 97 genes were found to be differentially expressed. 68 of these genes were up-regulated in response to exposure to cigarette smoke; these tended to be involved in the regulation of oxidant stress, glutathione and xenobiotic metabolism, and secretion, and included cytochrome P450 subfamily 1 polypeptide 1 (*CYP1B1*), glutathione peroxidase 2 (*GPX2*), aldehyde dehydrogenase family 3 subfamily A (*ALDH3A1*), carcinoembryonic antigen-related cell adhesion molecule 6 (*CEACAM6*), claudin 10 (*CLDN10*), pirin (*PIR*) and carbonic anhydrase XII (*CA12*). By contrast, the 29 genes that were found to be down-regulated appeared to be involved in the regulation of inflammation although the expression of several putative tumour suppressor genes was also found to be decreased; these included matrix metalloproteinase 10 (*MMP10*), downregulated in renal cell carcinoma 1 (*TU3A*), hepatic leukaemia factor (*HLF*), chemokine C-X3-C motif ligand 1 (*CX3CL1*), Slit homologues 1 and 2 (*SLIT1* & *SLIT2*) and growth arrest-specific 6 (*GAS6*). Changes in gene expression were confirmed by real-time PCR.

Spira *et al.* (2004) performed a 2D hierarchical clustering analysis of the data from the current and never-smokers based on the 97 genes that were differentially expressed between the two groups. Intriguingly, the expression of a subset of genes in three smokers more closely resembled that of never-smokers than of smokers in that the expression of a number of xenobiotic and redox-related genes was not increased. Since these individuals would have failed to increase the expression of several genes encoding proteins with detoxifying or antioxidant functions, Spira *et al.* (2004) speculated that they might be at increased risk of smoking-related genetic damage. That one of these apparently non-reactive individuals subsequently developed lung cancer during a 1-year follow-up was taken as being consistent with the view that there might indeed be a relationship between the unusual gene expression signature and the risk of lung cancer. These authors also noted a never-smoker who expressed a subset of genes characteristic of current smokers. On the face of it, these results are exciting because it has long been suspected that there may be individuals who smoke who are especially sensitive to smoke-related genetic damage, as well as individuals who do not smoke who may nevertheless be at increased risk of lung cancer. The *caveat* of course is that the smoking status of all the individuals studied must have been correctly determined whereas in reality, smoking status is notoriously difficult to ascertain with any accuracy. This type of approach nevertheless holds out the eventual promise of identifying individuals at increased risk in the hope that risk reduction measures might then be instituted (e.g. smoking avoidance/cessation, clinical monitoring etc).

Spira *et al.* (2004) have also claimed that changes in the expression of some genes correlates with cumulative smoking exposure (in pack-years). Among those genes deemed to be significantly over-expressed were cystatin 6 (*CST6*), eukaryotic translation initiation factor 2C 3 (*EIF2C3*), bromodomain-containing 2 (*BRD2*), fibroblast growth factor binding protein 1 (*FGFBP1*) and protocadherin γC3 (*PCDHGC3*) whereas replication protein A1 (*RPA1*) and phosphatidic acid phosphatase type 2B (*PPAP2B*) genes were under-expressed. Analysis of variance provided some evidence that race (but not age or gender) influenced the effect of smoking on bronchial epithelial cell gene expression. It must however be bourne in mind that when fibreoptic bronchoscopy is used to obtain sequential samples of airway epithelial cells, the expression of monitored genes may well be altered in response to the stress of the procedure (Heguy *et al.*, 2003).

Finally, Spira *et al.* (2004) compared the expression profiles of former smokers who discontinued smoking <2 years before the start of the study with those of former smokers who discontinued the habit >2 years before the study began. The expression level of many 'smoking-induced genes' (including those serving metabolic and antioxidant functions) among former smokers was found to resemble that of never-smokers after 2 years of smoking cessation. However, Spira *et al.* (2004) found that the expression of 13 genes did not return to normal levels (defined as levels characteristic of never-smokers) in former smokers, even in those former smokers who had apparently discontinued smoking some 20-30 years before being tested. These genes include *TU3A*, *CX3CL1* and metallothioneins 1F and 1X (*MT1F* & *MT1X*) whose expression was permanently decreased, and *CEACAM6* and haematological and neurological expressed 1 (*HN1*) whose expression was permanently increased. Spira *et al.* (2004) suggested that the persistence of abnormal expression of a specific set of genes after smoking cessation might have provided a growth advantage to bronchial epithelial cells, allowing them to expand in clonal fashion thereby perpetuating the gene expression signature many years after smoking had been discontinued. Whether the bronchial epithelial cell gene expression profiles of smokers could serve as a biomarker for lung cancer risk still remains an open question.

Microarray analysis of gene expression in the murine lung as a means to identify candidate genes

Microarrays have also proven useful in studying murine lung gene expression as an indirect means to identify novel genetic markers of human lung cancer. Thus, expression profiling has been employed to examine lung gene expression in mouse strains with varying susceptibility to lung tumour development with a view to finding quantitative trait loci that could lead to the identification of genes responsible for either susceptibility or resistance to lung cancer (Lemon *et al.*, 2002; Gariboldi *et al.*, 2003). Bonner *et al.* (2003; 2004) performed similar experiments using lung tissue from mice at different developmental stages and noted that some of the genes identified as being involved in lung development were also known tumour suppressors (e.g. *Rassf1*, *Robo1*).

Problems and pitfalls in the use of microarrays in studies of lung cancer

> "It isn't that they can't see the solution. It is that they can't see the problem".
> GK Chesterton (1935) *The Scandal of Father Brown.*

Microarray analyses are potentially very powerful but we must not be blind to their limitations nor to the problems and pitfalls in data interpretation (Russo *et al.*, 2003b). This notwithstanding, many of the differences in gene expression picked up in studies to date are not tumour type-specific and are often likely to be a consequence of the higher metabolic rate of the dividing tumour cells rather than a cause of the tumorigenesis *per se*. Another potential problem is that an expression profile in cultured cells might not reflect the profile of the tumour from which the cells originated although a recent study of SCLC cells/tumour tissue using oligonucleotide microarrays has suggested that expression profiles may not change as radically as has been feared (Pedersen *et al.*, 2003). Also encouraging have been the findings of Ross *et al.* (2000) who have shown that the most important factor responsible for variation in gene expression patterns, among 60 cell lines from diverse types of tumour, was the identity of the tissue from which each cell line was derived. It would thus appear that the adaptation of cells to culture conditions and the accompanying selection for growth in an *in vitro* environment, is not sufficient for those cells to override their basic tissue-specific gene expression programmes that were established during differentiation *in vivo*.

The results of large-scale studies of lung cancer gene expression performed to date are summarized in Table 6; details are given of specific genes that were found to be significantly over- or under-expressed by comparison with normal lung tissue. It is clear from inspection of this Table that a considerable number of different studies of lung cancer gene expression have been performed, generating a profusion of data. In principle, one way to address the twin questions of the authenticity and validity of these data is to perform a meta-analysis to identify those genes whose over- or under-expression is frequently or even invariably associated with a particular type of lung cancer. Unfortunately, in practice, such a meta-analysis is fraught with difficulties owing to the very considerable technical and methodological differences between studies (Li *et al.*, 2002; Järvinen *et al.*, 2004). At the very least, such studies may differ in terms of the quality of tumour identification/sub-classification and tissue preparation (Sorensen *et al.*, 1993).

Another problem is that lung tumour tissue often consists of a heterogeneous collection of different types of tumour cell as well as non-tumour cells (Yesner 2001; Bombi *et al.*, 2002). Further, the comparison of a given lung tumour, arising from a specific cell type, with 'normal lung tissue' may well be inherently problematic because the preneoplastic cell type may not be appropriately represented in the control tissue sample. Intra-tumoral heterogeneity, whether due to mutational or epimutational differences (Ohlsson *et al.*, 2003), may therefore lead to the generation of mixed expression profiles. Techniques such as laser capture microdissection (Kikuchi *et al.*, 2003) promise to be useful in separating out tumour cells from other contaminating cell types (e.g. stromal and surrounding cells) but could well suffer from the drawback of generating only small yields of mRNA.

It is also apparent from Table 6 that screening methodologies vary quite widely. Different studies utilizing cDNA and oligonucleotide microarrays may not be directly comparable either on the grounds of differences in the numbers of genes screened, the differential fidelity of the microarrays employed or the lack of probe specificity (Kothapalli *et al.*, 2002) or because methods of statistical evaluation of the generated data may differ markedly (Nadon and Shoemaker 2002). It may nevertheless be that similarities noted between the results of these studies can help us to discern trends and hence inform and guide our thinking. At this stage, however, only a relatively small number of genes appear to be over-expressed in two or more studies (e.g. pulmonary surfactants A2 & B, insulin-like growth factor binding protein 2, DNA topoisomerase II, ribonucleotide reductase, glutathione peroxidase 2, glutathione-*S*-transferases M1 & M3, pituitary tumour transforming gene 1, hepatoma-derived growth factor, cyclin B1, polo-like kinase, cell division cycle proteins cdc2 and cdc25A, and thyroid transcription factor 1). Further, some results appear to be inconsistent between studies with, for example, glutathione peroxidase being variously reported as either over- or under-expressed. Somewhat more encouragingly, however, Parmigiani *et al.* (2004) performed a cross-study comparison of several microarray studies that sought to refine lung cancer classification and concluded that there was significant agreement across studies in terms of gene expression patterns such that certain genes could be used reliably as markers to predict clinical outcomes.

In practice, relatively few of the individually studied genes in Table 5 showed up as significantly over-expressed (e.g. dihydrodiol dehydrogenase, wingless-type MMTV integration site, cyclins B1, D1 and E1, polo-like kinase, α-prothymosin and transcription factor E2F) in large-scale microarray studies (Table 6). Similarly, very few of the individually studied genes known to be inactivated in lung cancer by LOH, mutation or promoter hypermethylation (Tables 2 and 3) or whose expression is known to be down-regulated in lung cancer (Table 5), showed up as significantly under-expressed in microarray studies. This may be a question of the thresholds set in the studies involving large numbers of genes or it may reflect the common failure to confirm the initial results of microarray studies by, for example, techniques such as real-time PCR. Alternatively, perhaps the studies of the expression of individual genes have been better validated prior to publication. This notwithstanding, for now it would appear as if relatively few trends are apparent through comparison of the results generated by different studies.

As was the case for expression studies of individual genes at the mRNA level, microarray studies are blind to post-transcriptional processing. If post-transcriptional changes in mRNA stability were to turn out to be frequent, then the relatively crude mRNA levels measured *en masse* in microarray studies would not necessarily represent the levels of the encoded protein products in lung cancer cells and tissues. In any case, since expression profiling data may be regarded as "noisy", it would seem sensible to validate all data not only by providing adequate replicates (Yang and Speed 2002) but also by confirming all microarray-based conclusions independently by means of quantitative PCR (Bangur *et al.* 2002; Agathanggelou *et al.*, 2003b; Cho *et al.*, 2004; Whiteside *et al.*, 2004). Another approach to validating the results of any classificatory study based upon expression data is to repeat the process, blind, on an independent dataset. Even so, since conventional classification is still something of a hit-and-miss affair, the extent of misclassification is going to be difficult to assess and it is likely that the conventional and molecular classificatory tools will co-evolve in iterative fashion.

Some of the differences in expression observed in microarray experiments may be a consequence of the evolution of the tumour cells which may in turn have been driven by cellular changes brought about by radiotherapy or chemotherapy. Care must therefore be taken to compare like with like. It might thus be inappropriate to compare a profile derived from a primary tumour with a profile derived from cultured cells. More difficult are cases in which no obvious control is available. Thus, for example, it is unclear what control would be appropriate in the case of lung neuroendocrine tumours because the lung precursor cell for such tumours is unknown.

Probably the most serious problem facing those adopting the microarray approach is the sheer mass of data generated that makes the detection of truly significant changes in gene expression rather difficult. If the sensitivity is set too high, then the emerging data will be rendered meaningless by the subsequent number of positive results generated. If on the other hand the specificity is set too high, then subtle changes in the expression of key genes that may be critical for understanding lung tumorigenesis (Yan *et al.*, 2002a) will be missed and may even be undetectable. This notwithstanding, this type of approach holds out the very real promise not only of distinguishing SCLC from NSCLC at the level of gene expression (Petersen and Petersen 2001) but also of formulating a rational and empirical scheme for classifying the different sub-types of these very different forms of lung cancer (Sattler and Salgia 2003).

Finally, microarray studies have sometimes been criticized for being, by their very nature, "non-hypothesis driven". This is both unfair and incorrect for, as we have seen, a multitude of different hypotheses have already been tested by means of microarrays in the context of lung cancer. Indeed, the results of these studies aid the generation of new hypotheses every bit as much as they aid the testing of pre-existing hypotheses. Thus, as Greenberg (2001) opines, "as microarray technology develops, it will be important that disdain for 'nonhypothesis-driven' research does not prevent its potential benefits from being realised".

Gene expression profiling using approaches other than microarrays

Several other approaches to gene expression profiling that do not use microarrays have also been deployed in studies of human lung cancer (Liang and Pardee 2003).

Hibi *et al.* (1998) were the first to use *serial analysis of gene expression* (SAGE) to identify differentially expressed genes in a comparison of squamous cell carcinomas and normal bronchial tracheal epithelial cells. SAGE is a powerful method that allows the simultaneous quantitative analysis of a large number of different transcripts in a given cell type. In this technique, a large number of short DNA sequences (tags) derived from cDNA molecules and uniquely identifying transcripts from individual genes, are ligated together to form long concatemers and these concatemers are then sequenced. The relative representation of each transcript is indicated by the number of times a particular tag is encountered. Fujii *et al.* (2002) have attempted to make sense of NSCLC gene expression profiles generated by SAGE. Expressed genes were positioned onto chromosomes and a total of 43 and 55 clusters of differentially expressed genes were noted in squamous cell lung carcinoma and lung adenocarcinoma respectively. Gene numbers in each cluster varied between 18 and 78 (squamous cell carcinoma) and between 20 and 165 (adenocarcinoma) whilst cluster size varied from 1-65 megabases (Mb) [squamous cell carcinoma] and from 1.6-98 Mb (adenocarcinoma). The clusters of genes with increased expression in tumours manifested a 3-4 fold increase in expression as compared with a normal control whilst clusters of genes with reduced expression exhibited 50-65 % of the normal expression level. As far as the squamous cell carcinoma gene clusters were concerned, 9/15 with over-expressed genes corresponded to regions previously reported to be involved in lung cancer either by cytogenetic or LOH studies. The corresponding proportion was 13/28 for clusters with under-expressed genes. Fujii *et al.* (2002) concluded that a sizeable proportion of the gene clusters identified using this *transcriptome mapping* approach corresponded to regions of allelic imbalance (i.e. gene loss or amplification) in the original tumours. Inspection of Table 3 indicates that this correspondence may well be extended by consideration of gene inactivation through hypermethylation.

Differential display is another technique that is designed to allow the rapid and sensitive detection of altered gene expression by identifying cDNA sequences that are differentially expressed between two samples. Labelled cDNA fragments are generated by reverse transcription-PCR using an oligo-dT primer with a 2bp overhang and one arbitrary primer. The use of different primer combinations generates pools of cDNA fragments that represent subfractions of the transcribed sequences. PCR amplification products from equivalent samples are then compared and differentially expressed cDNAs identified. Manda *et al.* (1999) used differential display to isolate genes expressed at different levels in SCLC, NSCLC and normal lung cells.

Suppression subtractive hybridization (SSH) is a technique that combines normalization and subtraction in a single procedure; the normalization step serves to equalize the abundance of cDNAs within the target population whilst the subtraction step excludes the common cDNA sequences between the target and driver populations. SSH has been successfully used to identify differentially expressed genes in comparisons of a squamous cell lung carcinoma, a metastatic SCLC tumour and a metastatic lung adenocarcinoma cell line with normal bronchial epithelial cells (Petersen *et al.*, 2000a; Difilippantonio *et al.*, 2003; Sun *et al.*, 2004b). An *et al.* (2003)

used the same technique to identify differentially expressed genes in a bronchial epithelial cell line.

Finally, Wang *et al.* (2002a) used *representational difference analysis* (a subtractive hybridization method designed to permit the identification of differences between two genomes) on pooled mRNA derived from lung adenocarcinomas and their corresponding non-neoplastic lung tissues to identify genes that are overexpressed in this type of lung cancer.

Proteomic approaches to protein profiling

The proteome is much more complex than the genome since in excess of 500,000 proteins may be encoded by some 30,000 – 40,000 genes. The disparity may be accounted for in terms of the use of alternative promoters, splice sites and polyadenylation sites as well as variable post-translational modification and processing. The dissection of this complexity is a daunting task but has at its disposal a battery of new proteomic tools for protein separation, isolation, identification and quantification as well as techniques for determining protein structure, function, regulation and interactions. Such approaches include 2D polyacrylamide electrophoresis, mass spectrometry, multidimensional chromatography and protein or antibody arrays (Hanash 2003; Wulfkuhle *et al.*, 2003). Immunochemistry has been employed to compare the protein expression profiles of adenocarcinomas and squamous cell carcinomas (Volm *et al.*, 2002a; Cuezva *et al.*, 2004; Ullmann *et al.*, 2004), squamous cell lung carcinoma with and without lymph node involvement (Volm *et al.*, 2002b), lung adenocarcinoma from long-term survivors (Volm *et al.*, 2002c; Chen *et al.*, 2003h) and profiles indicative of drug resistance in NSCLC (Volm *et al.*, 2002d).

Campa *et al.* (2003) used matrix-assisted laser desorption/ionisation time-of-flight mass spectrometry (MALDI-TOF MS) to examine the protein expression profile in NSCLC and reported macrophage migration inhibitory factor (*MIF*) and cyclophilin A (*PPIA*) to be over-expressed. MALDI MS is capable of accurately determining the masses of large proteins due to its ability to ionise the molecules to be analysed without fragmentation. A time-of-flight analyser measures the time taken by ions to travel down a flight tube to a detector, the precise time depending on the mass/charge ratio. MALDI-TOF MS has also been used to generate protein profiles from small amounts of lung tumour tissue with the aim of identifying molecular markers, improving tumour classification, distinguishing between normal and neoplastic tissue, and between primary tumours and metastases, and making prognostic predictions (Yanagisawa *et al.*, 2003a; Li *et al.*, 2003c; Howard *et al.*, 2003; Alfonso *et al.*, 2004). Other workers have employed surface-enhanced laser desorption/ionization (SELDI) technology to generate protein expression profiles in both lung tumours and pre-malignant lung lesions (Zhukov *et al.*, 2003; Xiao *et al.*, 2004). For the future, protein microarrays hold out the promise of studying the post-translational modification of cellular proteins in lung cancer tissue on the scale of the entire proteome (Liotta *et al.*, 2003).

Hanash and colleagues have studied >1000 lung cancer cell/tissue samples by a combination of 2D polyacrylamide gel electrophoresis and MALDI-TOF MS and oligonucleotide microarrays (Hanash *et al.*, 2001) and constructed a lung proteomic database that integrates RNA and protein expression data (Oh *et al.*, 2001). This parallel transcriptomics/proteomics approach has demonstrated that only ~20%

of the proteins examined exhibited a significant correlation between protein and mRNA expression (Chen *et al.*, 2002c). This notwithstanding, the general approach employed by this group may still prove useful in identifying novel markers of diagnostic/prognostic significance in lung cancer as well as novel targets for therapeutic intervention (Beer *et al.*, 2002; Gharib *et al.*, 2002; Chen *et al.*, 2002b).

Finally, Zhong *et al.* (2004) have described 'antibody profiling' using *phage display* to identify proteins associated with NSCLC. Antibodies to tumour-associated proteins may have predictive value, particularly if used combinatorially. Protein microarrays also promise to be useful in this context (Qiu *et al.*, 2004).

Lung tumour cell ontogeny may be determined by gene expression pathways that recapitulate lung development

The currently dominant paradigm holds that lung tumours arise from the transformation of pleuripotent stem cells that are capable of differentiation into one or a number of different histological cell types (Reya *et al.*, 2001). Thus, lung tumour cell ontogeny is thought to be a consequence of the net effect of the activity of different genes whose activation or repression together serve to recapitulate the key events of embryonic lung development. The human lung is derived from the endodermal foregut by branching morphogenesis and alveolization concomitant with angiogenesis and vasculogenesis (Warburton *et al.*, 2000; *see* Chapter 2; Human Lung Development). The pathways involved in epithelial-mesenchymal-inductive signaling and epithelial cell differentiation (hedgehog/ patched/Gli/FGF/FGFR/ IGF/EGF/TGF) and some of the genes encoding the key transcription factors involved in these pathways (e.g. forkhead, GATA, POU, HOX, HLH, RAR) are now beginning to be identified (Warburton *et al.*, 2000; Cardoso 2001; *see* Chapter 2; Human Lung Development).

Borczuk *et al.* (2003) set out to test the hypothesis that genes expressed specifically in certain NSCLC histological classes would be associated with developmentally regulated genes and pathways. To this end, they employed oligonucleotide microarrays to generate gene expression profiles from 32 microdissected NSCLC tumours. The 100 top-ranked genes were identified for each tumour type viz. adenocarcinoma, squamous cell carcinoma, large cell carcinoma and carcinoid. The expression of murine orthologues of these genes during lung development was then determined by reference to publically available microarray-derived data. The adenocarcinoma gene set was found to be associated with gene expression in the terminal sac (E17.4 to P5) and alveolar (P5 to P30) stages of murine embryonic and post-natal lung development. By contrast, the large cell carcinoma gene set was associated with genes expressed in the rather earlier pseudoglandular (E9.5 to E16.6) and canalicular (E16.6 to E17.4) stages. These findings provide the basis with which to explain the distinction between the different types of NSCLC. The adenocarcinoma gene set encoded many proteins with roles in differentiation, signal transduction and cell adhesion including surfactant protein B (*SFTPB*), thyroid transcription factor (*TITF1*), and two retinoic acid-responsive proteins (*RAI3* and *RARRES3*). On the other hand, the large cell carcinoma gene set tended to be associated with undifferentiated cell growth and proliferation and cell cycle regulation. Some of these genes appear to be functionally associated e.g. *E2F3*, the E2F-responsive gene *MYBL2* and the upstream effectors of E2F3, *HDAC2, CDK4* and *PCNA*). In carcinoid, most genes encoded proteins associated with synaptic transmission and neurotransmitter maintenance e.g. *CPE, DTNA, PRKA1* and *KCNK3*. Finally, the squamous cell carcinoma genes encoded proteins involved in cell cycle maintenance, signal transduction and cell adhesion e.g. *EGFR* and *MAPK1* from the Ras/MAPK signal transduction pathway and *WNT5A* and *DVL3* from the Wnt (wingless)/β-catenin pathway. The identification of developmentally regulated pathways that are active during lung tumorigenesis and which serve to distinguish between histologically different types of NSCLC may go some way toward explaining prognostic differences between distinct types of NSCLC and is a prerequisite for the identification of histological type-specific biomarkers and targeted therapeutics.

CHAPTER 8

Lung Cancer Pathogenesis and Future Prospects for Treatment and Prevention

Towards an understanding of the pathogenesis of lung cancer at the molecular level

It might be thought that the search for genes involved in lung tumorigenesis will have been hampered by the comparative rarity of a familial predisposition to lung cancer. On the contrary, however, with the attention of geneticists largely focussed on the genetic, epigenetic and cytogenetic abnormalities to be found in lung tumours, a considerable number of genes have now been shown to be associated with lung cancer by virtue of their frequent (and sometimes very frequent) somatic mutation/inactivation in tumour tissue. These include oncogenes, tumour suppressor genes, genes involved in the regulation of the cell cycle or in the processes of apoptosis and DNA repair.

This notwithstanding, after two decades of research into the molecular genetics of lung cancer, we are still not yet in a position where we can be sure that we have identified all the key genes that play a role in the tumorigenic process. This is true regardless of whether these genes are defined in terms of the somatic mutations they harbour that promote cellular growth or resistance to apoptosis, or in terms of the inherited polymorphic variants that confer increased susceptibility to damage from specific exogenous mutagens (Wu *et al.*, 2004b). Whilst enormous progress has been made, there are still more questions than answers.

Losses of ~40 different chromosomal regions have been identified as being frequently associated with human lung tumorigenesis (Table 1). Somatic mutations in or involving at least 120 different human genes have so far been reported in lung cancer (Table 2); some of these genes are very commonly mutated and are therefore likely to be causal, others are much less frequently mutated and these lesions could therefore be consequential (although still potentially contributory). With relatively few exceptions, most subtle gene mutations characterized in lung cancer are solitary reports. Although those same tumours probably contain a sizeable number of other gene mutations in different combinations and permutations, such lesions have usually not been sought and therefore almost invariably still remain to be characterized. A proper understanding of the molecular and cellular biology of lung cancer will require much more complete descriptions of the underlying chromosomal abnormalities and gene mutations in each individual tumour or cell line as well as of the downstream consequences for gene expression on a genome-wide basis (Weber 2002; Strausberg *et al.*, 2003).

Epimutation, in the form of promoter hypermethylation, is known to be responsible for inactivating in excess of 50 different genes in human lung tumour tissue (Table 3) but this is likely to be a serious underestimate of the total number. Indeed, Costello *et al.* (2000) estimated that approximately 600 CpG islands (of the putative ~45,000 total in the human genome, most but not all of which are associated with gene promoters) are aberrantly methylated in the average tumour. An unknown proportion of these hypermethylated CpG islands will, in a given lung tumour, serve to inactivate genes that would normally encode proteins that promote genome stability, serve as growth regulators or which fulfil a proapoptotic function.

There are of course many more candidate genes still to be examined for evidence of mutation and/or epimutation in lung cancer: thus, of the ~30,000 human genes annotated to date, there are thought to be of the order of 900 proto-oncogenes, 500-800 genes encoding protein kinases, > 500 genes encoding molecules involved in cell adhesion, > 1800 genes encoding transcription factors and > 2300 genes encoding nucleic acid enzymes (Venter *et al.*, 2001; Manning *et al.*, 2002). Many of these could yet prove to be involved in lung tumorigenesis whether in cellular transformation or in tumour growth, tissue invasion, angiogenesis, evasion of host immunity or metastasis. A recent 'census of human cancer genes' has very conservatively counted 291 known 'cancer genes' (Futreal *et al.*, 2004) and a proportion of these (over and above those already recognized) may well turn out to play a role in lung tumorigenesis.

For those genes that have already been either associated with, or implicated in, lung tumorigenesis, a variety of different types of somatic mutation are apparent: missense and nonsense mutations, splicing mutations, promoter mutations, micro-deletions and micro-insertions, multi-exon deletions, gross deletions and gene amplification among others. We have also come to understand, at least in broad outline, some of the pathological mechanisms through which the transforming influence of these mutations is transduced: the signalling pathways by which lung cells are induced to grow abnormally and enabled to evade their usual growth constraints. It is, however, still unclear in many instances if the mutations found are causal rather than being merely consequential, and if causal, whether they are actually necessary for the tumorigenic process rather than merely contributory. It is also sometimes unclear whether the mutations detected have originated in the primary tumour, or whether they occurred subsequently during cell culture, or conceivably in some instances only after chemotherapeutic treatment (e.g. Kubo *et al.*, 1996; Mirski *et al.* 2000). Moreover, since it is very unusual for a particular tumour sample to be screened for mutations at more than two or three different gene loci (except rather indirectly by low-resolution loss of heterozygosity analysis), it follows that all examples of lung tumour tissue so far examined by molecular genetic methods must have been very inadequately described in terms of their net load of pathological lesions.

Whilst the identity of the loci mutated and the nature of the detected mutations (and/or epimutations) at these loci may be important for the rate of tumour progression, so may be the order in which these loci have been mutated or inactivated during tumorigenesis. Although very little information is as yet available about mutational order in lung tumorigenesis, studies of mutational order in other tumour types such as colorectal cancer (Kinzler and Vogelstein 1996; Tarafa *et al.*, 2003) and Barrett oesophagus (Barrett *et al.*, 1999) have provided important indicators as to the likely nature of the underlying principles as well as a glimpse of our eventual goal of acquiring a unified and unifying molecular description/explanation of lung cancer pathogenesis.

There are currently >260 different genes whose expression is known to be dysregulated in lung cancer, as evidenced by expression studies on individual genes (Table 5). This number is set to increase dramatically as new data from microarray studies emerge (Table 6) and the results are validated and confirmed. The investigation of the potential role in lung tumorigenesis of non-coding RNA molecules including microRNAs - very small non-coding RNA molecules that are thought to play key roles in regulatory networks - is eagerly awaited (McManus 2003; Takamizawa *et al.*, 2004). Gene expression profiling using oligonucleotide or cDNA microarrays holds out the promise of being able to determine the downstream consequences of specific mutations (whether chromosomal or gene) for cellular gene expression and ultimately in so doing, to define the process of lung tumorigenesis much more precisely in molecular terms (Luo *et al.*, 2003; Simon *et al.*, 2003; Macgregor 2003).

SCLC and NSCLC have been known for some time to be rather different in terms of their associated cytogenetic abnormalities as well as the genes involved and their underlying mutational spectra (Table 7). It is also now clear that both SCLC and NSCLC will turn out to be highly heterogeneous in terms of the downstream effects on the expression of cellular genes (see Tables 5 & 6). Microarray studies are thus likely to be instrumental in exploring and attempting to delineate the differences between SCLC and NSCLC and their subtypes at the level of gene expression, and in identifying the key pathognomonic features specific to these forms of lung cancer that will be so important for micro-classification. These new molecular genetic classificatory tools promise in turn to greatly increase the power of molecular epidemiological studies which are a prerequisite for the identification of a new generation of prognostic markers.

Prevention

> "Since, both in importance and in time, health precedes disease, so we ought to consider first how health may be preserved, and then how one may best cure disease".
>
> Claudius Galen ~180 AD *De Sanitate Tuenda*

Prevention, although clearly a topic well outwith the remit of this volume, must still receive a mention. Although in many contexts, 'prevention is better than cure' has become no more than a trite aphorism, it does still provide an apt description of the situation pertaining in lung cancer. As Dragnev *et al.* (2003) have succinctly summarized it: "Smoking prevention and smoking cessation represent an essential approach to reduce the societal impact of tobacco carcinogenesis". This notwithstanding, some workers are also exploring possibilities for chemoprevention in smokers, by dietary and/or pharmaceutical means (De Flora *et al.*, 2003; Petty *et al.*, 2003; Szabo and Kalebic 2003). We must remember that lung cancer also occurs in individuals who have never smoked and the disease would not therefore be completely eradicated even if the smoking of tobacco ceased immediately, completely and indefinitely.

Molecular genetics may prove directly useful in a preventive context in a number of different ways. Gene therapeutics (discussed below) is clearly one approach. Prevention might of course be interpreted to mean the early detection of lung cancer to ensure that opportunities for a beneficial outcome of clinical intervention are optimised (Mulshine 1999). Prevention may also refer to the preclinical identification of risk of lung cancer in order to potentiate the screening of those most at risk – this eventually reduces to the question of identifying reliable markers of genetic susceptibility (Marcy *et al.*, 2002). Finally, prevention could also refer to situations in which an individual who has received treatment for lung cancer is assessed in order to determine the risk of recurrence (Baksh *et al.*, 2003).

Novel treatment modalities offered by gene therapy

> "The art of medicine is to follow Nature, to imitate and assist her in the cure of diseases".
> Thomas Reid (1785) *Inquiry into the Human Mind on the Principles of Common Sense*

> "It is customary, in the study of any clinical subject, to conclude with a careful discussion of the treatment. The treatment of primary malignant growths of the lung has not required much discussion in the textbooks up to date. The diagnosis of a cancer of the lung was the death-warrant of the patient".
> Isaac Adler (1912) *Primary Malignant Growths of the Lungs and Bronchi*

The mainstays of lung cancer therapy have long been surgery, radiation and chemotherapy (Hoffman *et al.*, 2000). However, recent advances in our understanding of the molecular biology of lung cancer have guided the design and application of a number of new molecularly targeted therapeutic agents (Hoang *et al.*, 2002) and gene therapeutic approaches (Davies *et al.*, 2001; Albelda *et al.*, 2002; Daniel and Smythe 2003; Haura *et al.*, 2003a; 2003b; Roth and Grammer 2004) that are currently being explored in the search for the next generation of treatments (Giaccone 2002). A variety of new pharmaceuticals such as matrix metalloproteinase inhibitors, cyclin-dependent kinase inhibitors, farnesyltransferase inhibitors, epidermal growth factor inhibitors and modulators of protein phosphatases and kinases are undergoing testing but discussion of these is beyond the remit of this volume.

One aim of gene therapy has been to restore normal gene function, either by providing the wild-type gene to replace the properties of the wild-type product lost through mutation (as in the case of tumour suppressor genes) or by inactivating a gene that has become inappropriately active (as in the case of an activated oncogene). The potential of various other gene therapeutic approaches has also been explored e.g. the introduction of pro-apoptotic genes, or genes that stimulate the immune system such as interleukins 2 and 4 (Tan *et al.*, 1996; Swisher and Roth 2000; Davies *et al.*, 2001; Haura *et al.*, 2003a; 2003b; Park *et al.*, 2003a). Finally, the use of 'suicide genes' such as the herpes simplex virus thymidine kinase gene, which is introduced so as to confer sensitivity to ganciclovir, appears to be very promising. The SCLC tumour-specific expression of the suicide gene has been attempted by the use of the gastrin-releasing peptide gene (*GRP*) promoter (Inase *et al.*, 2000; Morimoto *et al.*, 2001).

The first attempt at gene replacement in lung cancer was reported by Roth *et al.* (1996). These workers injected a retroviral vector containing a wild-type *TP53* tumour suppressor gene into the NSCLC tumours of nine patients with advanced lung cancer. Tumour regression was noted in three cases whilst tumour growth was stabilized in three other cases. Post-treatment biopsies were found to contain vector-*TP53* sequences and displayed more apoptosis than pre-treatment biopsies. No vector-related toxic effects were noted. Further reports of the introduction of the *TP53* gene into NSCLC tumours using adenoviral vectors have since appeared (Schuler *et al.*, 1998; Swisher *et al.*, 1999; Weill *et al.*, 2000) whilst the adenoviral-mediated transduction of a number of other genes has been shown to induce apoptosis in lung cancer cells (Pataer *et al.*, 2003; Kim *et al.*, 2003a). Such studies have met with some degree of success in terms of anti-tumour activity and minimal toxicity but have the disadvantage of an omnipresent risk of an immune reaction to the adenoviral vector (Swisher and Roth 2002). Adeno-associated viral (AAV) vectors appear to exhibit a strong tropism for NSCLC cells and the transduction of a recombinant AAV-p53 vector into NSCLC cells has been shown to reduce tumour cell growth and to increase apoptosis (Rohr *et al.*, 2003).

TP53 gene transfer can induce radiation sensitization in previously radiation-resistant tumours. For this reason, a combination of adenoviral-mediated *TP53* gene transfer and radiation therapy has been attempted and lung tumour regression was successfully induced in a majority of cases (Swisher *et al.*, 2003). Importantly, several p53-regulated genes including *CDKN2A*, *BAK* and *MDM2* were up-regulated in response to the treatment. Adenoviral-mediated *TP53* gene transfer in conjunction with chemotherapeutic (e.g. cisplatin) administration has also been explored (Nemunaitis *et al.*, 2000; Schuler *et al.*, 2001) although Schuler *et al.* (2001) concluded that their phase II trial provided no additional clinical benefit to patients receiving chemotherapy for advanced NSCLC. In similar vein, the liposome-mediated transduction of the *FHIT* tumour suppressor gene was reportedly successful in suppressing growth in both primary and metastatic lung tumours (Ramesh *et al.*, 2001).

Countering the deleterious effects of an activated oncogene is potentially more difficult than replacing a wild-type gene sequence, but is in principle achievable using similar means. Lee *et al.* (2003c) have constructed adenoviruses that express dominant negative forms of the soluble extracellular domain of insulin-like growth factor I receptor (IGF-IR) containing engineered stop codons. Dominant negative effects on growth were noted in lung cancer cell lines transduced with these constructs as a consequence of competition between the defective receptor and the wild-type IGF-IR product.

Down-regulation of specific genes is also possible at the RNA level. Thus, Alemany *et al.* (1996) introduced an adenoviral vector containing a *KRAS* oncogene fragment in antisense orientation into lung cancer cells bearing a codon 61 *KRAS* mutation. The antisense *KRAS* gene fragment was efficiently expressed and served not only to reduce significantly the level of *KRAS* protein produced by the tumour cells, but also to reduce growth. Antisense genes work by blocking transcription or translation of the endogenous target gene by sequence-specific hybridisation to the cognate mRNA species. Much the same effect can be achieved by using antisense oligonucleotides. Indeed, a plethora of antisense oligodeoxynucleotides, ribozymes and double-stranded RNAs have now been designed for use as tools to target the biological pathways that are essential for lung tumorigenesis, blocking abnormal signal transduction or inhibiting growth factors or angiogenesis.

Antisense oligodeoxynucleotides designed specifically to inhibit the expression in human lung cancer cells of the thrombomodulin (*THBD*; Fujiwara *et al.*, 2002), RIα subunit of cAMP-dependent protein kinase (*PRKAR1A*; Wang *et al.*, 2002d), gastrin-releasing peptide receptor (*GRPR*; Langer *et al.*, 2002), *ERBB2* (Casalini *et al.*, 1997), *EGFR* (Washio *et al.*, 2003), *MYCL1* (Dosaka-Akita *et al.*, 1995), *SKP2* (Yokoi *et al.*, 2002), *IGF1R* and *IGF2* (Pavelic *et al.*, 2002) genes, have been reported to suppress cell growth. Further studies in human lung cancer cells have employed antisense oligodeoxynucleotides/constructs to target the *BCL2* (Ziegler *et al.*, 1997; Zangemeister-Wittke *et al.*, 2000), *MYC* (Akie *et al.*, 2000), checkpoint kinase 1 (*CHEK1*; Luo *et al.*, 2001), *MET* (Stabile *et al.*, 2004), XIAP (*BIRC4*; Hu *et al.*, 2003), *SKP2* (Yokoi *et al.*, 2003a), and survivin (*BIRC5*; Olie *et al.*, 2000) genes, thereby inducing apoptosis. Finally, the antisense-mediated down-regulation of some mRNAs has been employed to enhance the sensitivity of human lung cancer cells to chemotherapeutic agents (Kinzel *et al.*, 2002). Thus, the antisense-mediated suppression of the DNA repair gene Ku70 (*G22P1*) has also been shown to increase the radio- and chemo-sensitivity of cultured human lung cancer cells (Omori *et al.*, 2002). Qi and Martinez (2003) found that lung cancer cells could be induced to apoptose when exposed to ionising radiation by treatment with antisense 14-3-3 zeta (*YWHAZ*) implying that inhibition of 14-3-3 could be a useful therapeutic approach to sensitising lung tumours to radiation. Although such studies have been encouraging in terms of their minimal toxicity, the efficacy of the down-regulation of target mRNAs has been more limited (Flaherty *et al.*, 2001; Davies *et al.*, 2003).

Ribozymes are RNA molecules with catalytic activity that represent an alternative to antisense therapy by offering the advantage of their renewable capability for site-specific RNA cleavage. A ribozyme specific for RNA bearing the *KRAS* codon 12 mutation has been found to be a potent inhibitor of the growth of lung tumour cells (Zhang *et al.*, 2000c; 2000d; Tong *et al.*, 2001b). *Trans*-splicing ribozymes that can simultaneously reduce mutant p53 expression and restore wild-type p53 activity to cancer cells have also been reported (Watanabe and Sullenger 2000). Finally, a ribozyme has been successfully used to down-regulate vascular endothelial growth factor in a NSCLC cell line and appears to be capable of reducing vascularization of xenotransplanted lung tumours (Oshika *et al.*, 2000).

RNA interference (RNAi) is a cellular RNA degradation process that is activated when a double-stranded RNA (dsRNA) enters the cell, leading to the destruction of both the dsRNA molecule and its single-stranded endogenous counterpart (Shuey *et al.*, 2002). Since the artificial introduction of cognate double-stranded RNA molecules can be used to transcriptionally silence genes with key roles in tumorigenesis, a number of applications in the sphere of cancer therapy have become apparent (Borkhardt 2002; Agami 2002; Paddison and Hannon 2002). RNAi has already been used to target the polo-like kinase 1 (*PLK*) gene in a lung tumour cell line in order to reduce cellular proliferation and increase apoptosis (Spankuch-Schmitt *et al.*, 2002). RNAi has also been successfully used to inhibit both telomerase reverse transcriptase and human telomerase RNA, leading to reduced telomerase activity and decreased telomere length (Sarvesvaran *et al.*, 1999; Kosciolek *et al.*, 2003). Since RNAi may turn out to be more potent than antisense RNA in terms of reducing target gene expression in human cancer cell lines, its potential therapeutic importance cannot be understated. It is envisaged that combinations of many different

RNAi molecules will soon be deployed that will serve to inhibit both the growth and proliferation of lung cancer cells (Yin *et al.*, 2003).

Future prospects

> "We are at a crossroads at the turn of the century: the molecular detail of an entire cancer cell genome seems to be within our reach".
> D.P.Cahill, K.W.Kinzler, B.Vogelstein & C.Lengauer (1999) *Trends Cell Biol.* 9: M57-M59.

To be comprehensive, the future mutational profiling of lung cancer must combine seamlessly the battery of current techniques in molecular cytogenetics (including SNP and CGH arrays) with the mutation screening methodologies available to the molecular geneticist to ensure that the entire spectrum of somatic mutations in a given tumour is detectable, from changes in ploidy, gross chromosomal rearrangements and gene amplifications to the more subtle intragenic lesions (Popescu 2000; Garnis *et al.*, 2004; Pihan and Doxsey 2003; Zhou *et al.*, 2003c; Masayesva *et al.*, 2004). Since genetic and epigenetic events may well act in concert (Macaluso *et al.*, 2003), such analyses should also include 'methylation profiling': thus, we might envisage that the promoter regions of an appropriate set of tumour suppressor genes would be screened for hypermethylation to ensure that the epigenetic gene inactivation profile for a given lung tumour was also established (Yanagawa *et al.*, 2003; Feltus *et al.*, 2003; Chen *et al.*, 2003i; Fazzari and Greally 2004).

It is probably too early to speculate as to the likely combinations and permutations of different genetic lesions in lung cancer cells and their role in tumorigenesis. This will in any case be critically dependent upon tumour type and stage of progression, and the available data are extremely heterogeneous with respect to these parameters. This notwithstanding, loss of heterozygosity on specific chromosome regions is probably frequent enough in NSCLC (3p, 50-80%; 9p, 50-75%; 13q, 40-60%; 17p, ~70%) for any given NSCLC tumour to have a fairly high probability of loss of two or more of these regions (each probably containing at the very least many hundreds of genes). Since subtle mutations of the 17p-encoded *TP53* gene (40-60%) and the 9p-encoded *CDKN2A* gene (10-40%) are not uncommon in NSCLC, it also follows that a fair proportion of NSCLC tumours will possess two null alleles at these tumour suppressor loci. Some of the other gene mutations known to characterize NSCLC (e.g. amplification of a *MYC* gene family member in 5-10% of tumours, *PTEN* gene mutations in 8-17% of tumours) may of course play a facilitatory role as may, for example, the epimutational inactivation of the *RASSF1* and *RARB* genes (~40% of NSCLC tumours). Further, it is important to remember that these mutations and epimutations may often be superimposed upon a cellular background of increased tumorigenic potential conferred by the high frequency expression of telomerase in combination with genomic instability as well as compromised apoptotic and DNA repair machinery (Pihan and Doxsey 2003). The vast majority of the genes listed in Tables 2 and 3 are however inactivated in only a small proportion of lung tumours. Indeed, from the data to hand, it would appear that whereas there are relatively few genes that are very commonly mutated in NSCLC, there are probably a much larger number that are infrequently mutated. If the latter do indeed play a causal role in lung tumorigenesis as opposed to being merely consequential, then they could, by virtue of their number, still make a substantial net contribution.

In order to make sense of the very many different combinations of molecular lesions in lung cancer, we shall have to acquire a much more detailed knowledge of the basic mitogenic signaling, cell cycle control, DNA repair and apoptosis regulatory pathways, and the intricate complexities of their inter-relationships. Links between these pathways are potentiated by the myriad of self-regulating multi-functional proteins that act as caretakers, gatekeepers, and signal transducers. In some cases, the molecular lesions may exert their effects in additive fashion; in others the interactions will probably prove to be non-additive, perhaps even synergistic. It is not unreasonable to assume that mutations in proteins that play a role in the same biological pathway may, due to epistasis (Cordell 2002; Moore 2003), often be mutually exclusive in a particular tumour since no further selective advantage will accrue to neoplastic cells from the mutational inactivation of more than one gene per pathway. To what extent empirical data will confirm or refute this postulate is going to be crucially dependent upon the degree of multi-pathway functionality exhibited by the component proteins.

Microarray technology could eventually be applied to each and every tumour to provide expression profiles not only for purposes of tumour classification but also for individualized outcome prediction (Tomida *et al.*, 2004). Further, advances in microarray technology may even make it feasible to perform expression profiling down to the level of the single cell (Kawasaki 2004). Perhaps in the not too distant future, the lung cancer 'transcriptome' will be analysed by microarrays containing not only the ~40,000 or so basic gene sequences that make up the human genomic complement (or at least the lung cancer-relevant portion of them), but also the myriad of differentially spliced, initiated and polyadenylated mRNA (cDNA) products, many of which could yet prove to be clinically important (Bracco and Kearsey 2003; Lee and Roy 2004). In parallel, protein profiling (the 'proteomics' approach) will be necessary to ensure that all relevant post-translational protein modifications are covered and that the mRNA expression data are properly validated at the level of protein expression (Wulfkuhle *et al.*, 2003; Meyerson *et al.*, 2004). Finally, 'metabolomic' approaches, using mass spectrometry and high-resolution magnetic resonance imaging to study the metabolic phenotype (metabolites and low molecular weight intermediates) of tumours and tumour cells, should greatly improve our understanding of the molecular mechanisms underlying cancer (Griffin and Shockcor 2004). Taken together, such studies are essential for the development of new diagnostic and prognostic markers (Gunn and Smith 2004; Ornstein and Petricoin 2004) and new methods of risk assessment (Imyanitov *et al.*, 2004), and should open up new avenues for prevention, early detection and treatment (Etzioni *et al.*, 2003; Ntzami and Ioannidis 2003). However, the success of these initiatives hinges critically upon the efficacy of new computational/bioinformatics approaches that are being designed to analyse expression data from large numbers of genes or proteins simultaneously, and to look for higher order structure in the datasets in terms of both the clustering of functionally related genes and the identification of networks of interacting proteins (Kanehisa and Bork 2003; Rhodes and Chinnaiyan 2004).

The totality of the somatic mutational and epimutational changes experienced by the genome of cells in a particular tumour (an entity that could perhaps be termed the 'mutome'), or ultimately even a specific clonal cell population within the tumour, will serve not only to define the magnitude of departure from the original chromosomal and genomic architecture, but could also account for, and potentially even predict, its aberrant gene expression signature. I would argue that, in the future, knowledge of the 'mutome' and its cell-specific genome-wide consequences for gene/protein expression will not only be essential for identifying novel prognostic markers of lung cancer but should also inform and guide the development of new therapies, including bespoke personalized therapies tailor-made to fit the patient, and based upon their own individual genetic profiles. We are already beginning to see the emergence of targeted therapies that take into account the mutational profile of a given lung tumour. In the future, therapies may also be developed that anticipate the acquisition of further specific mutational lesions and their biological consequences in those residual cells that still survive in a post-treatment milieu, with a view to exploiting their unique properties and targeting their vulnerabilities.

The search for inherited polymorphic variants that could confer individual susceptibility to lung cancer will continue apace. Allelic variants in nearly 50 different genes have so far been found to display statistically significant associations with either an increased or a decreased risk of lung cancer. The majority of these genes encode xenobiotic metabolising enzymes or DNA repair enzymes. For these proteins, it is usually a fairly straightforward task to derive plausible hypotheses to account for the observed associations. However, whether or not these hypotheses are eventually supported by study replication and by subsequent functional analysis of the different protein isoforms *in vitro*, remains to be seen. For the remainder of the identified polymorphic variants, the mechanistic pathways through which they exert their influence on lung cancer risk are still unclear and remain to be elucidated. It may be that the magnitude of the risk conferred by each marker will turn out to be individually quite small. Nevertheless, the overall impact on disease prevalence will be influenced both by the population frequency of the different variants and by potential synergy between different non-allelic variants. Once validated, such markers of lung cancer susceptibility could be used to construct genetic risk profiles that would potentiate targeted intervention for those most at risk through, for example, recruitment to smoking prevention and cessation programmes.

Animal models promise to make a major contribution to the continuing search for lung cancer susceptibility genes (Malkinson 1998; Tuveson and Jacks 1999; Zhao *et al.*, 2000; Tripodis *et al.*, 2001). For example, human lung tumour cells may be transplanted into immunodeficient ('nude') mice to form tumours whose gene expression profiles can then be examined (Creighton *et al.*, 2003b). Linkage analyses using specific mouse strains can lead to the identification of novel lung cancer susceptibility loci (Devereux and Kaplan 1998; Tripodis *et al.*, 2001; Wang *et al.*, 2002e). Transgenic mouse models can allow the investigation of the phenotypic consequences both of the experimental over-expression of oncogenes ('knock-ins') and the loss of expression of tumour suppressor genes ('knock-outs'). Mutant genes may even be conditionally inactivated in murine lung epithelial cells, using the Cre/*lox*P site-directed recombination system, to produce mouse models of lung cancer ('conditional knock-outs') that can be induced in a tissue-specific and temporally controlled fashion [Meuwissen *et al.*, 2001; Jackson *et al.*, 2001]. These experimental models of human lung cancer promise to be invaluable tools in determining the pathways of lung tumour initiation and progression as well as exploring the potential for therapeutic intervention (Liu *et al.*, 1999; Fisher *et al.*, 2001b; Moorehead *et al.*, 2003; Kwak *et al.*, 2004).

Over the next decade, new molecular genetic imaging technologies are likely to come into their own in studies of tumorigenesis by permitting the spatial and temporal *in vivo* visualization of cellular processes at the molecular or genetic level. Such non-invasive imaging techniques may be applied to endogenous cellular processes such as signal transduction, protein-protein interactions and gene expression as well as to gene therapy monitoring, and will serve to complement existing *ex vivo* molecular biological assays that are currently dependent upon tissue sampling (Sullivan 2002; Contag and Bachmann 2002; Blasberg and Tjuvajev 2003). The ability to perform molecular imaging of the malignant phenotype in a particular patient at the molecular level and to monitor changes in that phenotype over time (including response to treatment; Tomida *et al.*, 2004), promises to revolutionize our ability to track the molecular and cellular consequences of mutation during lung tumorigenesis.

In "single gene" inherited disorders, we have come to realise that the path from mutant genotype to clinical phenotype progresses through multiple levels of control, each defined by dynamic yet self-regulating systems of interacting proteins. Lessons learned from the study of these monogenic conditions are now playing an important part in the unravelling of the molecular basis of cancer. In the process, na<i¨ve deterministic views have been replaced by an appreciation that genetic diseases are highly complex context-dependent entities whose expression is heavily modified and modulated, rather than wholly dictated by, the formative influence of genetic variation. In the coming years, we nevertheless anticipate continued progress toward improving our understanding of lung cancer by dissecting its cellular, cytogenetic, molecular genetic, epigenetic, transcriptional and protein-based heterogeneity (Sakatani and Onyango 2003). In reality, however, this is likely to remain, at least for the foreseeable future, a slow process of knowledge accumulation and accretion: an exercise in painstakingly cataloguing common features, differences, trends and themes. There is unlikely to be a single genetic Rosetta stone that will dramatically unlock the cardinal secrets of the disease. Indeed, it may be that the complexity of the system will ensure that there will always be features or properties of the tumour state that will prove refractory to analysis. The corollary to this is that we must beware a situation in which the anticipated veritable flood of genetic information from DNA and protein microarray studies serves to hamper our understanding and interpretation rather than to enhance it. In other words, as the once data-poor discipline becomes transformed into a data-rich one, our comprehension may, if we are not careful, suffer as much from a surfeit of scientific detail as it once did from paucity. Ultimately, however, we should appreciate that every lung tumour cell is likely to be unique in terms of the particular combination of the molecular properties that it possesses and displays. It is perhaps fitting to leave the last words with Isaac Adler who concluded his 1912 volume, *Primary Malignant Growths of the Lungs and Bronchi*, thus:

> "the writer may be permitted to express the hope that malignant disease of the lungs, so disastrous in its results, may perhaps in the near future be summarily dealt with in its incipiency, or at least modified in its progress, so as in some measure to assist in diminishing the sufferings of humanity".

Relevant Websites

American Lung Association:
http://www.lungusa.org/site/pp.asp?c=dvLUK9O0E&b=22555
[The American Lung Association is the oldest voluntary health organization in the United States. Founded in 1904 to fight tuberculosis, the American Lung Association today fights lung disease in all its forms, with special emphasis on asthma, tobacco control and environmental health].

Arylamine *N*-Acetyltransferase (NAT)**Nomenclature:**
http://www.louisville.edu/medschool/pharmacology/NAT.html

Atlas of Genetics and Cytogenetics in Oncology and Haematology:
http://www.infobiogen.fr/services/chromcancer/

Cancer Chromosomes
http://www.ncbi.nlm.nih.gov/entrez/query.fcgi?db=CancerChromosomes
Three databases, the NCI/NCBI SKY/M-FISH & CGH Database, the NCI Mitelman Database of Chromosome Aberrations in Cancer, and the NCI Recurrent Aberrations in Cancer, are now integrated into NCBI's Entrez system as "Cancer Chromosomes".

Cancer Genome Anatomy Project:
http://cgap.nci.nih.gov
[Includes genomic data for humans and mice, including transcript sequence, gene expression patterns, single nucleotide polymorphisms, clone resources and cytogenetic information. An extensive suite of informatics tools facilitates queries and analysis by the scientific community; Schaefer *et al.*, 2001].

CancerGene:
http://caroll.vjf.cnrs.fr/cancergene/HOME.html
[A relational database of genes involved in cancer including cellular genes associated with hereditary predisposition to cancer].

Cancer Genetics*Web*:
http://www.cancerindex.org/geneweb/X1501.htm
[Contains data on the molecular biology and genetics of lung cancer, specifically mutated genes and abnormal protein expression, recurrent chromosome abnorma-

lities, telomerase activity, smoking and lung cancer susceptibility, and clinical and epidemiological data].

Cancer Research UK:
http://www.cancerresearchuk.org
[Cancer Research UK supports over 500 research groups throughout the UK, through a variety of funding mechanisms including research institutes, clinical centres, programme and project grants. Website provides in-depth information on lung cancer, disease statistics and contact details of Cancer Research UK information nurses. It also provides information for lung cancer patients and their families about prevention, treatments, clinical trials, living with lung cancer and links for help and support].

Genetic Association Database:
http://geneticassociationdb.nih.gov
The Genetic Association Database is an archive of human genetic association studies of complex diseases and disorders including lung cancer (Becker *et al.*, 2004). This database is designed so as to allow the user to identify known medically relevant polymorphic variants from the published scientific literature. Study data are 'gene-centred' and recorded in the context of official human gene nomenclature with additional molecular reference numbers and links.

House of Commons Health Select Committee's Second Report on the Tobacco Industry and the Health Risks of Smoking, 5th June 2000:
http://www.parliament.the-stationery-office.co.uk/pa/cm199900/cmselect/cmhealth/27/2702.htm

Human Cytochrome P450 (*CYP*) Allele Nomenclature Committee:
http://www.imm.ki.se/CYPalleles/

Human DNA repair genes:
http://www.cgal.icnet.uk/DNA_Repair_Genes.html

Human Gene Mutation Database:
http://www.hgmd.org
[HGMD constitutes a comprehensive core collection of data on germ-line mutations in nuclear genes underlying or associated with human inherited disease (Stenson *et al.*, 2003). Data catalogued include single base-pair substitutions in coding, regulatory and splicing-relevant regions; micro-deletions and micro-insertions; indels; triplet repeat expansions as well as gross deletions; insertions; duplications; and complex rearrangements. By July 2004, the database contained in excess of 48,000 different lesions detected in $>$ 1,900 different nuclear genes, with new entries currently accumulating at a rate exceeding 5,000 per annum].

IARC *TP53* Mutation Database:
http://www.iarc.fr/p53/
[Database of somatic and germline mutations in the human *TP53* gene. The latest update (R9) from June 2004 contains 19,809 somatic mutations and ~2500 mutations pertaining to lung cancer].

Lung Cancer Resources Directory:
http://www.cancerindex.org/clinks2l.htm

Mitelman Database of Chromosome Aberrations in Cancer:
http://cgap.nci.nih.gov/Chromosomes/Mitelman
[Contains data that relate chromosomal aberrations to specific tumour characteristics in individual patient cases. Chromosomal aberrations and tumour histologies are related to genomic sequence data, typically genes rearranged as a consequence of structural chromosome changes. Clinical associations that relate chromosomal aberrations and/or gene rearrangements and tumour histologies to clinical variables, such as prognosis, tumour grade, and patient characteristics are highlighted].

Mouse Models of Human Cancers Consortium, National Cancer Institute (Murine lung cancer models):
http://emice.nci.nih.gov/emice/mouse_models/organ_models/lung_models/murinecancer/murinemodels

National Cancer Institute (Lung Cancer Home Page):
http://www.cancer.gov/cancerinfo/types/lung
[Information on lung cancer treatment, prevention, genetic causes, screening, testing, clinical trials, the scientific literature, and statistics].

National Cancer Institute (Surveillance, Epidemiology and End Results):
http://seer.cancer.gov
[The Surveillance, Epidemiology, and End Results (SEER) Program of the National Cancer Institute is an important source of information on cancer incidence and survival in the United States. The SEER Program currently collects and publishes cancer incidence and survival data from 14 population-based cancer registries and three supplemental registries covering ~26% of the US population. Information on more than 3 million *in situ* and invasive cancer cases is included in the SEER database, and approximately 170,000 new cases are added each year within the SEER coverage areas. The SEER Registries [http://seer.cancer.gov/registries] routinely collect data on patient demographics, primary tumour site, morphology, stage at diagnosis, first course of treatment, and follow-up for vital status].

National Center for Biotechnology Information (NCBI) Human Genome Resources:
http://www.ncbi.nlm.nih.gov/genome///guide/human/
[NCBI's Web site provides "an integrated, one-stop, genomic information infrastructure" for biomedical researchers].

Online Mendelian Inheritance in Man (OMIM):
http://www3.ncbi.nlm.nih.gov/omim
[This database is a catalogue of human genes and genetic disorders authored and edited by Dr. Victor A. McKusick and his colleagues at Johns Hopkins and elsewhere, and developed for the World Wide Web by the National Center for Biotechnology Information (NCBI). The database contains textual information and references. It also contains links to PubMed, sequence records in the Entrez system, and additional related resources. OMIM is intended for use primarily by physicians and other professionals concerned with genetic disorders, by genetics researchers, and by advanced students in science and medicine].

p53 Mutation Database Analysis & Search:
http://p53.genome.ad.jp/

PubMed (NCBI):
http://www.ncbi.nlm.nih.gov/entrez/query.fcgi
[PubMed is an online service of the National Library of Medicine which allows the user to search for specific keywords among the references/abstracts of over 14 million biomedical articles going back to the 1950s].

Roy Castle Lung Cancer Foundation:
http://www.roycastle.org/charity/
[The Centre is dedicated to the study and eventual elimination of lung cancer. Its principal activity is the Liverpool Lung Project (Field and Youngson 2002; http://www.roycastle.org/research/lungproject.htm), a 20-year programme designed to identify and eradicate the risk factors associated with lung cancer and also to develop screening programs in high-risk individuals which would enable earlier detection of the disease. The project intends to use genetic and epidemiological risk factors to identify populations and individuals most at risk of developing lung cancer].

Surgeon General's Reports (US Department of Health & Human Services, Centers for Disease Control and Prevention):
http://www.cdc.gov/tobacco/sgr/index.htm
[Contains all Surgeon General's Reports including "Smoking and Health" (1964 and 1979), "The Health Consequences of Smoking" (1982), "Reducing the Health Consequences of Smoking" (1989), "Smoking and Health: A National Status Report" (1990), "The Health Benefits of Smoking Cessation" (1990), "Reducing Tobacco Use" (2000), "Women and Smoking" (2001) and "The Health Consequences of Smoking" (2004).

Tumor Gene Database:
http://condor.bcm.tmc.edu/ermb/tgdb/tgdb.html
[Designed as a tool for biomedical researchers and contains information about genes that are targets for cancer-causing mutations and their encoded proteins].

UK Department of Health Report of the Scientific Committee on Tobacco and Health (1998):
http://www.archive.official-documents.co.uk/document/doh/tobacco/contents.htm

UK Office of National Statistics:
http://www.statistics.gov.uk
[Official UK statistics, including cancer incidence, survival and mortality; latest data from 2001].

UMD-p53 TP53 **database:**
http://p53.curie.fr

World Health Organization (WHO):
http://www.who.int/cancer/en
The World Health Organization, established in 1948, is the United Nations specialized agency for health. The website includes information on national cancer control programmes, prevention and treatment, screening, early detection and palliative care, as well as statistics on cancer incidence, mortality and survival.

Glossary

Adduct: the product formed by the chemical addition of one substance to another

Allele: one of several alternative forms of a gene or DNA sequence at a specific chromosomal location.

Alternative splicing: the use of different sets of splice junction sequences to produce more than one mRNA product from a given gene

Amino acid: one of the monomeric units of a protein molecule.

Anaphase: phase of mitosis in which the chromatids separate and move to opposite poles

Aneuploidy: a chromosome constitution with one or more chromosomes extra or missing by comparison with the normal karyotype

Angiogenesis: the growth of new blood vessels

Anoikis: apoptosis of cells after loss of contact with extracellular matrix.

Antisense: (RNA) a transcript that is complementary to a normal mRNA based on the non-template strand of a gene, (DNA) strand of a gene which is used during transcription as a template for mRNA synthesis.

Apoptosis: programmed cell death.

Association: the tendency of two characteristics (disease, marker, allele) to occur together at a frequency that cannot be attributed to chance alone.

Autocrine: action of a substance that binds to a surface receptor of the same cell that secreted it.

Autosome: chromosome other than a sex chromosome

Base-pair: hydrogen bonded structure formed by two complementary nucleotides in a double stranded DNA molecule.

cDNA: double stranded DNA copy of an mRNA molecule

Cell cycle: the series of events occurring in a cell between one division and the next.

Cell cycle checkpoint: period before entry into S or M phase of the cell cycle during which regulation may be exerted

Centromere: constricted region of a chromosome where the pair of chromatids is held together

Centrosome: a cytoplasmic region surrounding a pair of centrioles that are required for somatic cells to progress through G_1 and into the S phase of mitosis.

Chromatin: complex of DNA and histones found in chromosomes

***Cis*-acting:** linked on the same DNA molecule

Codominance: relationship between a pair of alleles which both contribute to the phenotype in a heterozygote.

Codon: nucleotide triplet that specifies an amino acid or a Stop signal.

Comparative genomic hybridisation (CGH): technique that involves the use of competitive fluorescence *in situ* hybridisation to detect chromosomal regions that are amplified or deleted.

Compound heterozygous: an individual is compound heterozygous at a locus if they have two different mutations, one on each allele

CpG island: GC-rich DNA region located upstream of many human genes.

Cre/*lox*P: a technique for generating pre-defined chromosomal deletions using the Cre protein that facilitates recombination between *lox*P sequences.

Cryptic splice site: a sequence in pre-mRNA that has some homology with a splice site and which is used by default in the absence (through mutation) of the normal splice site.

Cyclins: type of cell cycle regulatory protein

Cytokines: proteins involved in the regulation of cellular proliferation.

Cytokinesis: cytoplasmic division

Desmosome: one of the strong points anchoring cells to each other

Differential display: a technique that is designed to allow the rapid and sensitive detection of altered gene expression by identifying cDNA sequences that are differentially expressed between two samples.

Differentiation: the process by which cells become committed to being mature and specific cell types with specialized functions

Diploid: having two copies of each type of autosome.

EMAST: elevated microsatellite alterations at selected tetranucleotide repeats

Enhancer: sequence element that stimulates transcription of a gene and whose function is not critically dependent upon its position or orientation.

Epimutation: heritable change but not involving a change in DNA sequence.

Epistasis: the non-reciprocal interaction of non-allelic genes allowing one gene to mask the expression of another.

ERKs: extracellular-regulated kinases

Exon: a segment of a gene that is represented in the mature mRNA product.

Fluorescence *in situ* hybridisation: a form of molecular hybridisation which uses fluorescently labelled DNA as a probe to examine the target nucleic acid which is denatured DNA within chromosome preparations

Fragile site: a location in a chromosome that is prone to frequent breakage often because it contains an expanded trinucleotide repeat.

Frameshift mutation: mutation resulting from insertion or deletion of nucleotides that is not a multiple of three and which therefore changes the translational reading frame

Fusion protein: protein that consists of a fusion of two polypeptides normally encoded by separate distinct genes

Gene therapy: addition of a functional gene to a cell, whether temporally or permanently, to correct an hereditary or acquired genetic defect

Genome: the entire genetic complement

Genotype: a description of the genetic composition of a particular individual in terms of genetic variation at one or more loci.

Germline: sperm and egg cells and those cells which give rise to them.

Haploinsufficiency: a locus exhibits haploinsufficiency if a requirement of a normal phenotype is more gene product than a single copy of that locus can produce.

Haplotype: a specific combination of alleles inherited together on the same chromosome

Hemizygous: an individual is hemizygous at a locus if they have only one copy of an allele at that locus

Heterodimer: protein made up of two different polypeptide chains

Heterozygous: an individual is heterozygous at a locus if they have two different alleles at that locus

Homodimer: protein made up of identical polypeptide chains

Homozygous: an individual is heterozygous at a locus if they have two identical alleles at that locus

Hotspot: a sequence associated with an abnormally high frequency of recombination or mutation

Hybridization: the attachment to one another by base-pairing, of two complementary polynucleotides

Hyperplasia: an increase in the amount of tissue, produced by an increase in the number of cells

Immunocytochemistry: use of an antibody probe to identify or locate a specific protein

Imprinting: determination of the expression of a gene by its parental origin

In *cis*: on the same chromosome in a given individual

In *trans*: on the other chromosome in a given individual

Interphase: all the time in the cell cycle when the cell is not dividing

Intron: non-coding region within an otherwise discontinuous gene

Isoform: alternative form of a protein

Karyotype: a summary of the chromosome constitution of a cell or a person

Kinase: enzymes that catalyse the transfer of a phosphate group to a substrate

Knock-in: a targeted mutation that replaces the activity of one gene by that of an introduced gene

Knock-out: the targeted inactivation of a gene within the cell/organism

Ligase: an enzyme that synthesizes a phosphodiester bond at the ends of two DNA molecules to join them together

Linkage analysis: statistically-based study of the tendency for two genes or markers that lie on the same chromosome to co-segregate in a family pedigree/ procedure used to assign map positions to genes by genetic crosses.

Linkage disequilibrium: a statistical association between particular alleles at separate yet linked loci, normally a consequence of an ancestral haplotype being common in the population under study.

Locus: a specific chromosomal location of a genetic/DNA marker

Loss of heterozygosity: loss of alleles at numerous linked loci

Matrix-assisted laser desorption/ionisation (MALDI): a mass spectrophotometric method used for identifying protein molecules.

Marker, genetic: any polymorphic character that can be used to follow a chromosomal segment through a multi-generational pedigree.

Megabase-pair: 1,000 kilobases or 1,000,000 base-pairs

Messenger RNA: transcript of a protein-coding gene

Metabolomics: study of the metabolic phenotype e.g. metabolites and low molecular weight intermediates

Metaphase: stage of cell division (mitosis or meiosis) when chromosomes are maximally contracted and aligned on the 'metaphase plate' within the cell.

Metastasis: the spread of malignant neoplastic cells from the original site to another part of the body

Methylation: post-synthetic modification of cytosine in DNA by addition of a methyl group

Microarray: thousands of oligonucleotides or cDNA sequences deposited robotically onto a solid support which can be hybridized with mRNA to determine patterns of gene expression on a genome-wide basis.

Microsatellite: short run of tandem repeats of a very simple (1-4 bp) DNA sequence

Microsatellite instability: phenomenon of high frequency random changes in microsatellite repeat copy number

Microtubule: long hollow cylinders made from tubulin that form part of the cytoskeleton

Minisatellite: an intermediate size array of short tandemly repeated (0.1-20 kb) DNA sequences

Mismatch repair: enzymatic process that replaces a mis-paired nucleotide in a DNA duplex

Missense mutation: nucleotide substitution that results in an amino acid exchange

Modifier gene: a gene whose expression can influence a phenotype resulting from mutation at another locus.

Mutation: an alteration in the nucleotide sequence of a DNA molecule

Mutational spectrum: representative collection of mutations observed in a given disease state

Neoplasia: process by which a localized population of cells becomes independent of normal cellular growth controls and proliferates

Nonsense mutation: nucleotide substitution that introduces a premature stop codon.

Northern analysis: a membrane bearing size-fractionated RNA molecules used as a target in a hybridisation assay to determine the size or relative amount of a specific mRNA transcript

Nucleosome: structural unit of chromatin

Null allele: a mutant allele that produces no protein product

Odds ratio (OR): ratio of odds of disease among individuals with a specific allele or genotype compared to those without.

Oncogene: a gene involved in the control of cellular proliferation which when overactive, can help to transform a normal cell into a tumour cell.

Orthologue: member of a set of homologous genes in different species

Paracrine: action of a secreted substance that binds to a surface receptor of a nearby cell.

Penetrance: the frequency with which a given genotype manifests itself as a given phenotype

Phage display: expression cloning method in which foreign genes are inserted into a phage vector and are expressed to yield polypeptides that are displayed on the surface of the phage.

Phenotype: the observable characteristics of a cell or organism

Point mutation: single base-pair substitution

Polyadenine (polyA) tract: series of adenine residues

Polymerase chain reaction (PCR): technique which permits the selective amplification of a specific target DNA sequence from a heterogeneous collection of DNA sequences.

Polymorphism: Mendelian trait that exists in the population in at least two phenotypes, neither of which occurs at a frequency of less than 1%

Primer: a short oligonucleotide that base-pairs specifically to a target sequence to allow a polymerase to initiate synthesis of a complementary strand.

Probe: a known DNA or RNA fragment (or collection of different fragments) used in an hybridisation assay to identify closely related DNA or RNA sequences within a complex mixture of nucleic acid species.

Promoter: regulatory region of a gene

Prophase: phase of mitosis in which the chromosomes become visible within the nucleus as they coil up.

Proteome: the total set of different proteins in a cell, tissue or organism.

Proteomics: techniques and approaches used to study the proteome.

Pseudogene: a DNA sequence which shows a high degree of sequence homology to a non-allelic functional gene but which is itself non-functional.

Purine: guanine or adenine

Pyrimidine: cytosine or thymidine

Real-time PCR: A kind of quantitative PCR which amplifies nucleic acid sequences and simultaneously measures their concentrations.

Recessive: a character is recessive if it is manifest only in the homozygote

Relative risk (RR): an estimate of the magnitude of the association between exposure and disease that indicates the likelihood of developing the disease in the exposed group relative to those who are not exposed.

Representational difference analysis: a subtractive hybridization method designed to permit the identification of differences between two genomes.

Reverse transcriptase: an enzyme that can synthesize a DNA strand using an RNA template.

Reverse transcript PCR: Polymerase chain reaction that utilizes RNA as starting material with an initial reverse transcription step to produce cDNA.

Ribozyme: an RNA molecule with catalytic activity

RITEs: regions of increased tumour expression

RNA interference: the use of small interfering RNA to reduce the expression of a specified target gene.

Senescence: process of cellular aging

Serial analysis of gene expression (SAGE): a method that allows the simultaneous quantitative analysis of a large number of different transcripts in a given cell type.

Somatic: cells other than the gametes/germline

Splice site: junction between the end (3' end) of an intron and the start of the next exon (acceptor splice site) or the junction between the end of an exon and the start (5' end) of the next intron (donor splice site)

Splicing: removal of introns from primary transcript of a discontinuous gene

Stop codon: mRNA codon that signals the end of the polypeptide.

Suppression subtractive hybridization (SSH): a technique that combines normalization and subtraction in a single procedure; the normalization step serves to equalize the abundance of cDNAs within the target population whilst the subtraction step excludes the common cDNA sequences between the target and driver populations.

Telomerase: Ribonucleoprotein enzyme with reverse transcriptase activity that synthesizes the G-rich strands of telomeres, thereby sealing the ends of chromosomes.

Telomere: end of a eukaryotic chromosome

Telophase: stage of mitosis during which the spindle disappears and nuclear envelopes reform around the sets of progeny chromosomes

Topoisomerase: enzyme that introduces or removes turns from the DNA double helix by breakage and reunion of one or both polynucleotides

Toxicogenomics: Genome-wide search for changes in gene expression associated with exposure to toxic substances.

Transactivation domain: domain in transcription factor that is involved in transcriptional activation

Transcription: the synthesis of RNA on a DNA template by RNA polymerase

Transcriptome: the entire genetic complement represented at the mRNA level

Transcriptome mapping: A method which attempts to identify chromosomal regions in which clusters of genes exhibit non-random increased expression in tumour samples as compared to normal samples.

Transgenic: an animal in which an artificially introduced foreign DNA sequence becomes stably incorporated into the germline

Transition: a purine to purine or pyrimidine to pyrimidine substitution

Translocation: transfer of a chromosomal region between non-homologous chromosomes

Transversion: a purine to pyrimidine substitution or *vice versa*

Tumour suppressor: gene whose normal function is to inhibit or control cell division

Two-hit hypothesis: theory that hereditary cancers require two successive mutations within the same tumour cell.

Ubquitination: attachment of ubiquitin which marks a protein for degradation

Variable number tandem repeat (VNTR): array of tandemly repeated sequences

Vector: self-replicating DNA molecule that transfers a DNA segment between host cells

Xenobiotic: biochemical whose origin is from outside of the body

Tables

Table 1. Cytogenetic alterations in lung cancer (adapted from Kaye and Kubo 2001)

Small cell lung cancer		Non-small cell lung cancer		
1p		1p10-p13	1q11	
2q11-q13	2q33	2p	2q	
3p12-p23		3p14-p25	3q23-q27	
4p	4q	4p	4q	
5q13-q21	5q33-q35	5p13	5q11-q14	5q21
6p	6q	6p	6q15-q27	
		7p12-p15	7q	
8p21-p23		8p21-p23		
9p	9q	9p21	9q32	
10q		10q23-q26		
		11p13-p15	11q12-q23	11q23-q24
12p		12p		
13q14		13p	13q	
		14p	14q	
15q		15p	15q	
16q		16q24		
17p13		17p11-p13	17q11	
18q		18q		
		19p	19q13	
		21p	21q11-q12	
		22q		
		Yq		

Table 2. Genetic loci known to harbour somatic mutations in human lung cancer

	Gene/protein	Gene symbol	Chromosomal location	Types of lung cancer examined	Mutational lesions detected	References
Oncogenes and putative oncogenes	N-Ras	*NRAS*	1p13	Missense mutation in SW-1271 lung carcinoma cell line	Arg61Gln	Yuasa *et al.* (1984)
				Amplification in 2/137 NSCLC tumours		Shiraishi *et al.* (1989)
				Missense mutation in adenocarcinoma	Ser65Arg	Brose *et al.* (2002)
	L-*myc*	*MYCL1*	1p34	Gene amplification of a *myc* gene family member in < 10 % NSCLC and 20-50 % SCLC tumours, and in 50-83 % SCLC cell lines In some cases of SCLC, amplification of the *MYCL1* gene is associated with fusion of *MYCL1* gene to *RLF* gene which yields a novel chimeric transcript	~4-35 fold amplification	Nau *et al.* (1985) Johnson *et al.* (1987) Shiraishi *et al.* (1989) Gazzeri *et al.* (1990) Brennan *et al.* (1991) Mäkelä *et al.* (1992) Mäkelä *et al.* (1995) Harder *et al.* (1995) Kim *et al.* (1998c) Dai *et al.* (2003)
	Paired box protein PAX7	*PAX7*	1p36.2-p36.12	*PAX7* gene amplified in 5/37(13 %) cases of squamous cell carcinoma		Racz *et al.* (2000)
	N-*myc*	*MYCN*	2p24	Gene amplification of a *myc* gene family member in < 10 % NSCLC and 20-50 % SCLC tumours, and in 50-83 % SCLC cell lines	~5-170 fold amplification	Nau *et al.* (1986) Saksela *et al.* (1986) Kiefer *et al.* (1987) Noguchi *et al.* (1990) Gazzeri *et al.* (1990)
	Raf-1	*RAF1*	3p25	*RAF1* gene amplified in 1/42 NSCLC cell lines.	3-fold amplification	Hajj *et al.* (1990)
				Missense mutations	Amino acid residue 533	Storm and Rapp (1993)
	Tumour protein p73-like, p73L (p63/p40/p51)	*TP73L*	3q27	Frameshift mutation in 1/44 lung cancer cell lines.		Tani *et al.* (1999)
				Missense mutations in 2 % NSCLC tumours		Sunahara *et al.* (1999)
				Gene amplification in 10/10 squamous cell carcinomas and 2/13 (15 %) lung adenocarcinomas		Hibi *et al.* (2000)
				Gene amplification in 88 % squamous cell carcinomas, 42 % large cell carcinomas		Massion *et al.* (2003)

	Gene/protein	Gene symbol	Chromosomal location	Types of lung cancer examined	Mutational lesions detected	References
Oncogenes and putative oncogenes	Macrophage stimulating 1 receptor (RON protein tyrosine kinase)	*MST1R*	3p21.1	Missense mutation in adenocarcinoma	Leu296Pro	Angeloni *et al.* (2000)
	Phosphatidylinositol 3-kinase, catalytic, alpha subunit	*PIK3CA*	3q26.3	*PIK3CA* gene amplification in squamous cell lung carcinomas.		Racz *et al.* (1999) Massion *et al.* (2002)
				Missense mutation (Glu545Lys) in 1/24 'lung cancers'	Missense mutation occurs in evolutionarily conserved residue within helical domain of PIK3CA. Thought to increase kinase activity.	Samuels *et al.* (2004)
	Phosphatidylinositol 3-kinase, catalytic, beta subunit	*PIK3CB*	3q21-qter	*PIK3CB* gene amplification in squamous cell lung carcinomas		Massion *et al.* (2002)
	Translation initiation factor eIF4γ1	*EIF4G1*	3q27	*EIF4G1* gene amplification in NSCLC		Brass *et al.* (1997) Bauer *et al.* (2001) Rosenwald *et al.* (2001)
	Tyrosine kinase receptor, Kit	*KIT*	4q12	Heterozygous missense mutation in 1/28 SCLC tumours/cell lines (neutral polymorphism?)	Met541Leu.	Sekido *et al.* (1993)
				Heterozygous *KIT* missense mutations detected in 5/60 SCLC tumour samples. Unclear if of functional significance (polymorphisms?)	Asn495Ile (2) and Asn567Lys (3) in extracellular domain.	Boldrini *et al.* (2004)

Table 2. Genetic loci known to harbour somatic mutations in human lung cancer (continued)

	Gene/protein	Gene symbol	Chromosomal location	Types of lung cancer examined	Mutational lesions detected	References
Oncogenes and putative oncogenes	S-phase kinase-associated protein 2 (p45)	*SKP2*	5p13	*SKP2* gene amplified in 7/16 (44%) primary SCLC tumours. *SKP2* gene over-expressed in 83% tumours with *SKP2* gene amplification		Yokoi *et al.* (2002)
				SKP2 gene aberrations found in 88% NSCLC cell lines and 65% primary NSCLC tumours; in adenocarcinoma, LOH around *SKP2* locus found in 35% tumours whilst *SKP2* gene amplification noted in 13% tumours		Zhu *et al.* (2004a)
				SKP2 gene amplified in 5/25 (20%) NSCLC cell lines		Yokoi *et al.* (2004)
	Cadherin 6	*CDH6*	5p14-p15.1	*CDH6* gene homozygously deleted in SK-LU-1 lung cancer cell line		Teng *et al.* (2001)
	MYB	*MYB*	6q22	3 *MYB* gene deletions and one *MYB* gene amplification in 27 NSCLC cell lines.	–	Cline and Battifora (1987)
				20-25-fold amplification in SCLC cell line		Kiefer *et al.* (1987)
	MET	*MET*	7q31	Identical missense mutation (Arg988Cys) in 2/10 SCLC cell lines. Glu168Asp and Thr1010Ile missense mutations in 2/32 SCLC tumours	Arg988Cys and Thr1010Ile substitutions occur in juxtamembrane region whilst Glu168Asp mutation occurs in Sema domain.	Ma *et al.* (2003)
	BRAF kinase	*BRAF*	7q34	Missense mutations (5) in NSCLC	Val458Leu, Lys438Thr, Thr439Pro, Leu596-Val, Val599Glu	Brose *et al.* (2002)
				Missense mutations (2) in adenocarcinoma	Gly465Val, Leu596Arg	Naoki *et al.* (2002)

	Gene/protein	Gene symbol	Chromosomal location	Types of lung cancer examined	Mutational lesions detected	References
Oncogenes and putative oncogenes	Epidermal growth factor receptor	*EGFR*	7p12	*EGFR* gene amplified in ~6% NSCLC tumours	Up to 20-fold amplification	Hunts *et al.* (1985) Cline and Battifora (1987) Shiraishi *et al.* (1989) Taguchi *et al.* (1997) Reissmann *et al.* (1999) Franklin *et al.* (2002)
				EGFRvIII, a variant found in a number of different cancers including NSCLC, is caused by the in-frame deletion of exons 2-7 of the *EGFR* gene corresponding to the extracellular ligand-binding domain	Probably caused by homologous recombination between an *Alu* repeat in intron 7 and *Alu*-like sequences in intron 1	Garcia de Palazzo *et al.* (1993) Okamoto *et al.* (2003)
				Heterozygous activating mutations (missense & in-frame deletions) in the tyrosine kinase domain in 2/25 NSCLC patients, 8/9 gefitinib-responsive patients, and 0/7 gefitinib non-responsive patients.		Lynch *et al.* (2004)
				Heterozygous activating mutations in 15/58 tumours from Japanese NSCLC patients and 1/61 American NSCLC patient tumours. Tyrosine kinase domain mutations found in 5/5 gefitinib-responsive patients as compared with 0/4 gefitinib non-responsive patients	Leu858Arg within activation loop. Multiple in-frame deletions of 9-24 bp spanning codons 746-759, within the kinase domain of EGFR	Paez *et al.* (2004)
	PTK protein tyrosine kinase (Fak)	*PTK2*	8q24-qter	Gene amplification (3->10 fold) in 8/11 'lung cancer' cell lines		Agochiya *et al.* (1999)

Table 2. Genetic loci known to harbour somatic mutations in human lung cancer (continued)

	Gene/protein	Gene symbol	Chromosomal location	Types of lung cancer examined	Mutational lesions detected	References
Oncogenes and putative oncogenes	C-*myc*	*MYC*	8q24	*MYC* gene translocations in 2/12 NSCLC tumours accompanied by over-expression of gene Gene amplification of a *MYC* gene family member in < 10 % NSCLC and 20-50 % SCLC tumours, and in 50-83 % SCLC cell lines	~3-76 fold amplification	Little *et al.* (1983) Taya *et al.* (1984) Saksela *et al.* (1985) Yoshimoto *et al.* (1986a) Cline and Battifora (1987) Kiefer *et al.* (1987) Shiraishi *et al.* (1989) Gazzeri *et al.* (1990) Hajj *et al.* (1990) Prins *et al.* (1993) Gazzeri *et al.* (1994) Lu *et al.* (1996b) Ozkara *et al.* (1999) Mitani *et al.* (2001) Gugger *et al.* (2002)
	MOS	*MOS*	8q11	Missense mutation in NSCLC tumour	Arg22Leu within region critical for MOS stability	Gorgoulis *et al.* (2001)
	RET	*RET*	10q11	Missense mutation in two SCLC cell lines LOH in 6/8 SCLC tumours	Ala664Asp	Futami *et al.* (1995) Futami *et al.* (2003)
	Cyclin D1	*CCND1*	11q13	*CCND1* gene amplified in 5-80 % NSCLC and over-expressed in 42-56 % primary NSCLC	2-5 fold amplification	Schauer *et al.* (1994) Betticher *et al.* (1996) Mate *et al.* (1996) Marchetti *et al.* (1998a) Reissmann *et al.* (1999)
	H-Ras	*HRAS*	11p15	? NSCLC ? NSCLC	Gln61→Leu Gln61→Leu Gene deletion Gene deletions gene amplification	Yuasa *et al.* (1983) Srivastava *et al.* (1985) Cline and Battifora (1987) Hajj *et al.* (1990)

	Gene/protein	Gene symbol	Chromosomal location	Types of lung cancer examined	Mutational lesions detected	References
Oncogenes and putative oncogenes	Baculoviral IAP repeat containing 2 and 3 (cIAP1, cIAP2)	*BIRC2, BIRC3*	11q22	'Low level' amplification of 11q22 in one patient		Dai *et al.* (2003)
	K-Ras-2	*KRAS*	12p12	Mutated in 10-50% NSCLC	Single base-pair substitutions, mostly in codon 12. *KRAS* gene amplified in < 5% NSCLC tumours	Capon *et al.* (1983) McCoy *et al.* (1983) Shimizu *et al.* (1983) Nakano *et al.* (1984) Santos *et al.* (1984) Miyaki *et al.* (1985) Rodenhuis *et al.* (1987) Shiraishi *et al.* (1989) Rudduck *et al.* (1993) Sugio *et al.* (1994) Li *et al.* (1994) Tsuchiya *et al.* (1995) Mills *et al.* (1995a) Hsu *et al.* (1996) Gao *et al.* (1997) Yanez *et al.* (1987) Sekine *et al.* (1998) Urban *et al.* (2000) Kovalchuk *et al.* (2001) Brose *et al.* (2002) Broermann *et al.* (2002) Petmitr *et al.* (2003) Toyooka *et al.* (2003c) Hilbe *et al.* (2003a) Li *et al.* (2003d) Uchiyama *et al.* (2003) Pelosi *et al.* (2004)

Table 2. Genetic loci known to harbour somatic mutations in human lung cancer (continued)

	Gene/protein	Gene symbol	Chromosomal location	Types of lung cancer examined	Mutational lesions detected	References
Oncogenes and putative oncogenes	Mouse double minute 2 homologue	*MDM2*	12q14-q15	*MDM2* gene amplification in 6-7% of NSCLC tumours	Gene amplification and consequent overexpression	Marchetti *et al.* (1995) Higashiyama *et al.* (1997) Miller *et al.* (2003a) Mori *et al.* (2004)
	Dual-specificity tyrosine phosphorylation-regulated kinase 2	*DYRK2*	12q14	*DYRK2* gene amplified in 5% lung adenocarcinomas		Miller *et al.* (2003a)
	Hepatocyte nuclear factor 3α	*FOXA1*	14q13	*FOXA1* gene amplification in 2/5 adenocarcinomas found to over-express HNFα		Lin *et al.* (2002)
	c-erbB-2/neu/HER2 (heregulin receptor)	*ERBB2*	17q21	*ERBB2* gene amplification in 3-22% NSCLC with consequent over-expression of *ERBB2* gene	6-20 fold	Cline and Battifora (1987) Schneider *et al.* (1989) Shiraishi *et al.* (1989) Weiner *et al.* (1990) Keith *et al.* (1992) Kern *et al.* (1994) Hirsch *et al.* (2002a) Hirsch *et al.* (2002b) Nakamura *et al.* (2003a) Pellegrini *et al.* (2003) Tan *et al.* (2003d)
	Notch3	*NOTCH3*	19p13.1-p13.2	Balanced t(15;19) translocation in metastatic lung carcinoma. Translocation breakpoint ~50 kb upstream of *NOTCH3* gene which was found to be highly expressed in the adenocarcinoma cells	Other *NOTCH3*-expressing NSCLC cell lines also manifested translocations involving chromosome 19	Dang *et al.* (2000)
	Transcription factor E2F1	*E2F1*	20q11	*E2F1* gene amplification in 9% NSCLC tumours	Missense mutation (Glu371Asp) in one NSCLC tumour	Gorgoulis *et al.* (2002)

	Gene/protein	Gene symbol	Chromosomal location	Types of lung cancer examined	Mutational lesions detected	References
Apoptosis regulatory genes	Caspase 3	*CASP3*	4q35	Heterozygous missense mutations in 3/181 (1.6%) NSCLC tumours	Arg93Leu in p17 protease subunit. His185Asn & Gln225Leu in p12 protease subunit	Soung *et al.* (2004)
	Caspase 8	*CASP8*	2q33-q34	Homozygous gene deletion in SCLC cell line		Shivapurkar *et al.* (2002b)
	Caspase 10	*CASP10*	2q33-q34	Mutations in 4/80 NSCLC tumours	Missense mutations at codons 236, 292, 410 & 420	Shin *et al.* (2002)
	Fas apoptotic inhibitory molecule	*FAIM*	3q23	Mutations in 4/80 NSCLC tumours	Nonsense mutation Gln228Term and 3 missense mutations at codons 66, 84 & 250	Shin *et al.* (2002)
	Death receptor 4/TNF-related apoptosis-inducing ligand receptor (DR4/TRAIL-R1)	*TNFRSF10A*	8p21	2/38 (5%) NSCLC tumours/cell lines possessed missense mutations	Missense mutations within ligand-binding domain	Fisher *et al.* (2001a)
	Death receptor 5/TNF-related apoptosis-inducing ligand receptor (DR5/TRAIL-R2)	*TNFRSF10B*	8p21-p22	11/104 (10.6%) NSCLC tumours exhibited micro-lesions in *TNFRSF10B* gene	Missense, nonsense and splicing mutations mainly in "death domain"	Lee *et al.* (1999b)

Table 2. Genetic loci known to harbour somatic mutations in human lung cancer (continued)

	Gene/protein	Gene symbol	Chromosomal location	Types of lung cancer examined	Mutational lesions detected	References
Apoptosis regulatory genes	Fas antigen (Apo-1/CD95)	*TNFRSF6*	10q24	5/65 (8%) NSCLC possessed missense substitutions	Missense mutations mainly found in "death domain" of protein	Lee *et al.* (1999a)
				16/79 (20%) NSCLC possessed promoter or exonic mutations		Boldrini *et al.* (2001)
				Missense mutation in adenocarcinoma	Gly→Lys (no location given)	Boldrini *et al.* (2002)
	Fas (TNFRSF6)-associated via death domain	*FADD*	11q13.3	Mutations in 4/80 NSCLC tumours	Missense mutations at codons 18, 29, 37 & 202	Shin *et al.* (2002)
	Caspase 5	*CASP5*	11q22.2-q22.3	3 mutations found in 2/30 primary NSCLC tumour tissues	Case 30 was compound heterozygous for 1bp and 2bp micro-deletions in an $(A)_{10}$ tract leading to a frameshift. Case 19 had a heterozygous 1bp deletion at the same location	Hosomi *et al.* (2003)
	Fas ligand decoy receptor 3 (DCR3)	*TNFRSF6B*	20q13	*TNFRSF6B* gene amplified in 44% of lung tumours. Decoy receptor blocks Fas ligand; gene amplification may enable tumours to escape FasL-dependent immune attack	2-18 fold amplification	Pitti *et al.* (1998)

	Gene/protein	Gene symbol	Chromosomal location	Types of lung cancer examined	Mutational lesions detected	References
Putative tumour suppressor genes	Tumour protein p73	*TP73*	1p36	LOH in 42 % lung tumours, most frequently in squamous cell carcinoma. 21 % NSCLC tumours exhibited LOH. Intragenic lesions	Loss of heterozygosity missense mutations, microdeletions	Nomoto *et al.* (1998) Yoshikawa *et al.* (1999) Naka *et al.* (2001) Nicholson *et al.* (2001) Huqun *et al.* (2003)
				Gly264Trp missense mutation and two micro-deletions (del418 and del603)	Naturally occurring mutations found in lung cancer cell lines	
	Protein tyrosine phosphatase, receptor type F	*PTPRF*	1p34	Co-amplification of *PTPRF* and *MYCL1* genes in SCLC cell line		Harder *et al.* (1995)
				Of 11 lung tumours screened, one tumour was found to exhibit compound heterozygosity for LOH on one allele and an Ala381Val missense substitution on the other		Wang *et al.* (2004b)
	Activating transcription factor 2	*ATF2*	2q32	Possible missense mutation	Val258Ile	Woo *et al.* (2002)
	Lipoprotein receptor-related protein 1B deleted in tumours	*LRP1B*	2q21.2	Homozygous *LRP1B* gene deletions in 17 % NSCLC cell lines. Expression abnormal in additional 30 % cell lines. Missense mutation	Arg3157Cys substitution: significance unknown	Liu *et al.* (2000)
	Mitotic checkpoint protein, BUB1	*BUB1*	2q14	Missense mutation in adenocarcinoma	His51Asp	Gemma *et al.* (2000a)
	Roundabout (axon guidance receptor, *Drosophila*) homologue 1 (DUTT1)	*ROBO1*	3p12	Homozygous *ROBO1* gene deletions LOH in 2/26 SCLC/NSCLC cell lines		Sundaresan *et al.* (1998) Dallol *et al.* (2002b)

Table 2. Genetic loci known to harbour somatic mutations in human lung cancer (continued)

	Gene/protein	Gene symbol	Chromosomal location	Types of lung cancer examined	Mutational lesions detected	References
Putative tumour suppressor genes	Fragile histidine triad (FHIT)	*FHIT*	3p14	LOH in 50-70% NSCLC/SCLC Aberrant splicing or allele loss in high proportion of both SCLC and NSCLC	Aberrant splicing involving loss of exons leading to frameshifts. Allele loss	Sozzi *et al.* (1997) Sozzi *et al.* (1998) Burke *et al.* (1998) Geradts *et al.* (2000) Pavelic *et al.* (2001) Sozzi *et al.* (1996) Fong *et al.* (1997) Garinis *et al.* (2001)
	BRCA1-associated protein 1 (ubiquitin carboxy-terminal hydrolase)	*BAP1*	3p21.3	Intragenic homozygous rearrangements and deletions		Jensen *et al.* (1998)
	Ras effector homologue	*RASSF1*	3p21.3	Missense mutations in 4/41 NSCLC tumours	Asp129Glu, Ile135Thr, Arg257Gln, Ala336Thr	Dammann *et al.* (2000)
	Retinoic acid receptor β	*RARB*	3p24	10% of lung cancer cell lines have *RARB* gene rearrangement. LOH in 60% tumours	?	Gebert *et al.* (1991) Zhang *et al.* (1994) Martinet *et al.* (2000)
	von Hippel-Lindau disease gene	*VHL*	3p25-p26	Solitary case of SCLC cell line	Single base-pair substitution leading to Gly→Asp exchange	Sekido *et al.* (1994)
	β-Catenin	*CTNNB1*	3p21.3	4/123 NSCLC tumours & cell lines contained missense mutations	Ser37Cys, Asp6Gly, Ser45Phe, Thr75Ala	Shigemitsu *et al.* (2001)
				Missense mutations in 1 cell line and 2 adenocarcinomas	Claimed to cause constitutive activation of Wnt signaling pathway	Sunaga *et al.* (2001)
				Missense mutations in low-grade adenocarcinoma of the fetal lung type/ well differentiated fetal adenocarcinoma	Gly34Val, Ser37Cys	Nakatani *et al.* (2002)

	Gene/protein	Gene symbol	Chromosomal location	Types of lung cancer examined	Mutational lesions detected	References
Putative tumour suppressor genes	Transforming growth factor beta II receptor	*TGFBR2*	3p22	Mutation in exon 3 poly(A) tract in RER^+ SCLC	Single base insertion	Tani *et al.* (1997)
				Mutation in poly(A) tract in absence of microsatellite instability studies (SCLC)	Indel/nonsense mutation	Hougaard *et al.* (1999)
	Tumour suppressor candidate 2 (Fus1)	*TUSC2*	3p21.3	Two identical homozygous 28bp deletions and one homozygous ~30kb deletion in 79 lung cancer cell lines		Lerman and Minna (2000)
	Putative tumour suppressor (Fus2)/*N*-acetyltransferase 6	*NAT6*	3p21.3	2 homozygous missense mutations in 78 lung cancer cell lines	Arg145Trp, Thr207Ser	Lerman and Minna (2000)
	Hyalurono-glucos-aminidase 1	*HYAL1*	3p21.3	2 homozygous missense mutations and one homozygous deletion (~30kb) deletion in 40 lung cancer cell lines	Gly196Arg, Ala227Ser	Lerman and Minna (2000)
	RNA binding motif protein 6	*RBM6*	3p21.3	1 homozygous missense mutation in 39 lung cancer cell lines	Ser353Phe	Lerman and Minna (2000)
	Zinc finger MYND domain-containing 10 (Blu)	*ZMYND10*	3p21.3	2 homozygous missense mutations in 61 lung cancer cell lines	Asp198Gln, Arg407Gln	Lerman and Minna (2000)
	Semaphorin 3B	*SEMA3B*	3p21.3	3 homozygous missense mutations in 39 lung cancer cell lines	Arg348Cys, Asp397His, Thr415Ile	Lerman and Minna (2000)
	Tumour suppressor candidate 4 (NPR2L)	*TUSC4*	3p21.3	2 homozygous missense mutations and one homozygous nonsense mutation in 38 lung cancer cell lines	Pro28Leu, Gly86Asp, Gln261Term	Lerman and Minna (2000)

Table 2. Genetic loci known to harbour somatic mutations in human lung cancer (continued)

	Gene/protein	Gene symbol	Chromosomal location	Types of lung cancer examined	Mutational lesions detected	References
Putative tumour suppressor genes	Carboxy-terminal domain, RNA polymerase II, polypeptide A, small phosphatase-like (HYA22/RBSP3)	*CTDSPL*	3p21.3	Homozygous deletion of *CTDSPL* gene in SCLC cell line (deletion breakpoints intragenic). His139Tyr missense mutation in SCLC cell line		Kashuba *et al.* (2004)
	Adenomatous polyposis coli	*APC*	5q21	LOH at *APC* locus in 80 % SCLC tumours/cell lines and in 40 % NSCLC tumours/cell lines		D'Amico *et al.* (1992b)
				LOH in 44 % adenocarcinomas and 81 % squamous cell carcinomas		Sanz-Ortega *et al.* (1999)
				Intragenic mutations in *APC* gene in 2/44 (4 %) of squamous cell lung carcinomas and 1/32 (3 %) SCLC tumours	Glu1317Gln and delGT (codon 1465) in squamous cell lung cancer and Glu1284-Lys in SCLC	Ohgaki *et al.* (2004)
	Insulin-like growth factor II receptor	*IGF2R*	6q26	58 % squamous cell carcinomas of lung had LOH involving *IGF2R* gene	55 % of these also exhibited either missense mutations or micro-insertions	Kong *et al.* (2000)
				Missense mutation in adenocarcinoma cell line resistant to TGFβ1		Gemma *et al.* (2000b)
	Parkin	*PARK2*	6q25.2-q27	Homozygous deletions of exon 2 in two lung adenocarcinoma cell lines. LOH around *PARK2*/FRA6E of up to 45 %		Cesari *et al.* (2003)
	Mitotic checkpoint protein, MAD1	*MAD1L1*	7p22	Missense mutation in 1/49 NSCLC tumours.	Thr299Ala	Nomoto *et al.* (1999)

	Gene/protein	Gene symbol	Chromosomal location	Types of lung cancer examined	Mutational lesions detected	References
Putative tumour suppressor genes	Mitogen-activated protein kinase kinase 2 (MKK2, MEK2, MAPKK4)	*MAP2K2*	7q32	Missense mutation in NSCLC tumour	Pro298Leu substitution, significance unknown	Bansal *et al.* (1997)
	Protein phosphatase 1, regulatory (inhibitor) subunit 3A	*PPP1R3A*	7q31	Mutations detected in 5/33 (15%) NSCLC cell lines, 2/38 (5%) NSCLC tumours and in 1 SCLC cell line		Kohno *et al.* (1999a)
	Platelet-derived growth factor-beta-like tumour suppressor (PRLTS)	*PDGFRL*	8p22-p21.3	Structural rearrangement in sporadic NSCLC		Fujiwara *et al.* (1995)
	Cyclin-dependent kinase inhibitor 2A ($p16^{INK4A}$)	*CDKN2A*	9p21	*CDKN2A* gene inactivated in ~30% NSCLC	Multiple types of lesion including missense mutations and deletions	Hayashi *et al.* (1994) Okamoto *et al.* (1994) Shimizu & Sekiya (1995) Nakagawa *et al.* (1995) Merlo *et al.* (1995) Rusin *et al.* (1996) Pollock *et al.* (1996) Marchetti *et al.* (1997) Gazzeri *et al.* (1998a) Hamada *et al.* (1998) Fujishita *et al.* (1998) Gazzeri *et al.* (1998a) Nicholson *et al.* (2001) Su *et al.* (2002)

Table 2. Genetic loci known to harbour somatic mutations in human lung cancer (continued)

	Gene/protein	Gene symbol	Chromosomal location	Types of lung cancer examined	Mutational lesions detected	References
Putative tumour suppressor genes	Cyclin-dependent kinase inhibitor 2B (*p15*^{INK4B})	*CDKN2B*	9p21	Commonly co-deleted with *CDKN2A* gene in NSCLC		Nakagawa *et al.* (1995) Okamoto *et al.* (1995) Xiao *et al.* (1995) Washimi *et al.* (1995) Hamada *et al.* (1998)
	Deleted in malignant brain tumours 1	*DMBT1*	10q25.3-q26.1	100% SCLC cell lines and 43% NSCLC cell lines lacked *DMBT1* expression. Note: No mutations/deletions involving the *DMBT*1 gene were found by Petersen *et al.* (2000b)	Homozygous deletions of *DMBT1* gene in 10% SCLC but not in NSCLC. Missense mutation in one NSCLC cell line	Wu *et al.* (1999b)
				53% primary 'lung cancers' exhibited LOH. *DMBT1* gene mutated in 4/23 (17%) lung cancer cell lines	2 homozygous deletions, one rearrangement involving exons 5 & 6, one missense mutation	Takeshita *et al.* (1999)
	Phosphatase and tensin homologue (PTEN)	*PTEN*	10q23.3	*PTEN* gene mutated in 16-40% SCLC tumours/cell lines *PTEN* gene mutated in 8-17% NSCLC tumours/cell lines	Missense, nonsense, splice site mutations, multi-exon deletions. Intragenic deletions, missense/nonsense mutations	Yokomizo *et al.* (1998) Forgacs *et al.* (1998) Kohno *et al.* (1998a) Okami *et al.* (1998)
	Cell proliferation-associated antigen Ki-67	*MK167*	10q25-qter	Single base deletion (1496delC) in exon 7 of *MK167* gene in A549 lung carcinoma cell line	Results in premature Stop codon at residue 446	Bubán *et al.* (2004)
	Growth/ differentiation factor 10 (bone morphogenetic protein 3B)	*GDF10*	10q11.22	LOH in 40% NSCLC tumours		Dai *et al.* (2004)

	Gene/protein	Gene symbol	Chromosomal location	Types of lung cancer examined	Mutational lesions detected	References
Putative tumour suppressor genes	Immunoglobulin superfamily member 4 (tumour suppressor in lung cancer-1, TSLC1)	*IGSF4*	11q23.2	LOH in 42 % primary NSCLC tumours. Unspecified missense mutation in NSCLC tumour	Micro-deletion as second hit in one tumour	Kuramochi *et al.* (2001) Yageta *et al.* (2002) Murakami (2002)
	Multiple endocrine neoplasia type 1 gene (menin)	*MEN1*	11q13	Lung carcinoid from MEN1 patient. *MEN1* gene inactivated in 36 % sporadic tumours	Micro-deletions and micro-insertions, single base-pair substitution affecting splice site	Debelenko *et al.* (1997)
				LOH in 7 % SCLC tumours/cell lines. Micro-deletion in large cell lung carcinoma		Debelenko *et al.* (2000)
				LOH more prevalent in large cell neuroendocrine tumours of the lung (50 %) than in SCLC (22 %)		Haruki *et al.* (2000b)
	Serine/threonine protein phosphatase 2A subunit A β isoform	*PPP2R1B*	11q22-q24	Alterations in 15 % primary lung tumours and 6 % of cell lines (SCLC and NSCLC)	Missense mutations, micro-deletions and multi-exon deletions	Wang *et al.* (1998a)
	Protein tyrosine phosphatase, receptor, type J	*PTPRJ*	11p11.2	LOH in 6/12 'lung tumours'		Ruivenkamp *et al.* (2002)
	Wilms' tumour protein	*WT1*	11p13	LOH at *WT1* locus in 9 % NSCLC tumours		Fong *et al.* (1995c)

Table 2. Genetic loci known to harbour somatic mutations in human lung cancer (continued)

	Gene/protein	Gene symbol	Chromosomal location	Types of lung cancer examined	Mutational lesions detected	References
Putative tumour suppressor genes	Solute carrier family 22, member 18 (tumour-suppressing subchromosomal transferable fragment; TSSC5)	*SLC22A18*	11p15	LOH/missense mutation	Ser233Phe	Lee *et al.* (1998b)
	Cyclin-dependent kinase inhibitor 1C ($p57^{KIP2}$)	*CDKN1C*	11p15.5	*CDKN1C* gene is imprinted with maternal expression and maternal alleles lost in 85 % lung cancer cases carrying 11p15 deletions	–	Kondo *et al.* (1996)
	Protein kinase C, delta binding protein	*PRKCDBP*	11p15.4	LOH, frameshift, missense and truncating mutations	Arg190Term (NSCLC), 177del21, Ala192Gly (SCLC)	Xu *et al.* (2001a)
	Checkpoint protein with FHA and ring finger domains	*CHFR*	12q24.3	Missense mutations in 3/53 (6 %) of NSCLC tumours LOH in 1/2 NSCLC tumours that were found to harbour missense mutations on the other alleles	Pro166Leu, Arg202-Pro, Phe536Ser. *In vitro* expression assays indicated absence of expression of Phe536Ser whilst all three mutants were defective in a (functional) checkpoint assay	Mariatos *et al.* (2003)
	Deleted in cancer 1 (Dead/H box polypeptide 26)	*DDX26*	13q14	Loss of gene or down-regulation of gene expression in majority of NSCLC	–	Wieland *et al.* (1999)

	Gene/protein	Gene symbol	Chromosomal location	Types of lung cancer examined	Mutational lesions detected	References
Putative tumour suppressor genes	Retinoblastoma gene	*RB1*	13q14	LOH at *RB1* locus and loss of *RB1* gene expression in 9/9 SCLC cell lines Internal deletions and other structural abnormalities of *RB1* gene in 18-33% SCLC cell lines and 1/8 SCLC (12%) tumours LOH and point mutations in *RB1* gene in 70-90% of NSCLC tumours *RB1* gene expression altered in >90% of SCLC tumours but in only 14-32% NSCLC tumours		Harbour *et al.* (1988) Yokota *et al.* (1989) Mori *et al.* (1990) Horowitz *et al.* (1990) Hensel *et al.* (1990) Murakami *et al.* (1991) Xu *et al.* (1991) Reissmann *et al.* (1993) Gouyer *et al.* (1994) Sachse *et al.* (1994) Shimizu *et al.* (1994) Shimizu *et al.* (1994) Tamura *et al.* (1997) Tamura *et al.* (1997) Marchetti *et al.* (1998a) Tanaka *et al.* (1998) Gouyer *et al.* (1998) Hiroshima *et al.* (2004)
	Retinoblastoma-like 2 (p130)	*RBL2*	16q12.2	Loss of p130 in 1/17 (6%) SCLC cell lines	Splice site mutation	Helin *et al.* (1997)
				Mutations in 11/14 (78%) mainly SCLC tumours	Missense & splicing mutations & microinsertions	Claudio *et al.* (2000)
	WW domain-containing oxidoreductase	*WWOX*	16q23.3-q24.1	Homozygous in-frame deletion of exons 6-8 in 2/11 SCLC cell lines. Heterozygous Lys182Glu substitution in 1/7 NSCLC cell lines and 1/11 SCLC lines plus homozygous Lys182Glu substitution in 1/11 SCLC cell lines		Paige *et al.* (2001)
				Missense mutation of possible functional significance. LOH in 37% tumours	Asp183Asn	Yendamuri *et al.* (2003)

Table 2. Genetic loci known to harbour somatic mutations in human lung cancer (continued)

	Gene/protein	Gene symbol	Chromosomal location	Types of lung cancer examined	Mutational lesions detected	References
Putative tumour suppressor genes	p53	*TP53*	17p13	Mutations in 50-60% NSCLC and in >90% SCLC tumours	LOH, gross gene deletions, nonsense and missense mutations, splicing mutations	Takahashi *et al.* (1989b) Chiba *et al.* (1990) Takahashi *et al.* (1991) Mitsudomi *et al.* (1992) Sameshima *et al.* (1992) Kondo *et al.* (1992) D'Amico *et al.* (1992a) Suzuki *et al.* (1992) Kishimoto *et al.* (1992) Miller *et al.* (1992) Lohmann *et al.* (1993) Reichel *et al.* (1994) Ryberg *et al.* (1994) Shipman *et al.* (1996) Guang *et al.* (1997) de Anta *et al.* (1997) Liloglou *et al.* (1997) Jassem *et al.* (2001) Mori *et al.* (2004) Fouquet *et al.* (2004) See text, reviews and web-based mutation databases for full literature.
	Mitogen-activated protein kinase kinase 4 (MKK4, MEK4, MAPKK4, SERK1)	*MAP2K4*	17p11.2	Homozygous deletion of *MAP2K4* gene	Multi-exon deletion	Teng *et al.* (1997)

	Gene/protein	Gene symbol	Chromosomal location	Types of lung cancer examined	Mutational lesions detected	References
Putative tumour suppressor genes	Tyrosine 3-monooxygenase/ tryptophan 5-monooxygenase activation protein, ε polypeptide (14-3-3ε)	*YWHAE*	17p13.3	Homozygous gene deletion in two SCLC cell lines of same patient		Konishi *et al.* (2002)
	γ-Catenin (junction plakoglobin)	*JUP*	17q21	Missense mutation in 1/95 squamous cell carcinoma cell line		Ueda *et al.* (2001)
	Neurofibromatosis type 1 (neurofibromin)	*NF1*	17q11.2	Missense mutation in 2/9 SCLC tumours but 0/28 NSCLC tumours	Glu1415Gly	Furukawa *et al.* (2003)
	Mitogen-activated protein kinase kinase 3	*MAP2K3*	17q11.2	Gene homozygously deleted in NCI-H774 lung cancer cell line		Teng *et al.* (2001)
	JV18-1 (SMAD2)	*MADH2*	18q21	Two point mutations, one in SCLC tumour, one in NSCLC tumour	Missense mutation (Asp450His), 9bp deletion	Uchida *et al.* (1996) Yanagisawa *et al.* (2000)
	Deleted in pancreatic carcinoma 4 gene (SMAD4)	*MADH4*	18q21.1	Point mutations in 10% of NSCLC tumours	One micro-deletion, 2 missense mutations One partial gene deletion and micro-insertion	Nagatake *et al.* (1996a) Yanagisawa *et al.* (2000) Yanaihara *et al.* (2001)
	Deleted in colorectal carcinoma	*DCC*	18q21.1	Missense mutation in NSCLC cell line		Kohno *et al.* (2000)

Table 2. Genetic loci known to harbour somatic mutations in human lung cancer (continued)

	Gene/protein	Gene symbol	Chromosomal location	Types of lung cancer examined	Mutational lesions detected	References
Putative tumour suppressor genes	Serine/threonine protein phosphatase 2A subunit A α isoform	*PPP2R1A*	19q13	Missense mutation in unspecified lung carcinoma	Glu64Asp	Calin *et al.* (2000) Rüdiger *et al.* (2001)
	α-Actinin-4	*ACTN4*	19q13	Lys122Asn missense mutation, close to actin-binding domain of α-actinin-4 protein, in large cell carcinoma of the lung	Mutation results in loss of ability of α-actinin-4 to associate with actin cytoskeleton and loss of cell migration capacity. However, the mutation also abrogates the tumour suppressor function of ACTN4 (Menez *et al.*, 2004)	Echchakir *et al.* (2001)
	Serine/threonine kinase 11 (Peutz-Jeghers syndrome, LKB1)	*STK11*	19p13.3	Mutations found in 7/42 NSCLC (adenocarcinomas) Nonsense mutation (Gln37Term), 1 bp micro-deletion and 2 multi-exon deletions in 11 lung adenocarcinoma cell lines	6 nonsense mutations and 1 micro-deletion	Sanchez-Cespedes *et al.* (2002) Carretero *et al.* (2004)
				Mutations found in 5/19 lung adenocarcinoma tumours	Tyr60Term, Glu120-Term, insG codon 279, Glu65Term, del exon 8	Fernandez *et al.* (2004)
	SWI/SNF-related, matrix-associated, actin-dependent regulator of chromatin, sub family a, member 4	*SMARCA4*	19p13.3	Gene homozygously deleted in A-427 lung cancer cell line. Frameshift mutation in another lung cancer cell line		Wong *et al.* (2000)
				Gene homozygously deleted in A-427 lung cancer cell line; 69 bp deletion in exon 10		Teng *et al.* (2001)

	Gene/protein	Gene symbol	Chromosomal location	Types of lung cancer examined	Mutational lesions detected	References
Putative tumour suppressor genes	Forkhead box A2 transcription factor (HNF3B)	*FOXA2*	20p11	Of 31 lung cancer cell lines tested, 1 missense mutation found in SCLC and 1 homozygous micro-deletion found in NSCLC	Gly92Asp DelC in codon 194	Halmos *et al.* (2004)
	Protein tyrosine phosphatase, receptor type T	*PTPRT*	20q12-q13	*PTPRT* gene missense mutations were identified in 2/11 (18%) lung tumours	Arg1346Leu and Arg790Ile	Wang *et al.* (2004b)
	Seizure 6-like gene	*SEZ6L*	22q12	Missense mutations in 3/46 (7%) lung cancer cell lines. Microdeletion in acceptor splice site in 1/46 primary lung cancers	Glu60Gln, Arg188Gln, Gly630Val	Nishioka *et al.* (2000)
	Checkpoint kinase 2	*CHEK2*	22q12	Missense mutation in SCLC tumour	Asp311Val	Haruki *et al.* (2000a) Matsuoka *et al.* (2001)
				Missense mutation in NSCLC tumour	Pro85Leu	Miller *et al.* (2002a)
	Myosin XVIIIB	*MYO18B*	22q12.1	Missense mutations in 14/75 (19%) of SCLC and NSCLC lung cancer cell lines and 6/46 (13%) of primary lung cancers		Nishioka *et al.* (2002)
Miscellan	Arginine-rich protein, mutated in early stage tumours	*ARMET*	3p21.1	Identical missense mutation found in 1/2 SCLC tumours and 6/18 NSCLC tumours	Met50Arg	Shridhar *et al.* (1996)
	Aminoacylase 1	*ACY1*	3p21.1	Missense mutations with loss of aminoacylase activity in SCLC cell lines	Missense mutations at codons 195 and 254	Cook *et al.* (1998)
	Interferon regulatory factor (IRF-1)	*IRF1*	5q23-q31	Missense mutation abolishes DNA-binding and transactivating activities of IRF-1	Trp11Arg	Eason *et al.* (1999)
	β2 microglobulin	*B2M*	15q21-q22	Missense mutation in translational initiation codon of *B2M* gene in adenocarcinoma cell line	Met1Val	Chen *et al.* (1996a)

Table 2. Genetic loci known to harbour somatic mutations in human lung cancer (continued)

	Gene/protein	Gene symbol	Chromosomal location	Types of lung cancer examined	Mutational lesions detected	References
Miscellan	Somatostatin receptor 2	*SSTR2*	17q24	Nonsense mutation	Trp188Term	Zhang *et al.* (1995)
	Methyl-CpG binding protein, MBD1	*MBD1*	18q21	Nonsense mutation found in 1/80 lung cancer cell lines	Gln370Term (in SCLC cell line)	Bader *et al.* (2003)
Telomerase-associated	Telomerase reverse transcriptase	*TERT*	5p15.33	Gene amplification in 2/5 NSCLC and 2/3 SCLC cell lines and 8/21 NSCLC & SCLC tumours		Zhang *et al.* (2000a) Saretzki *et al.* (2002)
	Telomerase RNA	*TERC*	3q26.3	Gene amplification in 1/9 NSCLC tumours Gene amplification in SCLC Gene amplification in 9/19 (47%) NSCLC tumours		Soder *et al.* (1997) Sugita *et al.* (2000) Yokoi *et al.* (2003b)
DNA repair genes	Mut Y homologue	*MUTYH*	1p32.1-p34.3	Missense mutation found in 1/47 cell lines	Significance unclear	Shinmura *et al.* (2001)
				Novel *MUTYH* missense mutations noted in a screen of 276 patients with 'lung carcinoma' and 106 controls	Gly382Asp, Val22, Met, Gln324His, Ser501Phe. Their significance is unclear since none were over-represented in the patient group	Al-Tassan *et al.* (2003)
	Excision repair cross-complementing repair deficiency (comp. group 3)	*ERCC3*	2q21-q22	LOH in 2/9 (22%) "lung cancer" tissue samples		Takebayashi *et al.* (2001)

	Gene/protein	Gene symbol	Chromosomal location	Types of lung cancer examined	Mutational lesions detected	References
DNA repair genes	8-oxoguanine DNA glycosylase	*OGG1*	3p25-p26	LOH in 62 % primary NSCLC tumours LOH in 43 % primary NSCLC tumours		Chevillard *et al.* (1998) Wikman *et al.* (2000)
				Missense mutation in one cell line	Arg46Gln. Significance unclear	Kohno *et al.* (1998b)
	DNA mismatch repair gene, MLH1	*MLH1*	3p21	Lost in 55 % of NSCLC tumours Heterozygous Val384Asp missense mutation in 1/8 lung cancers of Chinese origin	Deletions Val384Asp mutation is a putative functional polymorphism in East Asian populations (Wang *et al.*, 1998b)	Benachenhou *et al.* (1998) Shi *et al.* (2003)
	DNA mismatch repair gene, MSH3	*MSH3*	5q11-q13	Lost in 42 % of NSCLC tumours	Deletions	Benachenhou *et al.* (1998)

Table 3. Human genes inactivated by promoter hypermethylation in lung cancer

Gene	Gene symbol	Chromosomal location	Methylation status	Loss of gene expression demonstrated	Reference
Retinoblastoma protein-interacting zinc finger, RIZ1	*PRDM2*	1p36	Promoter methylated in lung cancer cell lines	Yes	Du *et al.* (2001)
Runt-related transcription factor 3	*RUNX3*	1p36	Promoter methylated in 20-46 % NSCLC tumours	No	Yanagawa *et al.* (2003) Kim *et al.* (2004a)
			Promoter methylated in 19 % lung cancer cell lines and 24 % primary lung tumours		Li *et al.* (2004)
Tyrosine 3-monooxygenase/ tryptophan 5-monooxygenase activation protein, sigma isoform (14-3-3σ/stratifin)	*SFN*	1p35.3	Promoter methylated in 69 % SCLC cell lines and 5 % primary NSCLC tumours	Yes	Osada *et al.* (2002)
DNA mismatch repair gene, MSH2	*MSH2*	2p16	Promoter methylated in 29 % NSCLC tumours	Yes	Wang *et al.* (2003b)
Transmembrane protein with EGF-like and two follistatin-like domains 2 (HPP1)	*TMEFF2*	2q32.3	Promoter methylated in 13 % NSCLC tumours	No	Hanabata *et al.* (2004)
Caspase 8	*CASP8*	2q33-q34	Promoter methylated in 47 % SCLC tumours	Yes	Shivapurkar *et al.* (2002b)
Ras association domain family protein 1	*RASSF1*	3p21.3	Promoter methylated in 30-55 % NSCLC tumours, 17 % NSCLC cell lines and 10-100 % SCLC tumours/cell lines	Yes	Dammann *et al.* (2000); (2001); (2003) Burbee *et al.* (2001) Agathanggelou *et al.* (2001) Tomizawa *et al.* (2002) Hesson *et al.* (2003)

Gene	Gene symbol	Chromosomal location	Methylation status	Loss of gene expression demonstrated	Reference
Ras association domain family protein 1	*RASSF1*	3p21.3	Promoter methylated in 30-55 % NSCLC tumours, 17 % NSCLC cell lines and 10-100 % SCLC tumours/cell lines	Yes	Honorio *et al.* (2003) Endoh *et al.* (2003) Kim *et al.* (2003b) Yanagawa *et al.* (2003) Li *et al.* (2003d) Maruyama *et al.* (2004) Topaloglu *et al.* (2004)
DNA mismatch repair gene, MLH1	*MLH1*	3p21.3	Promoter methylated in 7-56 % NSCLC tumours	Yes	Wang *et al.* (2003b) Yanagawa *et al.* (2003)
Zinc finger MYND domain-containing 10 (Blu)	*ZMYND10*	3p21.3	Promoter methylated in 19 % NSCLC tumours and in 14 % SCLC tumours	Yes	Agathanggelou *et al.* (2003a)
Semaphorin 3B	*SEMA3B*	3p21.3	Promoter methylated in 50 % NSCLC cell lines and 41 % primary NSCLC tumours	Yes	Kuroki *et al.* (2003)
Transforming growth factor beta receptor II	*TGFBR2*	3p22	Promoter methylated in 42 % of NSCLC tumours	Yes	Zhang *et al.* (2004a)
Fragile histidine triad	*FHIT*	3p14.2	Promoter methylated in 37-38 % NSCLC & SCLC tumours and 65 % lung cancer cell lines	Yes	Zöchbauer-Müller *et al.* (2001a) Maruyama *et al.* (2004) Kim *et al.* (2004d)
Roundabout, homologue of *Drosophila* gene	*ROBO1*	3p12	Promoter methylated in 4 % NSCLC tumours	Yes	Dallol *et al.* (2002a)
Retinoic acid receptor β	*RARB*	3p24	Promoter methylated in 72 % SCLC tumours and in 10-40 % NSCLC tumours & cell lines	Yes	Virmani *et al.* (2000) Lamy *et al.* (2001) Zöchbauer-Müller *et al.* (2001b) Maruyama *et al.* (2004) Topaloglu *et al.* (2004) Hanabata *et al.* (2004)

Table 3. Human genes inactivated by promoter hypermethylation in lung cancer (continued)

Gene	Gene symbol	Chromosomal location	Methylation status	Loss of gene expression demonstrated	Reference
Retinol binding protein 1, cellular	*RBP1*	3q21-q23	Promoter methylated in 15% NSCLC tumours	Yes	Esteller *et al.* (2002)
Retinoic acid receptor responder 1 (tazarotene-induced 1)	*RARRES1*	3q25.32	Promoter methylated to variable extent in 14 'lung cancer' cell lines tested; 4 cell lines displayed 64-81% methylation of the *RARRES1* gene promoter and no gene expression, 10 cell lines displayed 6-22% methylation of the *RARRES1* promoter and expressed the *RARRES1* gene	Yes	Youssef *et al.* (2004)
SLIT2, homologue of *Drosophila* gene	*SLIT2*	4p15.2	Promoter methylated in 53% NSCLC tumours and in 36% SCLC tumours	Yes	Dallol *et al.* (2002b)
Adenomatous polyposis coli	*APC*	5q21-q22	Promoter methylated in 38-96% NSCLC tumours/cell lines and in 26% SCLC cell lines	No	Brabender *et al.* (2001) Virmani *et al.* (2001) Usadel *et al.* (2002) Harden *et al.* (2003) Yanagawa *et al.* (2003) Maruyama *et al.* (2004) Topaloglu *et al.* (2004)
5-hydroxy tryptamine (serotonin) receptor 1B	*HTR1B*	6q13	Promoter methylated in 55% NSCLC tumours	Yes	Takai *et al.* (2001)
Insulin-like growth factor-binding protein 3	*IGFBP3*	7p12-p13	Promoter methylated in 51/83 (61%) NSCLC tumours	No	Chang *et al.* (2002)
			Promoter methylated in 54% NSCLC cell lines and in 70% stage I, 78% stage II, 73% stage III, and 100% stage IV NSCLC tumours	Yes	Chang *et al.* (2004d)
Caveolin-1	*CAV1*	7q31.1	Promoter methylated in 93% of SCLC tumours and 9% of NSCLC tumours	Yes	Sunaga *et al.* (2004)

Gene	Gene symbol	Chromosomal location	Methylation status	Loss of gene expression demonstrated	Reference
Deleted in liver cancer	*DLC1*	8p21.3-p22	Promoter methylated in 8/11 (73 %) NSCLC cell lines	Yes	Yuan *et al.* (2004)
Paired box transcription factor 5	*PAX5*	9p13	Promoter methylated in 55 % lung tumour cell lines, 40 % adenocarcinomas, 52 % squamous cell carcinomas, and 10 % of adjacent tissue	Yes	Palmisano *et al.* (2003)
Cyclin-dependent kinase inhibitor 2A (*p16*^{INK4A})	*CDKN2A*	9p21	Promoter methylated in 20-38 % NSCLC tumours	Yes	Merlo *et al.* (1995)
			Promoter methylated in 35 % (smokers) and 7 % (non-smokers) of NSCLC tumours		Gazzeri *et al.* (1998a) Tanaka *et al.* (1998) Hamada *et al.* (1998) Seike *et al.* (2000) Zöchbauer-Müller *et al.* (2001b) Soria *et al.* (2002a) Yanagawa *et al.* (2002) Harden *et al.* (2003) Yanagawa *et al.* (2003) Jarmalaite *et al.* (2003) Liu *et al.* (2003a) Liu *et al.* (2004) Hiroshima *et al.* (2004) Topaloglu *et al.* (2004) Hiroshima *et al.* (2004) Maruyama *et al.* (2004) Hanabata *et al.* (2004)
			p14^{ARF} gene methylated & inactivated in 45 % of NSCLC tumours		Shimamoto *et al.* (2004) Mori *et al.* (2004)
			61 % of NSCLC tumours lacked *CDKN2A* expression but only 37 % of these tumours exhibited *CDKN2A* promoter hypermethylation		Gonzalez-Quevedo *et al.* (2004)

Table 3. Human genes inactivated by promoter hypermethylation in lung cancer (continued)

Gene	Gene symbol	Chromosomal location	Methylation status	Loss of gene expression demonstrated	Reference
Cyclin-dependent kinase inhibitor 2B (*p15*^{INK4B})	*CDKN2B*	9p21	Promoter methylated in 15 % neuro-endocrine lung tumours and 11 % NSCLC tumours	Yes No	Chaussade *et al.* (2001) Kurakawa *et al.* (2001)
Death-associated protein kinase	*DAPK1*	9q34.1	Promoter methylated in 17-44 % NSCLC tumours	No	Esteller *et al.* (1999) Tang *et al.* (2000) Zöchbauer-Müller *et al.* (2001b) Kim *et al.* (2001b) Soria *et al.* (2002a) Harden *et al.* (2003) Yanagawa *et al.* (2003)
Phosphatase and tensin homologue (PTEN)	*PTEN*	10q23.3	Promoter methylated in 11/16 (69 %) NSCLC cell lines and in 7/20 (35 %) PTEN-negative NSCLC tumours	Yes	Soria *et al.* (2002b)
Growth/ differentiation factor 10 (bone morphogenetic protein 3B)	*GDF10*	10q11.2	Promoter methylated in 5/6 NSCLC tumours	No	Dai *et al.* (2001)
			Promoter methylated in 49 % NSCLC tumours	No	Dai *et al.* (2004)
O6-methyl-guanine DNA methyltransferase	*MGMT*	10q26	Promoter methylated in 1-51 % NSCLC tumours	No	Zöchbauer-Müller *et al.* (2001b) Esteller *et al.* (2001) Hayashi *et al.* (2002) Harden *et al.* (2003) Pulling *et al.* (2003) Brabender *et al.* (2003)
			Promoter methylated in 38 % of NSCLC tumours, in 18 % of matching normal lung tissues of patients, and 0 % of controls without lung cancer		Nakagawachi *et al.* (2003) Maruyama *et al.* (2004) Topaloglu *et al.* (2004)

Gene	Gene symbol	Chromosomal location	Methylation status	Loss of gene expression demonstrated	Reference
Protein kinase C, delta binding protein	*PRKCDBP*	11p15.4	Promoter methylated in 11/14 (79 %) lung tumours	Yes	Xu *et al.* (2001a)
Immunoglobulin superfamily member 4 (tumour suppressor in lung cancer-1, TSLC1)	*IGSF4*	11q23.2	Promoter hypermethylated in 40-78 % NSCLC tumours and ~20 % SCLC tumours	No	Kuramochi *et al.* (2001) Murakami (2002) Fukami *et al.* (2003)
Fanconi anaemia, complementation group F	*FANCF*	11p15	Promoter methylated in 14 % of NSCLC tumours	No	Marsit *et al.* (2004)
Dickkopf 3 (REIC/DKK-3)	*DKK3*	11p15	Promoter methylated in 21 % NSCLC tumours	Yes	Kobayashi *et al.* (2002)
Glutathione *S*-transferase Pi	*GSTP1*	11q13-qter	Promoter methylated in 1-10 % NSCLC tumours	No	Zöchbauer-Müller *et al.* (2001b) Esteller *et al.* (2001b) Soria *et al.* (2002a) Harden *et al.* (2003) Yanagawa *et al.* (2003) Topaloglu *et al.* (2004)
Cyclin D2	*CCND2*	12p13	Promoter methylated in 57 % SCLC cell lines, 22 % SCLC tumour tissues, 47 % NSCLC cell lines and 40 % NSCLC tumour tissues	Yes	Virmani *et al.* (2003b)
Checkpoint protein with FHA and ring finger domains	*CHFR*	12q24.3	Promoter methylated in 19 % primary lung cancer tumours Promoter methylated in 10 % NSCLC tumours	Yes	Mizuno *et al.* (2002) Corn *et al.* (2003)
Deleted in cancer 1 (Dead/H Box 26)	*DDX26*	13q14	Promoter methylated in lung cancer cell lines	Yes	Wieland *et al.* (2000)

Table 3. Human genes inactivated by promoter hypermethylation in lung cancer (continued)

Gene	Gene symbol	Chromosomal location	Methylation status	Loss of gene expression demonstrated	Reference
Mothers against decapentaplegic, *Drosophila*, homologue of, 9 (SMAD8)	*MADH9*	13q12-q14	Promoter methylated in 1/19 (5%) lung cancer cell lines (expression could be successfully reactivated by 5-azacytidine treatment)	Yes	Cheng *et al.* (2004)
'Target of methylation-induced silencing' (Tms1)	*PYCARD*	16p12-p11.2	Promoter methylated in 41% of SCLC tumours, 70% of SCLC cell lines, 40% of NSCLC tumours and 48% of NSCLC cell lines	Yes	Virmani *et al.* (2003a)
E-cadherin	*CDH1*	16q22.1	Promoter methylated in 18-87% NSCLC tumours	No	Zöchbauer-Müller *et al.* (2001b) Yanagawa *et al.* (2003) Maruyama *et al.* (2004) Topaloglu *et al.* (2004) Shimamoto *et al.* (2004)
Heparan sulphate D-glucosaminyl 3-O-sulphotransferase 2	*HS3ST2*	16p12	Promoter methylated in 7/10 (70%) lung tumours	No	Miyamoto *et al.* (2003a)
H-cadherin	*CDH13*	16q24.2	Promoter methylated in 18/42 (43%) NSCLC tumours, 15/30 (50%) NSCLC cell lines and 6/30 (20%) SCLC cell lines	Yes	Toyooka *et al.* (2001a)
			Promoter methylated in 27% NSCLC tumours		Maruyama *et al.* (2004)
			Promoter methylated in 20% NSCLC tumours		Hanabata *et al.* (2004)
LOcated at Seventeen-p-Thirteen point three 1 (LOST1)	*TUSC5*	17p13.3	Promoter methylated in 4/6 lung cancer cell lines that exhibited negligible gene expression (marked reduction of *TUSC5* gene expression in 11/16 lung tumours)	Yes	Konishi *et al.* (2003)
Breast cancer 1	*BRCA1*	17q21-q24	Promoter methylated in 4% NSCLC tumours	No	Esteller *et al.* (2001) Marsit *et al.* (2004)

Gene	Gene symbol	Chromosomal location	Methylation status	Loss of gene expression demonstrated	Reference
Suppressor of cytokine signaling 3	*SOCS3*	17q25.3	Promoter methylated in 3/4 (75 %) NSCLC cell lines and 14/18 (78 %) primary NSCLC tissues	Yes	He *et al.* (2003) He *et al.* (2004)
Bcl-2	*BCL2*	18q21	Promoter methylated in 23 % NSCLC cases but in 0 % SCLC cases. Hypermethylation more frequent in adenocarcinomas (40 %) and large cell carcinomas (38 %) than in squamous cell carcinomas (13 %)	No	Nagatake *et al.* (1996b)
Serine/threonine kinase 11 (Peutz-Jeghers syndrome)	*STK11*	19p13.3	Promoter methylated in 1/42 adenocarcinomas	No	Sanchez-Cespedes *et al.* (2002)
Laminin α5	*LAMA5*	20q13.2-q13.3	3 *LAMA5* genes analysed: promoters methylated in 58-77 % SCLC tumours and 22-42 % NSCLC tumours	Yes	Sathyanarayana *et al.* (2003)
Checkpoint kinase 2	*CHEK2*	22q12.1	Promoter methylated in NSCLC tumours and cell lines but not in control lung tissue	Yes	Zhang *et al.* (2004b)
Tissue inhibitor of metalloproteinase	*TIMP3*	22q12.1-q13.2	Promoter methylated in 19-26 % NSCLC tumours	No	Zöchbauer-Müller *et al.* (2001b) Esteller *et al.* (2001)
Myosin XVIIIB	*MYO18B*	22q11.2-q12.1	Promoter methylated in 17 % lung cancer cell lines and 35 % primary lung cancers	Yes	Nishioka *et al.* (2002)

Table 4. Studies claiming an association between possession of a polymorphic allele and risk of lung cancer

Gene	Gene symbol	Chromosomal localization	Polymorphism association	Reference
NAD(P)H dehydrogenase, quinone 1 (NAD(P)H: quinone oxidoreductase/ diaphorase 4)	*NQO1*	16q22.1	Pro/Ser187 (C609→T) polymorphism. Ser187 null allele with no detectable NQO1 activity has prevalence of 20% in Caucasians. Ser187 allele exhibits OR of 0.3 in homozygous form	Wiencke *et al.* (1997) Chen *et al.* (1999) Lewis *et al.* (2001) Xu *et al.* (2001b)
			Heterozygosity for Pro/Ser187 associated with an OR of 0.66 for risk of lung cancer	Hamajima *et al.* (2002)
			Association between possession of Pro/Pro187 genotype and lung cancer risk (OR=2.15)	Sunaga *et al.* (2002)
Epoxide hydrolase 1, microsomal	*EPHX1*	1p11-qter	His113 variant, associated with a 40% reduction in EPHX activity, yielded an OR of 0.50 as compared with Tyr113 (OR =1.0)	Benhamou *et al.* (1998)
			Taiwanese	Lin *et al.* (2000)
			Mexican-Americans, OR=3.6	Wu *et al.* (2001)
			Smokers carrying HYL1*2 allele had higher relative risk for lung cancer (OR=5.7)	Yin *et al.* (2001)
			Decreased risk associated with His113 (OR=0.1)	London *et al.* (2000)
			Decreased risk associated with His113 (OR=0.4) and His139 (0.5) variants	To-Figueras *et al.* (2001)
			Association between Arg139 variant and lung cancer (OR=1.6)	Zhao *et al.* (2002)
			Meta-analysis provides evidence for association only between His113 variant and decreased risk of lung cancer (OR=0.7) and not for Arg139 (OR=1.0)	Lee *et al.* (2002b)
			No evidence for association between either His113 or Arg139 variants and risk of lung cancer	Zhou *et al.* (2001d)
			Association between homozygosity for His113 variant and decreased risk of lung cancer (OR=0.38)	Gsur *et al.* (2003)
			Association between homozygosity for Arg139 variant and lung cancer (OR=6.3)	Cajas-Salazar *et al.* (2003a)
N-acetyltransferase 1	*NAT1*	8p21.3-23.1	Slow acetylator genotypes associated with higher risk of lung cancer e.g. NAT1*10 allele over-represented in lung cancer patients <60 years of age (OR=6.8)	Bouchardy *et al.* (1998) Abdel-Rahman *et al.* (1998) Wikman *et al.* (2001)

Gene	Gene symbol	Chromosomal localization	Polymorphism association	Reference
N-acetyltransferase 2	*NAT2*	8p21.3-23.1	Slow acetylator genotypes associated with higher risk of lung cancer	Martinez *et al.* (1995) Cascorbi *et al.* (1996) Oyama *et al.* (1997a) Seow *et al.* (1999) Hou *et al.* (2001) Wikman *et al.* (2001)
Glutathione *S*-transferase μ1	*GSTM1*	1p13.3	Risk conferred by homozygous deletions appears to be small but magnitude of risk may be increased when interactions with other factors are considered	Zhong *et al.* (1991) Hayashi *et al.* (1992) Nazar-Stewart *et al.* (1993) Brockmoller *et al.* (1993) Hirvonen *et al.* (1993a) McWilliams *et al.* (1995) Goto *et al.* (1996) Rebbeck *et al.* (1997) Sun *et al.* (1997) Kihara *et al.* (1999) Ford *et al.* (2000) Chen *et al.* (2001c) Stücker *et al.* (2002) Nazar-Stewart *et al.* (2003) Kiyohara *et al.* (2003) Pinarbasi *et al.* (2003) Reszka *et al.* (2003) Ruano-Ravina *et al.* (2003)
Glutathione *S*-transferase θ	*GSTT1*	22q11.2	Risk conferred by homozygous deletions appears to be small but magnitude of risk may be increased when interactions with other factors are considered	Rebbeck *et al.* (1997) Sunaga *et al.* (2002)
			Association between possession of null allele and risk of lung cancer (OR=2.4)	Sorensen *et al.* (2004)
			Association between possession of null allele and risk of lung cancer (OR=1.7)	Yang *et al.* (2004b)
			Association between possession of null allele and risk of lung cancer (OR=1.7)	Chan-Yeung *et al.* (2004)

Table 4. Studies claiming an association between possession of a polymorphic allele and risk of lung cancer (continued)

Gene	Gene symbol	Chromosomal localization	Polymorphism association	Reference
Glutathione *S*-transferase μ4	*GSTM4*	1p13.3	Association between T2517C allele and risk of lung cancer (OR=2.2)	Liloglou *et al.* (2002)
Glutathione *S*-transferase P1	*GSTP1*	11q13	Association between Ala/Val114 alleles and risk of lung cancer (OR=1.4) Association between Val/Val105 alleles and risk of lung cancer	Watson *et al.* (1998) Wang *et al.* (2003a) Ryberg *et al.* (1997) Stücker *et al* (2002) Yang et al. (2004b)
Glutathione peroxidase 1	*GPX1*	3p21.3	Association between Pro/Leu198 polymorphic alleles and risk of lung cancer (OR=1.8) Association with lung cancer confined to smokers (OR=3.3)	Ratnasinghe *et al.* (2000) Yang *et al.* (2004b)
Cytochrome CYP1A1	*CYP1A1*	15q22-q24	Carriership of *Msp*I allele is purportedly more frequent in lung cancer patients (ORs=1.5-4.7) but note negative result of Hirvonen *et al.* (1993b) Carriership of Val allele of Ile462/Val polymorphism is more frequent in lung cancer (ORs=1.5-3.0)	Kawajiri *et al.* (1990) Hayashi *et al.* (1992) Goto *et al.* (1996) Taioli *et al.* (1995) Xu *et al.* (1996) Kawajiri *et al.* (1996) Lin *et al.* (2000) Houlston *et al.* (2000) Song *et al.* (2001) Vineis *et al.* (2003) Taioli *et al.* (2003) Le Marchand *et al.* (2003) Taioli *et al.* (2003) Dialyna *et al.* (2003) Hung *et al.* (2003) Sobti *et al.* (2003)
Cytochrome CYP1B1	*CYP1B1*	2p21-p22	Association between Ala/Ser119 polymorphic alleles and squamous cell lung cancer	Watanabe *et al.* (2000)

Gene	Gene symbol	Chromosomal localization	Polymorphism association	Reference
Cytochrome CYP2A6	*CYP2A6*	19q13.2	Deletion allele under-represented in lung cancer patients as compared to controls. OR (homozygotes) = 0.23 v. 1.00 (wild-type)	Miyamoto *et al.* (1999) Kamataki *et al.* (1999) Oscarson (2001) Ariyoshi *et al.* (2002)
			Individuals harbouring at least one *CYP2A6**4 allele at 2-fold increased risk of lung cancer. Not replicated by French study	Tan *et al.* (2001) Loriot *et al.* (2001)
Cytochrome CYP2A13	*CYP2A13*	19q13.2	Association between heterozygous possession of an Arg257Cys polymorphic allele and a significantly reduced risk of lung adenocarcinoma (OR=0.41)	Wang *et al.* (2003h)
			Association between heterozygous possession of null allele (Arg101Term) and small cell lung cancer (OR=9.9)	Cauffiez *et al.* (2004)
Cytochrome CYP2C19	*CYP2C19*	10q24	Association between possession of poor metabolizer *CYP2C19* genotype and lung cancer (OR=3.23)	Shi and Chen (2004)
Cytochrome CYP2D6	*CYP2D6*	22q13.1	Association between possession of additional copy of *CYP2D6* gene and lung cancer. However, most published studies are not supportive of a link between *CYP2D6* polymorphisms and lung cancer susceptibility (see Tefre *et al.*, 1994; Caporaso *et al.*, 1995; Legrand *et al.*, 1996; Christensen *et al.*, 1997; Shaw *et al.*, 1998)	Bouchardy *et al.* (1996) London *et al.* (1997) LaForest *et al.* (2000) Agundez *et al.* (2001)
Cytochrome CYP2E1	*CYP2E1*	10q24.3-qter	Crude association study using *Dra*I SNP Studies differ with regard to the allele that is supposed to confer susceptibility to lung cancer: OR=1.3 for c1/c1 (Wu), OR=2.45 for c2/c2 (Oyama)	Uematsu *et al.* (1991) Oyama *et al.* (1997b) Wu *et al.* (1997)
			*Pst*I allele over-represented in lung cancer (OR=3.5)	El-Zein *et al.* (1997b)
Cytochrome CYP3A4	*CYP3A4*	7q22.1	CYP3A4*1B allele associated with elevated risk of SCLC (OR=2.25)	Dally *et al.* (2003)
Cytochrome CYP3A5	*CYP3A5*	7q21.1	CYP3A5*1 allele present at lower frequency in lung cancer patients (OR=1.5)	Yeh *et al.* (2003)

Table 4. Studies claiming an association between possession of a polymorphic allele and risk of lung cancer (continued)

Gene	Gene symbol	Chromosomal localization	Polymorphism association	Reference
Advanced glycosylation endproduct-specific receptor (RAGE)	*AGER*	6p21.3	Homozygosity for T→A variant in promoter region, -388 relative to ATG, occurred significantly more frequently in NSCLC patients than in controls	Schenk *et al.* (2001)
Cyclooxygenase 2 (COX-2)	*PTGS2*	1q25.2-q25.3	Carriers of C allele of 3'UTR polymorphism had increased risk of lung cancer; OR (homozygotes)=4.3, OR (heterozygotes)=2.1	Campa *et al.* (2004)
Sulphotransferase 1A1	*SULT1A1*	16p12.1	Association between Arg213His polymorphic alleles and risk of lung cancer (OR=1.4)	Wang *et al.* (2002c)
			Association between 638GA and AA genotypes of Arg213His polymorphism and risk of lung cancer (OR=1.8)	Liang *et al.* (2004)
Surfactant protein B	*SFTPB*	2p12-p11.2	Association between intron 4 CA repeat polymorphic alleles and risk of lung cancer (OR=3.2)	Seifart *et al.* (2002)
Metalloproteinase-1	*MMP1*	11q21-q22	Association between possession of 2G/2G promoter genotype and lung cancer risk (OR =1.8)	Zhu *et al.* (2001)
Metalloproteinase-2	*MMP2*	16q13	Association between possession of CC genotype of promoter (-1306) polymorphism and lung cancer risk (OR=2.2)	Yu *et al.* (2002)
p53	*TP53*	17p13	Higher proportion of heterozygotes for intron 6 *Msp*I RFLP in "lung cancer" patients (OR=1.83).	Biros *et al.* (2001a)
			Significant excess of Arg/Arg72 homozygotes among non-smoker lung cancers	Murata *et al.* (1996)
			Association claimed between heterozygosity for Arg/Pro 72 polymorphism and survival time	Wang *et al.* (1999b)
			Significant excess of Pro/Pro72 homozygotes among smoking NSCLC cases	Hiraki *et al.* (2003)
			Significant excess of Pro/Pro72 homozygotes among smoking 'lung cancer' cases (OR=3.88)	Irarrazabal *et al.* (2003)

Gene	Gene symbol	Chromosomal localization	Polymorphism association	Reference
X-ray cross complementing group 1 protein	*XRCC1*	19q13.2	Association between possession of Gln/Gln399 genotype and lung cancer risk (OR=2.45)	Divine *et al.* (2001)
			Association between possession of Arg280His allele and lung cancer risk (OR=1.8)	Ratnasinghe *et al.* (2001a)
			Increased frequency of Gln399 allele in lung cancer	Hou *et al.* (2003)
			Increased frequency of Trp194 allele in lung cancer (OR=3.1)	Chen *et al.* (2002a)
			Decreased frequency of Trp194 allele in lung cancer (OR=0.7)	Ratnasinghe *et al.* (2003)
			Association between possession of Gln/Gln399 genotype and lung cancer risk (OR=3.3)	Park *et al.* (2002b)
			Increased frequency of Gln399 allele in lung cancer (OR=1.3)	Zhou *et al.* (2003a)
Excision repair, complementation group 2, Xeroderma pigmentosum complementing group D (DNA repair protein XPD)	*ERCC2*	19q13.3	Increased frequency of Asn312 allele in lung cancer (OR=1.8)	Hou *et al.* (2003) Xing *et al.* (2002)
			Association between possession of Lys/Gln751genotype and lung cancer risk (OR=3.2)	Chen *et al.* (2002a) Butkiewicz *et al.* (2001)
			Association between possession of Asn/Asn312 genotype and lung cancer risk (OR=1.5)	Zhou *et al.* (2002b) Zhou *et al.* (2003a)
			Increased frequency of homozygosity for Asn312 allele (OR=10.3) and Gln751allele (OR=2.7) in lung cancer	Liang *et al.* (2003)
			In case-cohort study, female Gln751 homozygotes were at increased risk of lung cancer for age intervals 50-55 (RR=5.6) and 56-60 (RR=10.6)	Vogel *et al.* (2004)
Excision repair, complementing defective 5, Xeroderma pigmentosum complementing group G (DNA repair gene XPG)	*ERCC5*	13q33	Asp1104 allele associated with decreased risk of squamous cell carcinoma (OR=0.55) and SCLC (OR=0.44)	Jeon *et al.* (2003)
Xeroderma pigmentosum complementing group A (DNA repair gene XPA)	*XPA*	9q22.3	GG709 genotype associated with decreased risk of lung cancer (OR=0.56)	Park *et al.* (2002c)
			GG709 genotype associated with decreased risk of lung cancer (OR=0.69)	Wu *et al.* (2003a)

Table 4. Studies claiming an association between possession of a polymorphic allele and risk of lung cancer (continued)

Gene	Gene symbol	Chromosomal localization	Polymorphism association	Reference
Neutrophil elastase	*ELA2*	19p13.3	Associations between TT-903 (OR=2.3) and GG-741 (OR=1.4) alleles and lung cancer	Taniguchi *et al.* (2002)
α1-Antitrypsin	*SERPINA1*	14q32.1	Possible association between carriership of α1-antitrypsin deficiency alleles and squamous cell (15.9%) and bronchoalveolar (23.8%) carcinoma of the lung. By comparison, reported carrier rate for US Caucasians taken as 7%	Yang *et al.* (1999b)
Transferrin	*TF*	3q21	Disproportionately low frequency of transferrin C3 allele in smokers with lung tumours (OR=0.03)	Sikstrom *et al.* (1996) Beckman *et al.* (1999)
L-*myc*	*MYCL1*	1p34.3	Association between LL genotype and lung cancer Association between SS (OR=3.2) and LS (OR=2.3) genotypes and lung cancer but not LL (OR=0.9)	Shih *et al.* (2002) Kumimoto *et al.* (2002)
Cyclin D1	*CCND1*	11q13	Association between homozygosity for A870 allele and risk of NSCLC (OR=1.95)	Qiuling *et al.* (2003)
Cytosine DNA methyltransferase 3β	*DNMT3B*	20q11.2	Association between combined CT and TT promoter genotypes and lung cancer (OR=1.9)	Shen *et al.* (2002)
ADP-ribosyltransferase pseudogene [poly(ADP-ribose) polymerase pseudogene]	*ADPRTP1*	13q34	193bp deletion polymorphism (B allele) reportedly had higher prevalence in lung cancer patients but see Choi *et al.* (2003)	Bhatia *et al.* (1990) Wu *et al.* (1998)
Methylenetetrahydrofolate reductase	*MTHFR*	1p36.3	Both SCLC and NSCLC patients claimed to exhibit statistically higher frequency of 677 TT genotype than non-cancer controls 677 CT and 677 TT genotypes found to constitute 39% of lung cancer cases and 47% of controls (OR=0.71). 677T allele may reduce risk of lung cancer?	Siemianowicz *et al.* (2003) Jeng *et al.* (2003)
Transforming growth factor, beta receptor 1	*TGFBR1*	9q22	Individuals homozygous for A allele of intron 7 polymorphism have 3-fold greater risk of NSCLC than those who are homozygous for the common G allele	Zhang *et al.* (2003a)

Gene	Gene symbol	Chromosomal localization	Polymorphism association	Reference
Myeloperoxidase	*MPO*	17q21.3-q23	Association between homozygosity for the A allele of G/A –463 promoter polymorphism and decreased risk of lung cancer (OR 0.5). Base change reduces binding of SP1 transcription factor thereby lowering *MPO* gene expression, and (putatively) reducing metabolic conversion of procarcinogenic compounds to reactive metabolites.	Le Marchand *et al.* (2000) Schabath *et al.* (2000)
			Decreased risk of lung cancer with G/A and A/A genotypes (OR=0.66)	Schabath *et al.* (2002)
			Decreased risk of lung cancer with A/A genotype (OR=0.39)	Kantarci *et al.* (2002)
			Decreased risk of lung cancer with G/A and A/A genotypes (OR=0.55)	Feyler *et al.* (2002)
			Decreased risk of lung cancer with G/A and A/A genotypes (OR=0.75)	Dally *et al.* (2002)
			GG homozygotes at an increased risk of squamous cell carcinoma (OR=2.3) but not adenocarcinoma.	Lu *et al.* (2002)
			Increased risk of lung cancer with A/A genotype	Wu *et al.* (2003e)
Nitric oxide synthase, endothelial	*NOS3*	7q35-q36	Distribution of genotypes significantly different between lung cancer group and controls	Cheon *et al.* (2000)
Fas antigen (Apo-1/CD95)	*TNFRSF6*	10q24	Association between Fas –670 AG and GG genotypes and risk of lung cancer (OR=4.00) but not between AA genotype and lung cancer (OR=0.97)	Wang *et al.* (2003e)
H-Ras	*HRAS*	11p15	Association of rare alleles of the Variable Number of Tandem Repeats (VNTR) locus 1kb 3' to the *HRAS* gene and lung cancer risk (OR=1.68, 3.3, 2.2)	Ryberg *et al.* (1990) Sugimura *et al.* (1990) Weston and Godbold (1997) Rosell *et al.* (1999) Lindstedt *et al.* (1999) Pierce *et al.* (2000)
8-α-oxyguanine DNA glycosylase	*OGG1*	3p25-p26	Association between possession of Cys/Cys326 genotype and lung cancer risk (OR=3.01). Cys326 is less effective in repair of 8-hydroxyguanine than Cys allele. NB, Study results not confirmed by Wikman *et al.* (2000)	Sugimura *et al.* (1999)

Table 4. Studies claiming an association between possession of a polymorphic allele and risk of lung cancer (continued)

Gene	Gene symbol	Chromosomal localization	Polymorphism association	Reference
8-α-oxyguanine DNA glycosylase	*OGG1*	3p25-p26	Association between possession of Ser/Cys326 genotype and lung cancer risk (OR=1.86)	Ito *et al.* (2002a)
			Association between possession of Cys/Cys326 genotype and lung cancer risk (OR=2.1)	Le Marchand *et al.* (2002)
			Association between possession of polymorphic allele 3 (G→T transition at nucleotide −18) and lung adenocarcinoma (OR=3.15)	Ishida *et al.* (1999)
			Association between possession of Cys326 allele and lung cancer risk (OR=3.8)	Park *et al.* (2004)
Telomerase	*TERT*	5p15.33	Association between LL genotype of MNS16A polymorphic minisatellite downstream of the *TERT* gene and lung cancer risk (OR=2.18)	Wang *et al.* (2003f)
O6-methylguanine DNA methyltransferase	*MGMT*	10q26	Association between Ile/Val 143 and Arg/Lys 178 polymorphic alleles and lung cancer risk (OR=2.1). However, Yang *et al.* (2004a) failed to repeat this finding	Kaur *et al.* (2000) Cohet *et al.* (2004)
Heat shock 70 kD protein 8 (HSC70)	*HSPA8*	11q23.3-q25	Association between possession of intron 2 1541-1542delGT allele and decreased risk of lung cancer (OR=0.44). The variant is associated with weaker immunohistochemical staining for the HSC70 protein in NSCLC tumours and a 20 % reduction in gene expression in *in vitro* reporter gene assays. These findings are intriguing in the light of the putative role of HSC70 in the cellular response to oxidative stress (Dastoor and Dreyer 2000)	Rusin *et al.* (2004)
Interleukin 1-beta	*IL1B*	2q14	Over-representation of T allele at −31 C/T polymorphism and C allele at −511 C/T polymorphism in NSCLC patients; homozygous −31 T/T (OR=2.4) and homozygous −511 C/C (OR=2.5) genotypes are risk factor for NSCLC	Zienolddiny *et al.* (2004)
Cyclin-dependent kinase inhibitor 1A (p21; WAF1/CIP1)	*CDKN1A*	6p21.2	Association between possession of Arg31 polymorphic allele and lung cancer (OR=1.7)	Sjalander *et al.* (1996)
			No association found between possession of Arg31 allele and lung cancer (Shih *et al.*, 2000; Su *et al.*, 2003b)	

Table 5. Human genes whose expression is altered in lung cancer

Gene	Gene symbol	Chromosomal localization	Up- or down-regulated	Tumour type	Reference
Ras homologue family member, RhoC	*ARHC*	1p13-p21	↑	NSCLC	Shikada *et al.* (2003)[2]
Cysteine-rich protein 61	*CYR61*	1p22	↓	NSCLC	Tong *et al.* (2001a)[2]
Dihydropyrimidine dehydrogenase	*DPYD*	1p22	↑	NSCLC	Fukushima *et al.* (2003)[1]
Collagen type XI α1	*COL11A1*	1p21	↑	NSCLC	Wang *et al.* (2002a)[2]
Enolase-α	*ENO1*	1p36.2	↓	NSCLC	Chang *et al.* (2003)[1]
S100 calcium-binding protein A2	*S100A2*	1q21-q24	↓	NSCLC	Feng *et al.* (2001)[1,2]
Phosphatidylinositol 3-kinase class 2 subunit, β polypeptide	*PIK3C2B*	1q32	↑	?	Lin *et al.* (2001)[1,2]
Cyclooxygenase 2 (COX-2)	*PTGS2*	1q25.2-q25.3	↑	NSCLC	Hida *et al.* (1998)[1] Ochiai *et al.* (1999)[2] Hasturk *et al.* (2002)[1] Takahashi *et al.* (2002)[1] Tian *et al.* (2003)[1] Petkova *et al.* (2004)[1]
JUN oncogene	*JUN*	1p31-p32	↑↓	?	Levin *et al.* (1994a)[2]; (1995a)[2] Szabo *et al.* (1996)[2]
SCL interrupting locus (TAL1/TCL5)	*SIL*	1p32	↑	?	Erez *et al.* (2004)[1,2]
Peroxiredoxin 1	*PRDX1*	1p34.1	↑	?	Kim *et al.* (2003f)[1]
p73	*TP73*	1p36	↑	?	Tokuchi *et al.* (1999b)[2]
Angiopoietin-like 1 (angioarrestin)	*ANGPTL1*	1p36.13-q31.3	↓	?	Sasaki *et al.* (2003c)[2]
Small proline-rich protein 1	*SPRR1A*	1q21-q22	↑	?	Lau *et al.* (2000a)[2]
Ephrin A3	*EFNA3*	1q21-q22	↑	?	Hafner *et al.* (2004)[2]
S100 calcium-binding protein A2	*S100A2*	1q21	↓	NSCLC	Feng *et al.* (2001)[1,2]
Glut1	*SLC2A1*	1p31.3-p35	↑	NSCLC	Younes *et al.* (1997)[1]
Mucin 1	*MUC1*	1q21	↑ ↓	NSCLC	Nguyen *et al.* (1996)[1,2] Seregni *et al.* (1996)[2] Awaya *et al.* (2004)[1]
Cathepsin K	*CTSK*	1q21	↑	?	Buhling *et al.* (2000)[1,2]
Fas ligand	*TNFSF6*	1q23	↓ ↑	NSCLC SCLC	Viard-Leveugle *et al.* (2003)[1] Badillo-Almaraz *et al.* (2003)[1]
RNA helicase A	*DHX9*	1q25	↑	SCLC NSCLC	Wei *et al.* (2004)[2]

Table 5. Human genes whose expression is altered in lung cancer (continued)

Gene	Gene symbol	Chromosomal localization	Up- or down-regulated	Tumour type	Reference
Cytochrome CYP4B1	*CYP4B1*	1p12-p34	↓	?	Czerwinski *et al.* (1994)[2]
Akt/protein kinase B	*AKT3*	1q43-q44	↑	NSCLC	Lee *et al.* (2002a)[1]
Interleukin 10	*IL10*	1q31-q32	↑	NSCLC	Kamiya *et al.* (2003)[2]
UDP-N-acetyl-alpha-D-galactosamine: polypeptide N-acetyl-galactosaminyl transferase 3	*GALNT3*	2q24-q31	↓	NSCLC	Dosaka-Akita *et al.* (2002)[1]
Villin	*VIL1*	2q35-q36	↑	NSCLC	Nambu *et al.* (1998)[1,2]
Neuropilin 2 (VEGF receptor 2)	*NRP2*	2q34	↑	NSCLC	Lantuejoul *et al.* (2003)[1]
Reticulon 4 (ASY/Nogo-B)	*RTN4*	2p13-p14	↓	SCLC	Li *et al.* (2001b)[2]
DNA mismatch repair gene, MSH2	*MSH2*	2p16	↓	NSCLC	Zhu *et al.* (2003a)[1]
Ras homologue gene family, member B	*RHOB*	2pter-p12	↓	?	Mazieres *et al.* (2004)[1]
Pro-opiomelano-cortin	*POMC*	2p23	↑	Neuro-endo-crine	Black *et al.* (1993)[2]
Syndecan 1	*SDC1*	2p24.1	↑	NSCLC	Nanki *et al.* (2001)[2]
Phospholipase C	*PLCL1*	2q33	↓	SCLC/ NSCLC	Kohno *et al.* (1995)[2]
Transforming growth factor α	*TGFA*	2p13	↑	NSCLC	Tateishi *et al.* (1990)[1] Rusch *et al.* (1993)[2]
Dipeptidyl peptidase IV (CD26)	*DPP4*	2q23-qter	↓	NSCLC	Wesley *et al.* (2004)[1,2]
Insulin-like growth factor binding protein 2	*IGFBP2*	2q33-q34	↑	NSCLC	Jaques *et al.* (1992)[1,2] Reeve *et al.* (1992)[2]
Insulin-like growth factor binding protein 5	*IGFBP5*	2q33-q36	↑	NSCLC	Wegmann *et al.* (1993)[2]
Epidermal growth factor receptor 4	*ERBB4*	2q33-q34	↑	NSCLC	al Moustafa *et al.* (1999)[1]
Thymosin-α1	*PTMA*	2q35-q36	↑	NSCLC	Sasaki *et al.* (1997)[1] Sasaki *et al.* (2001f)[1]
Homeobox D3	*HOXD3*	2q31-q32	↑	?	Hamada *et al.* (2001)[2]
Wingless-type MMTV integration site, member 7A	*WNT7A*	3p25	↓	NSCLC	Calvo *et al.* (2000)[2]
Ring finger protein 7 (sensitive to apoptosis, SAG)	*RNF7*	3q22-q24	↑	NSCLC	Sasaki *et al.* (2001a)[2]

Table 5. Human genes whose expression is altered in lung cancer (continued)

Gene	Gene symbol	Chromo-somal localization	Up- or down-regulated	Tumour type	Reference
Transferrin receptor	*TFRC*	3q29	↑	NSCLC	Whitney *et al.* (1995)[1]
Forkhead box P1	*FOXP1*	3p14.1	↓	SCLC	Banham *et al.* (2001)[1,2]
Fragile histidine triad (FHIT)	*FHIT*	3p14.2	↓	NSCLC	Sozzi *et al.* (1998)[1] Pylkkanen *et al.* (2002)[1] Zhu *et al.* (2003a)[1] Mascaux *et al.* (2003)[1]
RNA-binding motif protein 5 (H37)	*RBM5*	3p21.3	↓	NSCLC	Oh *et al.* (2002)[1]
Axin 1-upregulated gene	*AXUD1*	3p22	↓	?	Ishiguro *et al.* (2001)[2]
Transforming growth factor beta receptor II	*TGFBR2*	3p22	↓	NSCLC	Colasante *et al.* (2003)[1,2] Zhang *et al.* (2004a)[1,2]
Peroxisome prolife-rator-activated receptor γ	*PPARG*	3p25	↑	NSCLC	Keshamouni *et al.* (2004)[1]
Retinoic acid receptor β	*RARB*	3p24	↓	NSCLC	Gebert *et al.* (1991)[2] Zhang *et al.* (1994)[2] Xu *et al.* (1997)[2] Inui *et al.* (2003)[2]
			↑	NSCLC	Chang *et al.* (2004c)[1]
RAS-associated protein, RAB5A	*RAB5A*	3p22-p24	↑	NSCLC	Li *et al.* (1999)[2]
Deleted in lung and oesophageal cancer 1	*DLEC1*	3p21.3-p22	↓	NSCLC	Daigo *et al.* (1999)[2]
Semaphorin 3B	*SEMA3B*	3p21.3	↓	NSCLC/ SCLC	Sekido *et al.* (1996)[2] Lerman and Minna (2000)[2] Lantuejoul *et al.* (2003)[1]
Semaphorin 3F	*SEMA3F*	3p21.3	↓	NSCLC	Brambilla *et al.* (2000)[1]
Carboxy-terminal domain, RNA polymerase II, polypeptide A, small phosphatase-like (HYA22/RBSP3)	*CTDSPL*	3p21.3	↓↑	SCLC	Kashuba *et al.* (2004)[2]
Ubiquitin-specific protease 4 (Unph)	*USP4*	3p21	↑	SCLC/ NSCLC	Gray *et al.* (1995)[2]
β-Catenin	*CTNNB1*	3p21	↓	NSCLC	Soo Choi *et al.* (2003)[1]
Integrin, alpha 9	*ITGA9*	3p21.3	↑	SCLC	Hibi *et al.* (1994)[2]
Ceruloplasmin	*CP*	3q23-q25	↑	NSCLC	Wang *et al.* (2002a)[2]
Mitofusin 1 (trans-membrane GTPase)	*MFN1*	3q27.1	↑	NSCLC/ SCLC	Chung *et al.* (2001)[2]

Table 5. Human genes whose expression is altered in lung cancer (continued)

Gene	Gene symbol	Chromosomal localization	Up- or down-regulated	Tumour type	Reference
Dishevelled, *Drosophila* homologue 3	*DVL3*	3q27	↑	NSCLC	Uematsu *et al.* (2003)[1,2]
Basic helix-loop-helix domain containing, class B, 2	*BHLHB2*	3p26	↓	NSCLC	Giatromanolaki *et al.* (2003)[1]
Telomerase RNA	*TERC*	3q26	↑	NSCLC	Yokoi *et al.* (2003b)[2]
Laminin receptor 1	*LAMR1*	3p21.3	↑	NSCLC	Fontanini *et al.* (1997a)[1]
Neuropilin-related gene, CLCP1/Discoidin, CUB & LCCL domain-cont. 2	*DCBLD2*	3q12.2	↑	?	Koshikawa *et al.* (2002)[2]
Cytochrome *c* oxidase assembly protein	*COX17*	3q13.1-q21	↑	NSCLC	Suzuki *et al.* (2003a)[2]
Mucin 4	*MUC4*	3q29	↑	NSCLC	Nguyen *et al.* (1996)[1,2] Seregni *et al.* (1996)[2]
Membrane metallo-endopeptidase (CD10)	*MME*	3q21-q27	↑↓	NSCLC	Tokuhara *et al.* (2001b)[2] Kristiansen *et al.* (2002)[1]
Protein tyrosine phosphatase, receptor type, G	*PTPRG*	3p14-p21	↓	NSCLC	Van Niekerk and Poels (1999)[1]
Lipopolysaccharide-responsive and *beige*-like anchor	*LRBA*	4	↑	?	Wang *et al.* (2004c)[2]
Collapsin response mediator protein-1	*CRMP1*	4p16.1-p15	↓	NSCLC	Shih *et al.* (2003)[2]
Cyclin A2	*CCNA2*	4q27	↑	NSCLC	Müller-Tidow *et al.* (2001)[2] Dobashi *et al.* (2003)[1]
SPARC-like 1 (hevin)	*SPARCL1*	4q22-q25	↓	NSCLC	Bendik *et al.* (1998)[1,2]
Breast cancer resistance protein (BCRP)	*ABCG2*	4q22-q23	↑↓	NSCLC	Kawabata *et al.* (2003)[2]
Alcohol dehydrogenase 1C, γ polypeptide	*ADH3*	4q22	↓	NSCLC	Inui *et al.* (2003)[2]
Lung cancer-associated gene Y	*Lagy*	4q11-q13.1	↓	NSCLC	Chen *et al.* (2003a)[2]
Osteopontin	*SPP1*	4q11-q21	↑	?	Schneider *et al.* (2004a)[2]

Table 5. Human genes whose expression is altered in lung cancer (continued)

Gene	Gene symbol	Chromosomal localization	Up- or down-regulated	Tumour type	Reference
Tyrosine kinase receptor, Kit	*KIT*	4q12	↑↓	NSCLC/ SCLC	Hibi *et al.* (1991)[2] Sekido *et al.* (1981)[1]; (1991)[2], (1993)[1] Rygaard *et al.* (1993)[2] Hida *et al.* (1994)[1] Krystal *et al.* (1996)[1] Pietsch *et al.* (1998)[1] Naeem *et al.* (2002)[1] Micke *et al.* (2003)[1] Arakai *et al.* (2003)[1] Potti *et al.* (2003)[1] Boldrini *et al.* (2004)[1]
Ubiquitin carboxyl-terminal esterase L1 (PGP9.5)	*UCHL1*	4p14	↑	NSCLC	Bittencourt Rosas *et al.* (2001)[1,2] Hibi *et al.* (1999)[1]
Amphiregulin	*AREG*	4q13-q21	↓↑	NSCLC	Rusch *et al.* (1993)[2] Rusch *et al.* (1997)[1] Fontanini *et al.* (1998)[1]
Basic fibroblast growth factor	*FGF2*	4q25-q27	↑	NSCLC	Kuhn *et al.* (2004)[1,2]
Vascular endothelial growth factor C	*VEGFC*	4q33-q34	↑	NSCLC	Li *et al.* (2003b)[1]
Caspase 3	*CASP3*	4q35	↑	NSCLC	Krepela *et al.* (2004)[1]
Dentin matrix protein 1	*DMP1*	4q21	↑	NSCLC	Chaplet *et al.* (2003)[1,2]
Checkpoint control gene, Rad17	*RAD17*	5q13	↑	NSCLC	Sasaki *et al.* (2001e)[2]
Cyclin B1	*CCNB1*	5q12	↑	NSCLC	Arinaga *et al.* (2003)[1]
Programmed cell death 6 (Alg-2)	*PDCD6*	5p15.2-pter	↑	SCLC/ NSCLC	La Cour *et al.* (2003)[1]
S-phase kinase-associated protein 2	*SKP2*	5p13	↑	NSCLC	Yokoi *et al.* (2004)[2]
TGFβ-induced gene	*TGFBI*	5q31	↑ ↓	? SCLC	Sasaki *et al.* (2002a)[2] de Jonge *et al.* (1997)[2]
Platelet-derived growth factor receptor beta	*PDGFRB*	5q31-q32	↑	?	Antoniades *et al.* (1992)[1]
DNA polymerase kappa	*POLK*	5q13	↑	NSCLC	O-Wang *et al.* (2001)[1,2] Wang *et al.* (2004d)[1]
Occludin	*OCLN*	5q13.1	↓	NSCLC/ SCLC	Tobioka *et al.* (2004)
α-Methylacyl-CoA racemase	*AMACR*	5p13.2-q11.1	↑	?	Zhou *et al.* (2002c)[1]
Granulocyte-macrophage colony-stimulating factor	*CSF2*	5q31	↑	NSCLC	Kamiya *et al.* (2003)[2]

Table 5. Human genes whose expression is altered in lung cancer (continued)

Gene	Gene symbol	Chromosomal localization	Up- or down-regulated	Tumour type	Reference
α-Catenin	*CTNNA1*	5q31	↓	NSCLC	Toyoyama *et al.* (1999)[1]
Glycine receptor α1	*GLRA1*	5q32	↑	SCLC	Gurrola-Diaz *et al.* (2003)[2]
Secreted protein, acid, cysteine-rich (osteonectin)	*SPARC*	5q31-q32	↑	NSCLC	Koukourakis *et al.* (2003)[1] Siddiq *et al.* (2004)[1]
Mitogen-activated protein kinase phosphatase 1 (MKP-1/ Dual specificity phosphatase 1)	*DUSP1*	5q34	↑	NSCLC	Vicent *et al.* (2004)[1]
Pituitary tumour transforming gene 1 (securin)	*PTTG1*	5q35.1	↑	NSCLC	Honda *et al.* (2003)[2]
Transcription factor TFIID	*TBP*	6q27	↑	NSCLC & SCLC	McDonald & Pilgram (1999)[1]
Cyclin-dependent kinase inhibitor 1A (p21)	*CDKN1A*	6p21.2	↑ ↓	NSCLC	Marchetti *et al.* (1996)[1] Takeshima *et al.* (1998)[1] Hayashi *et al.* (1997)[1] Kameyama *et al.* (2003)[2] Ito *et al.* (2004)[1,2]
γ-Glutamyl cysteine synthetase	*GCLC*	6p12	↑	NSCLC	Soini *et al.* (2001)[1]
Receptor for advanced glycosylation end products (RAGE)	*AGER*	6p21.3	↑	?	Hsieh et al. (2003)[2]
Tumour necrosis factor α	*TNF*	6p21.3	↓	?	Badillo-Almaraz *et al.* (2003)[1]
Tumour differentially expressed gene 2	*TDE2*	6q22.32	↑	NSCLC	Player *et al.* (2003)[2]
Superoxide dismutase 2	*SOD2*	6q25	↑	?	Chung-man Ho *et al.* (2001)[1,2]
Estrogen receptor αβ	*ESR1*	6q24-q27	↑	NSCLC	Stabile *et al.* (2002)[1,2]
5-hydroxytryptamine receptor 1B	*HTR1B*	6q13	↓	NSCLC	Takai *et al.* (2001)[2]
Cell surface antigen CD109	*CD109*	6q14.1	↑	NSCLC	Hashimoto *et al.* (2004)[2]
N-Oct 3 transcription factor	*POU3F2*	6q16	↑	SCLC	Schreiber *et al.* (1992)[1]
Endothelin 1	*EDN1*	6p23-p24	↓	NSCLC	Takai *et al.* (2001)[2]

Table 5. Human genes whose expression is altered in lung cancer (continued)

Gene	Gene symbol	Chromosomal localization	Up- or down-regulated	Tumour type	Reference
Vascular endothelial growth factor	*VEGF*	6p12	↑	NSCLC	Cheung *et al.* (1998)[1] Takahama *et al.* (1999)[2] O'Byrne *et al.* (2000)[1] Yuan *et al.* (2000)[1,2] Lantuejoul *et al.* (2003)[1] Stefanou *et al.* (2003b)[1]
Parkin	*PARK2*	6q25.2-q27	↓	?	Picchio *et al.* (2004)[2]
Thrombospondin 2	*THBS2*	6q27	↓	NSCLC	Oshika *et al.* (1998a)[2]
Hepatocyte growth factor/ scatter factor	*HGF*	7q21.1	↑	NSCLC	Olivero *et al.* (1996)[1] Harvey *et al.* (1996)[1]
Met	*MET*	7q31	↑	NSCLC	Rygaard *et al.* (1993)[1,2] Ichimura *et al.* (1996)[1]
Homeobox HOXB9	*HOXB9*	7p14-p15	↑	NSCLC	Calvo *et al.* (2000)[2]
Homeobox HOXA9	*HOXA9*	7p14-p15	↑	NSCLC	Calvo *et al.* (2000)[2]
Homeobox HOXA10	*HOXA10*	7p14-p15	↑	NSCLC	Calvo *et al.* (2000)[2]
Heterogeneous nuclear ribonucleoprotein A2/B1	*HNRPA2B1*	7p15	↑	?	Zhou *et al.* (2001c)[2] Wu *et al.* (2003b)[1] Pino *et al.* (2003)[1]
Aryl hydrocarbon receptor	*AHR*	7p15	↑	NSCLC	Lin *et al.* (2003)[1,2]
Insulin-like growth factor -binding protein 1	*IGFBP1*	7p12-p13	↑	NSCLC	Reeve *et al.* (1992)[2]
Insulin-like growth factor -binding protein 3	*IGFBP3*	7p12-p13	↑	NSCLC	Jaques *et al.* (1992)[1,2] Wang *et al.* (2003d)[1,2]
Dopa decarboxylase	*DDC*	7p11	↑	SCLC	Vachtenheim & Novotna (1997)[2]
Epidermal growth factor receptor	*EGFR*	7p12	↑	NSCLC	Tateishi *et al.* (1990)[1] Damstrup *et al.* (1992)[1] Rusch *et al.* (1993)[2]; (1997)[2] Rachwal *et al.* (1995)[1] Sekine *et al.* (1998)[2] Fontanini *et al.* (1998)[1] Mukohara *et al.* (2003)[1] Hilbe *et al.* (2003b)[1,2] Hirsch *et al.* (2003)[1]
ATP-binding cassette, sub-family B (MDR1)	*ABCB1*	7q21.1	↓	NSCLC	Abe *et al.* (1994)[2]
Wingless-type MMTV integration site, member 2	*WNT2*	7q31	↑	NSCLC	Uematsu *et al.* (2003)[1] You *et al.* (2004)[1,2]

Table 5. Human genes whose expression is altered in lung cancer (continued)

Gene	Gene symbol	Chromosomal localization	Up- or down-regulated	Tumour type	Reference
Caveolin-1	*CAV1*	7q31	↑↓	SCLC/ NSCLC	Yoo *et al.* (2003)[1] Wikman *et al.* (2004)[1,2] Sunaga *et al.* (2004)[2]
Caveolin-2	*CAV2*	7q31	↓	SCLC/ NSCLC	Wikman *et al.* (2004)[1,2]
Pleiotrophin	*PTN*	7q33	↓	NSCLC	Garver *et al.* (1993)[2]
Mucin 3B	*MUC3B*	7q22	↑	NSCLC	Nguyen *et al.* (1996)[1,2]
MOS oncogene	*MOS*	8q11	↑	NSCLC	Gorgoulis *et al.* (2001)[1]
Leucine zipper, putative tumour suppressor 1	*LZTS1*	8p22	↓	NSCLC/ SCLC	Toyooka *et al.* (2002)[1,2]
Eukaryotic translation IF3 subunit 6	*EIF3S6*	8q22-q23	↓	NSCLC	Marchetti *et al.* (2001)[2]
Tumour necrosis factor-related apoptosis-inducing ligand receptor, TRAIL-R2	*TNFRSF10B*	8p22	↑	NSCLC	Wu *et al.* (2000)[2]
BCL2/adenovirus E1B 19kDa interacting protein 3-like	*BNIP3L*	8p21	↓	?	Sun *et al.* (2004a)[1,2]
A disintegrin and metalloproteinase domain 9	*ADAM9*	8p11.22	↑	NSCLC	Shintani *et al.* (2004)[2]
Angiopoietin-1	*ANGPT1*	8q22	↑	NSCLC	Takahama *et al.* (1999)[2]
Angiopoietin-2	*ANGPT2*	8q23	↑	NSCLC	Xing *et al.* (2003)[2]
CDC28 protein kinase regulatory subunit 1B	*CKS1B*	8q21	↑	NSCLC	Inui *et al.* (2003b)[1,2]
Cathepsin B	*CTSB*	8p22	↓	SCLC	Kayser *et al.* (2003)[1]
Deleted in liver cancer	*DLC1*	8p21.3-p22	↓	NSCLC	Yuan *et al.* (2004)[2]
Carbonic anhydrase I	*CA1*	8q22	↓	NSCLC	Chiang *et al.* (2002)[1]
Carbonic anhydrase II	*CA2*	8q22	↓	NSCLC	Chiang *et al.* (2002)[1]
Oestrogen receptor binding site-associated antigen 9 (RCAS1)	*EBAG9*	8q23	↑	NSCLC	Izumi *et al.* (2001)[1]
Myc oncogene	*MYC*	8q24	↑	?	Yoshimoto *et al.* (1986b)[2] Sasaki *et al.* (2001f)[1]
Cyclin-dependent kinase inhibitor 2B (p15)	*CDKN2B*	9p21	↑↓	Neuroendocrine	Chaussade *et al.* (2001)[1]

Table 5. Human genes whose expression is altered in lung cancer (continued)

Gene	Gene symbol	Chromosomal localization	Up- or down-regulated	Tumour type	Reference
TEK tyrosine kinase, Tie2	*TEK*	9p21	↑	NSCLC	Takahama *et al.* (1999)[2]
Prostaglandin E synthase	*PTGES*	9q34.3	↑	NSCLC	Yoshimatsu *et al.* (2001)[1,2]
Prostaglandin-endoperoxide synthase 1	*PTGS1*	9q32-q33.3	↑	NSCLC	Hasturk *et al.* (2002)[1]
Transforming growth factor beta receptor I	*TGFBR1*	9q33-q34	↑	NSCLC	Colasante *et al.* (2003)[1,2]
Gelsolin	*GSN*	9q34	↓	NSCLC	Dosaka-Akita *et al.* (1998)[1,2]
Hyaluronan-binding protein 2	*HABP2*	10q	↑	NSCLC	Wang *et al.* (2002a)[2]
Phosphatase and tensin homologue (PTEN)	*PTEN*	10q23.3	↓	NSCLC	Mori *et al.* (2004)[1,2] Goncharuk *et al.* (2004)[1]
Cytochrome P450 2E1	*CYP2E1*	10q24.3-qter	↓	?	Botto *et al.* (1994)[1,2]
Deleted in malignant brain tumours 1	*DMBT1*	10q25-q26	↓ ↑	?	Takeshita *et al.* (1999)[2] Mollenhauer *et al.* (2002)[2]
Programmed cell death 4	*PDCD4*	10q24	↓	NSCLC	Chen *et al.* (2003c)[2]
Fas	*TNFRSF6*	10q24	↓	NSCLC/ SCLC	Viard-Leveugle *et al.* (2003)[1]
NFκB transcription factor	*NFKB2*	10q24	↑	NSCLC	Mukhopadhyaya *et al.* (1995)[1]
Growth/ differentiation factor 10 (bone morphogenetic protein 3B)	*GDF10*	10q11.22	↓	NSCLC	Dai *et al.* (2001)[2]
Neuropilin 1 (VEGF receptor 1)	*NRP1*	10p12	↑	NSCLC	Lantuejoul *et al.* (2003)[1]
Dihydrodiol dehydrogenase 1 (DDH1)	*AKR1C1*	10p14-p15	↑	NSCLC	Hsu *et al.* (2001)[1]
Krüppel-like factor 6	*COPEB*	10p15	↓	NSCLC	Ito *et al.* (2004)[1,2]
CD151 antigen	*CD151*	11p15.5	↑	NSCLC	Tokuhara *et al.* (2001a)[1,2]
CD44 antigen	*CD44*	11p13-pter	↑	NSCLC	Penno *et al.* (1994)[1,2]
Neurite growth-promoting factor 2 (midkine)	*MDK*	11p11.2	↑	NSCLC	Garver *et al.* (1993)[2]
Kangai 1	*KAI1*	11p11.2	↓	NSCLC	Tagawa *et al.* (1999)[2] Goncharuk *et al.* (2004)[1]

Table 5. Human genes whose expression is altered in lung cancer (continued)

Gene	Gene symbol	Chromosomal localization	Up- or down-regulated	Tumour type	Reference
Ets-1	*ETS1*	11q23.3	↑	NSCLC	Sasaki *et al.* (2001f)[2]; (2001 g)[2] Takanami *et al.* (2001b)[2]
Matrix metalloproteinase-1 (collagenase 1)	*MMP1*	11q21-q22	↑	NSCLC	Muller *et al.* (1991)[2] Nawrocki *et al.* (1997)[2]
Matrix metalloproteinase-7	*MMP7*	11q21-q22	↑	?	Muller *et al.* (1991)[2] Nawrocki *et al.* (1997)[2] Sasaki *et al.* (2001f)[2]; (2001f)[2]
Matrix metalloproteinase-10 (stromelysin 2)	*MMP10*	11q22.3	↑	NSCLC	Muller *et al.* (1991)[2] Cho *et al.* (2004)[1,2]
Matrix metalloproteinase-12	*MMP12*	11q22.3	↑	?	Cho *et al.* (2004)[1,2]
Phosphatidylinositol 3-kinase class 2 subunit, α polypeptide	*PIK3C2A*	11p15.1-14	↑	?	Lin *et al.* (2001)[1,2]
Catalase	*CAT*	11p13	↓	?	Chung-man Ho *et al.* (2001)[1,2]
Cyclin D1	*CCND1*	11q13	↑	?	Betticher *et al.* (1997a)[1] Tanaka *et al.* (1998)[1] McDonald & Pilgram (1999)[1] Bombi *et al.* (2002)[1] Ikehara *et al.* (2003)[1] Ratschiller *et al.* (2003)[1,2]
Wilms' tumour gene	*WT1*	11p13	↑	NSCLC	Oji *et al.* (2002)[1,2]
Baculoviral IAP repeat-containing 3	*BIRC3*	11q22	↑	NSCLC	Hofmann *et al.* (2002)[2]
Tumour suppressor in lung cancer-1	*IGSF4*	11q23.2	↓	NSCLC	Ito *et al.* (2003)[1] Uchino *et al.* (2003)[1]
Dickkopf 3 (REIC/DKK-3)	*DKK3*	11p15	↓	NSCLC	Nozaki *et al.* (2001)[2] Tsuji *et al.* (2001)[2] Kobayashi *et al.* (2002)[2]
Mucin 2	*MUC2*	11p15.5	↑	NSCLC	Awaya *et al.* (2004)[1]
Mucin 5AC	*MUC5AC*	11p15.5	↑	NSCLC	Awaya *et al.* (2004)[1]
Mucin 6	*MUC6*	11p15.5	↑	NSCLC	Awaya *et al.* (2004)[1]
Gastrin receptor	*CCKBR*	11p15.4-p15.5	↑	SCLC	Matsushima *et al.* (1994)[2]
Calcitonin	*CALCA*	11p15.4	↑	SCLC	Kelley *et al.* (1994)[1,2]
Apoptosis inhibitor 5	*API5*	11p12-q12	↑	NSCLC	Sasaki *et al.* (2001b)[2]

Table 5. Human genes whose expression is altered in lung cancer (continued)

Gene	Gene symbol	Chromosomal localization	Up- or down-regulated	Tumour type	Reference
Flap structure-specific endonuclease 1	*FEN1*	11q12	↑	SCLC & NSCLC	Sato *et al.* (2003)[1,2]
Mouse double minute 2 homologue	*MDM2*	12q14-q15	↑	NSCLC	Gorgoulis *et al.* (2000)[1] Eymin *et al.* (2002)[1] Mori *et al.* (2004)[1,2]
Dual-specificity tyrosine phosphorylation-regulated kinase 2	*DYRK2*	12q14	↑	NSCLC	Miller *et al.* (2003a)[2]
Cyclin-dependent kinase inhibitor 1B ($p27^{KIP1}$)	*CDKN1B*	12p13	↑ ↓	SCLC NSCLC	Yatabe *et al.* (1998)[1] Esposito *et al.* (1997a)[1] Catzavelos *et al.* (1999)[1] Kaurana *et al.* (1998)[1] Tsukamoto *et al.* (2001)[1] Hayashi *et al.* (2000)[1]; (2001b)[1] Hommura *et al.* (2000)[1] Tsoli *et al.* (2001)[1]
Neuron-specific enolase	*ENO2*	12p13	↑	SCLC	Schneider *et al.* (2003)[1] Ferrino *et al.* (2003)[1]
CD9 antigen	*CD9*	12p13	↓	SCLC	Funakoshi *et al.* (2003)[1]
Achaete-scute homologue 1	*ASCL1*	12q22-q23	↑	SCLC & NSCLC	Westerman *et al.* (2002)[2] Jiang *et al.* (2004)[2]
Apoptotic protease activating factor 1	*APAF1*	12q23	↑	NSCLC	Krepela *et al.* (2004)[1]
Diablo homoloue (*Drosophila*)	*DIABLO*	12q24.31	↓	NSCLC	Sekimura *et al.* (2004)[2]
Keratin 8	*KRT8*	12q13	↑	NSCLC	Fukunaga *et al.* (2002)[1,2]
Wingless-type MMTV integration site, member 1	*WNT1*	12q12-q13	↑	NSCLC	Uematsu *et al.* (2003)[1]
Epidermal growth factor receptor 3	*ERBB3*	12q13	↑	NSCLC	Yi *et al.* (1997)[1] al Moustafa *et al.* (1999)[1] Sithanandam *et al.* (2003)[1]
High mobility group protein IC	*HMGA2*	12q15	↑	NSCLC	Rogalla *et al.* (1998)[2]
Glut2	*SLC2A3*	12p13.3	↑	NSCLC	Younes *et al.* (1997)[1]
Integral membrane protein 2 (BRI)	*ITM2B*	13q12-q13	↑	NSCLC	Chen et al. (2004c)[2]

Table 5. Human genes whose expression is altered in lung cancer (continued)

Gene	Gene symbol	Chromosomal localization	Up- or down-regulated	Tumour type	Reference
Retinoblastoma	*RB1*	13q14	↑↓	NSCLC	Xu *et al.* (1994)[1] Dosaka-Akita *et al.* (1997)[1] Gorgoulis *et al.* (2000)[1] Sugio *et al.* (2001)[1]
Inhibitor of growth, family member 1	*ING1*	13q34	↓	NSCLC	Kameyama *et al.* (2003)[2]
Pancreatic cancer-derived (PCD1)/LIM domain only 7	*LMO7*	13q21	↑	?	Kang *et al.* (2000)[2]
Periostin	*POSTN*	13q13.3	↓ ↑	?	Sasaki *et al.* (2001c)[2] Yoshioka *et al.* (2002)[2]
Fos oncogene	*FOS*	14q24.3	↓	NSCLC	Levin *et al.* (1994a)[2]; (1995a)[2] Lee *et al.* (1998c)[2]
α1-antichymotrypsin	*SERPINA3*	14q32.1	↑	NSCLC	Higashiyama *et al.* (1995a)[1,2]
Hepatocyte nuclear factor 3α	*FOXA1*	14q13	↑	NSCLC	Lin *et al.* (2002)[2]
Thyroid transcription factor 1	*TITF1*	14q13	↓	NSCLC	Yatabe *et al.* (2002)[1]
Galectin 3	*LGALS3*	14q21-q22	↑	NSCLC	Yoshimura *et al.* (2003)[2] Buttery *et al.* (2004)[1]
Chromogranin A	*CHGA*	14q32	↑	SCLC	Hamid *et al.* (1991)[1,2]
Metastasis-associated gene 1	*MTA1*	14q32.3	↑	NSCLC	Sasaki *et al.* (2002b)[2]
Estrogen receptor β	*ESR2*	14q21-q22	↑	NSCLC	Stabile *et al.* (2002)[1,2]
Placental growth factor	*PGF*	14q22-q24.3	↑	SCLC	Woo *et al.* (2004)[1,2]
Hypoxia-inducible factor 1a	*HIF1A*	14q21-q24	↑	NSCLC	Lee *et al.* (2003d)[1]
Pinin, desmosome-associated protein	*PNN*	14q13.3	↓	NSCLC	Shimakage *et al.* (2002)[2]
BUB1, homologue of yeast gene	*BUB1B*	15q14-q21	↑	?	Seike *et al.* (2002)[2]
Acute promyelocytic leukemia, inducer of	*PML*	15q22	↓	SCLC NSCLC	Zhang *et al.* (2000b)[1,2] Gurrieri *et al.* (2004)[1,2]
Integrin α11	*ITGA11*	15q22.3-q23	↑	NSCLC	Wang *et al.* (2002a)[2]
Furin	*FURIN*	15q25-q26	↑	NSCLC	Lopez de Cicco *et al.* (2002)[1]
Polo-like kinase	*PLK1*	16p12.3	↑	NSCLC	Wolf *et al.* (1997)[2]

Table 5. Human genes whose expression is altered in lung cancer (continued)

Gene	Gene symbol	Chromosomal localization	Up- or down-regulated	Tumour type	Reference
E-cadherin	*CDH1*	16q22.1	↓	NSCLC	Toyoyama *et al* (1999)[1] Bohm *et al.* (1994)[1] Zhong *et al.* (2001)[1] Fei *et al.* (2002)[1] Soo Choi *et al.* (2003)[1]
H-cadherin	*CDH13*	16q24	↓	NSCLC	Sato *et al.* (1998a)[2] Zhong *et al.* (2001)[1]
Autocrine motility factor receptor	*AMFR*	16q21	↑	NSCLC	Takanami *et al.* (2001a)[1,2] Takanami & Takeuchi (2003)[1,2]
NAD(P)H: quinone oxidoreductase I	*NQO1*	16q12-q22	↑	?	Gasdaska *et al.* (1993)[2] Kolesar *et al.* (2002)[2]
Crystallin-mu	*CRYM*	16p12.3-p13	↑	NSCLC	Wang *et al.* (2002a)[2]
Retinoblastoma-related protein, p130	*RBL2*	16q12.2	↓	NSCLC & SCLC	Baldi *et al.* (1996)[1] Xue Jun *et al.* (2003)[2]
Matrix metalloproteinase-2	*MMP2*	16q13	↑	NSCLC	Nakagawa and Yagihashi (1994)[2] Nawrocki *et al.* (1997)[2]
p53	*TP53*	17p13	↑	NSCLC	Westra *et al.* (1993b)[1] Gosney *et al.* (1993)[1] McDonald & Pilgram (1999)[1] Gorgoulis *et al.* (2000)[1]
Crk oncogene homologue	*CRK*	17p13	↑	NSCLC	Miller *et al.* (2003b)[1,2]
Eukaryotic translation initiation factor 5A	*EIF5A*	17p12-p13	↑	NSCLC	Chen *et al.* (2003f)[1,2]
Hypermethylated in cancer-1	*HIC1*	17p13.3	↓	NSCLC	Hayashi *et al.* (2001b)[2]
Granulocyte colony stimulating factor	*CSF3*	17q11.2-q12	↑	NSCLC	Nakamura *et al.* (1997)[2]
γ-Catenin (junction plakoglobin)	*JUP*	17q21	↓	NSCLC	Winn *et al.* (2002)[1]
Transducer of ERBB2, 1 (Tob)	*TOB1*	17q21	↓	NSCLC	Iwanaga *et al.* (2003)[1,2]
Deleted in multiple human cancers (DMHC or DMC1)	*C17orf28*	17q25.1	↓	?	Mikami *et al.* (2001)[2] Harada *et al.* (2001)[2]
Retinoic acid receptor α	*RARA*	17q12	↓	NSCLC	Inui *et al.* (2003)[2]
Survivin	*BIRC5*	17q25	↑	NSCLC	Monzo *et al.* (1999a)[2] Escuin and Rosell (1999)[2] Falleni *et al.* (2003)[1,2] Hofmann *et al.* (2002)[2]

Table 5. Human genes whose expression is altered in lung cancer (continued)

Gene	Gene symbol	Chromo-somal localization	Up- or down-regulated	Tumour type	Reference
Topoisomerase IIα	*TOP2A*	17q21-q22	↑	NSCLC	Hasegawa *et al.* (1993)[2] Giaccone *et al.* (1995)[1,2]
Signal transducer and activator of transcription 3	*STAT3*	17q21	↑	?	Seki *et al.* (2004)[1]
Non-metastatic cells 1, protein NM23A, expressed in	*NME1*	17q22	↑ ↓	NSCLC	Huwer *et al.* (1997)[1] Goncharuk *et al.* (2004)[1]
Non-metastatic cells 2, protein NM23B, expressed in	*NME2*	17q21.3	↑	NSCLC	Ozeki *et al.* (1994)[1]
Platelet endothelial cell adhesion molecule 1	*CD31*	17q23	↑	NSCLC	Takahama *et al.* (1999)[2]
Epidermal growth factor receptor 2	*ERBB2*	17q21.1	↑	NSCLC	Weiner *et al.* (1990)[1] Tateishi *et al.* (1991)[1] Shi *et al.* (1992)[1] Noguchi *et al.* (1993)[1] Yu *et al.* (1994)[2] Rachwal *et al.* (1995)[1] Tsai *et al.* (1995)[1] Pfeifer *et al.* (1996)[1] al Moustafa *et al.* (1999)[1] Kristiansen *et al.* (2001)[1] Turken *et al.* (2003)[1]
Insulin-like growth factor binding protein 4	*IGFBP4*	17q12-q21.1	↑	NSCLC	Wegmann *et al.* (1993)[2]
Cdk5 and Abl enzyme substrate 1 (Cables)	*CABLES1*	18q11.2	↓	NSCLC	Tan *et al.* (2003a)[1,2]
Erythrocyte membrane protein band 4.1-like 3	*EPB41L3*	18p11.32	↓	NSCLC	Tran *et al.* (1999)[1]
Thymidylate synthase	*TYMS*	18p11	↑	NSCLC	Otake *et al.* (1999)[1] Fukushima *et al.* (2003)[1] Shintani *et al.* (2003)[2]
Gastrin-releasing peptide	*GRP*	18q21	↑	SCLC	Ide *et al.* (2001)[1] Uchida *et al.* (2002)[1,2] Oremek & Sapoutzis (2003)[1] Schneider *et al.* (2003)[1]

Table 5. Human genes whose expression is altered in lung cancer (continued)

Gene	Gene symbol	Chromosomal localization	Up- or down-regulated	Tumour type	Reference
Bcl-2	*BCL2*	18q21	↑	SCLC	Ikegaki *et al.* (1994)[1,2] Jiang *et al.* (1995)[1] Yan *et al.* (1996)[1] Reeve *et al.* (1996)[2] Shabnam *et al.* (2004)[1,2]
				NSCLC	Joseph *et al.* (1999)[2] Pezzella *et al.* (1993)[1] Walker *et al.* (1995)[1] Laudanski *et al.* (1999)[1]
Serine protease inhibitor B5 (maspin)	*SERPINB5*	18q21.3	↑	NSCLC	Smith *et al.* (2003)[1,2] Yatabe *et al.* (2004)[1]
Cyclin E1	*CCNE1*	19q13	↑	NSCLC	Müller-Tidow *et al.* (2001)[2] Dobashi *et al.* (2003)[1]
Carcinoembryonic antigen-related cell adhesion molecule 1	*CEACAM1*	19q13.2	↑	NSCLC	Ohwada *et al.* (1994)[2] Sienel *et al.* (2003)[1]
CCAAT/enhancer binding protein	*CEBPA*	19q13.1	↓	NSCLC	Halmos *et al.* (2002)[2]
Cytochrome CYP2B7	*CYP2B7*	19q13.2	↓	?	Czerwinski *et al.* (1994)[2]
Peroxiredoxin 2	*PRDX2*	19p13.2	↑	?	Kim *et al.* (2003f)[1]
SWI/SNF-related, matrix-associated, actin-dependent regulator of chromatin, sub family a, member 4	*SMARCA4*	19p13.3	↓	NSCLC	Reisman *et al.* (2003)[1]
Bcl2-associated X protein	*BAX*	19q13.3-q13.4	↑	NSCLC	Caputi *et al.* (1999)[1]
			↑	NSCLC/SCLC	Reeve *et al.* (1996)[2]
			↓	NSCLC	Shabnam *et al.* (2004)[1,2]
				?	Badillo-Almaraz *et al.* (2003)[1]
Bcl2-like 1 (Bcl-X)	*BCL2L1*	20pter-p12.1	↑	NSCLC/SCLC	Reeve *et al.* (1996)[2]
			↑	NSCLC	Shabnam *et al.* (2004)[1,2]
Forkhead box A2 transcription factor (HNF3B)	*FOXA2*	20p11	↑	Neuroendocrine	Khoor *et al.* (2004)[1]
			↓	SCLC/NSCLC	Halmos *et al.* (2004)[1,2]
Bone morphogenetic protein 2	*BMP2*	20p12	↑	NSCLC	Langenfeld *et al.* (2003)[2]
Proliferating cell nuclear antigen	*PCNA*	20p12	↑	?	Carey *et al.* (1992)[1]

Table 5. Human genes whose expression is altered in lung cancer (continued)

Gene	Gene symbol	Chromosomal localization	Up- or down-regulated	Tumour type	Reference
Complement component 1, q subcomponent, receptor 1	*C1QR1*	20p12.3	↑	NSCLC	Rubinstein *et al.* (2004)[1,2]
URP1, *C. elegans* UNC-112 homologue	*C20orf42*	20p13	↑	?	Weinstein *et al.* (2003)[2]
Insulinoma-associated 1	*INSM1*	20p11.2	↑	SCLC	Lan *et al.* (1993)[2]
E2F transcription factor	*E2F1*	20q11	↑ ↑	SCLC NSCLC	Eymin *et al.* (2001)[1] Imai *et al.* (2004)[1,2]
Palate, lung and nasal epithelium carcinoma-associated	*PLUNC*	20q11.2	↑	NSCLC	Iwao *et al.* (2001)[2] Mitas *et al.* (2003)[2]
Chromosome 20 open reading frame 97/tribbles homologue 3	*TRIB3*	20p12.2-p13	↑	?	Bowers *et al.* (2003)[2]
Matrix metalloproteinase- 9 (gelatinase B)	*MMP9*	20q12-q13	↑	NSCLC	Naskagawa & Yagihashi (1994)[2] Nawrocki *et al.* (1997)[2] Jumper *et al.* (2004)[1]
Collagen 18α1	*COL18A1*	21q22.3	↑	NSCLC	Chang *et al.* (2004b)[1]
Platelet-derived growth factor beta	*PDGFB*	22q12.3-q13.1	↑	?	Antoniades *et al.* (1992)[1]
Macrophage migration inhibitory factor	*MIF*	22q11.2	↑	NSCLC	Kamimura *et al.* (2000)[1,2] Tomiyasu *et al.* (2002)[1,2]
Matrix metalloproteinase-11 (stromelysin 3)	*MMP11*	22q11.2	↑	NSCLC	Nawrocki *et al.* (1997)[2]
Checkpoint kinase 2	*CHEK2*	22q12	↓	NSCLC	Zhang *et al.* (2004b)[1,2]
Gastrin-releasing peptide receptor	*GRPR*	Xp21.2-p22.3	↑	SCLC	Uchida *et al.* (2002)[1,2]
G antigen, family D, 2 (XAGE-1)	*GAGED2*	Xp11.21-p11.22	↑	NSCLC	Wang *et al.* (2001b)[1,2] Egland *et al.* (2002)[2] Ali Eldib *et al.* (2004)[1,2]
Tissue inhibitor of matrix metalloproteinase 1	*TIMP1*	Xp11.23-p11.3	↑	NSCLC	Jumper *et al.* (2004)[1] Aljada *et al.* (2004)[1]
Heparan sulphate proteoglycan GPC3	*GPC3*	Xq26	↓	NSCLC	Kim *et al.* (2003d)[1,2]
Baculoviral IAP repeat-contaning 4	*BIRC4*	Xq25	↑	NSCLC	Hofmann *et al.* (2002)[2]
Melanoma antigen, family A, 3	*MAGEA3*	Xq28	↑	?	Tajima *et al.* (2003)[2]

Table 5. Human genes whose expression is altered in lung cancer (continued)

Gene	Gene symbol	Chromosomal localization	Up- or down-regulated	Tumour type	Reference
Sarcoma antigen 1	*SAGE1*	Xq28	↑	?	Sasaki *et al.* (2003d)[2]
Taxol resistance-associated gene, TRAG-3	*CSAG2*	Xq28	↑	?	Yao *et al.* (2004)[2]
Cancer/testis antigen 1	*CTAG1*	Xq28	↑	NSCLC	Konishi *et al.* (2004)[2]
Synovial sarcoma, X breakpoint 4	*SSX4*	Xp11.2	↑	?	Tajima *et al.* (2003)[2]

Gene symbols given in lower case have not yet been formally ascribed official gene symbols by the *Human Gene Nomenclature Committee.*

[1]Expression measured by immunohistochemical methods including western blotting.

[2]Expression by mRNA analysis including northern blotting, microarrays or real-time reverse transcript PCR.

Table 6. Transcriptomics/proteomics approaches to expression profiling in human lung cancer

Cancer type	Genes up-regulated	Genes down-regulated	Reference
Bronchioloalviolar carcinoma (type of adenocarcinoma)[1]	Osteopontin, retinoic acid-binding protein, intestinal trefoil factor, secreted cement gland protein, α1(XI) collagen, complement factor B, CD24 signal transducer, Kruppel-related zinc protein, insulin-like growth factor-binding protein, MAP kinase phosphatase, β-catenin	TSC403 protein, RAGE receptor, G protein-coupled receptor kinase, carboxylesterase, interleukin 10 receptor, actin-binding protein p57	Goodwin *et al.* (2001)
Adenocarcinoma/ squamous cell carcinoma[2]	Cdc25A, ephrin type A receptor, LIM domain kinase 1, ras-related proteins RAB2 and RAB6, JAK3, JNK1, ephrin A3, transmembrane receptor PTK7, insulin receptor, IL2Rα, CD27L antigen receptor, HGF activator, PDGFA	Protein tyrosine kinase c-kit, c-jun, c-src, c-fgr, protein tyrosine kinase receptor tyro3, TR3 orphan receptor, adenosine A1 receptor, caspase 8, ICAM1, amphiregulin, vascular endothelial growth factor C, macrophage inflammatory protein 2α	McDoniels-Silvers *et al.* (2002)
Adenocarcinomas/ SCLC/ squamous cell tumours[3]	PGP 9.5 antigen, annexins I/II	–	Hanash *et al.* (2001)
Adenocarcinomas/ squamous cell carcinomas[4]	Glutathione peroxidase 2, glutathione *S*-transferases M1 & M3, carboxylesterase, aldoketoreductase, peroxiredoxin 1, small proline-rich protein 3, TNF receptor 18, MHC class I, CD71 (squamous cell carcinoma). Pulmonary surfactants A2 & B, pronapsin A, mucin 1, MHC class II, CD74 (adenocarcinoma)	P21 (SCC). S100 protein, keratins 5, 14, 16, 17, small proline-rich protein 1B, small proline-rich protein 3, 14-3-3σ, $p21^{waf1/CIP1}$ (AC)	Nacht *et al.* (2001)
Adenocarcinomas[1]	Pulmonary surfactants A1, A2, B & C, thyroid transcription factor 1, flavin-associated monooxygenase 2, matrix Gla protein, endothelial cell growth factor 1, aldoketoreductase 1 (C1), immunoglobulins heavy Cγ3, Vλ & Cκ, β2-integrin, fibronectin 1, MHC class II DQα1, paraoxonase 3	–	Giordano *et al.* (2001)
SCLC[2]	Sox proteins 2, 4 & 21, ubiquitin-conjugating enzyme E2 homologue, pituitary tumour transforming gene 1, homologue of yeast NUF2 gene, collapsing response mediator protein 1	–	Bangur *et al.* (2002)
Adenocarcinoma[3]	Antioxidant enzyme AOE372, ATP synthase subunit d, β1,4-galactosyltransferase, cytosolic inorganic pyrophosphatase, glucose-regulated 58 kDa protein, glutathione-*S*-transferase M4, prolyl 4-hydroxylase β subunit, triosephosphate isomerase, ubiquitin thiolesterase 1	–	Chen *et al.* (2002b)

Cancer type	Genes up-regulated	Genes down-regulated	Reference
Adenocarcinoma[5]	Lc19, hyaluronan binding protein 2, crystallin μ, caeruloplasmin, α11-integrin, α1(XI) collagen	–	Wang *et al.* (2002a)
NSCLC[2]	α1(III) collagen, osteopontin, MAGE-1 antigen, glutamate receptor 4, chlordecone reductase, polyA binding protein, glutathione peroxidase-like protein, MAD2-like protein, ubiquitin carrier protein E2-C, keratins 6A, 13, 14, 15, 17, 19, dihydrodiol dehydrogenase, connexin 26, protein regulator of cytokinesis 1, DNA topoisomerase II, ubiquitin carboxy-terminal hydrolase L1-like, type I interstitial collagenase, S100 calcium-binding protein, α1(X) collagen, cell cycle control gene cdc2, α1(I) collagen, ubiquitin-conjugating enzyme E2, thymidine kinase, thrombospondin 2, insulin-like growth factor binding protein 2, bilirubin UDP-glucuronosyltransferase isozyme 1, pituitary tumour transforming gene 1, ARF-activated phosphatidylcholine-specific phospholipase D1a, α2(I) collagen, G protein coupled receptor GPR87, ladinin, 20 α-hydroxysteroid dehydrogenase, melanoma antigens A3 & A9, cysteine/glutamate exchanger, prourokinase, gremlin, migration inhibitory factor-related protein 8, diubiquitin, neutrophil gelatinase-associated lipocalin, CK2 interacting protein, p-cadherin, basic transcription factor 2 p44, ubiquitin B, protein tyrosine phosphatase CIP2, aldehyde dehydrogenase 3 A1, maspin, ubiquitin carrier protein E2-EPF, squamous cell carcinoma antigen SCCA1, small proline-rich proteins sprII & 3, lactate dehydrogenase B, small subunit ribonucleotide reductase	Pulmonary surfactant-associated proteins B, D, 2 & 5, α2-macroglobulin, smooth muscle myosin heavy chain, adult folate-binding protein, channel-like integral membrane protein, glutathione peroxidase, slow skeletal troponin C, Na/Cl-dependent serotonin transporter, α2δ calcium channel subunit isoform II, 15-hydroxy prostaglandin dehydrogenase, desmin, dual oxidase, Na-dependent phosphate transporters Iib & 3b, cathepsin H, endothelin receptor, cytochrome P450 (IV) BI, alcohol dehydrogease β1, fructose–1,6-bisphosphatase, lipoprotein lipase, endothelial cell-selective adhesion molecule, mannose receptor, macrophage receptor, matrix Gla protein, oxidative 3α hydroxysteroid dehydrogenase, nectin-like protein 2, prostaglandin transporter, tetranectin, leucine zipper nuclear factor BLZF1, catalase, von Willebrand factor, DMBT1, HLA-DQ, protein tyrosine phosphatase CL100, prepromultimerin, phosphatidylethanolamine binding protein, episialin B, CDW52 antigen, haptoglobin, insulin-like growth factor I, α1-antitrypsin, complement C1Qβ & C7, flavin-containing monooxygenase 2, γ-glutamyl transpeptidase, chitinase, ribonuclease A, endothelial cell adhesion molecule PECAM-1, β2-adrenergic receptor, chloride channel protein 7, LDLR, γ-fibrinogen, aldehyde oxidase 1, mesothelin, angiotensin II type 2, VE-cadherin, monocyte chemoattractant protein 1, tryptase, decorin, FcER1, stomatin, retinal-binding protein, leptin receptor, mast cell carboxypeptidase A, heparin-binding EGF-like growth factor, migration inhibitory factor-related protein 8, bone morphogenetic protein 2A, vimentin, DNAse γ, folate-binding protein, PPARγ, SLIT-2, HNF3β, aldehyde dehydrogenase 2	Heighway *et al.* (2002)

Table 6 Transcriptomics/proteomics approaches to expression profiling in human lung cancer (continued)

Cancer type	Genes up-regulated	Genes down-regulated	Reference
NSCLC/SCLC[2]	SOX4, reticulon 1, reelin, kinesin heavy chain 2, stathmin, FGF12, catenin δ2, internexin neuronal intermediate filament protein α, deoxycytidine kinase, minichromosome maintenance deficient 3 & 4, CDC7-like, uracil DNA glycosylase, HMG17 (SCLC), glutathione *S*-transferases M1 & M3, glutathione reductase, glutathione peroxidase 2, bullous pemphigoid antigen 1, ataxia-telangiectasia group D-associated protein, RaP2 interacting protein 8, keratins 5, 6B, 12, 13, 15, 17, small proline-rich protein spr1, pirin, wingless-type MMTV integration site members 2 & 5A, fumarylacetoacetate hydrolase, heparin-binding growth factor binding protein, phosphatidylserine synthase 1, phosphogluconate dehydrogenase, phosphoglucomutase and prothrombin (squamous cell carcinoma). α2A amylase, occluding, class VI type iiA ATPase, glutathione *S*-transferase A2, carcinoembryonic antigen-related cell adhesion molecules 5 & 7, and thyroid transcription factor 1 (adenocarcinoma and large cell carcinoma).	–	Virtanen *et al.* (2002)
Adenocarcinoma[2]	High mobility group protein 1, desmoplakins I & II, nucleoside diphosphate kinase B, G_2/mitotic-specific cyclin B1, hepatoma-derived growth factor, cytokeratin 8, histone H4, matrix metalloproteinase 12, DNA topoisomerase IIα, tyrosine 3-monooxygenase/ tryptophan 5-monooxygenase activation protein sigma isoform (14-3-3σ/stratifin), polo-like kinase 1, insulin-like growth factor-binding protein 3, macrophage migration inhibitory factor, tenascin C, ribonucleotide reductase.	Tissue inhibitor of metalloproteinase 3, DNA-binding protein CPBP, gravin, DNAX acivation protein 12, caveolins 1 & 2, bone morphogenetic protein 4 type II receptor, aryl hydrocarbon receptor, chitinase, early growth response α, nuclear protein C-193.	Wikman *et al.* (2002)

Cancer type	Genes up-regulated	Genes down-regulated	Reference
SCLC[1]	Neutral amino acid transport B (*SLC1A5*), thiopurine S-methyltransferase, thyroid transcription factor, DEAD/H (*DDX11*), sodium channel non-voltage-gated 1α, hepatoma-derived growth factor, nuclear receptor co-repressor N-Cor, Achaete scute homologous protein (*ASCL1*), extra spindle poles-like (*ESPL1*), gastrin-releasing peptide, uncoupling protein homologue (*UCP2*), cell division cycle 25B, inhibitor of DNA binding 2 (*ID2*), neurite growth-promoting factor 2, programmed cell death 6, immunoglobulin superfamily 4, cofilin 1, calmodulin 1, tyrosine 3-tryptophan 5-monooxygenase activation protein ξ, nuclear receptor subfamily 2 (*NR2F1*), sex determining region Y-box 2 (*SOX2*), exostoses-like 3, inositol polyphosphate-5-phosphatase-like 1, tripartite motif-containing 28, tyrosine phosphatase type IVA 3, vascular endothelial growth factor B, transport secretion protein 2.2, lysophosphatidic acid acyltransferase α, protective protein for β-galactosidase	–	Pedersen *et al.* (2003)
Adenocarcinoma[1]	Cyclins A1, B1, D1 and E1, polo-like kinase, α-prothymosin, proliferating cell nuclear antigen, cell division cycle proteins cdc25A, cdc20 and cdc2, cyclin-dependent kinases cdk2, cdk4 and cdk7, histone deacetylase 1, cyclin F, Mdm2, E2F transcription factor, Dp-1 transcription factor, E1A-binding protein p300	Cyclin-dependent kinase inhibitors 2B (p15), 1A (p21), 1C (p57) and 2D (p19), Chk2 checkpoint kinase 2, SMAD4, CDK4-binding protein (p34), GADD45	Singhal *et al.* (2003a)
Adenocarcinoma[2]	DNA topoisomerase IIα (*TOP2A*), matrix metalloproteinase 15 (*MMP15*), Myxovirus resistance 2, murine, homologue (*MX2*), IGF2 mRNA-binding protein/KH domain-containing protein over-expressed in cancer (*Koc1*)	–	Kobayashi *et al.* (2004)

Table 6. Transcriptomics/proteomics approaches to expression profiling in human lung cancer (continued)

Cancer type	Genes up-regulated	Genes down-regulated	Reference
NSCLC[2]	Ornithine decarboxylase 1, pulmonary-associated surfactant protein A2, transmembrane 4 superfamily member 1, SHC transforming protein 1, solute carrier family 34 member 2, prostaglandin endoperoxide synthase 2, pronapsin A, dual specificity phosphatase 6, aspartate β-hydroxylase, chitinase 3-like 1, urokinase, reticulocalbin 1, osteonectin, solute carrier family 2 member 1, TAF4 RNA polymerase II, trefoil factor 3, tyrosyl-tRNA synthetase, tissue factor inhibitor 2, collagen type I $\alpha 2$, calcitonin, calumenin, cytochrome CYP1B1, chondroitin sulphate proteoglycan 2, collagen type III $\alpha 1$, cystatin B, serotonin receptor 2B, epiregulin, keratin 6A (adenocarcinoma). Keratins 5, 6A, 6B & 14, RAN, neurotrophic tyrosine kinase receptor 1, solute carrier family 2 member 1, bullous pemphigoid antigen 1, S100 calcium-binding protein 2, aldoketoreductase family 1 member C3, phosphoglycerate kinase 1, p53-induced protein PIGPC1, collagen type I $\alpha 1$, N-myc downstream regulated 1, sperm-specific antigen 2, desmoplakin, TP73-like, tripartite motif-containing 29, aldoketo reductase family 1 members B10, C1 and C2, glycoprotein nmb, osteoblast-specific factor 2, urokinase, translational activator GCN1, S100 calcium-binding protein A11, parathyroid hormone-like hormone, osteonectin, meltrin γa, vezatin, solute carrier family 5 member 6, peroxiredoxin 1, osteopontin, aspartate β-hydroxylase, laminin gamma 2, claudin 1, annexin A1, collagen type I $\alpha 2$, collagen type III $\alpha 1$, cystatin B, desmoglein 3, dual specificity phosphatase 5, glyoxylase I (squamous cell carcinoma)	–	Amatschek *et al.* (2004)
Squamous cell carcinoma[2]	Type II cytoskeletal 2 epidermal keratin, desmocollin 3A/3B, type I cytoskeletal 14 keratins 10 & 14, integrins alpha 6 & 7B, retinoic acid receptor γ 1, integrin β4, CD104 antigen, macrophage migration inhibitory factor, IGF-binding protein 5, Jagged 1, low affinity nerve growth factor receptor, collagen 2$\alpha 1$	Caveolin 1, core promoter element-binding protein	Kettunen *et al.* (2004)

Cancer type	Genes up-regulated	Genes down-regulated	Reference
NSCLC/SCLC cell lines[1]	Achaete-scute homologue 1, aldoketoreductase family 1 member B10, ATP-binding cassette sub-family C member 2, cancer/testis antigen CTAG1/LAGE2, claudin 10, dopa decarboxylase, insulinoma-associated 1, keratin hair basic 1, TRIM9, melanoma antigen family A members 2, 3, 6, 10 & 12, Na^+K^+-ATPase subunit αIII, neurofilament light polypeptide, prostaglandin E synthase, secretogranins I and II, ubiquitin COOH-terminal esterase L1	–	Sugita *et al.* (2002)

[1]Oligonucleotide microarray
[2]cDNA expression array
[3]2-D gel electrophoresis/mass spectrometry
[4]Serial analysis of gene expression (SAGE)
[5]Representational difference analysis

Table 7. Characteristics of SCLC and NSCLC; a comparison

Characteristic/property	SCLC	NSCLC
Frequency (% of lung cancer)	20-25 %	75-80 %
Chemosensitivity	High	Low
RAS gene mutation	< 1 %	20-30 %
MYC gene amplification/ over-expression	25-30 %	5-10 %
ERBB2 gene over-expression	< 10 %	~30 %
BCL2 gene expression	80-90 %	10-35 %
3p allele loss	> 90 %	50-80 %
TP53 gene mutation	80-95 %	40-60 %
17p LOH	80-90 %	~70 %
RB1 expression absent	~90 %	~15-30 %
13q LOH	~75 %	40-60 %
CDKN2A gene mutation	< 1 %	10-40 %
9p LOH	20-50 %	50-75 %
CDKN2A gene expression absent	< 10 %	30-70 %
Telomerase expression	~100 %	80-85 %
PTEN gene mutation	16-40 %	8-17 %
Microsatellite instability	16-100 %	29-58 %
RASSF1 gene promoter hypermethylation	> 90 %	40 %
RARB gene promoter hypermethylation	72 %	41 %
Pre-neoplastic LOH, multiple changes	90 %	30 %

References

Abdel-Rahman SZ *et al.* (1998) Association of the NAT1*10 genotype with increased chromosome aberrations and higher lung cancer risk in cigarette smokers. *Mutation Res.* 398: 43-54.
Abdel-Rahman SZ, el-Zein RA (2000) The 399Gln polymorphism in the DNA repair gene *XRCC1* modulates the genotoxic response induced in human lymphocytes by the tobacco-specific nitrosamine NNK. *Cancer Lett.* 159: 63-71.
Abe Y *et al.* (1994) Expression of the multidrug resistance gene (*MDR1*) in non-small cell lung cancer. *Jpn. J. Cancer Res.* 85: 536-541.
Abe T *et al.* (1996) The somatic mutation frequency of the transforming growth factor β receptor type II gene varies widely among different cancers with microsatellite instability. *Eur. J. Surg. Oncol.* 22: 474-477.
Abeysinghe SS *et al.* (2003) Translocation and gross deletion breakpoints in human inherited disease and cancer I. Nucleotide composition and recombination-associated motifs. *Hum. Mutation* 22: 229-244.
Abujiang P *et al.* (1998) Loss of heterozygosity (LOH) at 17q and 14q in human lung cancers. *Oncogene* 17: 3029-3033.
Adachi J *et al.* (1995) Microsatellite instability in primary and metastatic lung carcinomas. *Genes Chrom. Cancer* 14: 301-306.
Adachi J *et al.* (1996) Induction of apoptosis but not G1 arrest by expression of the wild-type p53 gene in small cell lung carcinoma. *Cell Growth Differ.* 7: 879-886.
Adachi M *et al.* (1998) Reduced integrin α3 expression as a factor of poor prognosis in patients with adenocarcinoma of the lung. *J. Clin. Oncol.* 16: 1060-1067.
Adler I (1912) *Primary Malignant Growths of the Lungs and Bronchi: A Pathological and Clinical Study.* Longmans, Green & Co, London.
Affatato AA *et al.* (2004) Effect of *XPD/ERCC2* polymorphisms on chromosome aberration frequencies in smokers and on sensitivity to the mutagenic tobacco-specific nitrosamine NNK. *Environ. Mol. Mutagen.* 44: 65-73.
Agami R (2002) RNAi and related mechanisms and their potential use for therapy. *Curr. Opin. Chem. Biol.* 6: 829-834.
Agathanggelou A *et al.* (2001) Methylation associated inactivation of *RASSF1A* from region 3p21.3 in lung, breast and ovarian tumours. *Oncogene* 20: 1509-1518.
Agathanggelou A *et al.* (2003a) Epigenetic inactivation of the candidate 3p21.3 suppressor gene *BLU* in human cancers. *Oncogene* 22: 1580-1588.
Agathanggelou A *et al.* (2003b) Identification of novel gene expression targets for the Ras association domain family 1 (*RASSF1A*) tumor suppressor gene in non-small cell lung cancer and neuroblastoma. *Cancer Res.* 63: 5344-5351.
Agochiya M *et al.* (1999) Increased dosage and amplification of the focal adhesion kinase gene in human cancer cells. *Oncogene* 18: 5646-5653.
Agundez JA *et al.* (2001) Functionally active duplications of the *CYP2D6* gene are more more prevalent among larynx and lung cancer patients. *Oncology* 61: 59-63.
Ahmed SI *et al.* (2000) Studies on the expression of endothelin, its receptor subtypes, and converting enzymes in lung cancer and in human bronchial epithelium. *Am. J. Respir. Cell Mol. Biol.* 22: 422-431.
Ahrendt SA *et al.* (1999) Molecular detection of tumor cells in bronchoalveolar lavage fluid from patients with early stage lung cancer. *J. Natl. Cancer Inst.* 91: 332-339.
Ahrendt SA *et al.* (2000a) Microsatellite instability at selected tetranucleotide repeats is associated with p53 mutations in non-small cell lung cancer. *Cancer Res.* 60: 2488-2491.
Ahrendt SA *et al.* (2000b) Alcohol consumption and cigarette smoking increase the frequency of p53 mutations in non-small cell lung cancer. *Cancer Res.* 60: 3155-3159.

Ahrendt SA *et al.* (2001) Cigarette smoking is strongly associated with mutation of the K-ras gene in patients with primary adenocarcinoma of the lung. *Cancer* 92: 1525-1530.

Ahrendt SA *et al.* (2002) Molecular assessment of lymph nodes in patients with resected stage I non-small cell lung cancer: preliminary results of a prospective study. *J. Thorac. Cardiovasc. Surg.* 123: 466-473.

Ahrendt SA *et al.* (2003) p53 mutations and survival in stage I non-small-cell lung cancer: results of a prospective study. *J. Natl. Cancer Inst.* 95: 961-970.

Ahsan H, Rundle AG (2003) Measures of genotype versus gene products: promise and pitfalls in cancer prevention. *Carcinogenesis* 24: 1429-1434.

Akie K *et al.* (2000) A combination of c-*myc* antisense DNA with all-trans-retinoic acid inhibits cell proliferation by downregulating c-*myc* expression in small cell lung cancer. *Antisense Nucleic Acid Drug Dev.* 10: 243-249.

al Moustafa AE *et al.* (1999) Expression of P185erbB-2, P160erbB-3, P180erbB-4, and heregulin alpha in human normal bronchial epithelial and lung cancer cell lines. *Anticancer Res.* 19(1A): 481-486.

Al-Tassan N *et al.* (2004) Inherited variants in MYH are unlikely to contribute to the risk of lung carcinoma. *Hum. Genet.* 114: 207-210.

Alavanja MCR (2002) Biologic damage resulting from exposure to tobacco smoke and from radon: implication for preventive interventions. *Oncogene* 21: 7365-7375.

Albanell J *et al.* (1997) High telomerase activity in primary lung cancers: association with increased cell proliferation rates and advanced pathologic stage. *J. Natl. Cancer Inst.* 89: 1609-1615.

Albelda SM *et al.* (2002) Gene therapy for lung neoplasms. *Clin. Chest Med.* 23: 265-277.

Alberg AJ, Samet JM (2003) Epidemiology of lung cancer. *Chest* 123 Suppl 1: 21S-49S.

Alberola V *et al.* (2004) Effect of the methylenetetrahydrofolate reductase C677T polymorphism on patients with cisplatin/gemcitabine-treated stage IV non-small-cell lung cancer. *Clin. Lung Cancer* 5: 360-365.

Alemany R *et al.* (1996) Growth inhibitory effect of anti-K-ras adenovirus on lung cancer cells. *Cancer Gene Ther.* 3: 296-301.

Alexandrie A-K *et al.* (1994) Genetic susceptibility to lung cancer with special emphasis on *CYP1A1* and *GSTM1*: a study on best factors in relation to age at onset, gender and histological cancer types. *Carcinogenesis* 15: 1785-1790.

Alexandrie A-K *et al.* (2004) Influence of *CYP1A1*, *GSTM1*, *GSTT1*, and *NQO1* genotypes and cumulative smoking dose on lung cancer risk in a Swedish population. *Cancer Epidemiol. Biomarkers Prev.* 13: 908-914.

Alexandrov K *et al.* (2002) *CYP1A1* and *GSTM1* genotypes affect benzo[*a*]pyrene DNA adducts in smokers' lung: comparison with aromatic/hydrophobic adduct formation. *Carcinogenesis* 23: 1969-1677.

Alfonso P *et al.* (2004) Proteomic analysis of lung biopsies: differential protein expression profile between peritumoral and tumoral tissue. *Proteomics* 4: 442-447.

Ali IU *et al.* (1999) Mutational spectra of *PTEN/MMAC1* gene: a tumor suppressor with lipid phosphatase activity. *J. Natl. Cancer Inst.* 91: 1922-1932.

Ali Eldib AM *et al.* (2004) Immunoscreening of a cDNA library from a lung cancer cell line using autologous patient serum: identification of XAGE-1b as a dominant antigen and its immunogenicity in lung adenocarcinoma. *Int. J. Cancer* 108: 558-563.

Aljada IS *et al.* (2004) Upregulation of the tissue inhibitor of metalloproteinase-1 protein is associated with progression of human non-small-cell lung cancer. *J. Clin. Oncol.* 22: 3218-3229.

Allan JM *et al.* (2001) Genetic alterations in bronchial mucosa and plasma DNA from individuals at high risk of lung cancer. *Int. J. Cancer* 91: 359-365.

Amatschek S *et al.* (2004) Tissue-wide expression profiling using cDNA subtraction and microarrays to identify tumor-specific genes. *Cancer Res.* 64: 844-856.

Ames BN, Gold LS (1990) Too many rodent carcinogens: mitogenesis increases mutagenesis. *Science* 249: 970-971.

Amos CI *et al.* (1999) Is there a genetic basis for lung cancer susceptibility? *Cancer Res.* 151: 3-12.

Amundson SA *et al.* (2002) A nucleotide excision repair master-switch: p53 regulated coordinate induction of global genomic repair genes. *Cancer Biol. Ther.* 1: 145-149.

An Q *et al.* (2002a) Deletion of tumor suppressor genes in Chinese non-small cell lung cancer. *Cancer Letts* 184: 189-195.

An Q *et al.* (2002b) Detection of p16 hypermethylation in circulating plasma DNA of non-small cell lung cancer patients. *Cancer Letts.* 188: 109-114.

An Q *et al.* (2003) Identification of differentially expressed genes in immortalized human bronchial epithelial cell line as a model for *in vitro* study of lung carcinogenesis. *Int. J. Cancer* 103: 194-204.

Anbazhagan R *et al.* (1999) Classification of small cell lung cancer and pulmonary carcinoid by gene expression profiles. *Cancer Res.* 59: 5119-5122.
Ando K *et al.* (2004) Polo-like kinase 1 (Plk1) inhibits p53 function by physical interaction and phosphorylation. *J. Biol. Chem.* 279: 25549-25561.
Andriani F *et al.* (2004) Detecting lung cancer in plasma with the use of multiple genetic markers. *Int. J. Cancer* 108: 91-96.
Angeloni D *et al.* (2000) Gene structure of the human receptor tyrosine kinase RON and mutation analysis in lung cancer samples. *Genes Chrom. Cancer* 29: 147-156.
Antonarakis SE, Krawczak M, Cooper DN (2001) The nature and mechanisms of human gene mutation. Chap. 13 in *The Metabolic and Molecular Bases of Inherited Disease.* 8th Ed. Eds: CR Scriver, AL Beaudet, WS Sly, D Valle. pp 343-377.
Antoniades HN *et al.* (1992) Malignant epithelial cells in primary human lung carcinomas coexpress *in vivo* platelet-derived growth factor (PDGF) and PDGF receptor mRNAs and their protein products. *Proc. Natl. Acad. Sci. USA* 89: 3942-3946.
Anttila S *et al.* (2000) An uncommon phenotype of poor inducibility of CYP1A1 in human lung is not ascribable to polymorphisms in the *AHR*, *ARNT*, or *CYP1A1* genes. *Pharmacogenetics* 10: 741-751.
Anttila S *et al.* (2003) Methylation of cytochrome P4501A1 promoter in the lung is associated with tobacco smoking. *Cancer Res.* 63: 8623-8628.
Aoyagi Y *et al.* (2001) Accumulation of losses of heterozygosity and multistep carcinogenesis in pulmonary adenocarcinoma. *Cancer Res.* 61: 7950-7954.
Arai T *et al.* (1998) Application of telomerase activity for screening of primary lung cancer in broncho-alveolar lavage fluid. *Oncol. Rep.* 5: 405-408.
Araki K *et al.* (2003) Frequent overexpression of the c-kit protein in large cell neuroendocrine carcinoma of the lung. *Lung Cancer* 40: 173-180.
Archavachotikul K *et al.* (2002) Thyroid hormone affects embryonic mouse lung branching morphogenesis and cellular differentiation. *Am. J. Physiol. Lung Cell Mol. Physiol.* 282: L359-L369.
Arinaga M et al. (2003) Clinical implication of cyclin B1 in non-small cell lung cancer. *Oncol. Rep.* 10: 1381-1386.
Ariyoshi N *et al.* (2002) Genetic polymorphism of CYP2A6 gene and tobacco induced lung cancer risk in male smokers. *Cancer Epidemiol. Biomarkers Prev.* 11: 890-894.
Arkin A, Wagner DH (1936) Primary carcinoma of the lung. *JAMA* 106: 587-591.
Arvanitis DA *et al.* (2003) Fractional allele loss is a valuable marker for human lung cancer detection in sputum. *Lung Cancer* 40: 55-66.
Asher G *et al.* (2002) Mdm-2 and ubiquitin-independent p53 proteasomal degradation regulated by NQO1. *Proc. Natl. Acad. Sci. USA* 99: 13125-13130.
Asher G *et al.* (2003) p53 hot-spot mutants are resistant to ubiquitin-independent degradation by increased binding to NAD(P)H:quinone oxidoreductase 1. *Proc. Natl. Acad. Sci. USA* 100: 15065-15070.
Ashman JN *et al.* (2002) Chromosomal alterations in small cell lung cancer revealed by multicolour fluorescence *in situ* hybridisation. *Int. J. Cancer* 102: 230-236.
Attisano L, Wrana JL (2002) Signal transduction by the TGF-beta superfamily. *Science* 296: 1646-1647.
Au WW *et al.* (2001) Usefulness of genetic susceptibility and biomarkers for evaluation of environmental health risk. *Environ. Mol. Mutagen.* 37: 215-225.
Auckley DH *et al.* (2001) Reduced DNA-dependent protein kinase activity is associated with lung cancer. *Carcinogenesis* 22: 723-727.
Audebert M *et al.* (2000a) Effect of single mutations in the *OGG1* gene found in human tumors on the substrate specificity of the Oggl protein. *Nucleic Acids Res.* 28: 2672-2678.
Audebert M *et al.* (2000b) Alterations of the DNA repair gene *OGG1* in human clear cell carcinomas of the kidney. *Cancer Res.* 60: 4740-4744.
Auerbach O *et al.* (1957) Changes in the bronchial epithelium in relation to smoking and cancer of the lung; a report of progress. *New Engl. J. Med.* 256: 97-104.
Auerbach O *et al.* (1961) Changes in bronchial epithelium in relation to cigarette smoking and in relation to lung cancer. *New Engl. J. Med.* 265: 253-267.
Auerbach O *et al.* (1962) Bronchial epithelium in former smokers. *New Engl. J. Med.* 267: 119-125.
Averbuch SD (2003) The impact of gefitinib on epidermal growth factor receptor signalling pathways in cancer. *Clin. Lung Cancer* 5 Suppl, 1: S5-S10.
Awaya H *et al.* (2004) Expression of MUC1, MUC2, MUC5AC, and MUC6 in atypical adenomatous hyperplasia, bronchioloalveolar carcinoma, adenocarcinoma with mixed subtypes, and mucinous bronchioloalveolar carcinoma of the lung. *Am. J. Clin. Pathol.* 121: 644-653.

Ayesh R *et al.* (1984) Metabolic oxidation phenotypes as markers for susceptibility to lung cancer. *Nature* 312: 169-170.
Azam N *et al.* (2001) Interaction of CR6 (GADD45gamma) with proliferating cell nuclear antigen impedes negative growth control. *J. Biol. Chem.* 276: 2766-2774.
Bachman KE *et al.* (2003) Histone modifications and silencing prior to DNA methylation of a tumor suppressor gene. *Cancer Cell* 3: 89-96.
Bader S *et al.* (2003) *MBD1*, *MBD2* and *CGBP* genes at chromosome 18q21 are infrequently mutated in human colon and lung cancers. *Oncogene* 22: 3506-3510.
Badillo-Almaraz I *et al.* (2003) Defective expression of FasL and Bax in human lung cancer. *Clin. Exp. Med.* 3: 106-112.
Bai L, Sun Y, Li S (1997) Relationship of p53 alteration and *myc* family gene overexpression with the clinical characteristics of lung cancer. *Chin. Med. J.* 110: 250-254.
Bailey-Wilson JE *et al.* (2004) A major lung cancer susceptibility locus maps to chromosome 6q23-25. *Am. J. Hum. Genet.* 75: 460-474.
Bain C *et al.* (2004) Lung cancer rates in men and women with comparable histories of smoking. *J. Natl. Cancer Inst.* 96: 826-834.
Baksh FK *et al.* (2003) Widespread molecular alterations present in stage I non-small cell lung carcinoma fail to predict tumor recurrence. *Mod. Pathol.* 16: 28-34.
Baldi A *et al.* (1996) Differential expression of the retinoblastoma gene family members pRb/p105, p107, and pRb2/p130. *Clin. Cancer Res.* 2: 1239-1245.
Baldi A *et al.* (1997) Differential expression of Rb2/p130 and p107 in normal human tissues and in primary lung cancer. *Clin. Cancer Res.* 3: 1691-1697.
Ballaz S, Mulshine JL (2003) The potential contributions of chronic inflammation to lung carcinogenesis. *Clin. Lung Cancer* 5: 46-62.
Ballinger SW *et al.* (1996) Mitochondrial genome damage associated with cigarette smoking. *Cancer Res.* 56: 5692-5697.
Balmain A *et al.* (2003) The genetics and genomics of cancer. *Nature Genet. Suppl.* 33: 238-244.
Balsara BF, Testa JR (2002) Chromosomal imbalances in human lung cancer. *Oncogene* 21: 6877-6883.
Balsara BR *et al.* (2004) Frequent activation of AKT in non-small cell lung carcinomas and preneoplastic bronchial lesions. Carcinogenesis [DOI: 10.1093/carcin/bgh226].
Bangur CS *et al.* (2002) Identification of genes over-expressed in small cell lung carcinoma using suppression subtractive hybridisation and cDNA microarray expression analysis. *Oncogene* 21: 3814-3825.
Banham AH *et al.* (2001) The FOXP1 winged helix transcription factor is a novel candidate tumor suppressor gene on chromsome 3p. *Cancer Res.* 61: 8820-8829.
Bansal A *et al.* (1997) Mutation analysis of the coding sequences of MEK-1 and MEK-2 genes in human lung cancer cell lines. *Oncogene* 14: 1231-1234.
Barbareschi M *et al.* (2003) CDX-2 homeobox gene expression is a reliable marker of colorectal adenocarcinoma metastases to the lungs. *Am. J. Surg. Pathol.* 27: 141-149.
Barker N *et al.* (2001) The chromatin remodelling factor Brg-1 interacts with beta-catenin to promote target gene activation. *EMBO J.* 20: 4935-4943.
Barnard WG (1926) The nature of the "oat-celled sarcoma" of the mediastinum. *J. Path. Bacteriol.* 29: 241-244.
Barr LF *et al.* (2000) c-*Myc* suppresses the tumorigenicity of lung cancer cells and down-regulates vascular endothelial growth factor expression. *Cancer Res.* 60: 143-149.
Barrett MT *et al.* (1999) Evolution of neoplastic cell lineages in Barrett oesophagus. *Nature Genet.* 22: 106-109.
Bartram U, Speer CP (2004) The role of transforming growth factor beta in lung development and disease. *Chest* 125: 754-765.
Bates S *et al.* (1996) Characterisation of human cyclin G1 and G2: DNA damage inducible genes. *Oncogene* 13: 1103-1109.
Batra V *et al.* (2003) The genetic determinants of smoking. *Chest* 123: 1730-1739.
Bauer C *et al.* (2001) Translation initiation factor eIF-4G is immunogenic, overexpressed, and amplified in patients with squamous cell lung carcinoma. *Cancer* 92: 822-829.
Baylin SB *et al.* (2001) Aberrant patterns of DNA methylation, chromatin formation and gene expression in cancer. *Hum. Mol. Genet.* 10: 687-692.
Baylin S, Bestor TH (2002) Altered methylation patterns in cancer cell genomes: cause or consequence? *Cancer Cell* 1: 299-305.
Beadsmore CJ, Screaton NJ (2003) Classification, staging and prognosis of lung cancer. *Eur. J. Radiol.* 45: 8-17.

Beau-Faller M *et al.* (2003a) Genetic heterogeneity in lung and colorectal carcinoma as revealed by microsatellite analysis in plasma or tumor tissue DNA. *Cancer* 97: 2308-2317.
Beau-Faller M *et al.* (2003b) Plasma DNA microsatellite panel as sensitive and tumor-specific marker in lung cancer patients. *Int. J. Cancer* 105: 361-370.
Beau-Faller M *et al.* (2003c) Allelic imbalance at loci containing FGFR, FGF, c-Met and HGF candidate genes in non-small cell lung cancer sub-types, implication for progression. *Eur. J. Cancer* 39: 2538-2547.
Beaulieu N *et al.* (2002) An essential role for DNA methyltransferase DNMT3B in cancer cell survival. *J. Biol. Chem.* 277: 28176-28181.
Becker KG *et al.* (2004) The Genetic Association Database. *Nature Genet.* 36: 431-432.
Beckman LE *et al.* (1999) Protective effect of transferrin C3 in lung cancer. *Oncology* 56: 328-331.
Bednarek AK *et al.* (2001) *WWOX*, the FRA16D gene, behaves as a suppressor of tumor growth. *Cancer Res.* 61: 8068-8073.
Bearzatto A *et al.* (2002) p16(INK4A) hypermethylation detected by fluorescent methylation-specific PCR in plasmas from non-small cell lung cancer. *Clin. Cancer Res.* 8: 3782-3787.
Beer DG *et al.* (2002) Gene-expression profiles predict survival of patients with lung adenocarcinoma. *Nature Med.* 8: 816-824.
Begueret H *et al.* (2002) Detection of circulating cells expressing chromogranin A gene transcripts in patients with lung neuroendocrine carcinoma. *Eur. J. Cancer* 38: 2325-2330.
Behn M *et al.* (2000) Facilitated detection of oncogene mutations from exfoliated tissue material by a PNA-mediated 'enriched PCR' protocol. *J. Pathol.* 190: 69-75.
Belinsky SA *et al.* (1998) Aberrant methylation of $p16^{(INK4a)}$ is an early event in lung cancer and a potential marker for early diagnosis. *Proc. Natl. Acad. Sci. USA* 95: 11891-11896.
Belinsky SA *et al.* (2002) Aberrant promoter methylation in bronchial epithelium and sputum from current and former smokers. *Cancer Res.* 62: 2370-2377.
Belinsky SA *et al.* (2003) Inhibition of DNA methylation and histone deacetylation prevents murine lung cancer. *Cancer Res.* 63: 7089-7093.
Belinsky SA *et al.* (2004) Plutonium targets the p16 gene for inactivation by promoter hypermethylation in human lung adenocarcinoma. *Carcinogenesis* 25: 1063-1067.
Bellusci S *et al.* (1997a) Fibroblast growth factor 10 (FGF10) and branching morphogenesis in the embryonic mouse lung. *Development* 124: 4867-4878.
Bellusci S *et al.* (1997b) Involvement of Sonic hedgehog (Shh) in mouse embryonic lung growth and morphogenesis. *Development* 124: 53-63.
Belogubova EV *et al* (2004) A novel approach for assessment of cancer predisposing roles of *GSTM1* and *GSTT1* genes: use of putatively cancer resistant elderly tumor-free smokers as the referents. *Lung Cancer* 43: 259-266.
Benachenhou N *et al.* (1998) High resolution deletion mapping reveals frequent allelic losses at the DNA mismatch repair loci *hMLH1* and *hMSH3* in non-small cell lung cancer. *Int. J. Cancer* 77: 173-180.
Benhamou S, Sarasin A (2002) ERCC2/XPD gene polymorphisms and cancer risk. *Mutagenesis* 17: 463-469.
Bendik I *et al.* (1998) Characterization of MAST9/Hevin a SPARC-like protein, that is down-regulated in non-small cell lung cancer. *Cancer Res.* 58: 626-629.
Benhamou S *et al.* (1998) Association between lung cancer and microsomal epoxide hydrolase genotypes. *Cancer Res.* 58: 5291-5293.
Benhamou S *et al.* (2002) Meta-and pooled analyses of the effects of glutathione *S*-transsferase M1 polymorphisms and smoking on lung cancer risk. *Carcinogenesis* 23: 1343-1350.
Bennett WP *et al* (1999) Molecular epidemiology of human cancer risk: gene-environment interactions and p53 mutation spectrum in human lung cancer. *J. Pathol.* 187: 8-18.
Bepler G, Garcia-Blanco MA (1994) Three tumor suppressor regions on chromosome 11p identified by high-resolution deletion mapping in human non-small-cell lung cancer. *Proc. Natl. Acad. Sci. USA* 91: 5513-5517.
Bepler G *et al.* (2002) Prognostic significance of molecular genetic aberrations on chromosome segment 11p15.5 in non-small-cell lung cancer. *J. Clin. Oncol.* 20: 1353-1360.
Bepler G *et al.* (2004) RRM1 and PTEN as prognostic parameters for overall and disease-free survival in patients with non-small-cell lung cancer. *J. Clin. Oncol.* 22: 1878-1885.
Bernasconi NL *et al.* (2000) Post-transcriptional deregulation of myc genes in lung cancer cell lines. *Am. J. Respir. Cell Mol. Biol.* 23: 560-565.
Béroud C, Soussi T (2003) The UMD-p53 database: new mutations and analysis tools. *Hum. Mutation* 21: 176-181.
Berrieman HK *et al.* (2004) Chromosomal analysis of non-small-cell lung cancer by multicolour fluorescent *in situ* hybridisation. *Br. J. Cancer* 90: 900-905.

Bertrand P *et al.* (2004) p53's double life: transactivation-independent repression of homologous recombination. Trends Genet. 20: 235-243.

Besaratinia A *et al.* (2002) Mutational signature of the proximate bladder carcinogen N-hydroxy-4-acetylaminobiphenyl: inconsistency with the p53 mutational spectrum in bladder cancer. *Cancer Res.* 62: 4331-4338.

Betticher DC *et al.* (1996) Prognostic significance of *CCND1* (cyclin D1) overexpression in primary resected non-small-cell lung cancer. *Br. J. Cancer* 73: 294-300.

Betticher DC *et al.* (1997a) Abnormal expression of *CCND1* and *RB1* in resection margin epithelia of lung cancer patients. *Br. J. Cancer* 75: 1761-1768.

Betticher DC *et al.* (1997b) G1 control gene status is frequently altered in resectable non-small cell lung cancer. *Int. J. Cancer* 74: 556-562.

Bharadwaj R, Yu H (2004) The spindle checkpoint, aneuploidy, and cancer. *Oncogene* 23: 2016-2027.

Bhatia KG *et al.* (1990) A deletion linked to a poly(ADP-ribose) polymerase gene on chromosome 13q33-qter occurs frequently in the normal black population as well as in multiple tumor DNA. *Cancer Res.* 50: 5406-5413.

Bhattacharjee A *et al.* (2001) Classification of human lung carcinomas by mRNA expression profiling reveals distinct adenocarcinoma subclasses. *Proc. Natl. Acad. Sci. USA* 98: 13790-13795.

Bhattacharyya N *et al.* (1999) Variant forms of DNA polymerase beta in primary lung carcinomas. *DNA Cell Biol.* 18: 549-554.

Bianchi NO *et al.* (2001) Mitochondrial genome instability in human cancers. *Mutation Res.* 488: 9-23.

Bigger CA *et al.* (2000) Mutational spectra for polycyclic aromatic hydrocarbons in the supF target gene. *Mutation Res.* 450: 75-93.

Binaschi M *et al.* (1992) Characterization of a topoisomerase II gene rearrangement in a human small-cell lung cancer cell line. *J. Natl. Cancer Inst.* 84: 1710-1716.

Biros E *et al.* (2001a) Germ line polymorphisms of the tumor suppressor gene p53 and lung cancer. *Lung Cancer* 31: 157-162.

Biros E *et al.* (2001b) Polymorphism of the p53 gene within the codon 72 in lung cancer patients. *Neoplasma* 48: 407-411.

Birrer MJ, Minna JD (1988) Molecular genetics of lung cancer. *Semin. Oncol.* 15: 226-235.

Bittencourt Rosas SL *et al.* (2001) Methylation status in the promoter region of the human PGP9.5 gene in cancer and normal tissues. *Cancer Letts.* 170: 73-79.

Bjorkqvist AM *et al.* (1998) DNA gains in 3q occur frequently in squamous cell carcinoma of the lung, but not in adenocarcinoma. *Genes Chrom. Cancer* 22: 79-82.

Black M *et al.* (1993) Expression of the pro-opiomelanocortin gene in lung neuroendocrine tumours: *in situ* hybridisation and immunohistochemical studies. *J. Pathol.* 169: 329-334.

Blandino G *et al.* (1999) Mutant p53 gain of function: differential effects of different p53 mutants on resistance of cultured cells to chemotherapy. *Oncogene* 18: 477-485.

Bloom JL, Falconer DS (1964) A gene with major effect on susceptibility to induced lung tumors in mice. *J. Natl. Cancer Inst.* 33: 607-617.

Bochar DA *et al.* (2000) BRCA1 is associated with a human SWI/SNF-related complex: linking chromatin remodelling to breast cancer. *Cell* 102: 257-265.

Bodner SM *et al.* (1992) Expression of mutant p53 proteins in lung cancer correlates with the class of p53 gene mutation. *Oncogene* 7: 743-749.

Boettger MB *et al.* (2003) *BRCA1/2* mutation screening and LOH analysis of lung adenocarcinoma tissue in multiple-cancer patient with a strong family history of breast cancer. *J. Carcinogen.* 2: 5. [http://www.Carcinogenesis.com/content/2/1/5]

Bogenrieder T, Herlyn M (2003) Axis of evil: molecular mechanisms of cancer metastasis. *Oncogene* 22: 6524-6536.

Bogliolo M *et al.* (2002) The Fanconi anaemia genome stability and tumour suppressor network. *Mutagenesis* 17: 529-538.

Bohm M *et al.* (1994) Differences in E-cadherin expression levels and patterns in human lung cancer. *Ann. Hematol.* 68: 81-83.

Boland CR, Ricciardiello L (1999) How many mutations does it take to make a tumor? *Proc. Natl. Acad. Sci. USA* 96: 14675-14677.

Boldrini L *et al.* (2001) Alterations of Fas (APO-1/CD 95) gene and its relationship with p53 in non small cell lung cancer. *Oncogene* 20: 6632-6637.

Boldrini L *et al.* (2002) Identification of Fas (APO-1/CD95) and p53 gene mutations in non-small cell lung cancer. *Int. J. Oncol.* 20: 155-159.

Boldrini L *et al.* (2004) Expression and mutational status of c-kit in small-cell lung cancer: prognostic relevance. *Clin. Cancer Res.* 10: 4101-4108.

Bombi JA *et al.* (2002) Ultrastructural and molecular heterogeneity in non-small cell lung carcinomas: study of 110 cases and review of the literature. *Ultrastruct. Pathol.* 26: 211-218.

Bonafè M *et al.* (1999) p53 variants predisposing to cancer are present in healthy centenarians. *Am. J. Hum. Genet.* 64: 292-295.

Bonner AE *et al.* (2003) Gene expression signatures identify novel regulatory pathways during murine lung development: implications for lung tumorigenesis. *J. Med. Genet.* 40: 408-417.

Bonner AE *et al.* (2004) Molecular profiling of mouse lung tumors: association with tumor progression, lung development, and human lung adenocarcinomas. *Oncogene* 23: 1166-1176.

Borczuk AC *et al.* (2003) Non-small-cell lung cancer molecular signatures recapitulate lung developmental pathways. *Am. J. Pathol.* 163: 1949-1960.

Borczuk AC *et al.* (2004) Molecular signatures in biopsy specimens of lung cancer. *Am. J. Respir. Crit. Care Med.* 170: 167-174.

Borkhardt A *et al.* (2002) Blocking oncogenes in malignant cells by RNA interference: new hope for a highly specific cancer treatment? *Cancer Cell* 2: 167-168.

Botto F *et al.* (1994) Hypomethylation and hypoexpression of human *CYP2E1* gene in lung tumors. *Biochem. Biophys. Res. Commun.* 205: 1086-1092.

Bouchardy C *et al.* (1996) The effect of tobacco on lung cancer risk depends on CYP2D6 activity. *Cancer Res.* 56: 251-253.

Bouchardy C *et al.* (1998) *N*-acetyltransferase *NAT1* and *NAT2* genotypes and lung cancer risk. *Pharmacogenetics* 8: 291-298.

Bowers AJ *et al.* (2003) *SKIP3*, a novel *Drosophila tribbles* ortholog, is overexpressed in human tumors and is regulated by hypoxia. *Oncogene* 22: 2823-2835.

Brabender J *et al.* (2001) Adenomatous polyposis coli gene promoter hypermethylation in non-small cell lung cancer is associated with survival. *Oncogene* 20: 3528-3532.

Brabender J *et al.* (2003) Quantitative O(6)-methylguanine DNA methyltransferase methylation analysis in curatively resected non-small cell lung cancer: associations with clinical outcome. *Clin. Cancer Res.* 9: 223-227.

Bracco L, Kearsey J (2003) The relevance of alternative RNA splicing to pharmacogenomics. *Trends Biotechnol.* 21: 346-353.

Brambilla E *et al.* (1999) Alterations of Rb pathway (Rb-p16INK4-cyclin D1) in preinvasive bronchial lesions. *Clin. Cancer Res.* 5: 243-250.

Brambilla E *et al.* (2000) Semaphorin SEMA3F localization in malignant human lung and cell lines: a suggested role in cell adhesion and cell migration. *Am. J. Pathol.* 156: 939-950.

Brambilla E *et al.* (2001) The new World Health Organization classification of lung tumours. *Eur. Respir. J.* 18: 1059-1068.

Brambilla C *et al.* (2003) Early detection of lung cancer: role of biomarkers. *Eur. Respir. J. Suppl.* 39: 36s-44s.

Brass N *et al.* (1997) Translation initiation factor eIF-4gamma is encoded by an amplified gene and induces an immune response in squamous cell lung carcinoma. *Hum. Mol. Genet.* 6: 33-39.

Brauch H *et al.* (1987) Molecular analysis of the short arm of chromosome 3 in small-cell and non-small-cell carcinoma of the lung. *New Engl. J. Med.* 317: 1109-1113.

Braun MM *et al.* (1994) Genetic component of lung cancer: cohort study of twins. *Lancet* 344: 440-441.

Breathnach OS *et al.* (2001) Twety-two years of phase II trials for patients with advanced non-small-cell lung cancer: sobering results. *J. Clin. Oncol.* 19: 1734-1742.

Brechot JM *et al.* (2002) 2000 standards, options and recommendations for prognostic value of oncogenes and tumor suppressor genes in non small cell lung cancer. *Bull. Cancer* 89: 857-867.

Breeze RG, Wheeldon EB (1977) The cells of the pulmonary airways. *Am. Rev. Respir. Dis.* 116: 705-777.

Bremnes RM *et al.* (2002a) The E-cadherin cell-cell adhesion complex and lung cancer invasion, metastasis, and prognosis. *Lung Cancer* 36: 115-124.

Bremnes RM *et al.* (2002b) High-throughput tissue microarray analysis used to evaluate biology and prognostic significance of the E-cadherin pathway in non-small-cell lung cancer. *J. Clin. Oncol.* 20: 2417-2428.

Brennan J *et al.* (1991) *myc* family DNA amplification in 107 tumors and tumor cell lines from patients with small cell lung cancer treated with different combination chemotherapy regimens. *Cancer Res.* 51: 1708-1712.

Brentani H *et al.* (2003) The generation and utilization of a cancer-oriented representation of the human transcriptome by using expressed sequence tags. *Proc. Natl. Acad. Sci. USA* 100: 13418-13423.

Brockmoller J *et al.* (1993) Genotype and phenotype of glutathione *S*-transferase class mu isoenzymes mu and psi in lung cancer patients and controls. *Cancer Res.* 53: 1004-1011.

Brockmoller J *et al.* (1998) Polymorphisms in xenobiotic conjugation and disease predisposition. *Toxicol. Lett.* 28: 173-183.
Broermann P *et al.* (2002) Trimodality treatment in Stage III nonsmall cell lung carcinoma: prognostic impact of K-ras mutations after neoadjuvant therapy. *Cancer* 94: 2055-2062.
Brognard J *et al.* (2001) Akt/protein kinase B is constitutively active in non-small cell lung cancer cells and promotes cellular survival and resistance to chemotherapy and radiation. *Cancer Res.* 61: 61: 3986-3997.
Bromen K *et al.* (2000) Aggregation of lung cancer in families: results from a population-based case-control study in Germany. *Am. J. Epidemiol.* 152: 497-505.
Brookes AJ (1999) The essence of SNPs. *Gene* 234: 177-186.
Brose MS *et al.* (2002) *BRAF* and *RAS* mutations in human lung cancer and melanoma. *Cancer Res.* 62: 6997-7000.
Brown KD *et al.* (2003) The mismatch repair system is require for S-phase checkpoint activation. *Nature Genet.* 33: 80-84.
Brownson RC *et al.* (1997) Family history of cancer and risk of lung cancer in lifetime non-smokers and long-term ex-smokers. *Int. J. Epidemiol.* 26: 256-263.
Brownson RC *et al.* (2002) Epidemiology of environmental tobacco smoke exposure. *Oncogene* 21: 7341-7348.
Brundage MD *et al.* (2002) Prognostic factors in non-small cell lung cancer. *Chest* 122: 1037-1057.
Brunet A *et al.* (2002) 14-3-3 transits to the nucleus and participates in dynamic nucleocytoplasmic transport. *J. Cell Biol.* 156: 817-828.
Bubán T *et al.* (2004) Detection of mutations in the cDNA of the proliferation marker Ki-67 protein in four tumor cell lines. *Cancer Genet. Cytogenet.* 149: 81-84.
Buhling F *et al.* (2000) Cathepsin K expression in human lung. *Adv. Exp. Med. Biol.* 477: 281-286.
Burbee DG *et al.* (2001) Epigenetic inactivation of *RASSF1A* in lung and breast cancers and kalignant phenotype suppression. *J. Natl. Cancer Inst.* 93: 691-699.
Burch PRJ (1964) Genetic carrier frequency for lung cancer. *Nature* 202: 711-71.
Burke L *et al.* (1998) Allelic deletion analysis of the *FHIT* gene predicts poor survival in non-small cell lung cancer. *Cancer Res.* 58: 2533-2536.
Butkiewicz D *et al.* (2000) Polymorphisms of the *GSTP1* and *GSTM1* genes and PAH DNA adducts in human mononuclear white blood cells. *Environ. Mol. Mutagen.* 35: 99-105.
Butkiewicz D *et al.* (2001) Genetic polymorphisms in DNA repair genes and risk of lung cancer. *Carcinogenesis* 22: 593-599.
Buttery R *et al.* (2004) Galectin-3: differential expression between small-cell and non-small-cell lung cancer. *Histopathology* 44: 339-344.
Caballero OL *et al.* (2001) Loss of chromosome arms 3p and 9p and inactivation of p16 (INK4a) in normal epithelium of patients with primary lung cancer. *Genes Chrom. Cancer* 32: 119-125.
Cahill DP *et al.* (1998) Mutations of mitotic checkpoint genes in human cancers. *Nature* 392: 300-303.
Cahill DP *et al.* (1999) Genetic instability and Darwinian selection in tumours. *Trends Cell Biol.* 9: M57-M60.
Cajas-Salazar N *et al.* (2003a) Effect of epoxide hydrolase polymorphisms on chromosome aberrations and risk for lung cancer. *Cancer Genet. Cytogenet.* 145: 97-102.
Cajas-Salazar N *et al.* (2003b) Combined effect of *MPO*, *GSTM1* and *GSTT1* polymorphisms on chromosome aberrations and lung cancer risk. *Int. J. Hyg. Environ. Health* 206: 473-483.
Caligo MA *et al.* (1998) Microsatellite alterations and p53, TGFβRII, IGFIIR and BAX mutations in sporadic non-small-cell lung cancer. *Int. J. Cancer* 78: 606-609.
Calin GA *et al.* (2000) Low frequency of alterations of the α (PPP2R1A) and β (PPP2R1B) isoforms of the subunit A of the serine-threonine phosphatase 2A in human neoplasms. *Oncogene* 19: 1191-1195.
Calvo R *et al.* (2000) Altered HOX and WNT7A expression in human lung cancer. *Proc. Natl. Acad. Sci. USA* 97: 12776-12781.
Cam H, Dynlacht BD (2003) Emerging roles for E2F: beyond the G1/S transition and DNA replication. *Cancer Cell* 3: 311-316.
Campa MJ *et al.* (2003) Protein expression profiling identifies macrophage migration inhibitory factor and cyclophilin as potential molecular targets in non-small cell lung cancer. *Cancer Res.* 63: 1652-1656.
Campa D *et al.* (2004) Association of a common polymorphism in the cyclooxygenase 2 gene with risk of non-small cell lung cancer. *Carcinogenesis* 25: 229-235.
Cannon-Albright LA *et al.* (1994) Familiality of cancer in Utah. *Cancer Res.* 54: 2378-2385.
Capon DJ *et al.* (1983) Activation of Ki-ras2 gene in human colon and lung carcinomas by two different point mutations. *Nature* 304: 507-513.

Caporaso N *et al.* (1995) Lung cancer and CYP2D6 (the debrisoquine polymorphism): sources of heterogeneity in the proposed association. *Pharmacogenetics* 5: S129-S134.

Caputi M *et al.* (1999) Frequent high expression of bax pro-apoptotic protein in non-small cell lung cancer. *Anticancer Res.* 19 (1B): 825-827.

Caputi M *et al.* (2002) Loss of pRb2/p130 expression is associated with unfavourable clinical outcome in lung cancer. *Clin. Cancer Res.* 8: 3850-3856.

Cardon LR, Palmer LJ (2003) Population stratification and spurious allelic association. *Lancet* 361: 598-604.

Cardoso WV (2001) Molecular regulation of lung development. *Annu. Rev. Physiol.* 63: 471-494.

Carew JS, Huang P (2002) Mitochondrial defects in cancer. *Mol. Cancer* 1: 9.

Carey FA *et al.* (1992) Expression of proliferating cell nuclear antigen in lung cancer: a systematic study and correlation with DNA ploidy. *Histopathology* 20: 499-503.

Carmeliet P (2003) Angiogenesis in health and disease. *Nature Medicine* 9: 653-660.

Carmelli D *et al.* (1992) Genetic influence on smoking – a study of male twins. *New Engl. J. Med.* 327: 829-833.

Caron de Fromentel C, Soussi T (1992) *TP53* tumor suppressor gene: a model for investigating human mutagenesis. *Genes, Chrom. Cancer* 4: 1-15.

Carretero J *et al.* (2004) Novel and natural knockout lung cancer cell lines for the *LKB1/STK11* tumor suppressor gene. *Oncogene* 23: 4037-4040.

Casalini P *et al.* (1997) Inhibition of tumorigenicity in lung adenocarcinoma cells by c-erbB-2 antisense expression. *Int. J. Cancer* 72: 631-636.

Cascorbi I (2003) Pharmacogenetics of cytochrome p4502D6: genetic background and clinical implication. *Eur. J. Clin. Invest.* 33 Suppl. 2: 17-22.

Cascorbi I *et al.* (1996) Homozygous rapid arylamine N-acetyltransferase (NAT2) genotype as a susceptibility factor for lung cancer. *Cancer Res.* 56: 3961-3966.

Casey G *et al.* (1996) DNA sequence analysis of exons 2 through 11 and immunohistochemical staining are required to detect all known p53 alterations in human malignancies. *Oncogene* 13: 1971-1981.

Casse C *et al.* (2003) The *XRCC1* codon 399 Gln allele is associated with adenine to guanine p53 mutations in non-small cell lung cancer. *Mutation Res.* 528: 19-27.

Castaldo G *et al.* (2003) Carcinoembryonic mRNA analysis detects micrometastatic cells in blood from lung cancer patients. *Eur. Respir. J.* 22: 418-421.

Castedo M *et al.* (2004) The cell cycle checkpoint kinase Chk2 is a negative regulator of mitotic catastrophe. *Oncogene* 23: 4353-4361.

Catzavelos C *et al.* (1999) Reduced expression of the cell cycle inhibitor $p27^{Kip}1$ in non-small cell lung carcinoma: a prognostic factor independent of *Ras*. *Cancer Res.* 59: 684-688.

Cauchi S *et al.* (2001) Polymorphisms of human aryl hydrocarbon receptor (AhR) gene in a French population: relationship with CYP1A1 inducibility and lung cancer. *Carcinogenesis* 22: 1819-1824.

Cauffiez C et al. (2004) Genetic polymorphism of the human cytochrome *CYP2A13* in a French population: implication in lung cancer susceptibility. *Biochem. Biophys. Res. Commun.* 317: 662-669.

Cave-Riant F *et al.* (2002) Association of genetic defects in primary resected lung adenocarcinoma revealed by targeted allelic imbalance analysis. *Am. J. Respir. Cell Mol. Biol.* 27: 495-502.

Center R *et al.* (1993) Y chromosome loss and rearrangement in non-small-cell lung cancer. *Int. J. Cancer* 55: 390-393.

Cerami C *et al.* (1997) Tobacco smoke is a source of toxic reactive glycation products. *Proc. Natl. Acad. Sci. USA* 94: 13915-13920.

Cesari R *et al.* (2003) Parkin, a gene implicated in autosomal recessive juvenile parkinsonism, is a candidate tumor suppressor gene on chromosome 6q25-q27. *Proc. Natl. Acad. Sci. USA* 100: 5956-5961.

Chan EC *et al.* (2002) Aberrant promoter methylation in Chinese patients with non-small cell lung cancer: patterns in primary tumors and potential diagnostic application in bronchoalveolar lavage. *Clin. Cancer Res.* 8: 3741-3746.

Chan-Yeung M *et al.* (2004) Lung cancer susceptibility and polymorphisms of glutathione-*S*-transferase genes in Hong Kong. *Lung Cancer* 45: 155-160.

Chang JW *et al.* (2000) Correlation of genetic instability with mismatch repair protein expression and p53 mutations in non-small cell lung cancer. *Clin. Cancer Res.* 6: 1639-1646.

Chang NS *et al.* (2001) Hyaluronidase induction of a WW domain-containing oxidoreductase that enhances tumor necrosis factor cytotoxicity. *J. Biol. Chem.* 276: 3361-3370.

Chang YS *et al.* (2002) Correlation between insulin-like growth factor-binding protein-3 promoter methylation and prognosis of patients with stage I non-small cell lung cancer. *Clin. Cancer Res.* 8: 3669-3675.

Chang YS *et al.* (2003) Enolase-alpha is frequently down-regulated in non-small cell lung cancer and predicts aggressive biological behavior. *Clin. Cancer Res.* 9: 3641-3644.

Chang CC *et al.* (2004a) Connective tissue growth factor and its role in lung adenocarcinoma invasion and metastasis. *J. Natl. Cancer Inst.* 96: 364-375.

Chang H *et al.* (2004b) Increased expression of collagen XVIII and its prognostic value in nonsmall cell lung cancer. *Cancer* 100: 1665-1672.

Chang YS *et al.* (2004c) Retinoic acid receptor-beta expression in Stage I non-small cell lung cancer and adjacent normal appearing bronchial epithelium. *Yonsei Med. J.* 435-442.

Chang YS *et al.* (2004d) Mechanisms underlying lack of insulin-like growth factor-binding protein-3 expression in non-small-cell lung cancer. *Oncogene* In press [DOI: 10.1038/sj.onc.1207882].

Chang HW *et al.* (2004e) Smoking and drinking can induce p15 methylation in the upper aerodigestive tract of healthy individuals and patients with head and neck squamous cell carcinoma. *Cancer* 101: 125-132.

Chaplet M *et al.* (2003) Dentin matrix protein 1 is expressed in human lung cancer. *J. Bone Miner. Res.* 18: 1506-1512.

Chaussade L *et al.* (2001) Expression of p15 and p15.5 products in neuroendocrine lung tumours: relationship with p15(INK4b) methylation status. *Oncogene* 20: 6587-6596.

Chen IT *et al.* (1995) Direct interaction of Gadd45 with PCNA and evidence for competitive interaction of Gadd45 and p21Waf1/Cip1 with PCNA. *Oncogene* 11: 1931-1937.

Chen HL *et al.* (1996a) Structural and functional analysis of β2 microglobulin abnormalities in human lung and breast cancer. *Int. J. Cancer* 67: 756-763.

Chen HL *et al.* (1996b) A functionally defective allele of TAP1 results in loss of MHC class I antigen presentation in a human lung cancer. *Nature Genet.* 13: 210-213.

Chen X *et al.* (1996c) Microsatellite alterations in plasma DNA of small cell lung cancer patients. *Nature Medicine* 2: 1033-1035.

Chen H *et al.* (1999) Association of the NAD(P)H:quinone oxidoreductase 609C→T polymorphism with a decreased lung cancer risk. *Cancer Res.* 59: 3045-3048.

Chen JT *et al.* (2000) Detection of p53 mutations in sputum smears precedes diagnosis of non-small cell lung carcinoma. *Anticancer Res.* 20: 2687-2690.

Chen JJ *et al.* (2001a) Global analysis of gene expression in invasion by a lung cancer model. *Cancer Res.* 61: 5223-5230.

Chen Q *et al.* (2001b) Study on telomerase activity and its clinical value in human non small cell lung cancer. *Hunan Yi Ke Da Xue Xue Bao* 26: 221-222.

Chen S *et al.* (2001c) Polymorphisms of the *CYP1A1* and *GSTM1* genes in relation to individual susceptibility to lung carcinoma in Chinese population. *Mutation Res.* 458: 41-47.

Chen S *et al.* (2002a) DNA repair gene *XRCC1* and *XPD* polymorphisms and risk of lung cancer in a Chinese population. *Carcinogenesis* 23: 1321-1325.

Chen G *et al.* (2002b) Proteomic analysis of lung adenocarcinoma: identification of a highly expressed set of proteins in tumors. *Clin. Cancer Res.* 8: 2298-2305.

Chen G *et al.* (2002c) Discordant protein and mRNA expression in lung adenocarcinomas. *Mol. Cell. Proteomics* 1: 304-313.

Chen W *et al.* (2002d) Expression of telomerase activity, telomerase RNA component and telomerase catalytic subunit gene in lung cancer. *Chin. Med. J.* 115: 290-292.

Chen JT *et al.* (2002e) Alterations of the p16(ink4a) gene in resected nonsmall cell lung tumors and exfoliated cells within sputum. *Int. J. Cancer* 98: 724-731.

Chen YQ *et al.* (2002f) Targeted expression of the receptor tyrosine kinase RON in distal lung epithelial cells results in multiple tumor formation: oncogenic potential of RON *in vivo*. *Oncogene* 21: 6382-6386.

Chen Y *et al.* (2003a) Identification of a novel homeobox-containing gene, LAGY, which is down-regulated in lung cancer. *Oncology* 64: 450-458.

Chen JJ *et al.* (2003b) Up-regulation of tumor interleukin-8 expression by infiltrating macrophages: its correlation with tumor angiogenesis and patient survival in non-small cell lung cancer. *Clin. Cancer Res.* 9: 729-737.

Chen Y *et al.* (2003c) Loss of *PDCD4* expression in human lung cancer correlates with tumour progression and prognosis. *J. Pathol.* 200: 640-646.

Chen SK *et al.* (2003d) Age-asoociated decrease of oxidative repair enzymes, human 8-oxoguanine DNA glycosylases (hOgg1), in human aging. *J. Radiat. Res.* 44: 31-35.

Chen J *et al.* (2003e) Xpd/Ercc2 regulates CAK activity and mitotic progression. *Nature* 424: 228-232.

Chen G *et al.* (2003f) Proteomic analysis of eIF-5A in lung adenocarcinomas. *Proteomics* 3: 496-504.
Chen JT *et al.* (2003 g) The correlation between aberrant connexin 43 mRNA expression induced by promoter methylation and nodal micrometastasis in non-small cell lung cancer. *Clin. Cancer Res.* 9: 4200-4204.
Chen G *et al.* (2003h) Protein profiles associated with survival in lung adenocarcinoma. *Proc. Natl. Acad. Sci. USA* 100: 13537-13542.
Chen CM *et al.* (2003i) Methylation target array for rapid analysis of CpG island hypermethylation in multiple tissue genomes. *Am. J. Pathol.* 163: 37-45.
Chen HW *et al.* (2004a) Anti-invasive gene expression profile of curcumin in lung adenocarcinoma based on a high throughput microarray analysis. *Mol. Pharmacol.* 65: 99-110.
Chen G *et al.* (2004b) Reduced selenium-binding protein 1 expression is associated with poor outcome in lung adenocarcinomas. *J. Pathol.* 202: 321-329.
Chen YZ *et al.* (2004c) Correlative study between expression of *BRI* gene and metastatic potential in human non-small cell lung cancer. *Zhonghua Bing Li Xue Za Zhi* 33: 62-66.
Chen W *et al.* (2004d) Identification of specific PP2A complexes involved in human cell transformation. *Cancer Cell* 5: 127-133.
Cheng LS *et al.* (2000) A genetic analysis of smoking behavior in family members of older adult males. *Addiction* 95: 427-435.
Cheng YW *et al.* (2001a) Gender difference in DNA adduct levels among non-smoking lung cancer patients. *Environ. Mol. Mutagen.* 37: 304-310.
Cheng L *et al.* (2001b) Reduced expression levels of nucleotide excision repair genes in lung cancer: a case-control analysis. *Carcinogenesis* 21: 1527-1530.
Cheng YW *et al.* (2001c) The association of human papillomavirus 16/18 infection with lung cancer among non-smoking Taiwanese women. *Cancer Res.* 61: 2799-2803.
Cheng K-h *et al.*(2004) Elucidation of epigenetic inactivation of *SMAD8* in cancer using targeted expressed gene display. *Cancer Res.* 64: 1639-1646.
Cheon KT *et al.* (2000) Gene polymorphisms of endothelial nitric oxide synthase and angiotensin-converting enzyme in patients with lung cancer. *Lung* 178: 351-360.
Cherian MG *et al.* (2003) Metallothioneins in human tumors and potential roles in carcinogenesis. *Mutation Res.* 533: 201-209.
Cheung N *et al.* (1998) Tissue-specific expression pattern of vascular endothelial growth factor isoforms in the malignant transformation of lung and colon. *Hum. Pathol.* 29: 910-914.
Cheung VG *et al.* (2003) Natural variation in human gene expression assessed in lymphoblastoid cells. *Nature Genet.* 33: 422-428.
Chevillard C *et al.* (1998) Mutations in *OGG1*, a gene involved in the repair of oxidative DNA damage, are found in human lung and kidney tumours. *Oncogene* 16: 3083-3086.
Chhieng DC *et al.* (2001) Use of thyroid transcription factor 1, PE-10, and cytokeratins 7 and 20 in discriminating between primary lung carcinomas and metastatic lesions in fine needle aspiration biopsy specimens. *Cancer* 93: 330-336.
Chiang WL *et al.* (2002) The aberrant expression of cytosolic carbonic anhydrase and its clinical significance in human non-small cell lung cancer. *Cancer Lett.* 188: 199-205.
Chiba I *et al.* (1990) Mutations in the p53 gene are frequent in primary resected non-small cell lung cancer. *Oncogene* 5: 1603-1610.
Chin L *et al.* (1999) p53 deficiency rescues the adverse effects of telomere loss and cooperates with telomere dysfunction to accelerate carcinogenesis. *Cell* 97: 527-538.
Chiou HL *et al.* (2003) The presence of human papillomavirus type 16/18 DNA in blood circulation may act as a risk marker of lung cancer in Taiwan. *Cancer* 97: 1558-1563.
Chizhikov V *et al.* (2001) Two consistently deleted regions within chromosome 1p32-pter in human non-small cell lung cancer. *Mol. Carcinogen.* 30: 151-158.
Cho JY *et al.* (1997) Correlation between K-*ras* gene mutation and prognosis of patients with nonsmall cell lung carcinoma. *Cancer* 79: 462-467.
Cho ES *et al.* (2002) Identification of tumor suppressor loci on the long arm of chromosome 4 in primary small cell lung cancers. *Yonsei Med. J.* 43: 145-151.
Cho NH *et al.* (2004) MMP expression profiling in recurred stage IB lung cancer. *Oncogene* 23: 845-851.
Choi JE *et al.* (2003) No association between haplotypes of three variants (codon 81, 284, and 762) in poly (ADP-ribose) polymerase gene and risk of primary lung cancer. *Cancer Epidemiol. Biomarkers Prev.* 12: 947-949.
Choma D *et al.* (2001) Aneuploidy and prognosis of non-small-cell lung cancer: a meta-analysis of published data. *Br. J. Cancer* 85: 14-22.
Christensen PM *et al.* (1997) The sparteine/debrisoquine (CYP2D6) oxidation polymorphism and the risk of lung cancer: a meta analysis. *Eur. J. Clin. Pharmacol.* 51: 389-393.

Chujo M *et al.* (2002) Comparative genomic hybridisation analysis detected frequent overrepresentation of chromsome 3q in squamous cell carcinoma of the lung. *Lung Cancer* 38: 23-29.

Chung GTY *et al.* (1995) Sequential molecular genetic changes in lung cancer development. *Oncogene* 11: 2591-2598.

Chung GT *et al.* (1996) Clonal evolution of lung tumors. *Cancer Res.* 56: 1609-1614.

Chung J-G *et al.* (2001) Novel transmembrane GTPase of non-small cell lung cancer identified by mRNA differential display. *Cancer Res.* 61: 8873-8879.

Chung CH *et al.* (2002) Molecular portraits and the family tree of cancer. *Nature Genet. Suppl.* 32: 533-540.

Chung-man Ho J *et al.* (2001) Differential expression of manganese superoxide dismutase and catalase in lung cancer. *Cancer Res.* 61: 8578-8585.

Churchill GA (2002) Fundamentals of experimental design for cDNA microarrays. *Nature Genet.* 32: 490-495.

Classon M, Harlow E (2002) The retinoblastoma tumour suppressor in development and cancer. *Nature Rev. Cancer* 2: 910-917.

Claudio PP, Giordano A (2000) *RB1* and *RB2*. Chap. 7 in *Lung Cancer. Principles and Practice.* 2nd Ed. HI Pass, JB Mitchell, DH Johnson, AT Turrisi, JD Minna (Eds.). Lippincott Williams & Williams, Philadelphia, pp 133-155.

Claudio PP *et al.* (2000) Mutations in the retinoblastoma-related gene *RB2/p130* in lung tumors and suppression of tumor growth *in vivo* by retrovirus-mediated gene transfer. *Cancer Res.* 60: 372-382.

Cline MJ, Battifora H (1987) Abnormalities of protooncogenes in non-small cell lung cancer: correlations with tumor type and clinical characteristics. *Cancer* 60: 2669-2674.

Cohen AJ (2000) Outdoor air pollution and lung cancer. *Environ. Health Perspect.* 108 Suppl 4: 743-750.

Cohen MM (2003) TGFβ/Smad signalling system and its pathologic correlates. *Am. J. Med. Genet.* 116A: 1-10.

Cohet C *et al.* (2004) Exon 5 polymorphisms in the O6-alkylguanine DNA alkyltransferase gene and lung cancer risk in non-smokers exposed to second-hand smoke. *Cancer Epidemiol. Biomarkers Prev.* 13: 320-323.

Colasante A *et al.* (2003) Gene expression of transforming growth factor beta receptors I and II in non-small-cell lung tumors. *Cytokine* 24: 182-189.

Coleman CM, Grimberg A (2003) Alterations in the expression of insulin-like growth factors and their binding proteins in lung cancer. Chap. 11 in *Methods in Molecular Medicine Vol. 74: Lung Cancer, Vol. 1: Molecular Pathology Methods and Reviews.* Ed. B. Driscoll. Humana Press, Totowa, New Jersey. pp 167-186.

Coller HA *et al.* (1998) Mutational spectra of a 100-base pair mitochondrial DNA target sequence in bronchial epithelial cells: a comparison of smoking and non-smoking twins. *Cancer Res.* 58: 1268-1277.

Contag CH, Bachmann MH (2002) Advances in *in vivo* bioluminescence imaging of gene expression. *Annu. Rev. Biomed. Eng.* 4: 235-260.

Cook RM *et al.* (1998) Mutational inactivation of aminoacylase-1 in a small cell lung cancer cell line. *Genes Chrom. Cancer* 21: 320-325.

Cooke MS *et al.* (2003) Oxidative DNA damage: mechanisms, mutation, and disease. *FASEB J.* 17: 1195-1214.

Cooper DN, Krawczak M (1993) *Human Gene Mutation.* BIOS Scientific, Oxford.

Cooper DN (1999) *Human Gene Evolution.* BIOS Scientific, Oxford.

Cooper DN (2002) Human gene mutation in pathology and evolution. *J. Inher. Metab. Dis.* 25: 157-182.

Cooper DN *et al.* (2002) Proposed guidelines for papers describing DNA polymorphism-disease associations. *Hum. Genet.* 110: 207-208.

Cooper CA *et al.* (1996) Loss of heterozygosity at 5q21 in non-small cell lung cancer: a frequent event but without evidence of *APC* mutation. *J. Pathol.* 180: 33-37.

Cooper CA *et al.* (1997) The pattern of K-*ras* mutation in pulmonary adenocarcinoma defines a new pathway of tumour development in the human lung. *J. Pathol.* 181: 401-404.

Corbin S *et al.* (2002) Identification of unstable sequences within the common fragile site at 3p14.2: implications for the mechanism of deletions within fragile histidine triad gene/common fragile site at 3p14.2 in tumors. *Cancer Res.* 62: 3477-3484.

Cordell HJ (2002) Epistasis: what it means, what it doesn't mean, and statistical methods to detect it in humans. *Hum. Mol. Genet.* 11: 2463-2468.

Corn PG *et al.* (2003) Frequent hypermethylation of the 5' CpG island of the mitotic stress checkpoint gene Chfr in colorectal and non-small cell lung cancer. *Carcinogenesis* 24: 47-51.

Cortez D *et al.* (2004) Minichromosome maintenance proteins are direct targets of the ATM and ATR checkpoint kinases. *Proc. Natl. Acad. Sci. USA* 101: 10078-10083.
Cory S, Adams JM (2002) The Bcl2 family: regulators of the cellular life-or-death switch. *Nature Rev. Cancer* 2: 647-656.
Costa RH *et al.* (2001) Transcription factors in mouse lung development and function. *Am. J. Physiol. Lung Cell Mol. Physiol.* 280: L823-L838.
Costello JF *et al.* (2000) Aberrant CpG-island methylation has non-random and tumour-type-specific patterns. *Nature Genet.* 24: 132-138.
Cottu PH *et al.* (1996) Inverse correlation between RER$^+$ status and *p53* mutation in colorectal cancer cell lines. *Oncogene* 13: 2727-2730.
Coulson JM *et al.* (2000) A splice variant of the neuron-restrictive silencer factor repressor is expressed in small cell lung cancer: a potential role in derepression of neuroendocrine genes and a useful clinical marker. *Cancer Res.* 60: 1840-1844.
Coulson JM *et al.* (2003) Neuroendocrine phenotype of small cell lung cancer. Chap. 4 in *Methods in Molecular Medicine Vol. 74: Lung Cancer, Vol. 1: Molecular Pathology Methods and Reviews.* Ed. B. Driscoll. Humana Press, Totowa, New Jersey. pp 61-73.
Courtois S *et al.* (2004) p53 protein variants: structural and functional similarities with p63 and p73 isoforms. *Oncogene* 23: 631-638.
Cowell JK (2001) Basic principles in cancer genetics. Chap. 1 in *Molecular Genetics of Cancer.* JK Cowell (Ed), 2nd Ed. pp 1-32. BIOS Scientific, Oxford.
Creighton C *et al.* (2003a) Gene expression patterns define pathways correlated with loss of differentiation in lung adenocarcinomas. *FEBS Letts.* 540: 167-170.
Creighton C *et al.* (2003b) Profiling of pathway-specific changes in gene expression following growth of human cancer cell lines transplanted into mice. *Genome Biol.* 4: R46.
Crespo P, León J (2000) Ras proteins in the control of the cell cycle and cell differentiation. *Cell. Mol. Life Sci.* 57: 1613-1636.
Croce CM *et al.* (1999) Role of *FHIT* in human cancer. *J. Clin. Oncol.* 17: 1618-1624.
Crofts F *et al.* (1994) Functional significance of different human CYP1A1 genotypes. *Carcinogenesis* 15: 2961-2963.
Crowe DL *et al.* (2003) Retinoic acid differentially regulates cancer cell proliferation via dose-dependent modulation of the mitogen-activated protein kinase pathway. *Mol. Cancer Res.* 1: 532-540.
Crowell RE *et al.* (1996) Detection of trisomy 7 in non-malignant bronchial epithelium from lung cancer patients and individuals at risk for cancer. *Cancer Epidemiol. Biomarkers Prev.* 5: 631-637.
Crowley-Weber CL *et al.* (2003) Nicotine increases oxidative stress, activates NF-kappaB and GRP78, induces apoptosis and sensitises cells to genotoxic/xenobiotic stresses by a multiple stress inducer, deoxycholate: relevance to colon carcinogenesis. *Chem. Biol. Interact.* 145: 53-66.
Cuenda A (2000) Mitogen-activated protein kinase kinase 4 (MKK4). *Int. J. Biochem. Cell Biol.* 32: 581-587.
Cuezva JM *et al.* (2004) The bioenergetic signature of lung adenocarcinoma is a molecular marker of cancer diagnosis and prognosis. *Carcinogenesis* 25: 1157-1163.
Czerwinski M *et al.* (1994) Quantification of CYP2B7, CYP4B1, and CYPOR messenger RNAs in normal human lung and lung tumors. *Cancer Res.* 54: 1085-1091.
D'Amico D *et al.* (1992a) High frequency of somatically acquired *p53* mutations in small-cell lung cancer cell lines and tumors. *Oncogene* 7: 339-346.
D'Amico D *et al.* (1992b) Polymorphic sites within the *MCC* and *APC* loci reveal very frequent loss of heterozygosity in human small cell lung cancer. *Cancer Res.* 52: 1996-1999.
D'Amico TA *et al.* (1999) A biologic risk model for stage I lung cancer: immunohistochemical analysis of 408 patients with the use of ten molecular markers. *J. Thorac. Cardivasc. Surg.* 117: 736-743.
D'Amico TA *et al.* (2001) Predicting the sites of metastases from lung cancer using molecular biologic markers. *Ann. Thorac. Surg.* 72: 1144-1148.
Dai Y *et al.* (2000) Application of the p53 and K-ras gene mutation patterns for cytologic diagnosis of recurrent lung carcinomas. *Cancer* 90: 258-263.
Dai Z *et al.* (2001) Global methylation profiling of lung cancer identifies novel methylated genes. *Neoplasia* 3: 314-323.
Dai Z *et al.* (2003) A comprehensive search for DNA amplification in lung cancer identifies inhibitors of apoptosis cIAP1 and cIAP2 as candidate oncogenes. *Hum. Mol. Genet.* 12: 791-801.
Dai Z *et al.* (2004) Bone morphogenetic protein 3B silencing in non-small-cell lung cancer. *Oncogene* 23: 3521-3529.

Daigo Y *et al.* (1999) Molecular cloning of a candidate tumor suppressor gene, *DLC1*, from chromosome 3p21.3. *Cancer Res.* 59: 1966-1972.
Dallol A *et al.* (2002a) Tumour specific promoter region methylation of the human homologue of the *Drosophila Roundabout* gene DUTT1 (*ROBO1*) in human cancers. *Oncogene* 21: 3020-3028.
Dallol A *et al.* (2002b) *SLIT2*, a human homologue of the *Drosophila Slit2* gene, has tumor suppressor activity and is frequently inactivated in lung and breast cancers. *Cancer Res.* 62: 5874-5880.
Dally H *et al.* (2002) Myeloperoxidase (*MPO*) genotype and lung cancer histologic types: the *MPO* –463 A allele is associated with reduced risk for small cell lung cancer in smokers. *Int. J. Cancer* 102: 530-535.
Dally H *et al.* (2003) The CYP3A4*1B allele increases risk for small cell cancer: effect of gender and smoking dose. *Pharmacogenetics* 13: 607-618.
Dammann R *et al.* (2000) Epigenetic inactivation of a RAS association domain family protein from the lung tumour suppressor locus 3p21.3. *Nature Genet.* 25: 315-319.
Dammann R *et al.* (2001) The CpG island of the novel tumor suppressor gene *RASSF1A* is intensely methylated in primary small cell lung carcinomas. *Oncogene* 20: 3563-3567.
Dammann R *et al.* (2003) Epigenetic inactivation of the RAS-association domain family 1 (*RASSF1A*) gene and its function in human carcinogenesis. *Histol. Histopathol.* 18: 665-677.
Damstrup L *et al.* (1992) Expression of the epidermal growth factor receptor in human small cell lung cancer cell lines. *Cancer Res.* 52: 3089-3093.
Damstrup L *et al.* (2002) Epidermal growth factor receptor mutation type III transfected into a small cell lung cancer cell line is predominantly localized at the cell surface and enhances the malignant phenotype. *Int. J. Cancer* 97: 7-14.
Dang TP *et al.* (2000) Chromsome 19 translocation, overexpression of Notch3, and human lung cancer. *J. Natl. Cancer Inst.* 92: 1355-1357.
Dang TP *et al.* (2003) Constitutive activation of Notch3 inhibits terminal epithelial differentiation in lungs of transgenic mice. *Oncogene* 22: 1988-1997.
Daniel JC, Smythe Wr (2003) Gene therapy of lung cancer. *Semin. Surg. Oncol.* 21: 196-204.
Darby S *et al.* (2001) Radon: a likely carcinogen at all exposures. *Ann. Oncol.* 12: 1341-1351.
Dastoor Z, Dreyer J (2000) Nuclear translocation and aggregate formation of heat shock cognate protein 70 (Hsc70) in oxidative stress and apoptosis. *J. Cell. Sci.* 113: 2845-2854.
David O (2001) Akt and PTEN: new diagnostic markers of non-small cell lung cancer? *J. Cell Mol. Med.* 5: 430-433.
David-Beabes GL, London SJ (2001) Genetic polymorphism of *XRCC1* and lung cancer risk among African-Americans and Caucasians. *Lung Cancer* 34: 333-339.
Davidson E, Bulkin W (1966) Long marker chromosome in bronchogenic carcinoma. *Lancet* 2: 227.
Davidson JF *et al.* (2002) Endogenous mutagenesis and cancer. *Mutation Res.* 509: 17-21.
Davidson JD *et al.* (2004) An increase in the expression of ribonucleotide reductase large subunit 1 is associated with gemcitabine resistance in non-small cell lung cancer cell lines. *Cancer Res.* 64: 3761-3766.
Davies JC *et al.* (2001) Prospects for gene therapy in lung disease. *Curr. Opin. Pharmacol.* 1: 272-277.
Davies AM *et al.* (2003) Antisense oligonucleotides in the treatment of non-small-cell lung cancer. *Clin. Lung Cancer* 4 Suppl. 2: S68-S73.
Davis PK, Dowdy SF (2001) p73. *Int. J. Biochem. Cell Biol.* 33: 935-939.
de Anta JM *et al.* (1997) *TP53* mutational pattern in Spanish and Polish non-small cell lung canceer patients: null mutations are associated with poor prognosis. *Oncogene* 15: 2951-2958.
de Boer JG (2002) Polymorphisms in DNA repair and environmental interactions. *Mutation Res.* 509: 201-210.
de Castro J *et al.* (2003) New insights in β-tubulin sequence analysis in non-small cell lung cancer. *Lung Cancer* 41: 41-48.
De Flora S *et al.* (2003) Modulation of cigarette smoke-related end-points in mutagenesis and carcinogenesis. *Mutation Res.* 523-524:237-252.
De Fusco PA *et al.* (1989) Cytogenetic studies in 11 patients with small cell carcinoma of the lung. *Mayo Clin. Proc.* 64: 168-176.
De Gregorio L *et al.* (1998) Prognostic value of loss of heterozygosity and *KRAS2* mutations in lung adenocacinoma. *Int. J. Cancer* 79: 269-272.
De Juan C *et al.* (1998) Prognostic value of genomic damage in non-small-cell lung cancer. *Br. J. Cancer* 77: 1971-1977.
de Jonge RR *et al.* (1997) Frequent inactivation of the transforming growth factor β type II receptor in small-cell lung carcinoma cells. *Oncol. Res.* 9: 89-98.
de Lange R *et al.* (2003) Identification of genes associated with enhanced metastasis of a large cell lung carcinoma cell line. *Anticancer Res.* 23(1A): 187-194.

de Vos S *et al.* (1995) Alterations of *CDKN2* (p16) in non-small cell lung cancer. *Genes Chrom. Cancer* 14: 164-170.
Deb SP (2003) Cell cycle regulatory functions of the human oncoprotein MDM2. *Mol. Cancer Res.* 1: 1009-1016.
Debelenko LV *et al.* (1997) Identification of *MEN1* gene mutations in sporadic carcinoid tumors of the lung. *Hum. Mol. Genet.* 6: 2285-2290.
Debelenko LV *et al.* (2000) *MEN1* gene mutation analysis of high-grade neuroendocrine lung carcinoma. *Genes Chrom. Cancer* 28: 58-65.
Decaussin M *et al.* (1999) Expression of vascular endothelial growth factor (VEGF) and its two receptors (VEGF-R1-Flt1 and VEGF-R2-Flk1/KDR) in non-small cell lung carcinomas (NSCLCs): correlation with angiogenesis and survival. *J. Pathol.* 188: 369-377.
Delebecq TJ *et al.* (2000) Overexpression level of stromelysin 3 is related to the lymph node involvement in non-small cell lung cancer. *Clin. Cancer Res.* 6: 1086-1092.
DeMarini DM *et al.* (2001) Lung tumor *KRAS* and *TP53* mutations in nonsmokers reflect exposure to PAH-rich coal combustion emissions. *Cancer Res.* 61: 6679-6681.
Demopoulos K *et al.* (2002) *MYCL1, FHIT, SPARC, p16(INK4)* and *TP53* genes associated to lung cancer in idiopathic pulmonary fibrosis. *J. Cell Mol. Med.* 6: 215-222.
Deng CX, Wang RH (2003) Roles of BRCA1 in DNA damage repair: a link between development and cancer. *Hum. Mol. Genet.* 12: R113-123.
Denissenko MF *et al.* (1996) Preferential formation of benzo[*a*]pyrene adducts at lung cancer mutational hotspots in *p53*. *Science* 274: 430-432.
Denissenko MF *et al.* (1997) Cytosine methylation determines hot spots of DNA damage in the human *p53* gene. *Proc. Natl. Acad. Sci. USA* 94: 3893-3898.
Denissenko MF *et al.* (1998) Slow repair of bulky DNA adducts along the nontranscribed strand of the human p53 gene may explain the strand bias of transversion mutations in cancer. *Oncogene* 16: 1241-1247.
Dennis TR, Stock AD (1999) A molecular cytogenetic study of chromosome rearrangements in small cell lung cancer: consistent involvement of chromosome band 3q13.2. *Cancer Genet. Cytogenet.* 113: 134-140.
DePinho RA (2000) The age of cancer. *Nature* 408: 248-254.
Desinan L *et al.* (1996) Non-small cell lung cancer. Morphology and DNA flow cytometry. *Anal. Quant. Cytol. Histol.* 18: 438-452.
Desmaze C *et al.* (2003) Telomere-driven genomic instability in cancer cells. *Cancer Lett.* 194: 173-182.
Devereux TR, Kaplan NL (1998) Use of quantitative trait loci to map murine lung tumor susceptibility genes. *Exp. Lung Res.* 24: 407-417.
Dherin C *et al.* (1999) Excision of oxidatively damaged DNA bases by the human alpha-hOgg1 protein and the polymorphic alpha-hOgg1(Ser326Cys) protein which is frequently found in human populations. *Nucleic Acids Res.* 27: 4001-4007.
Dialyna IA *et al.* (2003) Genetic polymorphisms of *CYP1A1, GSTM1* and *GSTT1* genes and lung cancer risk. *Oncol. Rep.* 10: 1929-1835.
Difilippantonio S *et al.* (2003) Gene expression profiles in human non-small and small-cell lung cancers. *Eur. J. Cancer* 39: 1936-1947.
Dikmen E *et al.* (2003) Detection of telomerase activity in bronchial lavage as an adjunct to cytological diagnosis in lung cancer. *Eur. J. Cardiothorac. Surg.* 23: 194-199.
Dingemans AM *et al.* (1999) Expression of DNA topoisomerase IIα and topoisomerase IIβ genes predicts survival and response to chemotherapy in patients with small cell lung cancer. *Clin. Cancer Res.* 5: 2048-2058.
Dingemans AC *et al.* (2001) Topoisomerase IIα and other drug resistance markers in advanced non-small cell lung cancer. *Lung Cancer* 32: 117-128.
Divine KK *et al.* (2001) The *XRCC1* 399 glutamine allele is a risk factor for adenocarcinoma of the lung. *Mutat. Res.* 461: 273-278.
Dobashi Y *et al.* (2003) Diversity in expression and prognostic significance of G1/S cyclins in human primary lung carcinomas. *J. Pathol.* 199: 208-220.
Doll R (1953) Mortality from lung cancer among non-smokers. *Br. J. Cancer* 7: 303-312.
Doll R, Hill AD (1950) Smoking and carcinoma of the lung. *Br. Med. J.* ii: 739-748.
Doll R *et al.* (2004) Mortality in relation to smoking: 50 years' observations on male British doctors. *Br. Med. J.* 328: 1519-1528.
Dosaka-Akita H *et al.* (1995) Inhibition of proliferation by L-myc antisense DNA for the translational initiation site in human small cell lung cancer. *Cancer Res.* 55: 1559-1564.
Dosaka-Akita H *et al.* (1997) Altered retinoblastoma protein expression in nonsmall cell lung cancer: its synergistic effects with altered ras and p53 protein status on progress. *Cancer* 79: 1329-1337.

Dosaka-Akita H *et al.* (1998) Frequent loss of gelsolin expression in non-small cell lung cancers of heavy smokers. *Cancer Res.* 58: 322-327.

Dosaka-Akita H *et al.* (2002) *N*-acetylgalactosaminyl transferase-3 is a potential new marker for non-small cell lung cancers. *Br. J. Cancer* 87: 751-755.

Dosaka-Akita H *et al.* (2004) Expression of *N*-acetylglucosaminyltransferase V is associated with prognosis and histology in non-small cell lung cancers. *Clin. Cancer Res.* 10: 1773-1779.

Dragnev KH *et al.* (2003) Lung cancer prevention: the guidelines. *Chest* 123: 60S-71S.

Drakoulis N *et al.* (1994) Polymorphisms in the human *CYP1A1* gene as susceptibility factors for lung cancer: exon 7 mutation (4889 A to G), and a T to C mutation in the 3'-flanking region. *Clin Invest.* 72: 240-248.

Du Y *et al.* (2001) Hypermethylation in human cancers of the *RIZ1* tumor suppressor gene, a member of a histone/protein methyltransferase superfamily. *Cancer Res.* 61: 8094-8099.

Duan W *et al.* (2002) Lung-specific expression of human mutant p53-273H is associated with a high frequency of lung adenocarcinoma in transgenic mice. *Oncogene* 21: 7831-7838.

Dumont P *et al.* (2003) The codon 72 polymorphic variants of p53 have markedly different apoptotic potential. *Nature Genet.* 33: 357-364.

Dunaief JL *et al.* (1994) The retinoblastoma protein and BRG1 form a complex and cooperate to induce cell cycle arrest. *Cell* 79: 119-130.

Dunn BK (2003) Hypomethylation: one side of a larger picture. *Ann N Y Acad. Sci.* 983: 28-42.

Duval A, Hamelin R (2002) Mutations at coding repeat sequences in mismatch repair-deficient human cancers: toward a new concept of target genes for instability. *Cancer Res.* 62: 2447-2454.

Dworakowska D *et al.* (2004) *MDM2* gene amplification: a new independent factor of adverse prognosis in non-small cell lung cancer (NSCLC). *Lung Cancer* 43: 285-295.

el-Zein RA *et al.* (1997a) Combined genetic polymorphism and risk for development of lung cancer. *Mutation Res.* 381: 189-200.

el-Zein RA *et al.* (1997b) Polymorphism of metabolising genes and lung cancer histology: prevalence of CYP2E1 in adenocarcinoma. *Cancer Lett.* 112: 71-78.

Eason DD *et al.* (1999) Interferon regulatory factor 1, tryptophan 11 to arginine point mutation abolishes DNA binding. *Biochim. Biophys. Acta* 1446: 140-144.

Echchakir H *et al.* (2001) A point mutation in the α-actinin-4 gene generates an antigenic peptide recognized by autologous cytolytic T lymphocytes on a human lung carcinoma. *Cancer Res.* 61: 4078-4083.

Eglund KA *et al.* (2002) Characterization of overlapping XAGE-1 transcripts encoding a cancer testis antigen expressed in lung, breast, and other types of cancers. *Mol. Cancer Ther.* 1: 441-450.

Elferink CJ *et al.* (2003) Aryl hydrocarbon receptor-mediated cell cycle control. *Prog. Cell Cycle Res.* 5: 261-267.

Endo C *et al.* (1998) Sequential loss of heterozygosity in the progression of squamous cell carcinoma of the lung. *Br. J. Cancer* 78: 612-615.

Endoh H *et al.* (2003) *RASSF1A* gene inactivation in non-small cell lung cancer and its clinical implication. *Int. J. Cancer* 106: 45-51.

Engbring JA, Kleinman HK (2003) The basement membrane matrix in malignancy. *J. Pathol.* 200: 465-470.

Engel E *et al.* (2004) Detection of tumor-specific mRNA in cell-free bronchial lavage supernatant in patients with lung cancer. *Ann. N Y Acad. Sci.* 1022: 140-146.

Erblich J *et al.* (2004) Stress-induced cigarette craving: effects of the *DRD2 Taq*I RFLP and *SLC6A3* VNTR polymorphisms. *Pharmacogenetics* 4: 102-109.

Erez A *et al.* (2004) *Sil* overexpression in lung cancer characterizes tumors with increased mitotic activity. *Oncogene* 23: 5371-5377.

Escuin D, Rosell R (1999) The anti-apoptosis survivin gene and its role in human cancer: an overview. *Clin. Lung Cancer* 1: 138-143.

Esposito V *et al.* (1997a) Prognostic role of the cyclin-dependent kinase inhibitor p27 in non-small cell lung cancer. *Cancer Res.* 57: 3381-3385.

Esposito V *et al.* (1997b) Prognostic value of p53 in non-small cell lung cancer: relationship with proliferating cell nuclear antigen and cigarette smoking. *Hum. Pathol.* 28: 233-237.

Esposito V *et al.* (2004) Analysis of cell cycle regulator proteins in non-small cell lung cancer. *J. Clin. Pathol.* 57: 58-63.

Esteller M, Herman JG (2004) Generating mutations but providing chemosensitivity: the role of O^6-methylguanine DNA methyltransferase in human cancer. *Oncogene* 23: 1-8.

Esteller M *et al.* (1999) Detection of aberrant promoter hypermethylation of tumor suppressor genes in serum DNA from non-small cell lung cancer patients. *Cancer Res.* 59: 67-70.

Esteller M *et al.* (2001a) A gene hypermethylation profile of human cancer. *Cancer Res.* 61: 3225-3229.

Esteller M *et al.* (2001b) DNA methylation patterns in hereditary human cancers mimic sporadic tumorigenesis. *Hum. Mol. Genet.* 10: 3001-3007.

Esteller M *et al.* (2002) Hypermethylation-associated inactivation of the cellular retinal-binding-protein 1 gene in human cancer. *Cancer Res.* 62: 5902-5905.

Esteller M, Herman JG (2004) Generating mutations but providing chemosensitivity: the role of O6-methylguanine DNA methyltransferase in human cancer. *Oncogene* 23: 1-8.

Etzel CJ *et al.* (2003) Risk for smoking-related cancer among relatives of lung cancer patients. *Cancer Res.* 63: 8531-8535.

Etzioni R *et al.* (2003) The case for early detection. *Nature Rev. Cancer* 3: 243-252.

Evan G, Littlewood T (1998) A matter of life and cell death. *Science* 281: 1317-1322.

Evan GI, Vousden KH (2001) Proliferation, cell cycle and apoptosis in cancer. *Nature* 411: 342-348.

Eymin B *et al.* (2001) Distinct pattern of E2F1 expression in human lung tumours: E2F1 is upregulated in small cell lung carcinoma. *Oncogene* 20: 1678-1687.

Eymin B *et al.* (2002) Mdm2 overexpression and $p14^{ARF}$ inactivation are two mutually exclusive events in primary human lung tumors. *Oncogene* 21: 2750-2761.

Fabbro D *et al.* (1996) TTF-1 gene expression in human lung tumours. *Eur. J. Cancer* 32A:512-517.

Falconer DS, Bloom JL (1961) Inheritance of susceptibility to induced pulmonary tumours in mice. *Nature* 191: 1070-1071.

Falleni M *et al.* (2003) Survivin gene expression in early-stage non-small cell lung cancer. *J. Pathol.* 200: 620-626.

Falor WH *et al.* (1969) Chromosomes in bronchoscopic biopsies from patients with bronchial adenoma, bronciogenic carcinoma, and from heavy smokers. *Cancer* 24: 198-209.

Farber SM *et al.* (1951) The early diagnosis of primary lung cancer by cytologic methods. *Dis. Chest* 20: 237-256.

Fazzari MJ, Greally JM (2004) Epigenomics: beyond CpG islands. *Nature Rev. Genet.* 5: 446-455.

Fearon ER (1997) Human cancer syndromes: clues to the origin and nature of cancer. *Science* 278: 1043-1050.

Fearon ER, Vogelstein B (1990) A genetic model for colorectal tumorigenesis. *Cell* 61: 759-767.

Fei Q *et al.* (2002) Defected expression of E-cadherin in non-small cell lung cancer. *Lung Cancer* 37: 147-152.

Feinberg AP, Vogelstein B (1983a) Hypomethylation of ras oncogenes in primary human cancers. *Biochem. Biophys. Res. Commun.* 111: 47-54.

Feinberg AP, Vogelstein B (1983b) Hypomethylation distinguishes genes of some human cancers from their normal counterparts. *Nature* 301: 89-92.

Feinberg AP *et al.* (2002) DNA methylation and genomic imprinting: insights from cancer into epigenetic mechanisms. *Semin. Cancer Biol.* 12: 389-398.

Feld R *et al.* (2000) Staging and prognostic factors for small cell lung cancer. Chap. 34 in *Lung Cancer. Principles and Practice.* 2nd Ed. HI Pass, JB Mitchell, DH Johnson, AT Turrisi, JD Minna (Eds.). Lippincott Williams & Williams, Philadelphia, pp 612-627.

Feltus FA *et al.* (2003) Predicting aberrant CpG island methylation. *Proc. Natl. Acad. Sci. USA* 100: 12253-12258.

Feng G *et al.* (2001) Diminished expression of S100A2, a putative tumor suppressor, at early stage of human lung carcinogenesis. *Cancer Res.* 61: 7999-8004.

Feng Z *et al.* (2002a) 4-aminobiphenyl is a major etiological agent of human bladder cancer: evidence from its DNA binding spectrum in human *p53* gene. *Carcinogenesis* 23: 1721-1727.

Feng Z *et al.* (2002b) Preferential DNA damage and poor repair determine ras gene mutational hotspot in human cancer. *J. Natl. Cancer Inst.* 94: 1527-1536.

Feng Y *et al.* (2004) A common haplotype of the nicotine acetylcholine receptor $\alpha 4$ subunit gene is associated with vulnerability to nicotine addiction in men. *Am. J. Hum. Genet.* 75: 112-121.

Fernandez P *et al.* (2004) Distinctive gene expression of human lung adenocarcinomas carrying *LKB1* mutations. *Oncogene* 23: 5084-5091.

Ferretti G *et al.* (2000) Detection by denaturant gradient gel electrophoresis of tumor-specific mutations in biopsies and relative bronchoalveolar lavage fluid from resectable non-small cell lung cancer. *Clin. Cancer Res.* 6: 2393-2400.

Ferrigno D *et al.* (2003) Neuron-specific enolase is an effective tumour marker in non-small cell lung cancer (NSCLC). *Lung Cancer* 41: 311-320.

Ferron PE *et al.* (1997) Combined and sequential expression of p53, Rb, Ras and Bcl-2 in bronchial preneoplastic lesions. *Tumori* 83: 587-593.

Feyler A *et al.* (2002) Myeloperoxidase –463 G→A polymorphism and lung cancer risk. *Cancer Epidemiol. Biomarkers Prev.* 11: 1550-1554.

Field JK, Youngson JH (2002) The Liverpool Lung Project: a molecular epidemiological study of early lung cancer detection. *Eur. Respir. J.* 20: 464-479.

Fielding P *et al.* (1999) Heterogeneous nuclear ribonuclearprotein A2/B1 up-regulation in bronchial lavage specimens: a clinical marker of early lung cancer detection. *Clin. Cancer Res.* 5: 4048-4052.

Fields WR *et al.* (1998) Overexpression of stably transfected human glutathione *S*-transferase P1-1 protects against DNA damage by benzo[*a*]pyrene diol-epoxide in human T47D cells. *Mol. Pharmacol.* 54: 298-304.

Fisher RA (1958a) Lung cancer and cigarettes? *Nature* 182: 108.

Fisher RA (1958b) Cancer and smoking. *Nature* 182: 596.

Fisher MJ *et al.* (2001a) Nucleotide substitution in the ectodomain of trail receptor DR4 associated with lung cancer and head and neck cancer. *Clin. Cancer Res.* 7: 1688-1697.

Fisher GH *et al.* (2001b) Induction and apoptotic regression of lung adenocarcinoma by regulation of a K-Ras transgene in the presence and absence of tumor suppressor genes. *Genes Dev.* 15: 3249-3262.

Flaherty KT *et al.* (2001) Antisense therapeutics: lessons from early clinical trials. *Curr. Opin. Oncol.* 13: 499-505.

Fleischhacker M *et al.* (2001) Detection of amplifiable messenger RNA in the serum of patients with lung cancer. *Ann. N Y Acad. Sci.* 945: 179-188.

Fletcher O *et al.* (2004) Lifetime risks of common cancers among retinoblastoma survivors. *J. Natl. Cancer Inst.* 96: 357-363.

Flint A, Lloyd RV (1992) Pulmonary metastases of colonic carcinoma. Distinction from pulmonary adenocarcinoma. *Arch. Pathol. Lab. Med.* 116: 39-42.

Fliss MS *et al.* (2000) Facile detection of mitochondrial DNA mutations in tumors and bodily fluids. *Science* 287: 2017-2019.

Flores ER *et al.* (2002) p63 and p73 are required for p53-dependent apoptosis in response to DNA damage. *Nature* 416: 560-564.

Florl AR, Schulz WA (2003) Peculiar structure and location of 9p21 homozygous deletion breakpoints in human cancer cells. *Genes Chrom. Cancer* 37: 141-148.

Folkman J, Kalluri R (2004) Cancer without disease. *Nature* 427: 787.

Fong KM *et al.* (1995a) Microsatellite instability and other molecular abnormalities in non-small cell lung cancer. *Cancer Res.* 55: 28-30.

Fong KM *et al.* (1995b) Tumor progression and loss of heterozygosity at 5q and 18q in non-small cell lung cancer. *Cancer Res.* 55: 220-223.

Fong KM *et al.* (1995c) Lung pathology: the molecular genetics of non-small cell lung cancer. *Pathology* 27: 295-301.

Fong KM *et al.* (1997) *FHIT* and FRA3B 3p14.2 allele loss are common in lung cancer and preneoplastic bronchial lesions and are associated with cancer-related *FHIT* cDNA splicing aberrations. *Cancer Res.* 57: 2256-2267.

Fong KM et al. (2002) The molecular basis of lung carcinogenesis. Chap. 17 in *The Molecular Basis of Human Cancer*. pp 379-405. Eds. WB Coleman, GJ Tsongalis. Human Press, Titowa, New Jersey.

Fontanini G *et al.* (1995) Bcl-2 protein: a prognostic factor inversely correlated to p53 in non-small-cell lung cancer. *Br. J. Cancer* 71: 1003-1007.

Fontanini G *et al.* (1997a) 67-kilodalton laminin receptor expression correlates with worse prognostic indicators in non-small cell lung carcinomas. *Clin. Cancer Res.* 3: 227-231.

Fontanini G *et al.* (1997b) Neoangiogenesis and p53 protein in lung cancer: their prognostic role and their relation with vascular endothelial growth factor (VEGF) expression. *Br. J. Cancer* 75: 1295-1301.

Fontanini G *et al.* (1998). Evaluation of epidermal growth factor-related growth factors and receptors and of neoangiogenesis in completely resected stage I-IIIA non-small-cell lung cancer: amphiregulin and microvessel count are independent prognostic indicators of survival. *Clin. Cancer Res.* 4: 241-249.

Fontanini G *et al.* (1999) Thrombospondins I and II messenger RNA expression in lung carcinoma: relationship with p53 alterations, angiogenic growth factors, and vascular density. *Clin. Cancer Res.* 5: 155-161.

Ford CH *et al.* (1981) HLA frequency and prognosis in lung cancer. *Br. J. Cancer* 43: 610-614.

Ford JG *et al.* (2000) Glutathione *S*-transferase M1 polymorphism and lung cancer risk in African-Americans. *Carcinogenesis* 21: 1971-1975.

Forgacs E *et al.* (1998) Mutation analysis of the *PTEN/MMAC1* gene in lung cancer. *Oncogene* 17: 1557-1565.

Forrester K *et al.* (1995) Effects of p53 mutants on wild-type p53-mediated transactivation are cell type dependent. *Oncogene* 10: 2103-2111.
Forsythe HL *et al.* (2002) Retroviral-mediated expression of telomerase in normal human cells provides a selective growth advantage. *Int. J. Oncol.* 20: 1137-1143.
Fortes C *et al.* (2003) The protective effect of the Mediterranean diet on lung cancer. *Nutr. Cancer* 46: 30-37.
Fortini P *et al.* (2003) 8-oxoguanine DNA damage: at the crossroad of alternative repair pathways. *Mutation Res.* 531: 127-139.
Fouquet C *et al.* (2004) Rapid and sensitive p53 alteration analysis in biopsies from lung cancer patients using a functional assay and a universal oligonucleotide array: a prospective study. *Clin. Cancer Res.* 10: 3479-3489.
Frank SA (2003) Somatic mutation: early cancer steps depend on tissue architecture. *Curr. Biol.* 13: R261-R263.
Franklin WA *et al.* (1997) Widely dispersed p53 mutation in respiratory epithelium. A novel mechanism for field carcinogenesis. *J. Clin. Invest.* 100: 2133-2137.
Franklin WA *et al.* (2002) Epidermal growth factor receptor family in lung cancer and premalignancy. *Semin. Oncol.* 29 (1 Suppl. 4): 3-14.
Frazier MW *et al.* (1998) Activation of c-myc gene expression by tumor-derived *TP53* mutants require a discrete C-terminal domain. *Mol. Cell. Biol.* 18: 3735-3743.
Frederick L *et al.* (2000) Analysis of genomic rearrangements associated with EGRFvIII expression suggests involvement of *Alu* repeat elements. *Neuro-oncology* 2: 159-163.
Freeman DJ *et al.* (2003) PTEN tumor suppressor regulates p53 protein levels and activity through phosphatase-dependent and -independent mechanisms. *Cancer Cell* 3: 117-124.
Fridman JS, Lowe SW (2003) Control of apoptosis by p53. *Oncogene* 22: 9030-9040.
Friedberg EC (2001) How nucleotide excision repair protects against cancer. *Nature Reviews Cancer* 1: 22-33.
Friedberg EC, Walker GC, Siede W (1995) *DNA Repair and Mutagenesis.* ASM Press, Washington D.C., pp. 39-41.
Friedberg EC *et al.* (2004) The role of endogenous and exogenous DNA damage and mutagenesis. *Curr. Opin. Genet. Dev.* 14: 5-10.
Friedlander P *et al.* (1996) A mutant p53 that discriminates between p53-responsive genes cannot induce apoptosis. *Mol. Cell. Biol.* 16: 4961-4971.
Fromigué O *et al.* (2003) Gene expression profiling of normal human pulmonary fibroblasts following coculture with non-small-cell lung cancer cells reveals alterations related to matrix degradation, angiogenesis, cell growth and survival. *Oncogene* 22: 8487-8497.
Frosst P *et al.* (1995) A candidate genetic risk factor for vascular disease: a common mutation in methylenetetrahydrofolate reductase. *Nature Genet.* 10: 111-113.
Fu J *et al.* (2003) Abnormality of p16 gene and its clinicopathological significance in non-small cell lung cancer. *Zhonghua Bing Li Xue Za Zhi* 32: 133-136.
Fujii T *et al.* (2002) A preliminary transcriptome map of non-small cell lung cancer. *Cancer Res.* 62: 3340-3346.
Fujishita T *et al.* (1998) Coincidental alterations of p16INK4A/CDKN2 and other genes in human lung cancer cell lines. *Anticancer Res.* 18(3A): 1537-1542.
Fujita Y *et al.* (2003) The diagnostic and prognostic relevance of human telomerase reverse transcriptase mRNA expression detected *in situ* in patients with nonsmall cell lung carcinoma. *Cancer* 98: 1008-1013.
Fujiwara Y *et al.* (1995) Isolation of a candidate tumor suppressor gene on chromosome 8p21.3-p22 that is homologous to an extracellular domain of PDGF receptor beta gene. *Oncogene* 10: 891-895.
Fujiwara M *et al.* (2002) Antisense oligodeoxynucleotides against thrombomodulin suppress the cell growth of lung adenocarcinoma cell line A549. *Pathol. Int.* 52: 204-213.
Fujiwara M *et al.* (2004) Expression and alternative splicing pattern of human telomerase reverse transcriptase in human lung cancer cells. *Int. J. Oncol.* 24: 925-930.
Fukami T *et al.* (2003) Promoter methylation of the *TSLC1* gene in advanced lung tumors and various cancer cell lines. *Int. J. Cancer* 107: 53-59.
Fukasawa K *et al.* (1996) Abnormal centrosome amplification in the absence of p53. *Science* 271: 1744-1747.
Fukunaga Y *et al.* (2002) Expression of cytokeratin 8 in lung cancer cell lines and measurement of serum cytokeratin 8 in lung cancer patients. *Lung Cancer* 38: 31-38.
Fukuoka J *et al.* (2004) Chromatin remodelling factors and BRM/BRG1 expression as prognostic indicators in non-small cell lung cancer. *Clin. Cancer Res.* 10: 4314-4324.

Fukushima M *et al.* (2003) Population study of expression of thymidylate synthase and dihydropyrimidine dehydrogenase in patients with solid tumors. *Int. J. Mol. Med.* 12: 839-844.
Fukuyama Y *et al.* (1997) K-*ras* and p53 mutations are an independent unfavourable prognostic indicator in patients with non-small-cell lung cancer. *Br. J. Cancer* 75: 1125-1130.
Funa K *et al.* (1987) Increased expression of N-myc in human small cell lung cancer biopsies predicts lack of response to chemotherapy and poor prognosis. *Am. J. Clin. Pathol.* 88: 216-220.
Funakoshi T *et al.* (2003) Expression of tetraspanins in human lung cancer cells: frequent down-regulation of CD9 and its contribution to cell motility in small cell lung cancer. *Oncogene* 22: 674-687.
Furmaga WB *et al.* (2003) *Alu* profiling of primary and metastatic nonsmall cell lung cancer. *Exp. Mol. Pathol.* 74: 224-229.
Furukawa K *et al.* (2003) Novel mutations of neurofibromatosis type 1 gene in small cell lung cancers. *Surg. Today* 33: 323-327.
Futami H *et al.* (1995) A novel somatic point mutation of the *RET* proto-oncogene in tumor tissues of small cell lung cancer patients. *Jp. J. Cancer Res.* 86: 1127-1130.
Futami H *et al.* (2003) Allelic loss of DNA locus of the RET proto-oncogene in small cell lung cancer. *Cancer Lett.* 195: 59-65.
Futreal PA *et al.* (2004) A census of human cancer genes. *Nature Rev. Cancer* 4: 177-183.
Gabrielson E (2003) Clinical and biological relevance of recently defined categories of pulmonary neoplasia. Chap. 2 in *Methods in Molecular Medicine Vol. 74: Lung Cancer, Vol. 1: Molecular Pathology Methods and Reviews*. Ed. B. Driscoll. Humana Press, Totowa, New Jersey. pp 31-42.
Gadgeel SM, Kalemkerian GP (2003) Racial differences in lung cancer. *Cancer Metastasis Rev.* 22: 39-46.
Gahmberg CG *et al.* (1979) Induction of aryl hydrocarbon hydroxylase activity and pulmonary carcinoma. *Int. J. Cancer* 23: 302-305.
Gan DD *et al.* (2003) How does a normal human cell become a cancer cell? *J. Exp. Clin. Cancer Res.* 22: 509-516.
Gao YT *et al.* (1996) Risk factors for lung cancer among nonsmokers with emphasis on lifestyle factors. *Lung Cancer* 14 Suppl 1: S39-S45.
Gao H-G *et al.* (1997) Distribution of p53 and K-ras mutations in human lung cancer tissues. *Carcinogenesis* 18: 473-478.
Gao N *et al.* (2001) The exogenous wild-type p14ARF gene induces growth arrest and promotes radiosensitivity in human lung cancer cell lines. *J. Cancer Res. Clin. Oncol.* 127: 359-367.
Gao WM *et al.* (2003a) Comparison of p53 mutations between adenocarcinoma and squamous cell carcinoma of the lung: unique spectra involving G to A transitions and G to T transversions in both histologic types. *Lung Cancer* 40: 141-150.
Gao WM *et al.* (2003b) Association of the DNA repair gene *XPD* Asp312Asn polymorphism with p53 gene mutations in tobacco-related non-small-cell lung cancer. *Carcinogenesis* 24: 1671-1676.
Garber ME *et al.* (2001) Diversity of gene expression in adenocarcinoma of the lung. *Proc. Natl. Acad. Sci. USA* 98: 13784-13789.
Garcia-Closas M *et al.* (1997) A case-control study of cytochrome P450 1A1, glutathione *S*-transferase M1, cigarette smoking and lung cancer susceptibility (Massachusetts, United States). *Cancer Causes Control* 8: 544-553.
Garcia de Palazzo IE *et al.* (1993) Expression of mutated epidermal growth factor receptor by non-small cell lung carcinomas. *Cancer Res.* 53: 3217-3220.
Gariboldi M *et al.* (2003) Gene expression profile of normal lungs predicts genetic predisposition to lung cancer in mice. *Carcinogenesis* 24: 1819-1826.
Garinis GA *et al.* (2001) Association of allelic loss at the *FHIT* locus and p53 alterations with tumour kinetics and chromosomal instability in non-small cell lung carcinomas (NSCLCs). *J. Pathol.* 193: 55-65.
Garinis GA *et al.* (2002) DNA hypermethylation: when tumour suppressor genes go silent. *Hum. Genet.* 111: 115-127.
Garnis C *et al.* (2004) Genetic alteration and gene expression modulation during cancer progression. *Mol. Cancer* 3: 9.
Garte S (1998) The role of ethnicity in cancer susceptibility gene polymorphisms: the example of *CYP1A1*. *Carcinogenesis* 19: 1329-1332.
Garte S *et al.* (2001) Metabolic gene polymorphism frequencies in control populations. *Cancer Epidemiol. Biomarkers Prev.* 10: 1239-1248.
Garver RI *et al.* (1993) Reciprocal expression of pleiotrophin and midkine in normal versus malignant lung tissues. *Am. J. Respir. Cell Mol. Biol.* 9: 463-466.
Gasdaska PY *et al.* (1993) Cigarette smoking is a determinant of DT-diaphorase gene expression in human non-small cell lung carcinoma. *Cancer Res.* 53: 5458-5461.

Gasperino J, Rom WN (2004) Gender and lung cancer. *Clin. Lung Cancer* 5: 353-359.

Gatenby RA, Vincent TL (2003) An evolutionary model of carcinogenesis. *Cancer Res.* 63: 6212-6220.

Gau HG *et al.* (1997) Distribution of p53 and K-ras mutations in human lung cancer tissues. *Carcinogenesis* 18: 473-478.

Gauderman WJ and Morrison (2000) Evidence for age-specific genetic relative risks in lung cancer. *Am. J. Epidemiol.* 151: 41-49.

Gazdar AF *et al.* (2000) Molecular and cytologic techniques of early detection. Chap. 23 in *Lung Cancer. Principles and Practice.* 2nd Ed. HI Pass, JB Mitchell, DH Johnson, AT Turrisi, JD Minna (Eds.). Lippincott Williams & Williams, Philadelphia, pp 407-424.

Gazzeri S *et al.* (1990) Analysis of the activation of the myc family oncogene and of its stability over time in xenografted human lung carcinomas. *Cancer Res.* 50: 1566-1570.

Gazzeri S *et al.* (1994) *p53* genetic abnormalities and *myc* activation in human lung carcinoma. *Int. J. Cancer* 58: 24-32.

Gazzeri S *et al.* (1998a) Mechanisms of $p16^{INK4A}$ inactivation in non small-cell lung cancers. *Oncogene* 16: 497-504.

Gazzeri S *et al.* (1998b) The human p19ARF protein encoded by the beta transcript of the $p16^{INK4A}$ gene is frequently lost in small cell lung cancers. *Cancer Res.* 58: 3926-3931.

Gealy R *et al.* (1999) Comparison of mutations in the p53 and K-ras genes in lung carcinomas from smoking and non-smoking women. *Cancer Epidemiol. Biomarkers Prev.* 8: 297-302.

Gebert JF *et al.* (1991) High frequency of retinoic acid receptor β abnormalities in human lung cancer. *Oncogene* 6: 1859-1868.

Gemignani F *et al.* (2002) A catalogue of polymorphisms related to xenobiotic metabolism and cancer susceptibility. *Pharmacogenetics* 12: 459-463.

Gemma A *et al.* (2000a) Somatic mutation of the *hBUB1* mitotic checkpoint gene in primary lung cancer. *Genes Chrom. Cancer* 29: 213-218.

Gemma A *et al.* (2000b) Mutation analysis of the gene encoding the human mannose 6-phosphate/insulin-like growth factor 2 receptor (M6P/IGF2R) in human cell lines resistant to growth inhibition by transforming growth factor β1 (TGF-β1). *Lung Cancer* 30: 91-98.

Gemma A *et al.* (2001) Altered expression of several genes in highly metastatic subpopulations of a human pulmonary adenocarcinoma cell line. *Eur. J. Cancer* 37: 1554-1561.

Geng Z *et al.* (2003) Expression of telomerase hTERT in human non-small cell lung cancer and its correlation with c-myc gene. *Chin. Med. J.* 116: 1467-1470.

Geradts J *et al.* (1993) Human lung cancer cell lines exhibit resistance to retinoic acid treatment. *Cell Growth Differ.* 4: 799-809.

Geradts J *et al.* (2000) Loss of Fhit expression in non-small-cell lung cancer: correlation with molecular genetic abnormalities and clinicopathological features. *Br. J. Cancer* 82: 1191-1197.

Gerson SL (2004) *MGMT*: its role in cancer aetiology and cancer therapeutics. *Nature Rev. Cancer* 4: 296-301.

Gerwin BI *et al.* (1992) Mutant p53 can induce tumorigenic conversion of human bronchial epithelial cells and reduce their responsiveness to a negative growth factor, transforming growth factor beta 1. *Proc. Natl. Acad. Sci. USA* 89: 2759-2763.

Gessner C *et al.* (2002) BAX and $p16^{INK4A}$ are independent positive prognostic markers for advanced tumour stage of nonsmall cell lung cancer. *Eur. Respir. J.* 19: 134-140.

Gessner C *et al.* (2004a) Detection of p53 gene mutations in exhaled breath condensate of non-small cell lung cancer patients. *Lung Cancer* 43: 215-222.

Gessner C *et al.* (2004b) Nuclear YB-1 expression as a negative prognostic marker in nonsmall cell lung cancer. *Eur. Respir. J.* 23: 14-19.

Ghadirian P *et al.* (1997) Carcinoma of the lung in three brothers and two sisters. *Lung Cancer* 18: 259-264.

Gharib TG *et al.* (2002) Proteomic analysis of cytokeratin isoforms uncovers association with survival in lung adenocarcinoma. *Neoplasia* 4: 440-448.

Giaccone G *et al.* (1995) Differential expression of DNA topoisomerases in non-small cell lung cancer and normal lung. *Biochim. Biophys. Acta* 1264: 337-346.

Giaccone G (2002) Clinical impact of novel treatment strategies. *Oncogene* 21: 6970-6981.

Giatromanolaki A *et al.* (2003) DEC1 (STRA13) protein expression relates to hypoxia-inducible factor 1-alpha and carbonic anhydrase-9 overexpression in non-small cell lung cancer. *J. Pathol.* 200: 222-228.

Giordano TJ *et al.* (2001) Organ-specific molecular classification of primary lung, colon, and ovarian adenocarcinomas using gene expression profiles. *Am. J. Pathol.* 159: 1231-1238.

Giles RH *et al.* (2003) Caught up in a Wnt storm: Wnt signalling in cancer. *Biochim. Biophys. Acta* 1653: 1-24.

Gilliland FD *et al.* (2002) Glutathione *S*-transferase P1 and NADPH quinone oxidoreductase polymorphisms are associated with aberrant promoter methylation of p16(INK4a) and O(6)-methylguanine-DNA methyltransferase in sputum. *Cancer Res.* 62: 2248-2252.
Girard L *et al.* (2000) Genome-wide allelotyping of lung cancer identifies new regions of allelic loss, differences between small cell lung cancer and non-small cell lung cancer, and loci clustering. *Cancer Res.* 60: 4894-4906.
Gisselsson D *et al.* (2003) Chromosome instability in cancer: how, when, and why? *Adv. Cancer Res.* 87: 1-29.
Goeze A *et al.* (2002) Chromosomal imbalances of primary and metastatic lung adenocarcinomas. *J. Pathol.* 196: 8-16.
Goldgar DE *et al.* (1994) Systematic population-based assessment of cancer risk in first-degree relatives of cancer probands. *J. Natl. Cancer Inst.* 86: 1600-1608.
Gomperts BD, Tatham PER, Kramer IM (2002) *Signal Transduction*. Academic Press, San Diego.
Goncharuk VN *et al.* (2004) Co-downregulation of PTEN, KAI-1, and nm23-H1 tumor/metastasis suppressor proteins in non-small cell lung cancer. *Ann. Diagn. Pathol.* 8: 6-16.
Gonzalez R *et al.* (2000) Microsatellite alterations and *TP53* mutations in plasma DNA of small-cell lung cancer patients: follow-up study and prognostic significance. *Ann. Oncol.* 11: 1097-1104.
Gonzalez-Quevedo R *et al.* (2002) Cooperative role of telomerase activity and p16 expression in the prognosis of non-small-cell lung cancer. *J. Clin. Oncol.* 20: 254-262.
Gonzalez-Quevedo R *et al.* (2004) Differential impact of p16 inactivation by promoter methylation in non-small cell lung and colorectal cancer: clinical implications. *Int. J. Oncol.* 24: 349-355.
Goode EL *et al.* (2002) Polymorphisms in DNA repair genes and associations with cancer risk. *Cancer Epidemiol. Biomarkers Prev.* 11: 1513-1530.
Goodwin LO *et al.* (2001) Gene expression patterns of paired bronchioloalveolar carcinoma and benign lung tissue. *Ann. Clin. Lab. Sci.* 31: 369-375.
Gordon GJ *et al.* (2003) A prognostic test for adenocarcinoma of the lung from gene expression profiling data. *Cancer Epidemiol. Biomarkers Prev.* 12: 905-910.
Gorgoulis VG *et al.* (2000) Altered expression of the cell cycle regulatory molecules pRb, p53 and MDM2 exert a synergistic effect on tumor growth and chromosomal instability in non-small cell lung carcinomas (NSCLCs). *Mol. Med.* 6: 208-237.
Gorgoulis VG *et al.* (2001) Deregulated expression of *c-mos* in non-small cell lung carcinomas: relationship with *p53* status, genomic instability, and tumor kinetics. *Cancer Res.* 61: 538-549.
Gorgoulis VG *et al.* (2002) Transcription factor E2F-1 acts as a growth-promoting factor and is associated with adverse prognosis in non-small cell lung carcinomas. *J. Pathol.* 198: 142-156.
Gosney JR *et al.* (1993) Exposure to cigarette smoke and expression of the protein encoded by the p53 gene in bronchial carcinoma. *Ann. NY Acad. Sci.* 686: 243-247.
Goto I *et al.* (1996) Prognostic significance of germ line polymorphisms of the *CYP1A1* and glutathione and glutathione *S*-transferase genes in patients with non-small cell lung cancer. *Cancer Res.* 56: 3725-3730.
Gotoh K *et al.* (1999) Frameshift mutations in TGFβRII, IGFIIR, BAX, hMSH3 and hMSH6 are absent in lung cancers. *Carcinogenesis* 20: 499-502.
Gouyer V *et al.* (1994) Loss of heterozygosity at the RB locus correlates with loss of RB protein in primary malignant neuro-endocrine lung carcinomas. *Int. J. Cancer* 58: 818-824.
Gouyer V *et al.* (1998) Mechanism of retinoblastoma gene inactivation in the spectrum of neuro-endocrine lung tumors. *Am. J. Respir. Cell Mol. Biol.* 18: 188-196.
Graham EA, Singer JJ (1933) Successful removal of entire lung for carcinoma of the bronchus *JAMA* 101: 1371-1374.
Granjeaud S *et al.* (1999) Expression profiling: DNA arrays in many guises. *BioEssays* 21: 781-790.
Graves RJ (1848) *Clinical Lectures on the Practice of Medicine*. New Sydenham Society, London.
Gray DA *et al.* (1995) Elevated expression of Unph, a proto-oncogene at 3p21.3, in human lung tumors. *Oncogene* 10: 2179-2183.
Green P *et al.* (2003) Transcription-associated mutational asymmetry in mammalian evolution. *Nature Genet.* 33: 514-517.
Greenberg DA (2001) Microarrays, markers of disease, and the myth of "nonhypothesis-driven" research. *Ann. Neurol.* 50: 695.
Greenberg AK *et al.* (2002) Preneoplastic lesions of the lung. *Respir. Res.* 3: 20.
Greenblatt MS *et al.* (1994) Mutations in the *p53* tumor suppressor gene: clues to cancer etiology and molecular pathogenesis. *Cancer Res.* 54: 4855-4878.
Greenblatt MS *et al.* (1996) Deletions and insertions in the *p53* tumor suppressor gene in human cancers: confirmation of the DNA polymerase slippage/misalignment model. *Cancer Res.* 56: 2130-2136.

Greenblatt MS *et al.* (2003) Detailed computational study of p53 and p16: using evolutionary sequence analysis and disease-associated mutations to predict the functional consequences of allelic variants. *Oncogene* 22: 1150-1163.

Greenlee RT *et al.* (2001) Cancer statistics, 2001. *CA Cancer J. Clin.* 51: 15-36.

Gregore V *et al.* (2003) Relevance of p53, bcl-2 and Rb expression on resistance to cisplatin-based chemotherapy in advanced non-small cell lung cancer. *Lung Cancer* 39: 41-48.

Griffin JL, Shockcor JP (2004) Metabolic profiles of cancer cells. *Nature Rev. Cancer* 4: 551-561.

Grindley JC *et al.* (1997) Evidence for the involvement of the Gli gene family in embryonic mouse lung development. *Dev. Biol.* 188: 337-348.

Groet J *et al.* (2000) Narrowing of the region of allelic loss in 21q11-21 in squamous non-small cell lung carcinoma and cloning of a novel ubiquitin-specific protease gene from the deleted segment. *Genes Chrom. Cancer* 27: 153-161.

Grosse H (1953) Kritische Gedanken zur Krebsstatistik auf Grund der Sektionen des Stadtkrankenhauses Dresden-Friedrichstadt 1852-1951. *Ztschr. Krebsforsch.* 59: 316-339.

Grote HJ *et al.* (2004) Aberrant methylation of the adenomatous polyposis coli promoter 1A in bronchial aspirates from patients with suspected lung cancer. *Int. J. Cancer* 110: 751-755.

Grzybowska E *et al.* (2000) The effect of the genetic polymorphisms of CYP1A1, CYP2D6, GSTM1 and GSTP1 on aromatic DNA adduct levels in the population of healthy women. *Mutation Res.* 469: 271-277.

Gschwind A *et al.* (2004) The discovery of receptor tyrosine kinases: targets for cancer therapy. *Nature Rev. Cancer* 4: 361-370.

Gsur A *et al.* (2003) Association of microsomal epoxide hydrolase polymorphism and lung cancer risk. *Br. J. Cancer* 89: 702-706.

Gu C *et al.* (2004) Low expression of polypeptide GalNac N-acetylgalactosaminyltransferase-3 in lung adenocarcinoma: impact on poor prognosis and early recurrence. *Br. J. Cancer* 90: 436-442.

Guang SG *et al.* (1997) Association between p53 mutation and clinicopathological features of non-small cell lung cancer. *Jpn. J. Clin. Oncol.* 27: 211-215.

Gugger M *et al.* (2002) Quantitative expansion of structural genomic alterations in the spectrum of neuroendocrine lung carcinomas. *J. Pathol.* 196: 408-415.

Gunn L, Smith MT (2004) Emerging biomarker technologies. *IARC Sci. Publ.* 157: 437-450.

Gurrieri C *et al.* (2004) Loss of the tumor suppressor *PML* in human cancers of multiple histologic origins. *J. Natl. Cancer Inst.* 96: 269-279.

Gurrola-Diaz C *et al.* (2003) Reduced expression of the neuron restrictive silencer factor permits transcription of glycine receptor alpha1 subunit in small-cell lung cancer cells. *Oncogene* 22: 5636-5645.

Gurubhagavatula S *et al.* (2004) *XPD* and *XRCC1* genetic polymorphisms are prognostic factors in advanced non-small-cell lung cancer patients treated with platinum chemotherapy. *J. Clin. Oncol.* 22: 2594-2601.

Haber D, Harlow E (1997) Tumour-suppressor genes: evolving definitions in the genomic age. *Nature Genet.* 16: 320-322.

Hackett JA *et al.* (2001) Telomere dysfunction increases mutation rate and genomic instability. *Cell* 106: 275-286.

Hackett NR *et al.* (2003) Variability of antioxidant-related gene expression in the airway epithelium of cigarette smokers. *Am. J. Respir. Cell Mol. Biol.* 29: 331-343.

Hafner C *et al.* (2004) Differential gene expression of Eph receptors and ephrins in various benign human tissues and cancers. *Clin. Chem.* 50: 490-499.

Hainaut P, Hollstein M (2000) p53 and human cancer: the first ten thousand mutations. *Adv. Cancer Res.* 77: 81-137.

Hainaut P, Pfeiffer GP (2001) Patterns of p53 G→T transversions in lung cancers reflect the primary mutagenic signature of DNA-damage by tobacco smoke. *Carcinogenesis* 22: 367-374.

Hainaut P *et al.* (2001) *TP53* mutation spectrum in lung cancers and mutagenic signature of components of tobacco smoke:lessons from the IARC *TP53* mutation database. *Mutagenesis* 16: 551-553.

Hajj C *et al.* (1990) DNA alterations at proto-oncogene loci and their clinical significance in operable non-small cell lung cancer. *Cancer* 66: 733-739.

Hake SB *et al.* (2004) Linking the epigenetic 'language' of covalent histone modifications to cancer. *Br. J. Cancer* 90: 761-769.

Hakkarainen J *et al.* (2004) *TP53* mutation detection by SSCP and sequencing. *Methods Mol. Med.* 99: 191-208.

Hall M, Peters G (1996) Genetic alterations of cyclins, cyclin-dependent kinases, and Cdk inhibitors in human cancer. *Adv. Cancer Res.* 68: 67-108.

Halmos B *et al.* (2002) Down-regulation and antiproliferative role of C/EBPα in lung cancer. *Cancer Res.* 62: 528-534.
Halmos B *et al.* (2004) A transcriptional profiling study of CCAAT/enhancer binding protein targets identifies hepatocyte nuclear factor 3β as a novel tumor suppressor in lung cancer. *Cancer Res.* 64: 4137-4147.
Hamada GS *et al.* (1995) The heme-binding region polymorphism of cytochrome P450IA1 (CypIA1), rather than the *Rsa*I polymorphism of IIE1 (CypIIE1), is associated with lung cancer in Rio de Janeiro. *Cancer Epidemiol. Biomark. Prevent.* 4: 63-67.
Hamada K *et al.* (1998) Association of CDKN2A(p16)/CDKN2B(p15) alterations and homozygous chromosome arm 9p deletions in human lung carcinoma. *Genes Chrom. Cancer* 22: 232-240.
Hamada J *et al.* (2001) Overexpression of homeobox gene *HOXD3* induces coordinate expression of metastasis-related genes in human lung cancer cells. *Int. J. Cancer* 93: 516-525.
Hamajima N *et al.* (2002) NAD(P)H: quinone oxidoreductase 1 (*NQO1*) C609T polymorphism and the risk of eight cancers for Japanese. *Int. J. Clin. Oncol.* 7: 103-108.
Hamid Q *et al.* (1991) Expression of chromogranin A mRNA in small cell carcinoma of the lung. *J. Pathol.* 163: 293-297.
Hanabata T *et al.* (2004) DNA methylation of multiple genes and clinicopathological relationship of non-small cell lung cancers. *Oncol. Rep.* 12: 177-180.
Hanahan D, Folkman J (1996) Patterns and emerging mechanisms of the angiogenic switch during tumorigenesis. *Cell* 86: 353-364.
Hanahan D, Weinberg RA (2000) The hallmarks of cancer. *Cell* 100: 57-70.
Hanaoka T *et al.* (2000) A preliminary evaluation of intra- and inter-individual variations of hOGG1 messenger RNA levels in peripheral blood cells as determined by a real-time polymerase chain reaction technique. *Cancer Epidemiol. Biomarkers Prev.* 9: 1255-1258.
Hanash S (2003) Disease proteomics. *Nature* 422: 226-232.
Hanash S *et al.* (2001) A proteomic approach to the identification of lung cancer markers. *Disease Markers* 17: 295-300.
Hanawalt PC *et al.* (2003) Functional characterization of global genomic DNA repair and its implications for cancer. *Mutation Res.* 544: 107-114.
Hansen LT *et al.* (2003) The role of mismatch repair in small-cell lung cancer cells. *Eur. J. Cancer* 39: 1456-1467.
Hara H *et al.* (2001) Clinicopathologic significance of telomerase activity and hTERT mRNA expression in non-small cell lung cancer. *Lung Cancer* 34: 219-226.
Harada H *et al.* (2001) Identification of DMC1, a novel gene in the TOC region on 17q25.1 that shows loss of expression in multiple human cancers. *J. Hum. Genet.* 46: 90-95.
Harbour JW *et al.* (1988) Abnormalities in structure and expression of the human retinoblastoma gene in SCLC. *Science* 241: 353-357.
Harden SV *et al.* (2003) Gene promoter hypermethylation in tumors and lymph nodes of stage I lung cancer patients. *Clin. Cancer Res.* 9: 1370-1375.
Harder KW *et al.* (1995) Coexisting amplifications of the chromosome 1p32 genes (*PTPRF* and *MYCL1*) encoding protein tyrosine phosphatases LAR and L-myc in a small cell lung cancer line. *Genomics* 27: 552-553.
Hardie LJ *et al.* (2000) The effect of hOGG1 and glutathione peroxidase I genotypes and 3p chromosomal loss on 8-hydroxyguanosine levels in lung cancer. *Carcinogenesis* 21: 167-172.
Harms C *et al.* (2004) Polymorphisms in DNA repair genes, chromsome aberrations, and lung cancer. *Environ. Mol. Mutagen.* 443: 74-82.
Harper PA *et al.* (2002) Polymorphisms in the human AH receptor. *Chem. Biol. Interact.* 141: 161-187.
Harris CC *et al.* (1976) Interindividual variation in binding of benzo[*a*]pyrene to DNA in cultured human bronchi. *Science* 194: 1067-1069.
Harris H (2004) Putting on the brakes. *Nature* 427: 201.
Härting FH, Hesse W (1879) Der Lungenkrebs die Bergkrankheit in den Schneebergen Gruben. *Viertelj F. Gerichtl. Med. U. Offen. Sanitats.* 30: 296-309 and 31: 102-132, 313-337.
Hartman AR, Ford JM (2002) BRCA1 induces DNA damage recognition factors and enhances nucleotide excision repair. *Nature Genet.* 32: 180-184.
Haruki N *et al.* (2000a) Histological type-selective, tumor-predominant expression of a novel CHK1 isoform and infrequent *in vivo* somatic *CHK2* mutation in small cell lung cancer. *Cancer Res.* 60: 4689-4692.
Haruki N *et al.* (2000b) Characterization of high-grade neuroendocrine tumors of the lung in relation to rennin mutations. *Jpn. J. Cancer Res.* 91: 317-323.
Haruki N *et al.* (2001) Persistent increase in chromosomal instability in lung cancer: possible indirect involvement of p53 inactivation. *Am. J. Pathol.* 159: 1345-1352.

Harvey P *et al.* (1996) Immunoreactivity for hepatocyte growth factor/scatter factor and its receptor, met, in human lung carcinomas and malignant mesotheliomas. *J. Pathol.* 180: 389-394.
Hasegawa T *et al.* (1993) Higher expression of topoisomerase II in lung cancers than normal lung tissues: different expression pattern from topoisomerase I. *Biochem. Biophys. Res. Commun.* 195: 409-414.
Hashimoto T *et al.* (2000) Different subtypes of human lung adenocarcinoma caused by different etiological factors. Evidence from *p53* mutational spectra. *Am. J. Pathol.* 157: 2133-2141.
Hashimoto M *et al.* (2004) Expression of CD109 in human cancer. *Oncogene* 23: 3716-3720.
Hassett C *et al.* (1994) Human microsomal epoxide hydrolase: genetic polymorphism and functional expression *in vitro* of amino acid variants. *Hum. Mol. Genet.* 3: 421-428.
Hasturk S *et al.* (2002) Expression of cyclooxygenase-1 and cyclooxygenase-2 in bronchial epithelium and nonsmall cell lung carcinoma. *Cancer* 94: 1023-1031.
Hatanaka H *et al.* (2001) Significant correlation between interleukin 10 expression and vascularization through angiopoietin/TIE2 networks in non-small cell lung cancer. *Clin. Cancer Res.* 7: 1287-1292.
Haugen A (2002) Women who smoke: are women more susceptible to tobacco-induced lung cancer? *Carcinogenesis* 23: 227-229.
Haura EB *et al.* (2003a) Gene therapy for lung cancer. *Meths. Mol. Med.* 75: 529-543.
Haura EB et al. (2003b) Gene therapy for lung cancer. *Mol. Biotechnol.* 25: 139-148.
Hawighorst T *et al.* (2001) Thrombospondin-2 plays a protective role in multistep carcinogenesis: a novel host anti-tumor defense mechanism. *EMBO J.* 20: 2631-2640.
Hayakawa K *et al.* (1998) The prognostic significance of immunohistochemically detected p53 protein expression in non-small cell lung cancer treated with radiation therapy. *Anticancer Res.* 18(5B): 3685-3688.
Hayashi S *et al.* (1991) Genetic linkage of lung cancer-associated *Msp*I polymorphisms with amino acid replacement in the heme binding region of the human cytochrome *P4501A1* gene. *J. Biochem.* 110: 407-411.
Hayashi S *et al.* (1992) High susceptibility to lung cancer analysed in terms of combined genotypes of *P450IA1* and Mu-class gltathione S-transferase genes. *Jpn. J. Cancer Res.* 83: 866-870.
Hayashi N *et al.* (1994) Somatic mutations of the MTS (multiple tumor suppressor)1/CDK4I (cyclin-dependent kinase-4 inhibitor) gene in human primary non-small cell lung carcinomas. *Biochem. Biophys. Res. Commun.* 202: 1426-1430.
Hayashi H *et al.* (1997) Analysis of p21Waf1/Cip1 expression in normal, premalignant, and malignant cells during the development of human lung adenocarcinoma. *Am. J. Pathol.* 151: 461-470.
Hayashi H *et al.* (2000) Reduced expression of p27/Kip1 is associated with the development of pulmonary adenocarcinoma. *J. Pathol.* 192: 26-31.
Hayashi H *et al.* (2001a) High cyclin E and low p27/Kip1 expressions are potentially poor prognostic factors in lung adenocarcinoma patients. *Lung Cancer* 34: 59-65.
Hayashi H *et al.* (2001b) Reduced HIC-1 gene expression in non-small cell lung cancer and its clinical significance. *Anticancer Res.* 21(1B): 535-540.
Hayashi H *et al.* (2002) Inactivation of O6-methylguanine-DNA methyltransferase in human lung adenocarcinoma relates to high-grade histology and worse prognosis among smokers. *Jpn. J. Cancer Res.* 93: 184-189.
Hazan RB *et al.* (2004) Cadherin switch in tumor progression. *Ann. N Y Acad. Sci.* 1014: 155-163.
He B *et al.* (2003) SOCS-3 is frequently silenced by hypermethylation and suppresses cell growth in human lung cancer. *Proc. Natl. Acad. Sci. USA* 100: 14133-14138.
He B *et al.* (2004) Activity of the suppressor of cytokine signalling-3 promoter in human non-small-cell lung cancer. *Clin. Lung Cancer* 5: 366-370.
Heath AC, Martin HG (1993) Genetic models for the natural history of smoking: evidence for a genetic influence on smoking persistence. *Addict. Behav.* 18: 19-34.
Heath AC *et al.* (1995) Personality and the inheritance of smoking behavior: a genetic perspective. *Behav. Genet.* 25: 103-117.
Hecht SS (1999) Tobacco smoke carcinogens and lung cancer. *J. Natl. Cancer Inst.* 91: 1194-1210.
Hecht SS (2002) Cigarette smoking and lung cancer: chemical mechanisms and approaches to prevention. *Lancet Oncology* 3: 461-469.
Hecht SS (2003) Tobacco carcinogens, their biomarkers and tobacco-induced cancer. *Nature Rev. Cancer* 3: 733-744.
Heeschen C *et al.* (2003) Nicotine promotes arteriogenesis. *J. Am. Coll. Cardiol.* 41: 489-496.
Heguy A *et al.* (2003) Sampling-dependent up-regulation of gene expression in sequential samples of human airway epithelial cells. *Mol. Med.* 9: 200-208.

Heighway J *et al.* (2002) Expression profiling of primary non-small cell lung cancer for target identification. *Oncogene* 21: 7749-7763.
Heighway J *et al.* (2003) The alleles of the DNA repair gene O6-alkylguanine-DNA alkyltransferase are expressed at different levels in normal human lung tissue. *Carcinogenesis* 24: 1691-1694.
Heijmans BT *et al.* (2003) A common variant of the methylenetetrahydrofolate reductase gene (1p36) is associated with an increased risk of cancer. *Cancer Res.* 63: 1249-1253.
Hein DW (2002) Molecular genetics and function of NAT1 and NAT2: role in aromatic amine metabolism and carcinogenesis. *Mutation Res.* 506-507: 65-77.
Helin K *et al.* (1997) Loss of the retinoblastoma protein-related p130 protein in small cell lung carcinoma. *Proc. Natl. Acad. Sci. USA* 94: 6933-6938.
Hellmann GM *et al.* (2001) Gene expression profiling of cultured human bronchial epithelial and lung carcinoma cells. *Toxicol. Sci.* 61: 154-163.
Hemminki K, Shields PG (2002) Skilled use of DNA polymorphisms as a tool for polygenic cancers. *Carcinogenesis* 23: 379-380.
Henness S *et al.* (2004) Changes in gene expression associated with stable drug and radiation resistance in small cell lung cancer cells are similar to those caused by a single X-ray dose. *Radiation Res.* 161: 495-503.
Hensel CH *et al.* (1990) Altered structure and expression of the human retinoblastoma susceptibility gene in small cell lung cancer. *Cancer Res.* 50: 3067-3072.
Herman JG, Baylin SB (2003) Gene silencing in cancer in association with promoter hypermethylation. *New Engl. J. Med.* 349: 2042-2054.
Hernandez-Boussard T, Rodriguez-Tome P, Montesano R, Hainaut P (1999a) IARC p53 mutation database: a relational database to compile and analyze p53 mutations in human tumors and cell lines. *Hum. Mutation* 14: 1-8.
Hernandez-Boussard T, Montesano R, Hainaut P (1999b) Sources of bias in the detection and reporting of *p53* mutations in human cancer: analysis of the IARC *p53* mutation database. *Genetic Analysis: Biomolecular Engineering* 14: 229-233.
Hesson L *et al.* (2003) *NORE1A*, a homologue of *RASSF1A* tumour suppressor gene is inactivated in human cancers. *Oncogene* 22: 947-954.
Heston WE (1942) Inheritance of susceptibility to spontaneous pulmonary tumors in mice. *J. Natl. Cancer Inst.* 3: 79-82.
Hibi K *et al.* (1991) Coexpression of the stem cell factor and the c-kit genes in small-cell lung cancer. *Oncogene* 6: 2291-2296.
Hibi K *et al.* (1994) Aberrant upregulation of a novel integrin alpha subunit gene at 3p21.3 in small cell lung cancer. *Oncogene* 9: 611-619.
Hibi K *et al.* (1998) Serial analysis of gene expression in non-small cell lung cancer. *Cancer Res.* 58: 5690-5694.
Hibi K *et al.* (1999) PGP9.5 as a candidate tumor marker for non-small-cell lung cancer. *Am. J. Pathol.* 155: 711-715.
Hibi K *et al.* (2000) AIS is an oncogene amplified in squamous cell carcinoma. *Proc. Natl. Acad. Sci. USA* 97: 5462-5467.
Hida T *et al.* (1994) Ectopic expression of c-kit in small-cell lung cancer. *Int. J. Cancer Suppl.* 8: 108-109.
Hida T *et al.* (1998) Increased expression of cyclooxygenase 2 occurs frequently in human lung cancers, specifically in adenocarcinomas. *Cancer Res.* 58: 3761-3764.
Higashiyama M *et al.* (1995a) Alpha-1-antichymotrypsin expression in lung adenocarcinoma and its possible association with tumor progression. *Cancer* 76: 1368-1376.
Higashiyama M *et al.* (1995b) Reduced motility related protein-1 (MRP-1/CD9) gene expression as a factor of poor prognosis in non-small cell lung cancer. *Cancer Res.* 55: 6040-6044.
Higashiyama M *et al.* (1997) *MDM2* gene amplification and expression in non-small-cell lung cancer: immunohistochemical expression of its protein is a favourable prognostic marker in patients without p53 protein accumulation. *Br. J. Cancer* 75: 1302-1308.
Higashiyama M *et al.* (1998) KAI1/CD82 expression in nonsmall cell lung carcinoma is a novel, favourable prognostic factor: an immunohistochemical analysis. *Cancer* 83: 466-474.
Hilbe W *et al.* (2003a) Automated real-time PCR to determine K-ras codon 12 mutations in non-small cell lung cancer: comparison with immunohistochemistry and clinico-pathological features. *Int. J. Oncol.* 23: 1121-1126.
Hilbe W *et al.* (2003b) Characterization and predictive value of epidermal growth factor receptor status using quantitative real-time PCR combined with immunohistochemistry on non-small cell lung cancer specimens. *Int. J. Oncol.* 23: 893-899.

Hirabayashi H *et al.* (2002) Prognostic significance of p27KIP1 expression in resected non-small cell lung cancers: analysis in combination with expressions of p16INK4A, pRB, and p53. *J. Surg. Oncol.* 81: 177-184.

Hiraki A *et al.* (2003) Different risk relations with smoking for non-small-cell lung cancer: comparison of *TP53* and *TP73* genotypes. *Asian Pac. J. Cancer Prev.* 4: 107-112.

Hirao T *et al.* (2001) Tobacco smoke-induced DNA damage and an early age of smoking initiation induce chromosome loss at 3p21 in lung cancer. *Cancer Res.* 61: 612-615.

Hirata T *et al.* (2002) Expression of E-cadherin and lymph node metastasis in resected non-small-cell lung cancer. *Clin. Lung Cancer* 3: 134-140.

Hiroshima K *et al.* (2004) Genetic alterations in early-stage pulmonary large cell neuroendocrine carcinoma. *Cancer* 100: 1190-1198.

Hirsch FR *et al.* (2002a) Evaluation of HER-2/neu gene amplification and protein expression in non-small cell lung carcinomas. *Br. J. Cancer* 86: 1449-1456.

Hirsch FR *et al.* (2002b) HER2/neu expression in malignant lung tumors. *Semin. Oncol.* 29 (1 Suppl 4): 51-58.

Hirsch FR *et al.* (2003) Epidermal growth factor receptor in non-small-cell lung carcinomas: correlation between gene copy number and protein expression and impact on prognosis. *J. Clin. Oncol.* 21: 3798-3807.

Hirschhorn JN *et al.* (2002) A comprehensive review of genetic association studies. *Genet. Med.* 4: 45-61.

Hirvonen A *et al.* (1992) Metabolic cytochrome P450 geneotypes and assessment of individual susceptibility to lung cancer. *Pharmacogenetics* 2: 259-263.

Hirvonen A *et al.* (1993a) The *GSTM1* null genotype as a potential risk modifier for squamous cell carcinoma of the lung. *Carcinogenesis* 14: 1479-1481.

Hirvonen A *et al.* (1993b) Polymorphism in *CYP1A1* and *CYP2D6* genes: possible association with susceptibility to lung cancer. *Environ. Health Perspect.* 101 Suppl. 3: 109-112.

Hiyama K *et al.* (1995a) Telomerase activity in small-cell and non-small-cell lung cancers. *J. Natl. Cancer Inst.* 87: 895-902.

Hiyama K *et al.* (1995b) Alterations in telomeric repeat length in lung cancer are associated with loss of heterozygosity in p53 and Rb. *Oncogene* 10: 937-944.

Ho WL *et al.* (2002a) Loss of heterozygosity at loci of candidate tumor suppressor genes in microdissected primary non-small cell lung cancer. *Cancer Detect. Prev.* 26: 343-349.

Ho CC *et al.* (2002b) Up-regulated caveolin-1 accentuates the metastasis capability of lung adenocarcinoma by inducing filopodia formation. *Am. J. Pathol.* 161: 1647-1656.

Hoang T *et al.* (2002) Novel therapies for lung cancer. *Surg. Oncol.* 11: 229-241.

Hoang CD *et al.* (2004) Expression profiling of non-small cell lung carcinoma identifies metastatic genotypes based on lymph node tumor burden. *J. Thorac. Cardiovasc. Surg.* 127: 1332-1342.

Hoeijmakers JHJ (2001) Genome maintenance mechanisms for preventing cancer. *Nature* 411: 366-374.

Hoffman PC *et al.* (2000) Lung cancer. *Lancet* 355: 479-485.

Hofmann HS *et al.* (2002) Expression of inhibitors of apoptosis (IAP) proteins in non-small cell human lung cancer. *J. Cancer Res. Clin. Oncol.* 128: 554-560.

Hofmann HS *et al.* (2004) Discrimination of human lung neoplasm from normal lung by two target genes. *Am. J. Respir. Crit. Care Med.* In press. [10.1164/rccm.200401-127OC]

Holland EC *et al.* (2004) Signaling control of mRNA translation in cancer pathogenesis. *Oncogene* 23: 3138-3144.

Hollstein M, Shomer B, Greenblatt M, Soussi T, Hovig E, Montesano R, Harris CC (1996) Somatic point mutations in the p53 gene of human tumors and cell lines: updated compilation. *Nucleic Acids Res.* 24: 141-146.

Holmila R *et al.* (2003) Splice mutations in the p53 gene: case report and review of the literature. *Hum. Mutation* 21: 101-102.

Hommura F *et al.* (2000) Prognostic significance of p27KIP1 protein and ki67 growth fraction in non-small cell lung cancers. *Clin. Cancer Res.* 6: 4073-4081.

Hommura F *et al.* (2002) Increased expression of beta-catenin predicts better prognosis in nonsmall cell lung carcinomas. *Cancer* 94: 752-758.

Honda S *et al.* (2003) A role for the pituitary tumor-transforming gene in the genesis and progression of non-small cell lung carcinomas. *Anticancer Res.* 23(5A): 3775-3782.

Honda K *et al.* (2004) Alternative splice variant of actinin-4 in small cell lung cancer. *Oncogene* 23: 5257-5262.

Hong TM *et al.* (2000) Profiling the downstream genes of tumor suppressor *PTEN* in lung cancer cells by complementary DNA microarray. *Am. J. Respir. Cell Mol. Biol.* 23: 355-363.

Honorio S *et al.* (2003) Detection of *RASSF1A* aberrant promoter hypermethylation in sputum from chronic smokers and ductal carcinoma *in situ* from breast cancer patients. *Oncogene* 22: 147-150.

Horio Y *et al.* (1993) Prognostic significance of p53 mutations and 3p deletions in primary resected non-small cell lung cancer. *Cancer Res.* 53: 1-4.

Horowitz JM *et al.* (1990) Frequent inactivation of the retinoblastoma anti-oncogene is restricted to a subset of human tumor cells. *Proc. Natl. Acad. Sci.USA* 87: 2775-2779.

Hosomi Y *et al.* (2003) Somatic mutation of the caspase-5 gene in human lung cancer. *Int. J. Mol. Med.* 12: 443-446.

Hou SM *et al.* (2002) The *XPD* variant alleles are associated with increased aromatic DNA adduct level and lung cancer risk. *Carcinogenesis* 23: 599-603.

Hou SM *et al.* (2001) Differential interactions between *GSTM1* and *NAT2* genotypes on aromatic DNA adduct level and HPRT mutant frequency in lung cancer patients and population controls. *Cancer Epidemiol. Biomarkers Prev.* 10: 133-140.

Hou SM *et al.* (2003) Influence of common *XPD* and *XRCC1* variant alleles on p53 mutations in lung tumors. *Environ. Mol. Mutagen.* 41: 37-42.

Hougaard S *et al.* (1999) Inactivation of the transforming growth factor beta type II receptor in human small cell lung cancer cell lines. *Br. J. Cancer* 79: 1005-1011.

Houlston RS (2000) *CYP1A1* polymorphisms and lung cancer risk: a meta-analysis. *Pharmacogenetics* 10: 105-114.

Howard BA *et al.* (2003) Identification and validation of a potential lung cancer serum biomarker detected by matrix-assisted laser desorption/ionisation-time of flight spectra analysis. *Proteomics* 3: 1720-1724.

Hsieh CC *et al.* (1998) Prognostic significance of HER-2/neu overexpression in stage I adenocarcinoma of lung. *Ann. Thorac. Surg.* 66: 1159-1163.

Hsieh HL *et al.* (2003) Expression analysis of S100 proteins and RAGE in human tumors using tissue microarrays. *Biochem. Biophys. Res. Commun.* 307: 375-381.

Hsu NY *et al.* (1996) Detection of K-ras point mutations in codons 12 and 13 in non-small cell lung cancers. *J. Form. Med. Assoc.* 95: 741-745.

Hsu NY *et al.* (2001) Overexpression of dihydrodiol dehydrogenase as a prognostic marker of non-small cell lung cancer. *Cancer Res.* 61: 2727-2731.

Hsu CP *et al.* (2003) Concordant expression of the telomerase-associated genes in non-small cell lung cancer. *Eur. J. Surg. Oncol.* 29: 594-599.

Hu X *et al.* (1999) Differential protection against benzo[*a*]pyrene-7,8-dihydrodiol-9,10-epoxide-induced DNA damage in HepG2 cells stably transfected with allelic variants of pi class human glutathione *S*-transferase. *Cancer Res.* 59: 2358-2362.

Hu S *et al.* (2000) Interaction between the serotonin transporter gene and neuroticism in cigarette smoking behavior. *Mol. Psychiatry* 5: 181-188.

Hu JJ *et al.* (2001) Amino acid substitution variants of *APE1* and *XRCC1* genes associated with ionising radiation sensitivity. *Carcinogenesis* 22: 917-922.

Hu YC *et al.* (2002) Molecular detection approaches for smoking associated tumors. *Oncogene* 21: 7289-7297.

Hu Y *et al.* (2003) Antisense oligonucleotides targeting XIAP induce apoptosis and enhance chemotherapeutic activity against human lung cancer cells *in vitro* and *in vivo*. *Clin. Cancer Res.* 9: 2826-2836.

Hu W *et al.* (2004) Preferential carcinogen-DNA adduct formation at codons 12 and 14 in the human *K-ras* gene and their possible mechanisms. *Biochemistry* 42: 10012-10023.

Huang S (2002) Histone methyltransferases, diet nutrients and tumour suprressors. *Nature Rev. Cancer* 2: 469-476.

Huang J *et al.* (1996) *APC* mutations in colorectal tumors with mismatch repair deficiency. *Proc. Natl. Acad. Sci. USA* 93: 9049-9054.

Huang C *et al.* (1998) Mutations in exon 7 and 8 of p53 as poor prognostic factors in patients with non-small cell lung cancer. *Oncogene* 16: 2469-2477.

Huang Q *et al.* (2002) Pulmonary neuroendocrine carcinomas. A review of 234 cases and a statistical analysis of 50 cases treated at one institution using a simple clinicopathologic classification. *Arch. Pathol. Lab. Med.* 126: 545-553.

Huang C *et al.* (2003) Chromatin remodelling and human disease. *Curr. Opin. Genet. Dev.* 13: 246-252.

Hui L *et al.* (2004) Identification of alternatively spliced mRNA variants related to cancers by genome-wide ESTs alignment. *Oncogene* 23: 3013-3023.

Huncharek M *et al.* (1999) K-ras oncogene mutation as a prognostic marker in non-small cell lung cancer: a combined analysis of 881 cases. *Carcinogenesis* 20: 1507-1510.

Huncharek M *et al.* (2000) Prognostic significance of p53 mutations in non-small cell lung cancer: a meta-analysis of 829 cases from eight published studies. *Cancer Lett.* 153: 219-226.
Hung J *et al.* (1995) Allele-specific chromosome 3p deletions occur at an early stage in the pathogenesis of lung carcinoma. *JAMA* 273: 558-563.
Hung RJ *et al.* (2003) *CYP1A1* and *GSTM1* genetic polymorphisms and lung cancer risk in Caucasian non-smokers: a pooled analysis. *Carcinogenesis* 24: 875-882.
Hunter T, Pines J (1994) Cyclins and cancer II: cyclin D and CDK inhibitors come of age. *Cell* 79: 573-582.
Hunter KW (2004) Host genetics and tumour metastasis. *Br. J. Cancer* 90: 752-755.
Hunts J *et al.* (1985) Hyperproduction and gene amplification of the epidermal growth factor receptor in squamous cell carcinomas. *Jpn. J. Cancer Res.* 76: 663-666.
Huqun *et al.* (2003) A naturally occurring p73 mutation in a p73-p53 double-mutant lung cancer cell line encodes p73a protein with a dominant-negative function. *Cancer Sci.* 94: 718-724.
Hurbin A *et al.* (2002) Inhibition of apoptosis by amphiregulin via an insulin-like growth factor-1 receptor-dependent pathway in non-small cell lung cancer cell lines. *J. Biol. Chem.* 277: 49127-49133.
Hurr K *et al.* (1996) Microsatellite alteration at chromosome 3p loci in neuroendocrine and non-neuroendocrine lung tumors. Histogenetic and clinical relevance. *Am. J. Pathol.* 149: 613-620.
Husgafvel-Pursiainen K, Kannio A (1996) Cigarette smoking and p53 mutations in lung cancer and bladder cancer. *Environ. Health Perspect.* 104 Suppl. 3: 553-556.
Husgafvel-Pursiainen K *et al.* (2000) p53 mutations and exposure to environmental tobacco smoke in a multicenter study on lung cancer. *Cancer Res.* 60: 2906-2911.
Hussain SP, Harris CC (1998) Molecular epidemiology of human cancer: contribution of mutation spectra studies of tumor suppressor genes. *Cancer Res.* 58: 4023-4037.
Hussain SP, Harris CC (2000) Molecular epidemiology and carcinogenesis: endogenous and exogenous carcinogens. *Mutation Res.* 462: 311-322.
Hussain SP *et al.* (2001) Mutability of p53 hotspot codons to benzo[*a*]pyrene diol epoxide (BPDE) and the frequency of p53 mutations in nontumorous human lung. *Cancer Res.* 61: 6350-6355.
Hussain SP *et al.* (2003) Radical causes of cancer. *Nature Rev. Cancer* 3: 276-285.
Hustert E et al. (2001) The genetic determinants of the *CYP3A5* polymorphism. *Pharmacogenetics* 11: 773-779.
Huwer H *et al.* (1997) Expression of the candidate tumor suppressor gene nm23 in the bronchial system of patients with squamous cell lung cancer. *Eur. J. Cardiothorac. Surg.* 11: 206-209.
Hwang JH *et al.* (2001) Apoptosis and bcl-2 expression as predictors of survival in radiation-treated non-small-cell lung cancer. *Int. J. Radiat. Oncol. Biol. Phys.* 50: 13-18.
Hwang SJ *et al.* (2003) Lung cancer risk in germline p53 mutation carriers: association between an inherited cancer predisposition, cigarette smoking, and cancer risk. *Hum. Genet.* 113: 238-243.
Hynes NE, Lane HA (2001) Myc and mammary cancer: Myc is a downstream effector of the ErbB2 receptor tyrosine kinase. *J. Mammary Gland Biol. Neoplasia* 6: 141-150.
Ichimura E *et al.* (1996) Expression of c-met/HGF receptor in human non-small cell lung carcinomas *in vitro* and *in vivo* and its prognostic significance. *Jpn. J. Cancer Res.* 87: 1063-1069.
Ide A *et al.* (2001) Elevation of serum pro-gastrin-releasing peptide in patients with medullary thyroid carcinoma and small cell lung carcinoma. *Thyroid* 11: 1055-1061.
Ikegaki N *et al.* (1994) Expression of *bcl*-2 in small cell lung carcinoma cells. *Cancer Res.* 54: 6-8.
Ikehara M *et al.* (2003) Expression of cyclin D1 but not of cyclin E is an indicator of poor prognosis in small adenocarcinomas of the lung. *Oncol. Rep.* 10: 137-139.
Ikehara M *et al.* (2004) Genome-wide cDNA microarray screening to correlate gene expression profile with survival in patients with advanced lung cancer. *Oncol. Rep.* 11: 1041-1044.
Imai MA *et al.* (2004) Overexpression of E2F1 associated with LOH at RB locus and hyperphosphorylation of RB in non-small cell lung carcinoma. *J. Cancer Res. Clin. Oncol.* 130: 320-326.
Imyanitov EN *et al.* (2004) Searching for cancer-associated gene polymorphisms: promises and obstacles. *Cancer Letts.* 204: 3-14.
Inase N *et al.* (2000) Use of gastrin-releasing peptide promoter for specific expression of thymidine kinase gene in small-cell lung carcinoma cells. *Int. J. Cancer* 85: 716-719.
Inbal B *et al.* (1997) DAP kinase links the control of apoptosis to metastasis. *Nature* 390: 180-184.
Iniesta P *et al.* (2004) Relationship between 3p deletions and telomerase activity in non-small-cell lung cancer: prognostic implications. *Br. J. Cancer* 90: 1983-1988.
Inoue H et al. (1997) Sequence of the FRA3B common fragile region: implications for the mechanism of *FHIT* deletion. *Proc. Natl. Acad. Sci. USA* 94: 14584-14589.
Inoue M *et al.* (1998) Lung cancer patients have increased 8-hydroxydeoxyguanosine levels in peripheral lung tissue DNA. *Jpn. J. Cancer Res.* 89: 691-695.

Inui N *et al.* (2003a) The loss of retinoic acid receptor alpha, beta and alcohol dehydrogenase 3 expression in non-small cell lung cancer. *Respirology* 8: 302-309.

Inui N *et al.* (2003b) High expression of Cks1 in human non-small cell lung cancer carcinomas. *Biochem. Biophys. Res. Commun.* 303: 978-984.

Ioannidis JPA *et al.* (2001) Replication validity of genetic association studies. *Nature Genet.* 29: 306-309.

Ioannidis JPA *et al.* (2003) Genetic associations in large versus small studies: an empirical assessment. *Lancet* 361: 567-571.

Ionov Y *et al.* (1993) Ubiquitous somatic mutations in simple repeat sequences reveal a new mechanism for colonic carcinogenesis. *Nature* 363: 558-561.

Irarrazabal CE *et al.* (2003) Chilean pilot study on the risk of lung cancer associated with codon 72 polymorphism in the gene of protein p53. *Toxicol. Lett.* 144: 69-76.

Irwin MS, Kaelin WG (2001) Role of the newer p53 family proteins in malignancy. *Apoptosis* 6: 17-29.

Ishida H *et al.* (1997) The prognostic significance of p53 and bcl-2 expression in lung adenocarcinoma and its correlation with Ki-67 growth fraction. *Cancer* 80: 1034-1045.

Ishida T *et al.* (1999) New DNA polymorphisms of human MMH/*OGG1* gene: prevalence of one polymorphism among lung-adenocarcinoma patients in Japanese. *Int. J. Cancer* 80: 18-21.

Ishiguro H *et al.* (2001) Identification of *AXUD1*, a novel human gene induced by AXIN1 and its reduced expression in human carcinomas of the lung, liver, colon and kidney. *Oncogene* 20: 5062-5066.

Ishihara S *et al.* (1999) The cyclin-dependent kinase inhibitor p27 in advanced non-small cell lung cancer: its immunohistochemical evaluation using biopsy specimens. *Lung Cancer* 26: 187-194.

Ishikawa F (1997) Telomere crisis, the driving force in cancer cell evolution. *Biochem. Biophys. Res. Commun.* 230: 1-6.

Ishikawa Y *et al.* (2002) Loss of heterozygosity and the smoking index increase with decrease in differentiation of lung adenocarcinomas: etiological implications. *Cancer Lett.* 187: 47-51.

Ito H *et al.* (2002a) A limited association of *OGG1* Ser326Cys polymorphism for adenocarcinoma of the lung. *J. Epidemiol.* 12: 258-265.

Ito H *et al.* (2002b) Expression of vascular endothelial growth factor and basic fibroblast growth factor in small adenocarcinomas. *Oncol. Rep.* 9: 119-123.

Ito A *et al.* (2003) Expression of the TSLC1 adhesion molecule in pulmonary epthelium and its down-regulation in pulmonary adenocarcinoma other than bronchioloalveolar carcinoma. *Lab Invest.* 83: 1175-1183.

Ito H *et al.* (2004a) Gene-environment interactions between the smoking habit and polymorphisms in the DNA repair genes, *APE1* Asp148Glu and *XRCC1* Arg399Gln, in Japanese lung cancer risk. *Carcinogenesis* 25: 1395-1401.

Ito G *et al.* (2004b) Krüppel-like factor 6 is frequently down regulated and induces apoptosis in non-small cell lung cancer cells. *Cancer Res.* 64: 3838-3843.

Iwanaga K *et al.* (2003) Alteration of expression or phosphorylation status of tob, a novel tumor suppressor gene product, is an early event in lung cancer. *Cancer Lett.* 202: 71-79.

Iwao K *et al.* (2001) Isolation of a novel human lung-specific gene, LUNX, a potential molecular marker for detection of micrometastasis in non-small-cell lung cancer. *Int. J. Cancer* 91: 433-437.

Izumi M *et al.* (2001) Expression of tumor-associated antigen *RCAS1* correlates significantly with poor prognosis in nonsmall cell lung carcinoma. *Cancer* 92: 446-451.

Jackson DG *et al.* (1994) Expression of alternatively spliced forms of the CD44 extracellular-matrix receptor on human lung carcinomas. *Int. J. Cancer Suppl.* 8: 110-115.

Jackson EL *et al.* (2001) Analysis of lung tumor initiation and progression using conditional expression of oncogenic K-ras. *Genes. Dev.* 15: 3243-3248.

Jaffrey RG *et al.* (2000) Genomic instability at the BUB1 locus in colorectal cancer, but not in non-small cell lung cancer. *Cancer Res.* 60: 4349-4352.

Jain RK (2001) Clearing the smoke on nicotine and angiogenesis. *Nature Med.* 7: 775-777.

Jaiswal AS *et al.* (1999) A correlation of APC and c-myc mRNA levels in lung cancer cell lines. *Oncol. Rep.* 6: 1253-1256.

Jallepalli PV, Lengauer C (2001) Chromosome segregation and cancer: cutting through the mystery. *Nature Rev. Cancer* 1: 109-117.

Jang SJ *et al.* (2001) Activation of melanoma antigen tumor antigens occurs early in lung carcinogenesis. *Cancer Res.* 61: 7959-7963.

Jänne PA, Johnson BE (2000) The role of *MYC, JUN*, and *FOS* oncogenes. Chap. 5 in *Lung Cancer. Principles and Practice.* 2nd Ed. HI Pass, JB Mitchell, DH Johnson, AT Turrisi, JD Minna (Eds.). Lippincott Williams & Williams, Philadelphia, pp 98-119.

Jänne PA *et al.* (2004) High resolution single-nucleotide polymorphism array and clustering analysis of loss of heterozygosity in human lung cancer cell lines. *Oncogene* 23: 2716-2726.

Janssen K *et al.* (2001) DNA repair activity of 8-oxoguanine DNA glycosylase 1 (*OGG1*) in human lymphocytes is not dependent on genetic polymorphism Ser326/Cys326. *Mutation Res.* 486: 207-216.

Janssen-Heijnen ML *et al.* (1998) Prevalence of co-morbidity in lung cancer patients and its relationship with treatment: a population-based study. *Lung Cancer* 21: 105-113.

Janssen-Heijnen ML, Coebergh JW (2003) The changing epidemiology of lung cancer in Europe. *Lung Cancer* 41: 245-258.

Jaques G *et al.* (1992) Differential expression of insulin-like growth factor binding proteins in human non-small cell lung cancer cell lines. *Eur. J. Cancer* 28A: 1899-1904.

Jarmalaite S *et al.* (2003) Aberrant p16 promoter methylation in smokers and former smokers with nonsmall cell lung cancer. *Int. J. Cancer* 106: 913-918.

Jaruga P *et al.* (1994) Oxidative DNA base damage and antioxidant enzyme activities in human lung cancer. *FEBS Letts.* 341: 59-64.

Järvinen A-K *et al.* (2004) Are data from different gene expression microarray platforms comparable? *Genomics* 83: 1164-1168.

Jasin M (2002) Homologous repair of DNA damage and tumorigenesis: the BRCA connection. *Oncogene* 21: 8981-8993.

Jassem E *et al.* (2001) Types and localization of p53 gene mutations: a report on 332 non-small cell lung cancer patients. *Lung Cancer* 34 Suppl. 2: S47-S51.

Jassem J *et al.* (2004) p53 and k-ras mutations are frequent events in microscopically negative surgical margins from patients with nonsmall cell lung carcinoma. *Cancer* 100: 1951-1960.

Jelinsky SA *et al.* (1995) The major, N2-Gua adduct of the (+)-anti-benzo[*a*]pyrene diol epoxide is capable of inducing G→A and G→C, in addition to G→T, mutations. *Biochemistry* 34: 13545-13553.

Jemal A *et al.* (2001) Recent trends in lung cancer mortality in the United States. *J. Natl. Cancer Inst.* 92: 277-283.

Jemal A *et al.* (2003) Cancer statistics, 2003. *CA Cancer J. Clin.* 53: 5-26.

Jen Y *et al.* (1996) Expression patterns of Id1, Id2, and Id3 are highly related but distinct from that of Id4 during mouse embryogenesis. *Dev. Dyn.* 207: 235-252.

Jeng YL *et al.* (2003) The methylenetetrahydrofolate reductase 677C→T polymorphism and lung cancer risk in a Chinese population. *Anticancer* Res. 23(6D): 5149-5152.

Jensen DE *et al.* (1998) BAP1: a novel ubiquitin hydrolase which binds to the BRC RING finger and enhances BRCA1-mediated cell growth suppression. *Oncogene* 16: 1097-1112.

Jeon HS *et al.* (2003) Relationship between *XPG* codon 1104 polymorphism and risk of primary lung cancer. *Carcinogenesis* 24: 1677-1681.

Jheon S *et al.* (2004) Lung cancer detection by a RT-nested PCR using MAGE A1-8 common primers. *Lung Cancer* 43: 29-37.

Ji L *et al.* (1999) Induction of apoptosis and inhibition of tumorigeneicity and tumor growth by adenovirus vector-mediated fragile histidine triad (*FHIT*) gene overexpression. *Cancer Res.* 59: 3333-3339.

Ji L *et al.* (2002) Expression of several genes in the human chromosome 3p21.3 homozygous deletion region by an adenovirus vector results in tumor suppressor activities *in vitro* and *in vivo*. *Cancer Res.* 62: 2715-2720.

Ji P *et al.* (2003) MALAT-1, a novel noncoding RNA, and thymosin β4 predict metastasis and survival in early-stage non-small cell lung cancer. *Oncogene* 22: 6087-6097.

Jiang S-X *et al.* (1995) Expression of *bcl*-2 oncogene protein is prevalent in small cell lung carcinomas. *J. Pathol.* 177: 135-138.

Jiang S-X *et al.* (2004) hASH1 expression is closely correlated with endocrine phenotype and differentiation extent in pulmonary neuroendocrine tumors. *Mod. Pathol.* 222-229.

Jimenez AI *et al.* (2003) Growth and molecular profile of lung cancer cells expressing ectopic *LKB1*: down-regulation of the phosphatidylinositol 3'-phosphate kinase/PTEN pathway. *Cancer Res.* 63: 1382-1388.

Jin X *et al.* (1995) Higher lung cancer risk for younger African-Americans with a Pro/Pro *p53* genotype. *Carcinogenesis* 16: 2205-2208.

Jin M *et al.* (2001) Cyclin D1, p16 and retinoblastoma gene product expression as a predictor for prognosis in non-small cell lung cancer at stages I and II. *Lung Cancer* 34: 207-218.

Jin XJ *et al.* (2002) Mitochondrial DNA mutations in lung cancer. *Ai Zheng* 21: 715-718.

Jin Z *et al.* (2004) Nicotine induces multi-site phosphorylation in association with suppression of apoptosis. *J. Biol. Chem.* 279: 23837-23844.

Johannsson O *et al.* (1996) Founding *BRCA1* mutations to hereditary breast and ovarian cancer in southern Sweden. *Am. J. Hum. Genet.* 58: 441-450.

Johnson BE *et al.* (1987) *myc* family oncogene amplification in tumor cell lines established from small cell lung cancer patients and its relationship to clinical status and course. *J. Clin. Invest.* 79: 1629-1634.

Johnson BE *et al.* (1996) *MYC* family DNA amplification in 126 tumor cell lines from patients with small cell lung cancer. *J. Cell. Biochem Suppl.* 24: 210-217.

Johnson L *et al.* (2001) Somatic activation of the *K-ras* oncogene causes early onset lung cancer in mice. *Nature* 410: 1111-1116.

Jones PA, Baylin SB (2002) The fundamental role of epigenetic events in cancer. *Nature Rev. Genet.* 3: 415-428.

Jones MH *et al.* (2004) Two prognostically significant subtypes of high-grade lung neuroendocrine tumours independent of small-cell and large-cell neuroendocrine carcinomas identified by gene expression profiles. *Lancet* 363: 775-781.

Jorm AF *et al.* (2000) Association of smoking and personality with a polymorphism of the dopamine transporter gene: results from a community survey. *Am. J. Med. Genet.* 96: 331-334.

Joseph B *et al.* (1999) Differences in expression of pro-caspases in small cell and non-small cell lung carcinoma. *Biochem. Biophys. Res. Commun.* 262: 381-387.

Joseph MG *et al.* (2001) Metallothionein expression in patients with small cell carcinoma of the lung: correlation with other molecular markers and clinical outcome. *Cancer* 92: 836-842.

Jost CA *et al.* (1997) p73 is a simian p53-related protein that can induce apoptosis. *Nature* 389: 191-194.

Jourenkova-Mironova N *et al.* (1998) Role of glutathione *S*-transferase *GSTM1*, *GSTM3*, *GSTP1* and *GSTT1* genotypes in modulating susceptibility to smoking-related lung cancer. *Pharmacogenetics* 8: 495-502.

Juin P *et al.* (2002) c-Myc functionally cooperates with Bax to induce apoptosis. *Mol. Cell. Biol.* 22: 6158-6169.

Jumper C *et al.* (2004) Determination of the serum matrix metalloproteinase-9 (MMP-9) and tissue inhibitor of matrix metalloproteinase (TIMP-1) in patients with either advanced small-cell lung cancer or non-small-cell lung cancer prior to treatment. *Respir. Med.* 98: 173-177.

Junker K *et al.* (2000) Pathology of small-cell lung cancer. *J. Cancer Res. Clin. Oncol.* 126: 361-368.

Junker N *et al.* (2003) Expression and regulation patterns of hyaluronidases in small cell lung cancer and glioma lines. *Oncol. Rep.* 10: 609-616.

Kadam S, Emerson BM (2003) Transcriptional specificity of human SWI/SNF BRG1 and BRM chromatin remodelling complexes. *Mol. Cell* 11: 377-389.

Kaiser U *et al.* (1996) Expression of bcl-2 protein in small cell lung cancer. *Lung Cancer* 15: 31-40.

Kakiuchi S *et al.* (2003) Genome-wide analysis of organ-preferential metastasis of human small cell lung cancer in mice. *Mol. Cancer Res.* 1: 485-499.

Kalemkerian GP (2000) Growth factors. Chap. 10 in *Lung Cancer. Principles and Practice.* 2nd Ed. HI Pass, JB Mitchell, DH Johnson, AT Turrisi, JD Minna (Eds.). Lippincott Williams & Williams, Philadelphia, pp 181-198.

Kalinichenko VV *et al.* (2004) Foxf1 haploinsufficiency reduces Notch-2 signaling during mouse lung development. *Am. J. Physiol. Lung Cell Mol. Physiol.* 286: L521-L530.

Kalomenidis I *et al.* (2001) Combined expression of p53, Bcl-2 and p21WAF-1 proteins in lung cancer and premalignant lesions: association with clinical characteristics. *Lung* 179: 265-278.

Kamataki T *et al.* (1999) Genetic polymorphism of *CYP2A6* in relation to cancer. *Mutation Res.* 428: 125-130.

Kamataki T *et al.* (2002) Role of human cytochrome P450 (CYP) in the metabolic activation of nitrosamine derivatives: application of genetically engineered *Salmonella* expressing human CYP. *Drug Metab. Rev.* 34: 667-676.

Kameyama K *et al.* (2003) Reduced *ING1b* gene expression plays an important role in carcinogenesis of non-small cell lung cancer patients. *Clin. Cancer Res.* 9: 4926-4934.

Kamimura A *et al.* (2000) Intracellular distribution of macrophage migration inhibitory factor predicts the prognosis of patients with adenocarcinoma of the lung. *Cancer* 89: 334-341.

Kamiya H *et al.* (1992) c-Ha-*ras* containing 8-hydroxyguanine at codon 12 induces point mutations at the modified and adjacent positions. *Cancer Res.* 52: 3483-3485.

Kamiya H *et al.* (1995) 8-Hydroxyguanine (7,8-dihydro-8-oxoguanine) in hot spots of the c-Ha-*ras* gene: effects of sequence contexts on mutation spectra. *Carcinogenesis* 16: 883-889.

Kamiya T *et al.* (2003) Interleukin-10 expression is closely correlated with the expression of granulocyte-macrophage colony-stimulating factor in non-small cell lung cancer. *Anticancer Res.* 23(3C): 2909-2913.

Kandasamy K, Srivastava RK (2002) Role of the phosphatidylinositol 3'-kinase/PTEN/Akt kinase pathway in tumor necrosis factor-related apoptosis-inducing ligand-induced apoptosis in non-small cell lung cancer cells. *Cancer Res.* 62: 4929-4937.
Kandioler D *et al.* (1996) Molecular genetic differentiation between primary lung cancers and lung metastases of other tumors. *J. Thorac. Cardiovasc. Surg.* 111: 827-831.
Kandioler-Eckersberger D *et al.* (1999) The *TP53* genotype but not immunohistochemical result is predictive of response to cisplatin-based neoadjuvant therapy in stage III non-small cell lung cancer. *J. Thorac. Cardiovasc. Surg.* 117: 744-750.
Kaneda M *et al.* (2004) Essential role for *de novo* DNA methyltransferase Dnmt3a in paternal and maternal imprinting. *Nature* 429: 900-903.
Kanehisa M, Bork P (2003) Bioinformatics in the post-sequence era. *Nature Genet. Suppl.* 33: 305-310.
Kanematsu T *et al.* (2003) Phosphorylation, but not overexpression, of epidermal growth factor receptor is associated with poor prognosis of non-small cell lung cancer patients. *Oncol. Res.* 13: 289-298.
Kang S *et al.* (2000) *PCD1*, a novel gene containing PDZ and LIM domains, is overexpressed in several human cancers. *Cancer Res.* 60: 5296-5302.
Kang Y *et al.* (2003) A multigenic program mediating breast cancer metastasis to bone. *Cancer Cell* 3: 537-549.
Kannan K *et al.* (2000) Profile of gene expression regulated by induced p53: connection to the TGF-beta family. *FEBS Letts.* 470: 77-82.
Kannan K *et al.* (2001a) DNA microarrays: identification of primary and secondary target genes regulated by p53. *Oncogene* 20: 2225-2234.
Kannan K *et al.* (2001b) DNA microarray analysis of genes involved in p53 mediated apoptosis: activation of Apaf-1. *Oncogene* 20: 3449-3455.
Kano M *et al.* (2003) Expression imbalance map: a new visualization method for detection of mRNA expression imbalance regions. *Physiol. Genomics* 13: 31-46.
Kantarci OH *et al.* (2002) Myeloperoxidase –463 (G→A) polymorphism associated with lower risk of lung cancer. *Mayo Clin. Proc.* 77: 17-22.
Kao YR *et al.* (2003) Tumor-associated antigen L6 and the invasion of human lung cancer cells. *Clin. Cancer Res.* 9: 2807-2816.
Kaplan R *et al.* (2003) Monoallelic up-regulation of the imprinted *H19* gene in airway epithelium of phenotypically normal cigarette smokers. *Cancer Res.* 63: 1475-1482.
Karuman P *et al.* (2001) The Peutz-Jegher gene product LKB1 is a mediator of p53-dependent cell death. *Mol. Cell* 7: 1307-1319.
Kashiwagi H, Uchida K (2000) Genome-wide profiling of gene amplification and deletion in cancer. *Hum. Cell* 13: 135-141.
Kashuba VI *et al.* (2004) *RBSP3* (*HYA22*) is a tumor suppressor gene implicated in major epithelial malignancies. *Proc. Natl. Acad. Sci. USA* 101: 4906-4911.
Kato S *et al.* (1995) Human lung carcinogen-DNA adduct levels mediated by genetic polymorphisms *in vivo*. *J. Natl. Cancer Inst.* 87: 902-907.
Kaufman WG, Paules RS (1996) DNA damage and cell cycle checkpoints. *FASEB J.* 10: 238-247.
Kaur TB *et al.* (2000) Role of polymorphisms in codons 143 and 160 of the O6-alkylguanine DNA alkyltransferase gene in lung cancer risk. *Cancer Epidemiol. Biomarkers Prev.* 9: 339-342.
Kawabata K *et al.* (2002) The role of neutrophil elastase in acute lung injury. *Eur. J. Pharmacol.* 451: 1-10.
Kawabata S *et al.* (2003) Expression and functional analyses of breast cancer resistance protein in lung cancer. *Clin. Cancer Res.* 9: 3052-3057.
Kawajiri K *et al.* (1990) Identification of genetically high risk individuals to lung cancer by DNA polymorphisms of the cytochrome *P450IA1* gene. *FEBS Letts.* 263: 131-133.
Kawajiri K *et al.* (1993) Germ line polymorphisms of p53 and *CYP1A1* genes involved in human lung cancer. *Carcinogenesis* 14: 1085-1089.
Kawajiri K *et al.* (1996) Association of *CYP1A1* germ line polymorphisms with mutations of the p53 gene in lung cancer. *Cancer Res.* 56: 72-76.
Kawakami T *et al.* (2002) Neuropilin 1 and neuropilin 2 co-expression is significantly correlated with increased vascularity and poor prognosis in nonsmall cell lung carcinoma. *Cancer* 95: 2196-2201.
Kawana H *et al.* (1998) Role of p27Kip1 and cyclin-dependent kinase 2 in the proliferation of non-small cell lung cancer. *Am. J. Pathol.* 153: 505-513.
Kawanishi M *et al.* (1997) Allelotype and replication error phenotype of small cell lung carcinoma. *Carcinogenesis* 18: 2057-2062.

Kawasaki M *et al.* (1998) Immunohistochemically detected p53 and P-glycoprotein predict the response to chemotherapy in lung cancer. *Eur. J. Cancer* 34: 1352-1357.

Kawasaki H *et al.* (2001) p53 gene alteration in atypical epithelial lesions and carcinoma in patients with idiopathic pulmonary fibrosis. *Hum. Pathol.* 32: 1043-1049.

Kawasaki ES (2004) Microarrays and the gene expression profile of a single cell. *Ann. N Y Acad. Sci.* 1020: 92-100.

Kaya A *et al.* (2004) The prognostic significance of vascular endothelial growth factor levels in sera of non-small cell lung cancer patients. *Respir. Med.* 98: 632-636.

Kaye FJ (2002) RB and cyclin dependent kinase pathways: defining a distinction between RB and p16 loss in lung cancer. *Oncogene* 21: 6908-6914.

Kaye FJ, Kubo A (2001) Molecular genetics of lung cancer. Chap. 17 in *Molecular Genetics of Cancer.* 2nd Ed. Ed. JK Cowell, BIOS Scientific, Oxford. pp 387-413.

Kayser K *et al.* (2003) Expression, proliferation activity and clinical significance of cathepsin B and cathepsin L in operated lung cancer. *Anticancer Res.* 23(3C): 2767-2772.

Kayyali US *et al.* (2003) Upregulation of xanthine oxidase by tobacco smoke condensate in pulmonary endothelial cells. *Toxicol. Appl. Pharmacol.* 188: 59-68.

Kazazian HH (1976) A geneticist's view of lung disease. *Am. Rev. Resp. Dis.* 113: 261-266.

Kee HJ *et al.* (2003) Identification of tumor suppressor loci on the long arm of chromsome 15 in primary small cell lung cancer. *Yonsei Med. J.* 44: 65-74.

Keith WN *et al.* (1992) Amplification of the topoisomerase II α gene in a non-small cell lung cancer cell line and characterisation of polymorphisms at the human topoisomerase II α and β loci in normal tissue. *Genes Chrom. Cancer* 4: 169-175.

Kellermann G *et al.* (1973) Aryl hydrocarbon hydroxylase inducibility and bronchiogenic carcinoma. *New Engl. J. Med.* 289: 934-939.

Kelley MJ *et al.* (1994) Small cell lung carcinoma cell lines express mRNA for calcitonin and alpha- and beta-calcitonin gene related peptides. *Cancer Lett.* 81: 19-25.

Kelley MJ *et al.* (1995) Differential inactivation of CDKN2 and Rb protein in non-small-cell and small-cell lung cancer cell lines. *J. Natl. Cancer Inst.* 87: 756-761.

Kelley MJ *et al.* (2001) Genetic analysis of the β-tubulin gene, *TUBB*, in non-small-cell lung cancer. *J. Natl. Cancer Inst.* 93: 1886-1888.

Kelsey KT *et al.* (1997) Polymorphisms in the glutathione *S*-transferase class mu and theta genes interact and increase susceptibility to lung cancer in minority populations (Texas, United States). *Cancer Causes Control* 8: 554-559.

Kemkemer R *et al.* (2002) Increased noise as an effect of haploinsufficiency of the tumor-suppressor gene neurofibromatosis type 1 *in vitro. Proc. Natl. Acad. Sci. USA* 99: 13783-13788.

Kenneway NM, Kenneway EL (1936) A study of the incidence of cancer of the lung and larynx. *J. Hygiene* 36: 236-237.

Kenneway EL, Kenneway NM (1947) A further study of the incidence of cancer of the lung and larynx. *Br. J. Cancer* 1: 260-298.

Keohavong P *et al.* (2001) Topographic analysis of K-ras mutations in histologically normal lung tissues and tumours of lung cancer patients. *Br. J. Cancer* 85: 235-241.

Keohavong P *et al.* (2004) Detection of K-ras and p53 mutations in sputum samples of lung cancer patients using laser capture microdissection microscopy and mutation analysis. *Anal. Biochem.* 324: 92-99.

Kern JA *et al.* (1990) p185neu expression in human lung adenocarcinomas predicts shortened survival. *Cancer Res.* 50: 5184-5187.

Kern JA *et al.* (1994) C-*erb*B-2 expression and codon 12 K-*ras* mutations both predict shortened survival for patients with pulmonary adenocarcinomas. *J. Clin. Invest.* 93: 516-520.

Kerr KM (2001) Pulmonary preinvasive neoplasia. *J. Clin. Pathol.* 54: 257-271.

Kersting M *et al.* (2000) Differential frequencies of p16(INK4a) promoter hypermethylation, p53 mutation, and K-ras mutation in exfoliative material mark the development of lung cancer in symptomatic chronic smokers. *J. Clin. Oncol.* 18: 3221-3229.

Keshamouni VG *et al.* (2004) Peroxisome proliferator-activated receptor-gamma activation inhibits tumor progression in non-small-cell lung cancer. *Oncogene* 23: 100-108.

Kettunen E *et al.* (2004) Differentially expressed genes in nonsmall cell lung cancer: expression profiling of cancer-related genes in squamous cell lung cancer. *Cancer Genet. Cytogenet.* 149: 98-106.

Khan S *et al.* (2004) Genetic abnormalities in plasma DNA of patients with lung cancer and other respiratory diseases. *Int. J. Cancer* 110: 891-895.

Khokhlatchev A *et al.* (2002) Identification of a novel Ras-regulated proapoptotic pathway. *Curr. Biol.* 12: 253-265.

Khoor A *et al.* (1997) Expression of surfactant protein B precursor and surfactant protein B mRNA in adenocarcinoma of the lung. *Mod. Pathol.* 10: 62-67.

Khoor A *et al.* (2004) Forkhead box A2 transcription factor is expressed in all types of neuroendocrine lung tumors. *Hum. Pathol.* 35: 560-564.

Khuri FR *et al.* (2000) Retinoic acid receptor-beta as a prognostic indicator in stage I non-small-cell lung cancer. *J. Clin. Oncol.* 18: 2798-2804.

Kiefer PE (1987) Amplification and expression of protooncogenes in human small cell lung cancer cell lines. *Cancer Res.* 47: 6236-6242.

Kihara M *et al.* (1995) Risk of smoking for squamous and small cell carcinomas of the lung modulated by combinations of *CYP1A1* and *GSTM1* gene polymorphisms in a Japanese population. *Carcinogenesis* 16: 2331-2336.

Kihara M *et al.* (1999) Lung cancer risk of the *GSTM1* null genotype is enhanced in the presence of the *GSTP1* mutated genotype in male Japanes smokers. *Cancer Lett.* 137: 53-60.

Kijima T *et al.* (2003) Molecular alterations in lung cancer. Impact on prognosis. *Methods Mol. Med.* 75: 29-38.

Kikuchi A (2003) Tumor formation by genetic mutations in the components of the Wnt signalling pathway. *Cancer Sci.* 94: 225-229.

Kikuchi T *et al.* (2003) Expression profiles of non-small cell lung cancers on cDNA microarrays: identification of genes for prediction of lymph-node metastasis and sensitivity to anti-cancer drugs. *Oncogene* 22: 2192-2205.

Kim CH *et al.* (1998a) Genetic instability of microsatellite sequences in non-small cell lung cancers. *Lung Cancer* 21: 21-25.

Kim SK *et al.* (1998b) Identification of two distinct tumor-suppressor loci on the long arm of chromosome 10 in small cell lung cancer. *Oncogene* 17: 1749-1753.

Kim JO *et al.* (1998c) Co-amplification of a novel cyclophilin-like gene (PPIE) with L-myc in small cell lung cancer cell lines. *Oncogene* 17: 1019-1026.

Kim WS *et al.* (2000) Microsatellite instability (MSI) in non-small cell lung cancer (NSCLC) is highly associated with transforming growth factor-beta type II receptor (TGF-beta RII) frameshift mutation. *Anticancer Res.* 20: 1499-1502.

Kim D-H *et al.* (2001a) $p16^{INK4a}$ and histology-specific methylation of CpG islands by exposure to tobacco smoke in non-small cell lung cancer. *Cancer Res.* 61: 3419-3424.

Kim D-H *et al.* (2001b) Promoter methylation of DAP-kinase: association with advanced stage in non-small cell lung cancer. *Oncogene* 20: 1765-1770.

Kim PK *et al.* (2003a) Fas-associating death domain protein overexpression induces apoptosis in lung cancer cells. *J. Thorac. Cardiovasc. Surg.* 125: 1336-1342.

Kim DH *et al.* (2003b) Hypermethylation of *RASSF1A* promoter is associated with the age at starting smoking and a poor prognosis in primary non-small cell lung cancer. *Cancer Res.* 63: 3743-3746.

Kim DH *et al.* (2003c) Relationship of ras association domain family 1 methylation and k-ras mutation in primary non-small cell lung cancer. *Cancer Res.* 63: 6206-6211.

Kim H *et al.* (2003d) The heparan sulfate proteoglycan GPC3 is a potential lung tumor suppressor. *Am. J. Respir. Cell Mol. Biol.* 29: 694-701.

Kim HS *et al.* (2003e) Correlation between cyclooxygenase-2 and tumor angiogenesis in non-small cell lung cancer. *Lung Cancer* 42: 163-170.

Kim HS *et al.* (2003f) Preferential elevation of Prx I and Trx expression in lung cancer cells following hypoxia and in human lung cancer tissues. *Cell Biol. Toxicol.* 19: 285-298.

Kim SY *et al.* (2003 g) Skp2 regulates Myc protein tability and activity. *Mol. Cell* 11: 1177-1188.

Kim TY *et al.* (2004a) Methylation of *RUNX3* in various types of human cancers and premalignant stages of gastric carcinoma. *Lab. Invest.* 84: 479-484.

Kim JS *et al.* (2004b) B-Raf is dispensible for K-Ras-mediated oncogenesis in human cancer cells. *Cancer Res.* 64: 1932-1937.

Kim H *et al.* (2004c) Tumor-specific methylation in bronchial lavage for the early detection of non-small-cell lung cancer. *J. Clin. Oncol.* 22: 2363-2370.

Kim JS *et al.* (2004d) Aberrant methylation of the *FHIT* gene in chronic smokers with early stage squamous cell carcinoma of the lung. *Carcinogenesis* In press [DOI: 10.1093/carcin/bgh217].

Kimura T *et al.* (2004) Mutant DNA in plasma of lung cancer patients: potential for monitoring response to therapy. *Ann. N Y Acad. Sci.* 1022: 55-60.

Kinnula VL, Crapo JD (2003) Superoxide dismutases in the lung and human lung diseases. *Am. J. Respir. Crit. Care Med.* 167: 1600-1619.

Kinnula VL *et al.* (2004) Antioxidant enzymes and redox regulating thiol proteins in malignancies of human lung. *FEBS Letts.* 569: 1-6.

Kinzel B *et al.* (2002) Downregulation of *Hus1* by antisense oligonucleotides enhances the sensitivity of human lung carcinoma cells to cisplatin. *Cancer* 94: 1808-1814.

Kinzler KW, Vogelstein B. (1996) Lessons from hereditary colorectal cancer. *Cell* 87: 159-170.
Kinzler KW, Vogelstein B. (2002) Introduction. Chap. 1 in *The Genetic Basis of Human Cancer.* 2nd Ed. Eds, B Vogelstein, KW Kinzler. McGraw-Hill, New York. Pp3-6.
Kishimoto Y *et al.* (1992) Aberrations of the *p53* tumor suppressor gene in human non-small cell carcinomas of the lung. *Cancer Res.* 52: 4799-4804.
Kishimoto Y *et al.* (1995) Allele-specific loss in chromosome 9p loci in preneoplastic lesions accompanying non-small-cell lung cancers. *J. Natl. Cancer Inst.* 87: 1224-1229.
Kitagawa Y *et al.* (1996) Overexpression of Bcl-2 and mutations in p53 and K-ras in resected human non-small cell lung cancers. *Am. J. Respir. Cell Mol. Biol.* 15: 45-54.
Kitaguchi S *et al.* (1998) Proliferative activity, p53 expression and loss of heterozygosity on 3p, 9p and 17p in atypical adenomatous hyperplasia of the lung. *Hiroshima J. Med. Sci.* 47: 17-25.
Kiyohara C *et al.* (2002) Genetic polymorphisms and lung cancer susceptibility: a review. *Lung Cancer* 37: 241-256.
Kiyohara C *et al.* (2003) Risk modification by *CYP1A1* and *GSTM1* polymorphisms in the association of environmental tobacco smoke and lung cancer: a case-control study in Japanese nonsmoking women. *Int. J. Cancer* 107: 139-144.
Kiyohara C *et al.* (2004) Genetic polymorphisms and environmental risk of lung cancer: a review. *Rev. Environ Health* 19: 15-38.
Knudson AG (2001) Two genetic hits (more or less) to cancer. *Nature Rev. Cancer* 1: 157-162.
Ko JL *et al.* (2000) *MDM2* mRNA expression is a favorable prognostic factor in non-small-cell lung cancer. *Int. J. Cancer* 89: 265-270.
Ko JL *et al.* (2002) A novel p53 mutant retained functional activity in lung carcinomas. *DNA Repair* 1: 755-762.
Kobayashi K *et al.* (2002) Reduced expression of the *REIC/Dkk-3* gene by promoter-hypermethylation in human tumor cells. *Gene* 282: 151-158.
Kobayashi K *et al.* (2004) Identification of genes whose expression is upgegulated in lung adenocarcinoma cells in comparison with type II alveolar cells and bronchiolar epithelial cells *in vivo*. *Oncogene* 23: 3089-3096.
Kohda M *et al.* (2001) Frequent loss of imprinting of *IGF2* and *MEST* in lung adenocarcinoma. *Mol. Carcinogen.* 31: 184-191.
Kohn KW (1999) Molecular interaction map of the mammalian cell cycle control and DNA repair systems. *Mol. Biol. Cell* 10: 2703-2734.
Kohno T, Yokota J (1999) How many tumor suppressor genes are involved in human lung carcinogenesis? *Carcinogenesis* 20: 1403-1410.
Kohno T *et al.* (1995) Identification of a novel phospholipase C family gene at chromosome 2q33 that is homozygously deleted in human small cell lung carcinoma. *Hum. Mol. Genet.* 4: 667-674.
Kohno T *et al.* (1998a) Inactivation of the *PTEN/MMAC1/TEP1* gene in human lung cancers. *Genes Chrom. Cancer* 22: 152-156.
Kohno T *et al.* (1998b) Genetic polymorphisms and alternative splicing of the *hOGG1* gene, that is involved in the repair of 8-hydroxyguanine in damaged DNA. *Oncogene* 16: 3219-3225.
Kohno T *et al.* (1998c) Homozygous deletion and frequent allelic loss of the 21q11.1-q21.1 region including the ANA gene in human lung carcinoma. *Genes Chrom. Cancer* 21: 236-243.
Kohno T *et al.* (1999a) Alterations of the *PPP1R3* gene in human cancer. *Cancer Res.* 59: 4170-4174.
Kohno H *et al.* (1999b) p53 mutation and allele loss of chromosome 3p, 9p of preneoplastic lesions in patients with nonsmall cell lung carcinoma. *Cancer* 85: 341-347.
Kohno T *et al.* (2000) Mutation and expression of the *DCC* gene in human lung cancer. *Neoplasia* 2: 300-305.
Kolesar JM *et al.* (2002) Evaluation of *NQO1* gene expression and variant allele in human NSCLC tumors and matched normal lung tissue. *Int. J. Oncol.* 21: 1119-1124.
Kolomietz E *et al.* (2002) The role of *Alu* repeat clusters as mediators of recurrent chromosomal aberrations in tumors. *Genes Chrom. Cancer* 35: 97-112.
Kondo M *et al.* (1992) Mutations in the p53 tumour suppressor gene in primary lung cancer in Japan. *Biochem. Biophys. Res. Commun.* 183: 1139-1146.
Kondo M *et al.* (1995) Frequent loss of imprinting of the *H19* gene is often associated with its overexpression in human lung cancers. *Oncogene* 10: 1193-1198.
Kondo M *et al.* (1996) Selective maternal-allele loss in human lung cancers of the maternally expressed $p57^{KIP2}$ gene at 11p15.5. *Oncogene* 12: 1365-1368.
Kondo M *et al.* (2001) Overexpression of candidate tumor suppressor gene *FUS1* isolated from the 3p21.3 homozygous deletion region leads to G1 arrest and growth inhibition of lung cancer cells. *Oncogene* 20: 6258-6262.
Kong F-M *et al.* (2000) *M6P/IGF2R* is mutated in squamous cell carcinoma of the lung. *Oncogene* 19: 1572-1578.

Konishi H *et al.* (2002) Identification of frequent G2 checkpoint impairment and a homozygous deletion of 14-3-3ε at 17p13.3 in small cell lung cancers. *Cancer Res.* 62: 271-276.
Konishi H *et al.* (2003) Detailed characterization of a homozygously deleted region corresponding to a candidate tumor suppressor locus at distal 17p13.3 in human lung cancer. *Oncogene* 22: 1892-1905.
Konishi J *et al.* (2004) The relationship between NY-ESO-1 mRNA expression and clinicopathological features in non-small cell lung cancer. *Oncol. Rep.* 11: 1063-1067.
Kopelovich L *et al.* (2003) The epigenome as a target for cancer chemoprevention. *J. Natl. Cancer Inst.* 95: 1747-1757.
Kops GJ, Burgering BM (1999) Forhead transcription factors: new insights into protein kinase B (c-akt) signalling. *J. Mol. Med.* 77: 656-665.
Korrapati V *et al.* (2001) Effect of HER2/neu expression on survival in non-small-cell lung cancer. *Clin. Lung Cancer* 2: 216-219.
Kosciolek BA *et al.* (2003) Inhibition of telomerase activity in human cancer cells by RNA interference. *Mol. Cancer Ther.* 2: 209-216.
Koshikawa K *et al.* (2002) Significant up-regulation of a novel gene, *CLCP1*, in a highly metastatic lung cancer subline as well as in lung cancers *in vivo. Oncogene* 21: 2822-2828.
Kost-Alimova M *et al.* (2003) Coincidence of synteny breakpoints with malignancy-related deletions on human chromosome 3. *Proc. Natl. Acad. Sci. USA* 100: 6622-6627.
Kothapalli R et al. (2002) Microarray results: how accurate are they? *BMC Bioinformatics* 3: 22.
Koukourakis MI *et al.* (1997) Platelet-derived endothelial cell growth factor expression correlates with tumour angiogenesis and prognosis in non-small-cell lung cancer. *Br. J. Cancer* 75: 477-481.
Koukourakis MI *et al.* (2003) Enhanced expression of SPARC/osteonectin in the tumor-associated stroma of non-small cell lung cancer is correlated with markers of hypoxia/acidity and with poor prognosis of patients. *Cancer Res.* 63: 5376-5380.
Kouri RE *et al.* (1982) Positive correlation between aryl hydrocarbon hydroxylase activity and primary lung cancer as analysed in cryopreserved lymphocytes. *Cancer Res.* 42: 5030-5037.
Kovalchuk O *et al.* (2001) K-ras codon 12 mutations detected with enriched PCR method in operable non-small cell lung cancer. *Folia Histochem. Cytobiol.* 39 Suppl. 2: 68-69.
Kramata P *et al.* (2003) A single site-specific *trans*-opened 7,8,9,10-tetrahydrobenzo[*a*]pyrene 7,8-diol 9,10-epoxide N2-deoxyguanosine adduct induces mutations at multiple sites in DNA. *J. Biol. Chem.* 278: 14940-14948.
Kraut M, Wozniak A (2000) Clinical presentation. Chap. 29 in *Lung Cancer. Principles and Practice.* 2nd Ed. HI Pass, JB Mitchell, DH Johnson, AT Turrisi, JD Minna (Eds.). Lippincott Williams & Williams, Philadelphia, pp 521-534.
Krawczak M *et al.* (1995) The somatic spectrum of cancer-associated single basepair substitutions in the *TP53* gene is determined mainly by endogenous mechanisms of mutation and by selection. *Hum. Mutation* 5: 48-57.
Krawczak M *et al.* (1998) Neighboring nucleotide effects on the rates of meiotic single base-pair substitution in human genes. *Am. J. Hum. Genet.* 63: 474-488.
Krawczak M, Cooper DN (1998) *p53* mutations, benzo[*a*]pyrene and lung cancer. *Mutagenesis* 13: 319-320.
Krawczak M *et al.* (2001) Correcting for multiple testing in genetic association studies: the legend lives on. *Hum. Genet.* 109: 566-567.
Krepela E *et al.* (2004) Increased expression of Apaf-1 and procaspase-3 and the functionality of intrinsic apoptosis apparatus in non-small cell lung carcinoma. *Biol. Chem.* 385: 153-168.
Kreuzer M *et al.* (1998) Risk factors for lung cancer in young adults. *Am. J. Epidemiol.* 147: 1028-1037.
Kriek E *et al.* (1998) Polycyclic aromatic hydrocarbon-DNA adducts in humans: relevance as biomarkers for exposure and cancer risk. *Mutation Res.* 400: 215-231.
Kristiansen G *et al.* (2001) Overexpression of c-erbB2 protein correlates with disease-stage and chromosomal gain at the c-erbB2 locus in non-small cell lung cancer. *Eur. J. Cancer* 37: 1089-1095.
Kristiansen G *et al.* (2002) CD10 expression in non-small cell lung cancer. *Anal. Cell Pathol.* 24: 41-46.
Krystal GW *et al.* (1996) Autocrine growth of small cell lung cancer mediated by coexpression of c-kit and stem cell factor. *Cancer Res.* 56: 370-376.
Kubo A *et al.* (1996) Point mutations of the topoisomerase IIα gene in patients with small cell lung cancer treated with etoposide. *Cancer Res.* 56: 1232-1236.
Kubokura H *et al.* (2001) Relations of the c-*myc* gene and chromosome 8 in non-small cell lung cancer: analysis by fluorescence *in situ* hybridisation. *Ann. Thorac. Cardiovasc. Surg.* 7: 197-203.

Kuehl P *et al.* (2001) Sequence diversity in *CYP3A* promoters and characterization of the genetic basis of polymorphic *CYP3A5* expression. *Nature Genet.* 27: 383-391.
Kuhn H *et al.* (2004) Influence of basic fibroblast growth factor on the proliferation of non-small cell lung cancer cell lines. *Lung Cancer* 44: 167-174.
Kumaki F *et al.* (2001) Telomerase activity and expression of human telomerase RNA component and human telomerase reverse transcriptase in lung carcinomas. *Hum. Pathol.* 32: 188-195.
Kumar Biswas S *et al.* (2004) Down-regulation of Bcl-2 is associated with cisplatin resistance in human small cell lung cancer H69 cells. *Mol. Cancer Ther.* 3: 327-334.
Kumari A *et al.* (2004) p53 protects from replication-associated DNA double-strand breaks in mammalian cells. *Oncogene* 23: 2324-2329.
Kumimoto H *et al.* (2002) L-myc genotype is associated with different susceptibility to lung cancer in smokers. *Jpn. J. Cancer Res.* 93: 1-5.
Kurakawa E *et al.* (2001) Hypermethylation of p16(INK4a) and p15(INK4b) genes in non-small cell lung cancer. *Int. J. Oncol.* 19: 277-281.
Kuramochi M *et al.* (2001) *TSLC1* is a tumor-suppressor gene in human non-small-cell lung cancer. *Nature Genet.* 27: 427-430.
Kure EH *et al.* (1996) *p53* mutations in lung tumours: relationship to gender and lung DNA adduct levels. *Carcinogenesis* 17: 2201-2205.
Kurimoto F *et al.* (2001) Unchanged frequency of loss of heterozygosity and size of the deleted region at 8p21-23 during metastasis of lung cancer. *Int. J. Mol. Med.* 8: 89-93.
Kuroki T *et al.* (2003) Allelic loss on chromosome 3p21.3 and promoter hypermethylation of semaphorin 3B in non-small cell lung cancer. *Cancer Res.* 63: 3352-3355.
Kurusu Y *et al.* (1999) Detection of circulating tumor cells by reverse transcriptase polymerase chain reaction in patients with resectable non-small-cell lung cancer. *Surgery* 126: 820-826.
Kwabi-Addo B *et al.* (2001) Haploinsufficiency of the Pten tumor suppressor gene promotes prostate cancer progression. *Proc. Natl. Acad. Sci. USA* 98: 11563-11568.
Kwak I *et al.* (2004) Genetically engineered mouse models for lung cancer. *Annu. Rev. Physiol.* 66: 647-663.
La Cour JM *et al.* (2003) Up-regulation of ALG-2 in hepatomas and lung cancer tissue. *Am. J. Pathol.* 163: 81-89.
Lacroix J *et al.* (2001) Sensitive detection of rare cancer cells in sputum and peripheral blood samples of patients with lung cancer by preproGRP-specific RT-PCR. *Int. J. Cancer* 92: 1-8.
Ladanyi M, Gerald WL (Eds) (2003) *Expression Profiling of Human Tumors.* Humana Press, Totowa, New Jersey.
Lader AS *et al.* (2004) Identification of a transcriptional profile associated with *in vitro* invasion in non-small cell lung cancer cell lines. *Cancer Biol. Ther.* 3: In press. [DOI?. PMID: 15153803]
Laforest L *et al.* (2000) *CYP2D6* gene polymorphism in Caucasian smokers: lung cancer susceptibility and phenotype-genotype relationships. *Eur. J. Cancer* 36: 1825-1832.
Laird PW (1997) Oncogenic mechanisms mediated by DNA methylation. *Mol. Med. Today* 3: 223-229.
Lamy A *et al.* (2001) Promoter methylation and silencing of the retinoic acid receptor-beta gene in lung carcinomas. *J. Natl. Cancer Inst.* 93: 66-68.
Lamy A *et al.* (2002) Aberrant methylation of the CDKN2a/p16INK4a gene promoter region in preinvasive bronchial lesions: a prospective study in high-risk patients without invasive cancer. *Int. J. Cancer* 100: 189-193.
Lan MS *et al.* (1993) IA-1, a new marker for neuroendocrine differentiation in human lung cancer cell lines. *Cancer Res.* 53: 4169-4171.
Landi S *et al.* (2003) Evaluation of a microarray for genotyping polymorphisms related to xenobiotic metabolism and DNA repair. *Biotechniques* 35: 816-827.
Langdon JA, Armour JAL (2003) Evolution and population genetics of the H-*ras* minisatellite and cancer predisposition. *Hum. Mol. Genet.* 12: 891-900.
Langendijk JA *et al.* (1995) The prognostic significance of accumulation of p53 protein in stage III non-small cell lung cancer treated by radiotherapy. *Radiother. Oncol.* 36: 218-224.
Langenfeld EM *et al.* (2003) The mature bone morphogenetic protein-2 is aberrantly expressed in non-small cell lung carcinomas and stimulates tumor growth of A549 cells. *Carcinogenesis* 24: 1445-1454.
Langer DA *et al.* (2002) Inhibition of proliferation of human small cell lung cancer cells expressing an autocrine system for gastrin releasing peptide by antisense oligodeoxynucleotides to gastrin releasing peptide receptor. *Lung Cancer* 36: 235-242.
Langhans T (1871) Primärer Krebs der Trachea und Bronchien. *Virchows Arch.* 53: 470-484.
Lantuejoul S *et al.* (2003) Expression of VEGF, semaphorin SEMA3F, and their common receptors neuropilins NP1 and NP2 in preinvasive bronchial lesions, lung tumours, and cell lines. *J. Pathol.* 200: 336-347.

Lantuejoul S *et al.* (2004) Differential expression of telomerase reverse transcriptase (hTERT) in lung tumours. *Br. J. Cancer* 90: 1222-1229.
Largaespada DA (2001) Haploinsufficiency for tumor suppression: the hazards of being single and living a long time. *J. Exp. Med.* 193: F15-F18.
Lau DH *et al.* (1997) Clonal origin of multiple lung cancers: K-ras and p53 mutations determined by nonradioisotopic single-strand conformation polymorphism analysis. *Diagn. Mol. Pathol.* 6: 179-184.
Lau D *et al.* (2000a) Expression and regulation of a molecular marker, SPR1, in multistep bronchial carcinogenesis. *Am. J. Respir. Cell Mol. Biol.* 22: 92-96.
Lau CL *et al.* (2000b) Clinical and molecular prognostic factors and models for non-small cell lung cancer. Chap. 33 in *Lung Cancer. Principles and Practice.* 2nd Ed. HI Pass, JB Mitchell, DH Johnson, AT Turrisi, JD Minna (Eds.). Lippincott Williams & Williams, Philadelphia, pp 602-611.
Laudanski J *et al.* (1999) Expression of *bcl*-2 protein in non-small cell lung cancer: correlation with clinicopathology and patient survival. *Neoplasma* 46: 25-30.
Le Marchand L *et al.* (2000) Association of the myeloperoxidase –463G→A polymorphism with lung cancer risk. *Cancer Epidemiol. Biomarkers Prev.* 9: 181-184.
Le Marchand L *et al.* (2002) Association of the hOGG1 Ser326Cys polymorphism with lung cancer risk. *Cancer Epidemiol. Biomarkers Prev.* 11: 409-412.
Le Marchand L *et al.* (2003) Pooled analysis of the *CYP1A1* exon 7 polymorphism and lung cancer (United States). *Cancer Causes Control* 14: 339-346.
Le Page F *et al.* (1995) Mutagenicity of a unique 8-oxoguanine in human Ha-*ras* sequence in mammalian cells. *Carcinogenesis* 16: 2779-2784.
Le Page F *et al.* (2000) Transcription coupled repair of 8-oxoguanine in murine cells: the ogg1 protein is required for repair in non-transcribed sequences but not in transcribed sequences. *Proc. Natl. Acad. Sci. USA* 97: 8397-8402.
LeRoith D, Roberts CT (2003) The insulin-like growth factor system and cancer. *Cancer Lett.* 195: 127-137.
Lechner JF, Fugaro JM (2000) Chap. 4 in *Lung Cancer. Principles and Practice.* 2nd Ed. HI Pass, JB Mitchell, DH Johnson, AT Turrisi, JD Minna (Eds.). Lippincott Williams & Williams, Philadelphia, pp 89-97.
Lechner JF *et al.* (2002) Human lung cancer cells and tissues partially recapitulate the homeobox gene expression profile of embryonic lung. *Lung Cancer* 37: 41-47.
Lee JC *et al.* (1998a) Telomerase activity in lung cancer cell lines and tissues. *Lung Cancer* 21: 99-103.
Lee MP *et al.* (1998b) Somatic mutation of *TSSC5*, a novel imprinted gene from human chromosome 11p15.5. *Cancer Res.* 58: 4155-4159.
Lee HY *et al.* (1998c) Suppression of c-Fos gene transcription with malignant transformation of human bronchial epithelial cells. *Oncogene* 16: 3039-3036.
Lee SH *et al.* (1999a) Alterations of Fas (Apo-1/CD95) gene in non-small cell lung cancer. *Oncogene* 18: 3754-3760.
Lee SH *et al.* (1999b) Alterations of the *DR5/TRAIL* receptor 2 gene in non-small cell lung cancers. *Cancer Res.* 59: 5683-5686.
Lee SH *et al.* (2002a) Non-small cell lung cancers frequently express phosphorylated Akt; an immunohistochemical study. *APMIS* 110: 587-592.
Lee WJ *et al.* (2002b) Microsomal epoxide hydrolase polymorphisms and lung cancer risk: a quantitative review. *Biomarkers* 7: 230-241.
Lee SH *et al.* (2002c) Non-small cell lung cancers frequently express phosphorylated Akt; an immunohistochemical study. *APMIS* 110: 587-592.
Lee HY *et al.* (2003a) Evidence that phosphatidylinositol 3-kinase- and mitogen-activated protein kinase kinase-4/cJun N-terminal kinase-dependent pathways cooperate to maintain lung cancer cell survival. *J. Biol. Chem.* 278: 23630-23638.
Lee EB *et al.* (2003b) Loss of heterozygosity on the long arm of chromosome 21 in non-small cell lung cancer. *Ann. Thorac. Surg.* 75: 1597-1600.
Lee CT *et al.* (2003c) Recombinant adenoviruses expressing dominant negative insulin-like growth factor-I receptor demonstrate antitumor effects on lung cancer. *Cancer Gene Ther.* 10: 57-63.
Lee CH *et al.* (2003d) Differential expression of hypoxia inducible factor-1 alpha and tumor cell proliferation between squamous cell carcinomas and adenocarcinomas among operable non-small cell lung carcinomas. *J. Korean Med. Sci.* 18: 196-203.
Lee C, Roy M (2004) Analysis of alternative splicing with microarrays: successes and challenges. *Genome Biol.* 5(7): 231.
Lee YC *et al.* (2004) Frequent allelic deletion at the *FHIT* locus associated with p53 overexpression in squamous cell carcinoma subtype of Taiwanese non-small-cell lung cancers. *Br. J. Cancer* 90: 2378-2383.

Legrand M *et al.* (1996) Influence of a mutation reducing the catalytic activity of the cytochrome P450 CYP2D6 on lung cancer susceptibility. *Carcinogenesis* 17: 2267-2269.

Lehman TA *et al.* (1996) Detection of K-ras oncogene mutations by polymerase chain reaction-based ligase chain reaction. *Anal. Biochem.* 239: 153-159.

Lei YC *et al.* (2002) Effects on sister chromatid exchange frequency of polymorphisms in DNA repair gene *XRCC1* in smokers. *Mutation Res.* 519: 93-101.

Lemon WJ *et al.* (2002) Identification of candidate lung cancer susceptibility genes in mouse using oligonucleotide arrays. *J. Med. Genet.* 39: 644-655.

Lengauer C *et al.* (1998) Genetic instabilities in human cancers. *Nature* 396: 643-649.

Lerebours F *et al.* (1999) Fine deletion mapping of chromosome 8p in non-small-cell lung carcinoma. *Int. J. Cancer* 81: 854-858.

Lerman C, Niaura R (2002) Applying genetic approaches to the treatment of nicotine dependence. *Oncogene* 21: 7412-7420.

Lerman C, Berrettini W (2003) Elucidating the role of genetic factors in smoking behavior and nicotine dependence. *Am. J. Med. Genet.* 118B: 48-54.

Lerman C *et al.* (1999) Evidence suggesting the role of specific genetic factors in cigarette smoking. *Health Psych.* 18: 14-20.

Lerman C *et al.* (2000) Interacting effects of the serotonin transporter gene and neuroticism in smoking practices and nicotine dependence. *Mol. Psychiatry* 5: 189-192.

Lerman MI, Minna JD (2000) The 630-kb lung cancer homozygous deletion region on human chromosome 3p21.3: identification and evaluation of the resident candidate tumor suppressor genes. The International Lung Cancer Chromosome 3p21.3 Tumor Suppressor Gene Consortium. *Cancer Res.* 60: 6116-6133.

Levin ML *et al.* (1950) Cancer and tobacco smoking: a preliminary report. *JAMA* 143: 336-338.

Levin WJ *et al.* (1994a) Tumor suppressor and immediate early transcription factor genes in non-small cell lung cancer. *Chest* 106 (6 Suppl.): 372-376S.

Levin NA *et al.* (1994b) Identification of frequent novel genetic alterations in small cell lung carcinoma. *Cancer Res.* 54: 5086-5091.

Levin WJ *et al.* (1995a) Expression patterns of immediate early transcription factors in human non-small cell lung cancer. The Lung Cancer Study Group. *Oncogene* 11: 1261-1269.

Levin NA *et al.* (1995b) Identification of novel regions of altered DNA copy number in small cell lung tumors. *Genes Chrom. Cancer* 13: 175-185.

Levine AJ (1993) The tumor suppressor genes. *Annu. Rev. Biochem.* 62: 623-651.

Levine AJ (1997) *p53*, the cellular gatekeeper for growth and division. *Cell* 88: 323-331.

Lewis SJ *et al.* (2001) Polymorphisms in the NAD(P)H: quinone oxidoreductase gene and small cell lung cancer risk in a UK population. *Lung Cancer* 34: 177-183.

Lewis PD, Parry JM (2004) *In silico* p53 mutation hotspots in lung cancer. *Carcinogenesis* 25: 1099-1107.

Li GM (1999) The role of mismatch repair in DNA damage-induced apoptosis. *Oncol. Res.* 11: 393-400.

Li J *et al.* (1994) K-*ras* gene point mutation: a stable tumor marker in non-small cell lung carcinoma. *Lung Cancer* 11: 19-27.

Li Y *et al.* (1999) Over-expression of the *RAB5* gene in human lung adenocarcinoma cells with high metastatic potential. *Chin. Med. Sci. J.* 14: 96-101.

Li D *et al.* (2001a) Sensitivity to DNA damage induced by benzo[*a*]pyrene diol epoxide and risk of lung cancer: a case-control analysis. *Cancer Res.* 61: 1445-1450.

Li Q *et al.* (2001b) Link of a new type of apoptosis-inducing gene ASY/Nogo-B to human cancer. *Oncogene* 20: 3929-3936.

Li X *et al.* (2002) DNA microarrays: their use and misuse. *Microcirculation* 9: 13-22.

Li J *et al.* (2003a) LOH of chromosome 12p correlates with Kras2 mutation in non-small cell lung cancer. *Oncogene* 22: 1243-1246.

Li Q *et al.* (2003b) Clinical significance of co-expression of VEGF-C and VEGFR-3 in non-small cell lung cancer. *Chin. Med. J.* 116: 727-730.

Li C *et al.* (2003c) Comparative proteomics analysis of human lung squamous carcinoma. *Biochem. Biophys. Res. Commun.* 309: 253-260.

Li J *et al.* (2003d) RASSF1A promoter methylation and Kras2 mutations in non small cell lung cancer. *Neoplasia* 5: 362-366.

Li Y *et al.* (2003e) A common polymorphism in *XRCC1* as a biomarker of susceptibility for chemically induced genetic damage. *Biomarkers* 8: 408-414.

Li QL *et al.* (2004) Transcriptional silencing of the *RUNX3* gene by CpG hypermethylation is associated with lung cancer. *Biochem. Biophys. Res. Commun.* 314: 223-228.

Liang P, Pardee AB (2003) Analysing differential gene expression in cancer. *Nature Rev. Cancer* 3: 869-876.

Liang G *et al.* (2003) Sequence variations in the DNA repair gene *XPD* and risk of lung cancer in a Chinese population. *Int. J. Cancer* 105: 669-673.

Liang G *et al.* (2004) A functional polymorphism in the *SULT1A1* gene (G638A) is associated with risk of lung cancer in relation to tobacco smoking. *Carcinogenesis* 25: 773-778.

Lichtenstein P *et al.* (2000) Environmental and heritable factors in the causation of cancer – analyses of cohorts of twins from Sweden, Denmark and Finland. *New Engl. J. Med.* 343: 78-85.

Lickint F (1929) Tabak und Tabakrauch als ätiologischer Faktor des Carcinoms. *Z. Krebsforsch.* 30: 349-365.

Lickint F (1935) Der Bronchialkrebs der Raucher. *München. Med. Wochenschr.* 82: 1232-1234.

Lieber MR *et al.* (2003) Mechanism and regulation of human non-homologous DNA end-joining. *Nature Rev. Mol. Cell Biol.* 4: 712-720.

Liebow A *et al.* (1948) Cytological studies of sputum and bronchial secretions in the diagnosis of carcinoma of the lung. *Cancer* 1: 223-229.

Liloglou T *et al.* (1997) *p53* aberrations in non-small-cell lung carcinomas from a smoking population. *Br. J. Cancer* 75: 1119-1124.

Liloglou T *et al.* (2001) Cancer-specific genomic instability in bronchial lavage: a molecular tool for lung cancer detection. *Cancer Res.* 61: 1624-1628.

Liloglou T *et al.* (2002) A T2517C polymorphism in *GSTM4* gene is associated with risk of developing lung cancer. *Lung Cancer* 37: 143-146.

Lim M, Jablons DM (2003) Matrix metalloproteinase expression in lung cancer. Chap. 25 in *Methods in Molecular Medicine Vol. 74: Lung Cancer, Vol. 1: Molecular Pathology Methods and Reviews.* Ed. B. Driscoll. Humana Press, Totowa, New Jersey. pp 349-356.

Lim EH *et al.* (2003) Feasibility of using low-volume tissue samples for gene expression profiling of advanced non-small cell lung cancers. *Clin. Cancer Res.* 9: 5980-5987.

Lim W *et al.* (2004) Relative frequency and morphology of cancers in *STK11* mutation carriers. *Gasteroenterology* 126: 1788-1794.

Lin P *et al.* (2000) Association of *CYP1A1* and microsomal epoxide hydrolase polymorphisms with lung squamous cell carcinoma. *Br. J. Cancer* 82: 852-857.

Lin X *et al.* (2001) Overexpression of phosphatidylinositol 3-kinase in human lung cancer. *Langenbecks Arch. Surg.* 386: 293-301.

Lin L *et al.* (2002) The hepatocyte nuclear factor 3 α gene, HNF3α (*FOXA1*), on chromosome band 14q13 is amplified and overexpressed in esophageal and lung adenocarcinomas. *Cancer Res.* 62: 5273-5279.

Lin P *et al.* (2003) Overexpression of aryl hydrocarbon receptor in human lung carcinomas. *Toxicol. Pathol.* 31: 22-30.

Lindblad-Toh K *et al.* (2000) Loss-of-heterozygosity analysis of small-cell lung carcinomas using single-nucleotide polymorphism arrays. *Nature Biotechnol.* 18: 1001-1005.

Lindstedt BA *et al.* (1999) *Hras1* VNTR alleles as susceptibility markers for lung cancer: relationship to microsatellite instability in tumors. *Anticancer Res.* 19: 5523-5527.

Lindstrom I *et al.* (2002) DNA copy number changes in lung adenocarcinoma in younger patients. *Mod. Pathol.* 15: 372-378.

Ling D *et al.* (2004) Association between polymorphism of the dopamine transporter gene and early smoking onset: an interaction risk on nicotine dependence. *J. Hum. Genet.* 49: 35-39.

Liotta LA *et al.* (2003) Protein microarrays: meeting analytical challenges for clinical applications. *Cancer Cell* 3: 317-325.

Little CD *et al.* (1983) Amplification and expression of the c-*myc* oncogene in human lung cancer cell lines. *Nature* 306: 194-196.

Little MP (2002) Comparisons of lung tumour mortality risk in the Japanese A-bomb survivors and in the Colorado Plateau uranium miners: support for the ICRP lung model. *Int. J. Radiat. Biol.* 78: 145-163.

Little NA, Jochemsen AG (2002) p63. *Int. J. Biochem. Cell. Biol.* 34: 6-9.

Liu L *et al.* (1999) Reduced lung tumorigenesis in human methylguanine DNA methyltransferase transgenic mice achieved by expression of transgene within the target cell. *Carcinogenesis* 20: 279-284.

Liu C-X *et al.* (2000) *LRP-DIT*, a putative endocytic receptor gene, is frequently inactivated in non-small cell lung cancer cell lines. *Cancer Res.* 60: 1961-1967.

Liu D *et al.* (2001) E-cadherin expression associated with differentiation and prognosis in patients with non-small cell lung cancer. *Ann. Thorac. Surg.* 71: 949-954.

Liu Y *et al.* (2003a) Hypermethylation of p16INK4a in Chinese lung cancer patients: biological and clinical implications. *Carcinogenesis* 24: 1897-1901.

Liu L *et al.* (2003b) Control of microtubule stability by the RASSF1A tumor suppressor. *Oncogene* 22: 8125-8136.
Liu Y *et al.* (2003c) Role for ETS domain transcription factors Pea3/Erm in mouse lung development. *Dev. Biol.* 261: 10-24.
Liu JY *et al.* (2004) Hypermethylation of p16 gene in clinical specimens of patients with lung cancer. *Zhonghua Zhong Liu Za Zhi* 26: 75-77.
Lo HS *et al.* (2003) Allele variation in gene expression is common in the human genome. *Genome Res.* 13: 1855-1862.
Lockhart DJ, Winzeler EA (2000) Genomics, gene expression and DNA arrays. *Nature* 405: 827-836.
Loeb KR, Loeb LA (2000) Significance of multiple mutations in cancer. *Carcinogenesis* 21: 379-385.
Loeb LA *et al.* (2003) Multiple mutations and cancer. *Proc. Natl. Acad. Sci. USA* 100: 776-761.
Lohmann D *et al.* (1993) Mutational spectrum of the p53 gene in human small-cell lung cancer and relationship to clinicopathological data. *Am. J. Pathol.* 142: 907-915.
Lohr K *et al.* (2003) p21/CDKN1A mediates negative regulation of transcription by p53. *J. Biol. Chem.* 278: 32507-32516.
Loktionov A (2004) Common gene polymorphisms, cancer progression and prognosis. *Cancer Letts.* 208: 1-33.
Lombard HL, Doering CR (1928) Cancer studies in Massachussets 2. Habits, characteristics and environment of individuals with and without cancer. *New Engl. J. Med.* 198: 481-487.
Lonardo F, Albanell J (2000) Telomerase in lung cancer. Chap. 9 in *Lung Cancer. Principles and Practice.* 2nd Ed. HI Pass, JB Mitchell, DH Johnson, AT Turrisi, JD Minna (Eds.). Lippincott Williams & Williams, Philadelphia, pp 170-177.
Lonardo F *et al.* (1999) Overexpression of cyclins D1 and E is frequent in bronchial preneoplasia and precedes squamous cell carcinoma development. *Cancer Res.* 59: 2470-2476.
London SJ *et al.* (1997) Genetic polymorphism of *CYP2D6* and lung cancer risk in African-Americans and Caucasians in Los Angeles County. *Carcinogenesis* 18: 1203-1214.
London SJ *et al.* (2000) Lung cancer risk in relation to genetic polymorphisms of microsomal epoxide hydrolase among African-Americans and Caucasians in Los Angeles County. *Lung Cancer* 28: 147-155.
Lopez de Cicco R *et al.* (2002) Furin expression in squamous cell carcinomas of the oral cavity and other sites evaluated by tissue microarray technology. *Acta Odontol. Latinoam.* 15: 29-37.
Lorenzato A *et al.* (2002) Novel somatic mutations of the *MET* oncogene in human carcinoma metastases activating cell motility and invasion. *Cancer Res.* 62: 7025-7030.
Lorimer IA, Lavictoire SJ (2001) Activation of extracellular-regulated kinases by normal and mutant EGF receptors. *Biochim. Biophys. Acta* 1538: 1-9.
Loriot MA *et al.* (2001) Genetic polymorphisms of cytochrome P450 2A6 in a case-control study on lung cancer in a French population. *Pharmacogenetics* 11: 39-44.
Lu S-L *et al.* (1995) Mutations of the transforming growth factor-beta type II receptor gene and genomic instability in hereditary nonpolyposis colorectal cancer. *Biochem. Biophys. Res. Commun.* 216: 452-457.
Lu YJ *et al.* (1996a) 2q-, a non-random chromosomal abnormality in human non-small-cell lung cancer. *Carcinogenesis* 17: 1589-1593.
Lu Y *et al.* (1996b) C-myc activation by translocation in human primary non-small cell lung cancer. *Zhonghua Yi Xue Za Zhi* 76: 275-277.
Lu R *et al.* (1997) A mammalian DNA repair enzyme that excises oxidatively damaged guanines maps to a locus frequently lost in lung cancer. *Curr. Biol.* 7: 397-407.
Lu YJ *et al.* (1999) Chromosome 3 imbalances are the most frequent aberration found in non-small cell lung carcinoma. *Lung Cancer* 23: 61-66.
Lu W *et al.* (2002) Genetic polymorphism in myeloperoxidase but not GSTM1 is associated with risk of lung squamous cell carcinoma in a Chinese population. *Int. J. Cancer* 102: 275-279.
Lui WO *et al.* (2001) High level amplification of 1p32-33 and 2p22-24 in small cell lung carcinomas. *Int. J. Oncol.* 19: 451-457.
Luk C *et al.* (2001) Molecular cytogenetic analysis of non-small cell lung carcinoma by spectral karyotyping and comparative genomic hybridisation. *Cancer Genet. Cytogenet.* 125: 87-99.
Lukeis R *et al.* (1990) Cytogenetics of non-small cell lung cancer: analysis of consistent non-random abnormalities. *Genes Chrom. Cancer* 2: 116-124.
Lukeis R *et al.* (1993) Chromosome abnormalities in non-small cell lung cancer pleural effusions: cytogenetic indicators of disease subgroups. *Genes Chrom. Cancer* 8: 262-269.
Lunn RM *et al.* (2000) *XPD* polymorphisms: effects on DNA repair proficiency. *Carcinogenesis* 21: 551-555.
Luo Y *et al.* (2001) Blocking Chk1 expression induces apoptosis and abrogates the G2 checkpoint mechanism. *Neoplasia* 3: 411-419.

Luo J *et al.* (2003) Looking beyond morphology: cancer gene expression profiling using DNA microarrays. *Cancer Invest.* 21: 937-949.
Lynch CJ (1926) Studies on the relation between tumor susceptibility and heredity. III. Spontaneous tumors of the lung in mice. *J. Exp. Med.* 43: 339-355.
Lynch TJ *et al.* (2004) Activating mutations in the epidermal growth factor receptor underlying responsiveness of non-small-cell lung cancer to Gefitinib. *New Engl. J. Med.* 350: 2129-2139.
McConnell BB, Vertino PM (2000) Activation of a caspase-9-mediated apoptotic pathway by subcellular redistribution of the novel caspase recruitment domain protein *TMS1*. *Cancer Res.* 60: 6243-6247.
McCoy MS *et al.* (1983) Characterization of a human colon/lung carcinoma oncogene. *Nature* 302: 79-81.
McDonald JW, Pilgram TK (1999) Nuclear expression of p53, p21 and cyclin D1 is increased in bronchioloalveolar carcinoma. *Histopathology* 34: 439-446.
McDonald ER, El-Deiry W (2001) Checkpoint genes in cancer. *Ann. Med.* 33: 113-122.
McDoniels-Silvers AL *et al.* (2002) Differential expression of critical cellular genes in human lung adenocarcinomas and squamous cell carcinomas in comparison to normal lung tissues. *Neoplasia* 4: 141-150.
McDuffie HH (1991) Clustering of cancer in families of patients with primary lung cancer. *J. Clin. Epidemiol.* 44: 69-76.
McLemore TL *et al.* (1990) Expression of *CYP1A1* in patients with lung cancer: evidence for cigarette smoke-induced gene expression in normal lung tissue and for altered gene regulation in primary pulmonary carcinomas. *J. Natl. Cancer Inst.* 82: 1333-1339.
McManus MT (2003) MicroRNAs and cancer. *Semin. Cancer Biol.* 13: 253-258.
McNally WD (1932) The tar in cigarette smoke and its possible effects. *Am. J. Cancer* 16: 1502-1514.
McWilliams JE *et al.* (1995) Glutathione *S*-transferase M1 (GSTM1) deficiency and lung cancer risk. *Cancer Epidemiol. Biomarkers Prev.* 4: 589-594.
Ma Q *et al.* (2001) Induction of CYP1A1. The AhR/DRE paradigm: transcription, receptor regulation, and expanding biological roles. *Curr. Drug Metab.* 2: 149-164.
Ma PC *et al.* (2003) c-MET mutational analysis in small cell lung cancer: novel juxtamembrane domain mutations regulating cytoskeletal functions. *Cancer Res.* 63: 6272-6281.
Macaluso M (2003) Genetic and epigenetic alterations as hallmarks of the intricate road to cancer. *Oncogene* 22: 6472-6478.
Macgregor PF (2003) Gene expression in cancer: the application of microarrays. *Expert Rev. Mol. Diagn.* 3: 185-200.
Mackintosh C (2004) Dynamic interactions between 14-3-3 proteins and phosphoproteins regulate diverse cellular processes. *Biochem. J.* 381: 329-342.
Maehama T, Dixon JE (1999) PTEN: a tumour suppressor that functions as a phospholipid phosphatase. *Trends Cell Biol.* 9: 125-128.
Magewu AN, Jones PA (1994) Ubiquitous and tenacious methylation of the CpG site in codon 248 of the *p53* gene may explain its frequent appearance as a mutational hot spot in human cancer. *Mol. Cell. Biol.* 14: 4225-4232.
Mai S, Mushinski JF (2003) c-Myc-induced genomic instability. *J. Environ. Pathol. Toxicol. Oncol.* 22: 179-199.
Mai H *et al.* (2003) A functional role for nicotine in Bcl2 phosphorylation and suppression of apoptosis. *J. Biol. Chem.* 278: 1886-1891.
Mäkelä TP *et al.* (1992) Amplification and rearrangement of L-*myc* in human small-cell lung cancer. *Mutation Res.* 276: 307-315.
Mäkelä TP *et al.* (1995) The rearranged L-myc fusion gene (*RLF*) encodes an Zn-15 related zinc finger protein. *Oncogene* 11: 2699-2704.
Makos M *et al.* (1992) Distinct hypermethylation patterns occur at altered chromosome loci in human lung and colon cancer. *Proc. Natl. Acad. Sci. USA* 89: 1929-1933.
Maley CC *et al.* (2004) Selectively advantageous mutations and hitchhikers in neoplasms. *Cancer Res.* 64: 3414-3427.
Malkinson AM (1998) Molecular comparison of human and mouse pulmonary adenocarcinomas. *Exp. Lung Res.* 24: 541-555.
Malpel S *et al.* (2000) Regulation of retinoic acid signalling during lung morphogenesis. *Development* 127: 3057-3067.
Malumbres M, Barbacid M (2001) To cycle or not to cycle: a critical decision in cancer. *Nature Reviews Cancer* 1: 222-231.
Manda R *et al.* (1999) Identification of genes (SPON2 and C20orf2) differentially expressed between cancerous and noncancerous lung cells by mRNA differential display. *Genomics* 61: 5-14.

Manda R *et al.* (2000) Differential expression of the *LAMB3* and *LAMC2* genes between small cell and non-small cell lung carcinomas. *Biochem. Biophys. Res. Commun.* 275: 440-445.

Maniwa Y *et al.* (2001) Association of p53 gene mutation and telomerase activity in resectable non-small cell lung cancer. *Chest* 120: 589-594.

Manning G *et al.* (2002) The protein kinase complement of the human genome. *Science* 298: 1912-1934.

Mao L *et al.* (1994) Detection of oncogene mutations in sputum precedes diagnosis of lung cancer. *Cancer Res.* 54: 1634-1637.

Mao L *et al.* (1997) Clonal genetic alterations in the lungs of current and former smokers. *J. Natl. Cancer Inst.* 89: 857-862.

Mao L (2002) Recent advances in the molecular diagnosis of lung cancer. *Oncogene* 21: 6960-6969.

Mao X *et al.* (2003) The cytoplasmic domain is critical to the tumour suppressor activity of TSLC1 in non-small cell lung cancer. *Cancer Res.* 63: 7979-7985.

Mao X *et al.* (2004) Re-expression of *TSLC1* in a non-small-cell lung cancer cell line induces apoptosis and inhibits tumor growth. *Oncogene* 23: 5632-5642.

Marchetti A *et al.* (1995) *mdm2* gene amplification and overexpression in non-small cell lung carcinomas with accumulation of the p53 protein in the absence of *p53* gene mutations. *Diag. Molec. Pathol.* 4: 93-97.

Marchetti A *et al.* (1996) p21 RNA and protein expression in non-small cell lung carcinomas: evidence of p53-independent expression and association with tumoral differentiation. *Oncogene* 12: 1319-1324.

Marchetti A *et al.* (1997) Alterations of p16 (MTS1) in node-positive non-small cell lung carcinomas. *J. Pathol.* 181: 178-182.

Marchetti A *et al.* (1998a) Cyclin D1 and retinoblastoma susceptibility gene alterations in non-small cell lung cancer. *Int. J. Cancer* 75: 187-192.

Marchetti A *et al.* (1998b) Genetic analysis of lung tumours of non-smoking subjects: p53 gene mutations are constantly associated with loss of heterozygosity at the *FHIT* locus. *Br. J. Cancer* 78: 73-78.

Marchetti A *et al.* (1999) Telomerase activity as a prognostic indicator in stage I non-small cell lung cancer. *Clin. Cancer Res.* 5: 2077-2081.

Marchevsky AM *et al.* (2004) Classification of individual lung cancer cell lines based on DNA methylation markers: use of linear discriminant analysis and artificial neural networks. *J. Mol. Diag.* 6: 28-36.

Marcy TW *et al.* (2002) Genetic testing for lung cancer risk: if physicians can do it, should they? *J. Gen. Intern. Med.* 17: 946-951.

Marchetti A *et al.* (2001) Reduced expression of INT-6/eIF3-p48 in human tumors. *Int. J. Oncol.* 18: 175-179.

Mariatos G *et al.* (2003) Inactivating mutations targeting the *chfr* mitotic checkpoint gene in human lung cancer. *Cancer Res.* 63: 7185-7189.

Marignani PA *et al.* (2001) LKB1 associates with Brg1 and is necessary for Brg1-induced growth arrest. *J. Biol. Chem.* 276: 32415-32418.

Markowitz S *et al.* (1995) Inactivation of the type II TGF-beta receptor in colon cancer cells with microsatellite instability. *Science* 268: 1336-1339.

Marsin S *et al.* (2003) Role of XRCC1 in the coordination and stimulation of oxidative damage repair initiated by the DNA glycosylase hOGG1. *J. Biol. Chem.* 278: 44068-44074.

Marsit CJ *et al.* (2004) Inactivation of the Fanconi anemia/BRCA pathway in lung and oral cancers: implications for treatment and survival. *Oncogene* 23: 1000-1004.

Martin ACR *et al.* (2002) Integrating mutation data and structural analysis of the *TP53* tumor-suppressor protein. *Hum. Mutation* 19: 149-164.

Martin B *et al.* (2003) Role of Bcl-2 as a prognostic factor for survival in lung cancer: a systematic review of the literature with meta-analysis. *Br. J. Cancer* 89: 55-64.

Martinet N *et al.* (2000) Retinoic acid receptor and retinoid X receptor alterations in lung cancer precursor lesions. *Cancer Res.* 60: 2869-2875.

Martinez C *et al.* (1995) Lung cancer and mutations at the polymorphic *NAT2* gene locus. *Pharmacogenetics* 5: 207-214.

Maruyama R *et al.* (2004) Hypermethylation of *FHIT* as a prognostic marker in nonsmall cell lung cancer. *Cancer* 100: 1472-1477.

Marwick JA *et al.* (2002) Cigarette smoke-induced oxidative stress and TGF-beta1 increase p21waf1/cip1 expression in alveolar epithelial cells. *Ann. NY Acad. Sci.* 973: 278-283.

Masayesva BG *et al.* (2004) Gene expression alterations over large chromosomal regions in cancers include multiple genes unrelated to malignant progression. *Proc. Natl. Acad. Sci. USA* 101: 8715-8720.

Mascaux C *et al.* (2003) Fragile histidine triad protein expression in nonsmall cell lung cancer and correlation with Ki-67 and with p53. *Eur. Respir. J.* 21: 753-758.
Maser RS, DePinho RA (2002) Connecting chromosomes, crisis, and cancer. *Science* 297: 565-569.
Massion PP *et al.* (2002) Genomic copy number analysis of non-small cell lung cancer using array comparative genomic hybridisation: implications of the phosphatidylinositol 3-kinase pathway. *Cancer Res.* 62: 3636-3640.
Massion PP *et al.* (2003) Significance of p63 amplification and overexpression in lung cancer development and prognosis. *Cancer Res.* 63: 7113-7121.
Masters SC *et al.* (2002) Survival-promoting functions of 14-3-3 proteins. *Biochem. Soc. Trans.* 30: 360-365.
Masuya D *et al.* (2001) The intratumoral expression of vascular endothelial growth factor and interleukin 8-associated with angiogenesis in nonsmall cell lung carcinoma patients. *Cancer* 92: 2628-2638.
Matakidou A *et al.* (2003) *TP53* polymorphisms and lung cancer risk: a systematic review and meta-analysis. *Mutagenesis* 18: 377-385.
Mate J *et al.* (1996) Cyclin D1 overexpression in non-small cell lung carcinoma: correlation with Ki67 labelling index and poor cytoplasmic differentiation. *J. Pathol.* 180: 395-399.
Máthé E (2004) RASSF1A, the new guardian of mitosis. *Nature Genet.* 36: 117-118.
Matsuda A, Takahashi T (2002) Chromosome instability in human lung cancers: possible underlying mechanisms and potential consequences in the pathogenesis. *Oncogene* 21: 6884-6897.
Matsuki Y *et al.* (2003) Histidine decarboxylase expression as a new sensitive and specific marker for small cell lung carcinoma. *Mod. Pathol.* 16: 72-78.
Matsuoka S *et al.* (2001) Reduced expression and impaired kinase activity of a Chk2 mutant identified in human lung cancer. *Cancer Res.* 61: 5362-5365.
Matsushima Y *et al.* (1994) Gastrin receptor gene expression in several human carcinomas. *Jpn. J. Cancer Res.* 85: 819-824.
Matsuyama W *et al.* (2003) Mitochondrial DNA mutation correlates with stage progression and prognosis in non-small cell lung cancer. *Hum. Mutation* 21: 441-443.
Matsuzoe D *et al.* (1999a) Discrimination of double primary lung cancer from intrapulmonary metastasis by p53 gene mutation. *Br. J. Cancer* 79: 1549-1552.
Matsuzoe D *et al.* (1999b) p53 mutations predict non-small cell lung carcinoma response to radiotherapy. *Cancer Letts.* 135: 189-194.
Mattern J *et al.* (1996) Association of vascular endothelial growth factor expression with intratumoral microvessel density and tumor cell proliferation in human epidermoid lung carcinoma. *Br. J. Cancer* 73: 931-934.
Matullo G *et al.* (2001) *XRCC1, XRCC3, XPD* gene polymorphisms, smoking and $^{(32)}$P-DNA adducts in a sample of healthy subjects. *Carcinogenesis* 22: 1437-1445.
Matullo G *et al.* (2003) Combination of DNA repair gene single nucleotide polymorphisms and increased levels of DNA adducts in a population-based study. *Cancer Epidemiol. Biomarkers Prev.* 12: 674-677.
Matzke MA *et al.* (2003) Does the intrinsic instability of aneuploidy genomes have a causal role in cancer? Trends Genet. 19: 253-256.
Maya R *et al.* (2001) ATM-dependent phosphorylation of Mdm2 on serine 395: role in p53 activation by DNA damage. *Genes Dev.* 15: 1067-1077.
Mazieres J *et al.* (2004) Loss of RhoB expression in human lung cancer progression. *Clin. Cancer Res.* 10: 2742-2750.
Mead LJ *et al.* (1997) Frequent loss of heterozygosity in early non-small cell lung cancers at chromsome 9p21 proximal to the *CDKN2A* gene. *Int. J. Cancer* 71: 213-217.
Medina PP *et al.* (2003) Screening of homologous recombination gene polymorphisms in lung cancer patients reveals an association of the *NBS1*-185Gln variant and p53 mutations. *Cancer Epidemiol. Biomarkers Prev.* 12: 699-704.
Meert AP *et al.* (2003) The role of HER-2/neu expression on the survival of patients with lung cancer: a systematic review of the literature. *Br. J. Cancer* 89: 959-965.
Mendelson CR (2000) Role of transcription factors in fetal lung development and surfactant protein gene expression. *Annu. Rev. Physiol.* 62: 875-915.
Mendes-da-Silva P *et al.* (2000) Frequent loss of heterozygosity on chromosome 5 in non-small cell lung carcinoma. *Mol. Pathol.* 53: 184-187.
Mendoza C *et al.* (2000) Allelotype and loss of heterozygosity around the L-*myc* gene locus in primary lung cancers. *Lung Cancer* 28: 117-125.
Menez J *et al.* (2004) Mutant α-actinin-4 promotes tumorigenicity and regulates cell motility of a human lung carcinoma. *Oncogene* 23: 2630-2639.

Mercer KE, Pritchard CA (2003) Raf proteins and cancer: B-Raf is identified as a mutational target. *Biochim. Biophys. Acta* 1653: 25-40.

Merlo A *et al.* (1994) Frequent microsatellite instability in primary small cell lung cancer. *Cancer Res.* 54: 2098-2101.

Merlo A *et al.* (1995) 5' CpG island methylation is associated with transcriptional silencing of the tumor suppressor *p16/CDKN2/MTS1* in human cancers. *Nature Medicine* 1: 686-692.

Merrick BA *et al.* (1996) HSP binding and mitochondrial localization of p53 protein in human HT1080 and mouse C3H10T1/2 cell lines. *Biochim. Biophys. Acta* 1297: 57-68.

Metzger RJ, Krasnow MA (1999) Genetic control of branching morphogenesis. *Science* 284: 1635-1639.

Meuwissen R *et al.* (2001) Mouse model for lung tumorigenesis through Cre/*lox* controlled sporadic activation of the K-Ras oncogene. *Oncogene* 20: 6551-6558.

Meuwissen R *et al.* (2003) Induction of small cell lung cancer by somatic inactivation of both Trp53 and Rb1 in a conditional mouse model. *Cancer Cell* 4: 181-189.

Meyerson M *et al.* (2004) Molecular classification and molecular genetics of human lung cancers. *Semin. Oncol.* 31 (Suppl. 1): 4-19.

Michelland S et al. (1999) Comparison of chromosomal imbalances in neuroendocrine and non-small-cell lung carcinomas. *Cancer Genet. Cytogenet.* 114: 22-30.

Michor F *et al.* (2004) Dynamics of cancer progression. *Nature Rev. Cancer* 4: 197-205.

Micke P *et al.* (2003) Characterization of c-kit expression in small cell lung cancer: prognostic and therapeutic implications. *Clin. Cancer Res.* 9: 188-194.

Mikami I *et al.* (2001) Down-regulation in multiple human cancers of a novel gene, DMHC, from 17q25.1 that encodes an integral membrane protein. *Jpn. J. Cancer Res.* 92: 417-422.

Miles CP (1967) Chromosome analysis of solid tumors: II. Twenty six nonepithelial tumors. *Cancer* 20: 1274-1287.

Miller JA, Jones OR (1930) Primary carcinoma of the lung. *Am. Rev. Tuberc.* 21: 1-56.

Miller CW *et al.* (1992) p53 mutations in human lung tumors. *Cancer Res.* 52: 1695-1698.

Miller CW *et al.* (2002a) Mutations of the *CHK2* gene are found in some osteosarcomas, but are rare in breast, lung, and ovarian tumors. *Genes Chrom. Cancer* 33: 17-21.

Miller DP *et al.* (2002b) Combinations of the variant genotypes of *GSTP1*, *GSTM1*, and *p53* are associated with an increased lung cancer risk. *Cancer Res.* 62: 2819-2823.

Miller LD *et al.* (2002c) Optimal gene expression analysis by microarrays. *Cancer Cell* 2: 353-361.

Miller CT *et al.* (2003a) Amplification and overexpression of the dual-specificity tyrosine-(Y)-phosphorylation regulated kinase 2 (*DYRK2*) gene in esophageal and lung adenocarcinomas. *Cancer Res.* 63: 4136-4143.

Miller CT *et al.* (2003b) Increased *C-CRK* proto-oncogene expression is associated with an aggressive phenotype in lung adenocarcinomas. *Oncogene* 22: 6006-6013.

Miller DP *et al.* (2003c) Smoking and the risk of lung cancer: susceptibility with *GSTP1* polymorphisms. *Epidemiology* 14: 545-551.

Miller DP *et al.* (2003d) Association between self-reported environmental tobacco smoke exposure and lung cancer: modification by *GSTP1* polymorphism. *Int. J. Cancer* 104: 758-763.

Miller AB *et al.* (2004) Fruits and vegetables and lung cancer: findings from the European prospective investigation into cancer and nutrition. *Int. J. Cancer* 108: 269-276.

Mills NE *et al.* (1995a) Increased prevalence of K-*ras* oncogene mutations in lung adenocarcinoma. *Cancer Res.* 55: 1444-1447.

Mills NE *et al.* (1995b) Detection of K-ras oncogene mutations in bronchoalveolar lavage fluid for lung cancer diagnosis. *J. Natl. Cancer Inst.* 87: 1056-1060.

Minna JD (2003) Nicotine exposure and bronchial epithelial cell nicotinic acetylcholine receptor expression in the pathogenesis of lung cancer. *J. Clin. Invest.* 111: 31-33.

Minoo P (2000) Transcriptional regulation of lung development: emergence of specificity. *Respir. Res.* 1: 109-115.

Minowa O *et al.* (2000) Mmh1/Ogg1 gene inactivation results in accumulation of 8-hydroxyguanine in mice. *Proc. Natl. Acad. Sci. USA* 97: 4156-4161.

Miozzo M *et al.* (1996) Microsatellite alterations in bronchial and sputum specimens of lung cancer patients. *Cancer Res.* 56: 2285-2288.

Mirski SE *et al.* (2000) A truncated cytoplasmic topoisomerase II alpha in a drug-resistant lung cancer cell line is encoded by a *TOP2A* allele with a partial deletion of exon 34. *Int. J. Cancer* 85: 534-539.

Mirza A *et al.* (2002) Human survivin is negatively regulated by wild-type p53 and participates in p53-dependent apoptotic pathway. *Oncogene* 21: 2613-2622.

Mirza A *et al.* (2003) Global transcriptional program of p53 target genes during the process of apoptosis and cell cycle regulation. *Oncogene* 22: 3645-3654.

Mistry P, Herbert KE (2003) Modulation of hOGG1 DNA repair enzyme in human cultured cells in response to pro-oxidant and antioxidant challenge. *Free Radic Biol. Med.* 35: 397-405.

Mitani S *et al.* (2001) Analysis of c-myc DNA amplification in non-small cell lung carcinoma in comparison with small cell lung carcinoma using polymerase chain reaction. *Clin. Exp. Med.* 1: 105-111.

Mitas M *et al.* (2003) Lunx is a superior molecular marker for detection of non-small lung cell cancer in peripheral blood. *J. Mol. Diagn.* 5: 237-242.

Mitelman F *et al.* (1997) A breakpoint map of recurrent chromosomal rearrangements in human neoplasia. *Nature Genet.* Special issue (April 1997): 417-430.

Mitelman F (2000) Recurrent chromosome aberrations in cancer. *Mutation Res.* 462: 247-253.

Mitsudomi T *et al.* (1991a) Mutations in *ras* genes distinguish a subset of non-small-cell lung cancer cell lines from small-cell lung cancer cell lines. *Oncogene* 6: 1353-1362.

Mitsudomi T *et al.* (1991b) *ras* gene mutations in non-small cell lung cancers are associated with shortened survival irrespective of treatment intent. *Cancer Res.* 51: 4999-5002.

Mitsudomi T *et al.* (1992) *p53* gene mutations in non-small-cell lung cancer cell lines and their correlation with the presence of *ras* mutations and clinical features. *Oncogene* 7: 171-180.

Mitsudomi T *et al.* (1994) Mutations of the p53 gene as a predictor of poor prognosis in patients with non-small-cell lung cancer. *J. Natl. Cancer Inst.* 85: 2018-2023.

Mitsudomi T *et al.* (1996) Loss of heterozygosity at 3p in non-small cell lung cancer and its prognostic implication. *Clin. Cancer Res.* 2: 1185-1189.

Mitsudomi T *et al.* (2000) Prognostic significance of p53 alterations in patients with non-small cell lung cancer: a meta-analysis. *Clin. Cancer Res.* 6: 4055-4063.

Mitsuuchi Y, Testa JR (2002) Cytogenetics and molecular genetics of lung cancer. *Am. J. Med. Genet.* 115: 183-188.

Miura I *et al.* (1992) Chromosome alterations in small cell lung cancer: frequent involvement of 5q. *Cancer Res.* 52: 1322-1328.

Miura K *et al.* (2002) Laser capture microdissection and microarray expression analysis of lung adenocarcinoma reveals tobacco smoking- and prognosis-related molecular profiles. *Cancer Res.* 62: 3244-3250.

Miyaki M, Kuroki T (2003) Role of Smad4 (DPC4) inactivation in human cancer. *Biochem. Biophys. Res. Commun.* 306: 799-804.

Miyaki M *et al.* (1985) Amplification and enhanced expression of cellular oncogene c-Ki-ras-2 in a human epidermoid carcinoma of the lung. *Jpn. J. Cancer Res.* 76: 260-265.

Miyamoto M *et al.* (1999) *CYP2A6* gene deletion reduces susceptibility to lung cancer. *Biochem. Biophys. Res. Commun.* 261: 658-660.

Miyamoto K *et al.* (2003a) Methylation-associated silencing of heparan sulfate D-glucosaminyl 3-O-sulfotransferase-2 (3-OST-2) in human breast, colon, lung and pancreatic cancers. *Oncogene* 22: 274-280.

Miyamoto Y *et al.* (2003b) Oxidative stress caused by inactivation of glutathione peroxidase and adaptive responses. *Biol. Chem.* 384: 567-574.

Miyasaki YJ *et al.* (2002) HOXD3 enhances motility and invasiveness through the TGF-beta-dependent and -independent pathways in A549 cells. *Oncogene* 21: 798-808.

Miyoshi T *et al.* (1997) The expression of the CD44 variant exon 6 is associated with lymph node metastasis in non-small cell lung cancer. *Clin. Cancer Res.* 3: 1289-1297.

Mizuno K *et al.* (2002) Aberrant hypermethylation of the *CHFR* prophase checkpoint gene in human lung cancers. *Oncogene* 21: 2328-2333.

Mohr LC *et al.* (2003) Glutathione *S*-transferase M1 polymorphism and the risk of lung cancer. *Anticancer Res.* 23: 2111-2124.

Mohrenweiser HW *et al.* (2002) Identification of 127 amino acid substitution variants in screening 37 DNA repair genes in humans. *Cancer Epidemiol. Biomarkers Prev.* 11: 1054-1064.

Molinari M (2000) Cell cycle checkpoints and their inactivation in human cancer. *Cell Prolif.* 33: 261-274.

Mollenhauer J *et al.* (2002) Sequential changes of the *DMBT1* expression and location in normal lung tissue and lung carcinomas. *Gene Chrom. Cancer* 35: 164-169.

Mollerup S *et al.* (1999) Sex differences in lung *CYP1A1* expression and DNA adduct levels among lung cancer patients. *Cancer Res.* 59: 3317-3320.

Momand J *et al.* (1998) The *MDM2* gene amplification database. *Nucleic Acids Res.* 26: 3453-3459.

Monzo M *et al.* (1999a) A novel anti-apoptosis gene: re-expression of survivin messenger RNA as a prognosis marker in non-small-cell lung cancer. *J. Clin. Oncol.* 17: 2100-2104.

Monzo M *et al.* (1999b) Paclitaxel resistance in non-small-cell lung cancer associated with beta-tubulin gene mutations. *J. Clin. Oncol.* 17: 1786-1793.

Moody TW *et al.* (2003) Neuropeptides as autocrine growth factors in cancer cells. *Curr. Pharm. Des.* 9: 495-509.

Mooibroek H *et al.* (1987) Loss of heterozygosity for a chromosome 3 sequence presumably at 3p21 in small cell lung cancer. *Cancer Genet. Cytogenet.* 27: 361-365.

Moolenaar CE *et al.* (1992) Alternative splicing of neural-cell-adhesion molecule mRNA in human small-cell lung cancer cell line H69. *Int. J. Cancer* 51: 238-243.

Mooney LA *et al.* (1997) Contribution of genetic and nutritional factors to DNA damage in heavy smokers. *Carcinogenesis* 18: 503-509.

Moore JH (2003) The ubiquitous nature of epistasis in determining susceptibility to common human diseases. *Hum. Hered.* 56: 73-82.

Moorhead RA *et al.* (2003) Transgenic overexpression of IGF-II induces spontaneous lung tumors: a model for human lung adenocarcinoma. *Oncogene* 22: 853-857.

Morgagni GB (1761) *De Sedibus et Causis Morborum per Anatomen Indagatis.* Remondiniana, Helvetia.

Mori N *et al.* (1990) Variable mutations of the *RB* gene in small-cell lung carcinoma. *Oncogene* 5: 1713-1717.

Mori M *et al.* (2001) Atypical adenomatous hyperplasia of the lung: a probable forerunner in the development of adenocarcinoma of the lung. *Mod. Pathol.* 14: 72-84.

Mori S *et al.* (2004) p53 apoptotic pathway molecules are frequently and simultaneously altered in nonsmall cell lung carcinoma. *Cancer* 100: 1673-1682.

Morimoto E *et al.* (2001) Adenovirus-mediated suicide gene transfer to small cell lung carcinoma using a tumor-specific marker. *Anticancer Res.* 21(1A):329-331.

Moriyama S *et al.* (2002) Ratio of expression of p16(INK4a) to p14(ARF) correlates with the progression of non-small cell lung cancer. *Jpn. J. Cancer Res.* 93: 783-788.

Morkve O *et al.* (1993) Prognostic significance of p53 protein expression and DNA ploidy in surgically treated non-small cell lung carcinomas. *Anticancer Res.* 13: 571-578.

Morris SM (2002) A role for p53 in the frequency and mechanism of mutation. *Mutation Res.* 511: 45-62.

Morstyn G *et al.* (1987) Heterogeneous cytogenetic abnormalities in small cell lung cancer cell lines. *Cancer Res.* 47: 3322-3327.

Moscatello DK *et al.* (1998) Constitutive activation of phosphatidylinositol 3-kinase by a naturally occurring mutant epidermal growth factor receptor. *J. Biol. Chem.* 273: 200-206.

Moschos SJ, Mantzoros CS (2002) The role of the IGF system in cancer: from basic to clinical studies and clinical applications. *Oncology* 63: 317-332.

Moses RE (2001) DNA damage processing defects and disease. *Annu. Rev. Genomics Hum. Genet.* 2: 41-68.

Motoyama J *et al.* (1998) Essential function of Gli2 and Gli3 in the formation of lung, trachea and oesophagus. *Nature Genet.* 20: 54-57.

Mountain CF (2000) International staging system for lung cancer. Chap. 32 in *Lung Cancer. Principles and Practice.* 2nd Ed. HI Pass, JB Mitchell, DH Johnson, AT Turrisi, JD Minna (Eds.). Lippincott Williams & Williams, Philadelphia, pp 591-601.

Mucenski ML *et al.* (2003) β-Catenin is required for specification of proximal/distal cell fate during lung morphogenesis. *J. Biol. Chem.* 278: 40231-40238.

Mukhopadhyay T *et al.* (1995) Altered expression of the p50 subunit of the NF-kappa B transcription factor complex in non-small cell lung carcinoma. *Oncogene* 11: 999-1003.

Mukohara T *et al.* (2003) Expression of epidermal growth factor receptor (EGFR) and downstream-activated peptides in surgically excised non-small-cell lung cancer (NSCLC). *Lung Cancer* 41: 123-130.

Müller FH (1939) Tabakmißbrauch und Lungencarcinom. *Z. Krebsforsch.* 49: 57-85.

Muller D *et al.* (1991) Expression of collagenase-related metalloproteinase genes in human lung or head and neck tumours. *Int. J. Cancer* 48: 550-556.

Müller-Tidow C *et al.* (2001) Cyclin E is the only cyclin-dependent kinase 2-associated cyclin that predicts metastasis and survival in early stage non-small cell lung cancer. *Cancer Res.* 61: 647-653.

Mulshine JL (1999) Reducing lung cancer risk: early detection. *Chest* 116: 493S-496S.

Murakami Y *et al.* (1991) Inactivation of the retinoblastoma gene in a human lung carcinoma cell line detected by single-strand conformation polymorphism analysis of the polymerase chain reaction product of cDNA. *Oncogene* 6: 37-42.

Murakami Y (2002) Functional cloning of a tumor suppressor gene, *TSLC1*, in human non-small cell lung cancer. *Oncogene* 21: 6936-6948.

Murata M *et al.* (1996) Analysis of a germ line polymorphism of the p53 gene in lung cancer patients; discrete results with smoking history. *Carcinogenesis* 17: 261-264.

Murata M *et al.* (1998) Correlation of the mutation of p53 gene and the polymorphism at codon 72 in smoking-related non-small cell lung cancer patients. *Int. J. Oncol.* 12: 577-581.
Muratovska A *et al.* (2003) Paired box genes are frequently expressed in cancer and often required for cancer cell survival. *Oncogene* 22: 7989-7997.
Nacht M *et al.* (2001) Molecular characteristics of non-small cell lung cancer. *Proc. Natl. Acad. Sci. USA* 98: 15203-15208.
Nadon R, Shoemaker J (2002) Statistical issues with microarrays: processing and analysis. *Trends Genet.* 18: 265-271.
Naeem M *et al.* (2002) Analysis of c-*kit* protein expression in small-cell lung carcinoma and its implication for prognosis. *Hum. Pathol.* 33: 1182-1187.
Nafstad P *et al.* (2003) Lung cancer and air pollution: a 27 year follow up of 16209 Norwegian men. *Thorax* 58: 1071-1076.
Nagatake M *et al.* (1996a) Somatic *in vivo* alterations of the *DC4* gene at 18q21 in human lung cancers. *Cancer Res.* 56: 2718-2720.
Nagatake M *et al.* (1996b) Aberrant hypermethylation at the bcl-2 locus at 18q21 in human lung cancers. *Cancer Res.* 56: 1886-1891.
Nair U, Bartsch H (2001) Metabolic polymorphisms as susceptibility markers for lung and oral cavity cancer. *IARC Scientific Publications* 154: 271-290.
Naka M *et al.* (2001) Functional characterization of naturally occurring mutants (P405R and P425L) of p73α and p73β found in neuroblastoma and lung cancer. *Oncogene* 20: 3568-3572.
Nakachi K *et al.* (1993) Polymorphism of the *CYP1A1* and glutathione *S*-transferase genes associated with susceptibility to lung cancer in relation to cigarette dose in a Japanese population. *Cancer Res.* 53: 2994-2999.
Nakagawa H, Yagihashi S (1994) Expression of type IV collagen and its degrading enzymes in squamous cell carcinoma of lung. *Jpn. J. Cancer Res.* 85: 934-938.
Nakagawa K *et al.* (1995) Mechanism of inactivation of CDKN2 and MTS2 in non-small cell lung cancer and association with advanced stage. *Oncogene* 11: 1843-1851.
Nakagawa T *et al.* (2002) Prognostic value of thymidylate synthase expression in patients with p-stage I adenocarcinoma of the lung. *Lung Cancer* 35: 165-170.
Nakagawa T *et al.* (2004) Expression of thymidylate synthase is correlated with proliferative activity in non-small cell lung cancer (NSCLC). *Lung Cancer* 43: 145-149.
Nakagawachi T *et al.* (2003) Silencing effect of CpG island hypermethylation and histone modifications on O6-methylguanine-DNA methyltransferase (MGMT) gene expression in human cancer. *Oncogene* 22: 8835-8844.
Nakajima T *et al.* (1995) Expression and polymorphism of glutathione *S*-transferase in human lungs: risk factors in smoking-related lung cancer. *Carcinogenesis* 16: 707-711.
Nakajima E *et al.* (2001) *K-ras* mutation in sputum of primary lung cancer patients does not always reflect that of cancerous cells. *Int. J. Oncol.* 18: 105-110.
Nakamura M *et al.* (1997) Gene expression of granulocyte colony stimulating factor (G-CSF) in non-small cell lung cancer. *Anticancer Res.* 17(1B): 573-576.
Nakamura H *et al.* (2003a) Correlation between encoded protein overexpression and copy number of the *HER2* gene with survival in non-small cell lung cancer. *Int. J. Cancer* 103: 61-66.
Nakamura H *et al.* (2003b) Chromosomal instability detected by fluorescence *in situ* hybridisation in surgical specimens of non-small cell lung cancer is associated with poor survival. *Clin. Cancer Res.* 9: 2294-2299.
Nakamura H *et al.* (2003c) cDNA microarray analysis of gene expression in pathologic Stage IA nonsmall cell lung carcinomas. *Cancer* 97: 2798-2805.
Nakamura Y *et al.* (2004) Isolation of p53-target genes and their functional analysis. *Cancer Sci.* 95: 7-11.
Nakanishi Y *et al.* (1999) Expression of p53 and glutathione *S*-transferase-pi relates to clinical drug resistance in non-small cell lung cancer. *Oncology* 57: 318-323.
Nakanishi K *et al.* (2002) Expression of human telomerase RNA component and telomerase reverse transcriptase mRNA in atypical adenomatous hyperplasia of the lung. *Hum. Pathol.* 33: 697-702.
Nakano H *et al.* (1984) Isolation of transforming sequences of two human lung carcinomas: structural and functional analyses of the activated c-*K-ras* oncogenes. *Proc. Natl. Acad. Sci. USA* 81: 71-75.
Nakao M *et al.* (2001) Regulation of transcription and chromatin by methyl-CpG binding protein MBD1. *Brain Dev.* 23 Suppl. 1: S174-S176.
Nakashima T *et al.* (2003) Neural-cadherin expression associated with angiogenesis in non-small-cell lung cancer patients. *Br. J. Cancer* 88: 1727-1733.
Nakashima T *et al.* (2004) Expression of vascular endothelial growth factor-A and vascular endothelial growth factor-C as prognostic factors for non-small cell lung cancer. *Med. Sci. Monit.* 10: BR157-BR165.

Nakatani Y *et al.* (2002) Aberrant nuclear localization and gene mutation of β-catenin in low-grade adenocarcinoma of fetal lung type: up-regulation of the Wnt signalling pathway may be a common denominator for the development of tumors that form modules. *Mod. Pathol.* 15: 617-624.

Nambu Y *et al.* (1998) Unique expression patterns and alterations in the intestinal protein villin in primary and metastatic pulmonary adenocarcinomas. *Mol. Carcinogen.* 23: 234-242.

Nanki N *et al.* (2001) Expression of oncofetal fibronectin and syndecan-1 mRNA in 18 human lung cancer cell lines. *Tumour Biol.* 22: 390-396.

Naoki K *et al.* (2002) Missense mutations of the *BRAF* gene in human lung adenocarcinoma. *Cancer Res.* 62: 7001-7003.

Naruse I et al. (2000) Over-expression of p27kip1 growth arrest and apoptosis mediated by changes of pRb expression in lung cancer cell lines. *Int. J. Cancer* 88: 377-383.

Nau MM *et al.* (1985) L-*myc*, a new *myc*-related gene amplified and expressed in human small cell lung cancer. *Nature* 318: 69-73.

Nau MM *et al.* (1986) Human small-cell lung cancers show amplification and expression of the N-*myc* gene. *Proc. Natl. Acad. Sci. USA* 83: 1092-1096.

Nawrocki B *et al.* (1997) Expression of matrix metalloproteinases and their inhibitors in human bronchopulmonary carcinomas: quantitative and morphological analyses. *Int. J. Cancer* 72: 556-564.

Naylor SL *et al.* (1987) Loss of heterozygosity of chromsome 3p markers in small cell lung cancer. *Nature* 329: 451-454.

Nazar-Stewart V *et al.* (1993) The glutathione *S*-transferase mu polymorphism as a marker for susceptibility to lung carcinoma. *Cancer Res.* 53 (10 Suppl.): 2313-2318.

Nazar-Stewart V *et al.* (2003) A population-based study of glutathione *S*-transferase M1, T1 and P1 genotypes and risk for lung cancer. *Lung Cancer*: 247-258.

Nebert DW (2002) Transcription factors and cancer: an overview. *Toxicology* 181-182: 131-141.

Nelkin BD *et al.* (2001) Lung cancer. Chap. 58 in *The Metabolic & Molecular Bases of Inherited Disease.* 8th Ed. Eds. CR Scriver, AL Beaudet, WS Sly, D Valle. McGraw-Hill, New York. pp 1147-1155.

Nelson WJ, Nusse R (2004) Convergence of Wnt, beta-catenin, and cadherin pathways. *Science* 303: 1483-1487.

Nelson MA *et al.* (1996) Detection of K-*ras* gene mutations in non-neoplastic lung tissue and lung cancers. *Cancer Letts.* 103: 115-121.

Nelson HH *et al.* (1998) Chromosome 3p14 alterations in lung cancer: evidence that *FHIT* exon deletion is a target of tobacco carcinogens and asbestos. *Cancer Res.* 58: 1804-1807.

Nemunaitis J *et al.* (1998) Prognostic value of K-ras mutations, ras oncoprotein, and c-erb B oncoprotein expression in adenocarcinoma of the lung. *Am. J. Clin. Oncol.* 21: 155-160.

Nemunaitis J *et al.* (2000) Adenovirus-mediated p53 transfer in sequence with cisplatin to tumors of patients with non-small-cell lung cancer. *J. Clin. Oncol.* 18: 609-622.

Nephew KP, Huang TH (2003) Epigenetic gene silencing in cancer initiationand progression. *Cancer Lett.* 190: 125-133.

Neville EM *et al.* (1995) Loss of heterozygosity at 9p23 defines a novel locus in non-small cell lung cancer. *Oncogene* 11: 581-585.

Nevins JR (2001) The Rb/E2F pathway and cancer. *Hum. Mol. Genet.* 10: 699-703.

Newbold RF (2002) The significance of telomerase activation and cellular immortalization in human cancer. *Mutagenesis* 17: 539-550.

Ng CS *et al.* (2002) Tumor p16M is a possible marker of advanced stage in non-small cell lung cancer. *J. Surg. Oncol.* 79: 101-106.

Nguyen PL *et al.* (1996) Membrane-bound MUC1) and secretory (MUC2, MUC3, and MUC4) mucin) mucin gene expression in human lung cancer. *Tumour Biol.* 17: 176-192.

Nichols KE *et al.* (2001) Germ-line p53 mutations predispose to a wide spectrum of early-onset cancers. *Cancer Epidemiol. Biomarkers Prev.* 10: 83-87.

Nicholson SA *et al.* (2001) Alterations of *p14*^{ARF}, *p53*, and *p73* genes involved in the E2F-1-mediated apoptotic pathways in non-small cell lung carcinoma. *Cancer Res.* 61: 5636-5643.

Niehans GA *et al.* (1997) Human lung carcinomas express Fas ligand. *Cancer Res.* 57: 1007-1012.

Niklinska W *et al.* (2001a) Detection of p16 abnormalities in early stage non-small cell lung cancer. *Folia Histochem. Cytobiol.* 39 Suppl. 2: 30-32.

Niklinska W *et al.* (2001b) Expression of vascular endothelial growth factor (VEGF) in non-small cell lung cancer (NSCLC): association with p53 gene mutation and prognosis. *Lung Cancer* 34 Suppl. 2: S59-S64.

Niklinski J *et al.* (2001) Molecular genetic abnormalities in premalignant lung lesions: biological and clinical implications. *Eur. J. Cancer Prev.* 10: 213-226.

Nilsson JA, Cleveland JL (2003) Myc pathways provoking cell suicide and cancer. *Oncogene* 22: 9007-9021.

Nishio M *et al.* (1996) Prognostic significance of abnormal p53 accumulation in primary, resected non-small-cell lung cancers. *J. Clin. Oncol.* 14: 497-502.

Nishioka M *et al.* (2000) Identification of a 428-kb homozygously deleted region disrupting the *SEZ6L* gene at 22q12.1 in a lung cancer cell line. *Oncogene* 19: 6251-6260.

Nishioka M *et al.* (2002) *MYO18B*, a candidate tumor suppressor gene at chromosome 22q12.1, deleted, mutated, and methylated in human lung cancer. *Proc. Natl. Acad. Sci. USA* 99: 12269-12274.

Nishisaka T *et al.* (2000) Evaluation of p53 gene mutation and loss of heterozygosity of 3p, 9p and 17p in precancerous lesions of 29 lung cancer patients. *Hiroshima J. Med. Sci.* 49: 109-116.

Nishiumi N *et al.* (2003) Use of 11p15 mucins as prognostic factors in small adenocarcinoma of the lung. *Clin. Cancer Res.* 9: 5616-5619.

Noda N *et al.* (2001) K-ras gene mutations in non-small cell lung cancer in Japanese. *Oncol. Rep.* 8: 889-892.

Noguchi M *et al.* (1990) Heterogenous amplification of *myc* family oncogenes in small cell lung carcinoma. *Cancer* 66: 2053-2058.

Noguchi M *et al.* (1993) Biological consequences of overexpression of a transfected c-erbB-2 gene in immortalized human bronchial epithelial cells. *Cancer Res.* 53: 2035-2043.

Nomoto S *et al.* (1998) Search for mutations and examination of allelic expression imbalance of the p73 gene at 1p36.33 in human lung cancers. *Cancer Res.* 58: 1380-1383.

Nomoto S *et al.* (1999) Search for *in vivo* somatic mutations in the mitotic checkpoint gene, *hMAD1*, in human lung cancers. *Oncogene* 18: 7180-7183.

Norbury CJ, Zhivotovsky B (2004) DNA damage-induced apoptosis. *Oncogene* 23: 2797-2808.

Norgaard P *et al.* (1996) Expression and autoregulation of transforming growth factor beta receptor mRNA in small-cell lung cancer cell lines. *Br. J. Cancer* 73: 1037-1043.

Nowak MA *et al.* (2002) The role of chromosomal instability in tumor initiation. *Proc. Natl. Acad. Sci. USA* 99: 16226-16231.

Nozaki I *et al.* (2001) Reduced expression of REIC/Dkk-3 gene in non-small cell lung cancer. *Int. J. Oncol.* 19: 117-121.

Ntzani EE, Ioannidis JP (2003) Predictive ability of DNA microarrays for cancer outcomes and correlates: an empirical assessment. *Lancet* 362: 1439-1444.

Nyberg F *et al.* (1998) Glutathione *S*-transferase mu1 and *N*-acetyltransferase 2 genetic polymorphisms and exposure to tobacco smoke in non-smoking and smoking lung cancer patients and population controls. *Cancer Epidemiol. Biomarkers Prev.* 7: 875-883.

O'Byrne KJ *et al.* (2000) Vascular endothelial growth factor, platelet-derived endothelial cell growth factor and angiogenesis in non-small-cell lung cancer. *Br. J. Cancer* 82: 1427-1432.

O'Byrne KJ *et al.* (2003) Vascular endothelial growth factor expression in non-small cell lung cancer. Chap. 26 in *Methods in Molecular Medicine Vol. 74: Lung Cancer, Vol. 1: Molecular Pathology Methods and Reviews.* Ed. B. Driscoll. Humana Press, Totowa, New Jersey. pp 357-373.

O'Hagan RC *et al.* (2002) Telomere dysfunction provokes regional amplification and deletion in cancer genomes. *Cancer Cell* 2: 149-155.

O-Wang J *et al.* (2001) DNA polymerase kappa, implicated in spontaneous and DNA damage-induced mutagenesis, is overexpressed in lung cancer. *Cancer Res.* 61: 5366-5369.

Ochi K *et al.* (2002) Identification of semaphorin 3B as a direct target of p53. *Neoplasia* 4: 82-87.

Ochiai M *et al.* (1999) Cyclooxygenase-2 (*COX-2*) mRNA expression levels in normal lung tissues and non-small cell lung cancers. *Jpn. J. Cancer Res.* 90: 1338-1343.

Ochsner A, Debakey ME (1939) Primary pulmonary malignancy: Treatment by total pneumonectomy; analysis of 79 collected cases and presentation of 7 personal cases. *Surg. Gynecol. Obstet.* 48: 433-451.

Ochsner A, Debakey ME (1941) Carcinoma of the lung. *Arch. Surg.* 42: 209-258.

Ogawa J *et al.* (1996) Expression of α-1,3-fucosyltransferase type IV and VII genes in related to poor prognosis in lung cancer. *Cancer Res.* 56: 325-329.

Oguri T *et al.* (2000) Increased expression of the MRP5 gene is associated with exposure to platinum drugs in lung cancer. *Int. J. Cancer* 86: 95-100.

Oh JMC *et al.* (2001) A database of protein expression in lung cancer. *Proteomics* 1: 1303-1319.

Oh JJ *et al.* (2002) A candidate tumor suppressor gene, *H37*, from the human lung cancer tumor suppressor locus 3p21.3. *Cancer Res.* 62: 3207-3213.

Ohgaki H *et al.* (2004) APC mutations are infrequent but present in human lung cancer. *Cancer Lett.* 207: 197-203.

Ohira T *et al.* (2002) Up-regulated gene expression of angiogenesis factors in post-chemotherapeutic lung cancer tissues determined by cDNA macroarray. *Oncol. Rep.* 9: 723-728.

Ohira T *et al.* (2003) WNT7a induces E-cadherin in lung cancer cells. *Proc. Natl. Acad. Sci. USA* 100: 10429-10434.

Ohlsson R *et al.* (2003) Epigenetic variability and the evolution of human cancer. *Adv. Cancer Res.* 88: 145-168.

Ohmiya N *et al.* (2001) Germline and somatic mutations in hMSH6 and hMSH3 in gastrointestinal cancers of the microsatellite mutator phenotype. *Gene* 272: 301-313.

Ohta Y, Watanabe Y (2000) Current concepts of angiogenesis related to primary lung cancer. Chap. 11 in *Lung Cancer. Principles and Practice.* 2nd Ed. HI Pass, JB Mitchell, DH Johnson, AT Turrisi, JD Minna (Eds.). Lippincott Williams & Williams, Philadelphia, pp 199-210.

Ohta Y *et al.* (1997) Vascular endothelial growth factor and lymph node metastasis in primary lung cancer. *Br. J. Cancer* 76: 1041-1045.

Ohwada A *et al.* (1994) Biliary glycoprotein mRNA expression is increased in primary lung cancer, especially in squamous cell carcinoma. *Am. J. Respir. Cell Mol. Biol.* 11: 214-220.

Oizumi S *et al.* (2002) RCAS1 expression: a potential prognostic marker for adenocarcinomas of the lung. *Oncology* 62: 333-339.

Oji Y *et al.* (2002) Overexpression of the Wilms' tumor gene *WT1* in *de novo* lung cancers. *Int. J. Cancer* 100: 297-303.

Okami K *et al.* (1998) Analysis of PTEN/MMAC1 alterations in aerodigestive tract tumors. *Cancer Res.* 58: 509-511.

Okamoto A *et al.* (1994) Mutations and altered expression of $p16^{INK4}$ in human cancer. *Proc. Natl. Acad. Sci. USA* 91: 11045-11049.

Okamoto A *et al.* (1995) Mutations in the $p16^{INK4}/MTS1/CDKN2$, $p15^{INK4B}/MTS2$, and *p18* genes in primary and metastaic lung cancer. *Cancer Res.* 55: 1448-1451.

Okamoto I *et al.* (2003) Expression of constitutively activated EGFRvIII in non-small cell lung cancer. *Cancer Sci.* 94: 50-56.

Okano M *et al.* (1999) DNA methyltransferases Dnmt3a and Dnmt3b are essential for *de novo* methylation and mammalian development. *Cell* 99: 247-257.

Okudela K *et al.* (2004) K-ras gene mutation enhances motility of immortalized airway cells and lung adenocarcinoma cells via Akt activation: possible contribution to non-invasive expansion of lung adenocarcinoma. *Am. J. Pathol.* 164: 91-100.

Olaussen KA *et al.* (2003) Loss of PTEN expression is not uncommon, but lacks prognostic value in stage I NSCLC. *Anticancer Res.* 23(6C): 4885-4890.

Olie RA *et al.* (2000) A novel antisense oligonucleotide targeting surviving expression induces apoptosis and sensitises lung cancer cells to chemotherapy. *Cancer Res.* 60: 2805-2809.

Olinski R *et al.* (2003) Oxidative DNA damage in cancer patients: a cause or a consequence of the disease development. *Mutation Res.* 531: 177-190.

Olivero M *et al.* (1996) Overexpression and activation of hepatocyte growth factor/scatter factor in human non-small-cell lung carcinomas. *Br. J. Cancer* 74: 1862-1868.

Olivier M *et al.* (2002) The IARC *TP53* database: new online mutation analysis and recommendations to users. *Hum. Mutation* 19: 607-614.

Olivier M *et al.* (2004) *TP53* mutation spectra and load: a tool for generating hypotheses on the etiology of cancer. *IARC Sci. Publ.* 157: 247-270.

Olsson C *et al.* (2004) Association between dependent smoking and a polymorphism in the tyrosine hydroxylase gene in a prospective population-based study of adolescent health. *Behav. Genet.* 34: 85-91.

Omori S *et al.* (2002) Suppression of a DNA double-strand break repair gene Ku70, increases radio- and chemosensitivity in a human lung carcinoma cell line. *DNA Repair* 1: 299-310.

Ong ST *et al.* (1997) Precise localization of the *FHIT* gene to the common fragile site at 3p14.2 (FRA3B) and characterization of homozygous deletions within FRA3B that affect FHIT transcription in tumor cell lines. *Genes Chrom. Cancer* 20: 16-23.

Onn A, Herbst RS (2003) Angiogenesis, metastasis, and lung cancer. Chap. 24 in *Methods in Molecular Medicine Vol. 74: Lung Cancer, Vol. 1: Molecular Pathology Methods and Reviews.* Ed. B. Driscoll. Humana Press, Totowa, New Jersey. pp 329-348.

Onn A *et al.* (2004) Synchronous overexpression of epidermal growth factor receptor and HER2-neu protein is a predictor of poor outcome in patients with stage I non-small cell lung cancer. *Clin. Cancer Res.* 10: 136-143.

Ooi WL *et al.* (1986) Increased familial risk for lung cancer. *J. Natl. Cancer Inst.* 76: 217-222.

Ookawa K *et al.* (1993) Reconstitution of the RB gene suppresses the growth of small-cell lung carcinoma cells carrying multiple genetic alterations. *Oncogene* 8: 2175-2181.

Oremek GM, Sapoutzis N (2003) Pro-gastrin-releasing peptide (Pro-GRP), a tumor marker for small cell lung cancer. *Anticancer Res.* 23(2A): 895-898.

Ornstein DK, Petricoin EF (2004) Proteomics to diagnose human tumors and provide prognostic information. *Oncology* 18: 521-529.

Osada H *et al.* (2001) Heterogeneous transforming growth factor (TGF)-beta unresponsiveness and loss of TGF-beta receptor type II expression caused by histone deacetylation in lung cancer cell lines. *Cancer Res.* 61: 8331-8339.

Osada H, Takahashi T (2002) Genetic alterations of multiple tumor suppressors and oncogenes in the carcinogenesis and progression of lung cancer. *Oncogene* 21: 7421-7434.

Osada H *et al.* (2002) Frequent and histological type-specific inactivation of 14-3-3σ in human lung cancers. *Oncogene* 21: 2418-2424.

Oscarson M (2001) Genetic polymorphisms in the cytochrome P450 2A6 (*CYP2A6*) gene: implications for interindividual differences in nicotine metabolism. *Drug Metab. Dispos.* 29: 91-95.

Oshika Y *et al.* (1998a) Thrombospondin 2 gene expression is correlated with decreased vascularity in non-small cell lung cancer. *Clin. Cancer Res.* 4: 1785-1788.

Oshika Y *et al.* (1998b) Multidrug resistance-associated protein and mutant p53 protein expression in non-small cell lung cancer. *Mod. Pathol.* 11: 1059-1063.

Oshika Y *et al.* (2000) Ribozyme approach to downregulate vascular endothelial growth factor (VEGF) 189 expression in non-small cell lung cancer (NSCLC). *Eur. J. Cancer* 36: 2390-2396.

Oshita F *et al.* (2002) Increased expression of integrin β1 is a poor prognostic factor in small-cell lung cancer. *Anticancer Res.* 22(2B): 1065-1070.

Oshita F *et al.* (2003) Detection of occult tumor cells in peripheral blood from patients with small cell lung cancer by promoter methylation and silencing of the retinoic acid receptor-beta. *Oncol. Rep.* 10: 105-108.

Oshita F *et al.* (2004) High expression of integrin beta1 and p53 is a greater poor prognostic factor than clinical stage in small-cell lung cancer. *Am. J. Clin. Oncol.* 27: 215-219.

Ossipova O *et al.* (2003) LKB1 (XEEK1) regulates Wnt signalling in vertebrate development. *Nature Cell Biol.* 5: 889-894.

Otake Y *et al.* (1999) Expression of thymidylate synthase in human non-small cell lung cancer. *Jpn. J. Cancer Res.* 90: 1248-1253.

Otterson GA *et al.* (1994) Absence of $p16^{INK4}$ protein is restricted to the subset of lung cancer lines that retains wildtype RB. *Oncogene* 9: 3375-3378.

Otterson GA *et al.* (1995) *CDKN2* gene silencing in lung cancer by DNA hypermethylation and kinetics of $p16^{INK4}$ protein induction by 5-aza 2'deoxycytidine. *Oncogene* 11: 1211-1216.

Oyama T *et al.* (1997a) *N*-acetylation polymorphism in patients with lung cancer and its association with p53 gene mutation. *Anticancer Res.* 17(1B): 577-581.

Oyama T *et al.* (1997b) Cytochrome P450 2E1 polymorphism as a risk factor for lung cancer: in relation to p53 gene mutation. *Anticancer Res.* 17(1B): 583-587.

Ozawa S *et al.* (1999) Analyses of bronchial bulky DNA adduct levels and *CYP2C9*, *GSTP1* and *NQO1* genotypes in a Hungarian study population with pulmonary diseases. *Carcinogenesis* 20: 991-995.

Ozeki Y *et al.* (1994) Immunohistochemical analysis of nm23/NDP kinase expression in human lung adenocarcinoma: association with tumor progression in Clara cell type. *Jpn. J. Cancer Res.* 85: 840-846.

Ozkara HA *et al.* (1999) Amplification of the c-*myc* oncogene in non-small cell lung cancer. *Tumori* 85: 508-511.

Paddison PJ, Hannon GJ (2002) RNA interference: the new somatic cell genetics? *Cancer Cell* 2: 17-23.

Paez JG *et al.* (2004) *EGFR* mutations in lung cancer: correlation with clinical response to Gefitinib therapy. *Science* 304: 1497-1500.

Paige AJW *et al.* (2001) *WWOX*: a candidate tumor suppressor gene involved in multiple tumor types. *Proc. Natl. Acad. Sci. USA* 98: 11417-11422.

Palli D *et al.* (2001) DNA adduct levels and DNA repair polymorphisms in traffic-exposed workers and a general population sample. *Int. J. Cancer* 94: 121-127.

Palli D *et al.* (2004) DNA bulky adducts in a Mediterranean population correlate with environmental ozone concentration, an indicator of photochemical smog. *Int. J. Cancer* 109: 17-23.

Palmisano WA *et al.* (2000) Predicting lung cancer by detecting aberrant promoter methylation in sputum. *Cancer Res.* 60: 5954-5958.

Palmisano WA *et al.* (2003) Aberrant promoter methylation of the transcription factor genes PAX5 α and β in human cancers. *Cancer Res.* 63: 4620-4625.

Pandolfi PP (2004) Aberrant mRNA translation in cancer pathogenesis: an old concept revisited comes finally of age. *Oncogene* 23: 3134-3137.

Papadakis ED *et al.* (2002) Association of p53 codon 72 polymorphism with advanced lung cancer: the Arg allele is preferentially retained in tumours arising in Arg/Pro germline heterozygotes. *Br. J. Cancer* 87: 1013-1018.

Papadopoulos N, Lindblom A. (1997) Molecular basis of HNPCC: mutations of MMR genes. *Hum. Mutation* 10: 89-99.

Paramio JM *et al.* (1999) PTEN tumour suppressor is linked to the cell cycle control through the retinoblastoma protein. *Oncogene* 18: 7462-7468.

Parascandola M, Weed DL (2001) Causation in epidemiology. *J. Epidemiol. Community Health* 55: 905-912.

Park IW *et al.* (1999) Multiple clonal abnormalities in the bronchial epithelium of patients with lung cancer. *J. Natl. Cancer Inst.* 91: 1863-1868.

Park JY *et al.* (2000) Microsatellite alteration in histologically normal lung tissue of patients with non-small cell lung cancer. *Lung Cancer* 30: 83-89.

Park YJ *et al.* (2001) Genetic changes of hOGG1 and the activity of ohGua glycosylase in colon cancer. *Eur. J. Cancer* 37: 340-346.

Park C *et al.* (2002a) Effects of transforming growth factor beta (TGF-beta) receptor on lung carcinogenesis. *Lung Cancer* 38: 143-147.

Park JY *et al.* (2002b) Polymorphism of the DNA repair gene *XRCC1* and risk of primary lung cancer. *Cancer Epidemiol. Biomarkers Prev.* 11: 23-27.

Park JY *et al.* (2002c) Polymorphisms of the DNA repair gene xeroderma pigmentosum group A and risk of primary lung cancer. *Cancer Epidemiol. Biomarkers Prev.* 11: 993-997.

Park KH *et al.* (2003a) Gene therapy with GM-CSF, interleukin-4 and herpes simplex virus thymidine kinase shows strong antitumor effect on lung cancer. *Anticancer Res.* 23(2B): 1559-1564.

Park MJ *et al.* (2003b) Pathogenetic and biologic significance of $TP14^{ARF}$ alterations in nonsmall cell lung carcinoma. *Cancer Genet. Cytogenet.* 141: 5-13.

Park J *et al.* (2004) The human 8-oxoguanine DNA N-glycosylase 1 (hOGG1) DNA repair enzyme and its association with lung cancer risk. *Pharmacogenetics* 14: 103-109.

Parker AR, Eshleman JR (2003) Human MutY: gene structure, protein functions and interactions, and role in carcinogenesis. *Cell Mol. Life Sci.* 60: 2064-2083.

Parmigiani G *et al.* (2004) A cross-study comparison of gene expression studies for the molecular classification of lung cancer. *Clin. Cancer Res.* 10: 2922-2927.

Parsons R *et al.* (1995) Microsatellite instability and mutations of the transforming growth factor beta type II receptor gene in colorectal cancer. *Cancer Res.* 55: 5548-5550.

Paschke T (2000) Analysis of different versions of the IARC p53 database with respect to G→T transversion mutation frequencies and mutation hotspots in lung cancer of smokers and non-smokers. *Mutagenesis* 15: 457-458.

Paschke T (2001) Reply to the contribution of Hainaut, Olivier and Pfeifer (*TP53* mutation spectrum in lung cancers and mutagenic signature of components of tobacco smoke: lessons from the IARC *TP53* mutation database). *Mutagenesis* 16: 555-556.

Pastorelli R *et al.* (1998) Impact of inherited polymorphisms in glutathione *S*-transferase M1, microsomal epoxide hydrolase, cytochrome P450 enzymes on DNA, and blood protein adducts of benzo[*a*]pyrene-diolepoxide. *Cancer Epidemiol. Biomarkers Prev.* 7: 703-709.

Pastorelli R *et al.* (2002) Effect of DNA repair gene polymorphisms on BPDE-DNA adducts in human lymphocytes. *Int. J. Cancer* 100: 9-13.

Pataer A *et al.* (2003) Adenoviral melanoma differentiation-associated gene 7 induces apoptosis in lung cancer cells through mitochondrial permeability transition-independent cytochrome *c* release. *J. Thorac. Cardiovasc. Surg.* 125: 1328-1335.

Patel JD *et al.* (2004a) Lung cancer in US women: a contemporary epidemic. *JAMA* 291: 1763-1768.

Patel JH *et al.* (2004b) Analysis of genomic targets reveals complex functions of Myc. *Nature Rev. Cancer* 4: 562-568.

Paulovitch AG *et al.* (1997) When checkpoints fail. *Cell* 88: 315-321.

Pavelic K *et al.* (2001) Aberration of *FHIT* gene is associated with increased tumor proliferation and decreased apoptosis-clinical evidence in lung and head and neck carcinomas. *Mol. Med.* 7: 442-453.

Pavelic J *et al.* (2002) Insulin-like growth factor family and combined antisense approach in therapy of lung carcinoma. *Mol. Med.* 8: 149-157.

Paz MF *et al.* (2002) Germ-line variants in methyl-group metabolism genes and susceptilibity to DNA methylation in normal tissues and human primary tumors. *Cancer Res.* 62: 4519-4524.

Paz MF *et al.* (2003) A systematic profile of DNA methylation in human cancer cell lines. *Cancer Res.* 63: 1114-1121.

Paz-Elizur T *et al.* (2003) DNA repair activity for oxidative damage and risk of lung cancer. *J. Natl. Cancer Inst.* 95: 1312-1319.

Peace BE *et al.* (2001) Point mutations and overexpression of Ron induce transformation, tumor formation, and metastasis. *Oncogene* 20: 6142-6151.

Peake MD *et al.* (2003) Ageism in the management of lung cancer. *Age Ageing* 32: 171-177.

Pedersen MW *et al.* (2001) Profile of differentially expressed genes mediated by the type III epidermal growth factor receptor mutation expressed in a small-cell lung cancer cell line. *Br. J. Cancer* 85: 1211-1218.
Pedersen N *et al.* (2003) Transcriptional gene expression profiling of small cell lung cancer cells. *Cancer Res.* 63: 1943-1953.
Pei XH *et al.* (1999) Benzo[*a*]pyrene activates the human p53 gene through induction of nuclear factor κB activity. *J. Biol. Chem.* 274: 35240-35246.
Pei J *et al.* (2001) Genomic imbalances in human lung adenocarcinomas and squamous cell carcinomas. *Genes Chrom. Cancer* 31: 282-287.
Pekarsky Y *et al.* (2004) Fhit is a physiological target of the protein kinase Src. *Proc. Natl. Acad. Sci. USA* 101: 3775-3779.
Pellegrini S *et al.* (1999) Lung tumours from non-smoking subjects: a p53-related genetic instability in a subset of cases. *Int. J. Mol. Med.* 4: 419-424.
Pellegrini C *et al.* (2003) HER-2/Neu alterations in non-small cell lung cancer: a comprehensive evaluation by real time reverse transcription-PCR, fluorescence *in situ* hybridization, and immunohistochemistry. *Clin. Cancer Res.* 9: 3645-3652.
Pelosi G *et al.* (2004) K-ras gene mutational analysis supports a monoclonal origin of biphasic pleomorphic carcinoma of the lung. *Mod. Pathol.* 17: 538-546.
Peng Y *et al.* (2001) Stabilization of the MDM2 oncoprotein by mutant p53. *J. Biol. Chem.* 276: 6874-6878.
Peng ZM *et al.* (2004) Study of gene expression difference in lung carcinogenesis by cDNA microarray. *Ai Zheng* 23: 150-154.
Penno MB *et al.* (1994) Expression of CD44 in human lung tumors. *Cancer Res.* 54: 1381-1387.
Perera FP *et al.* (2002) Associations between carcinogen-DNA damage, glutathione *S*-transferase genotypes, and risk of lung cancer in the prospective Physicians' Health Cohort Study. *Carcinogenesis* 23: 1641-1646.
Perkins P *et al.* (1997) Carcinoid tumors of the lung and family history of cancer. *J. Clin. Epidemiol.* 50: 705-709.
Persson I *et al.* (1997) *In vitro* kinetics of two human CYP1A1 variant enzymes suggested to be associated with interindividual differences in cancer susceptibility. *Biochem. Biophys. Res. Commun.* 231: 227-230.
Perucho M. (1996) Microsatellite instability: the mutator that mutates the other mutator. *Nature Medicine* 2: 630-631.
Petersen I *et al.* (1997) Small-cell lung cancer is characterized by a high incidence of deletions on chromosomes 3p, 4q, 5q, 10q, 13q and 17p. *Br. J. Cancer* 75: 79-86.
Petersen S *et al.* (2000a) Gene expression profiling of advanced lung cancer. *Int. J. Cancer* 86: 512-517.
Petersen S *et al.* (2000b) Analysis of the *DMBT1* gene in carcinomas of the respiratory tract. *Int. J. Cancer* 88: 71-76.
Petersen I, Petersen S (2001) Towards a genetic-based classification of human lung cancer. *Anal. Cell Pathol.* 22: 111-121.
Petes TD (2001) Meiotic recombination hot spots and cold spots. *Nature Rev. Genet.* 2: 360-369.
Petkova DK *et al.* (2004) Overexpression of cyclooxygenase-2 in non-small cell lung cancer. *Respir. Med.* 98: 164-172.
Petmitr S *et al.* (2002) Microsatellite alterations in non-small cell lung cancer. *J. Exp. Clin. Cancer Res.* 21: 31-35.
Petmitr S *et al.* (2003) Mutational analysis of ras gene family in lung cancer in Thai. *Oncol. Rep.* 10: 1497-1501.
Petty TL, Miller YE (2000) Early diagnosis and intervention in lung cancer: clinical studies. Chap. 22 in *Lung Cancer. Principles and Practice.* 2nd Ed. HI Pass, JB Mitchell, DH Johnson, AT Turrisi, JD Minna (Eds.). Lippincott Williams & Williams, Philadelphia, pp 398-406.
Petty WJ *et al.* (2003) Cyclin D1 as a target for chemoprevention. *Lung Cancer* 41 Suppl. 1: S155-S161.
Petty RD *et al.* (2004) Gene expression profiling in non-small cell lung cancer: from molecular mechanisms to clinical application. *Clin. Cancer Res.* 10: 3237-3248.
Pezzella F *et al.* (1993) *bcl*-2 protein in non-small-cell lung carcinoma. *New Engl. J. Med.* 329: 690-694.
Pfeifer GP (2000) p53 mutational spectra and the role of methylated CpG sequences. *Mutation Res.* 450: 155-166.
Pfeifer GP, Hainaut P (2003) On the origin of G→T transversions in lung cancer. *Mutation Res.* 526: 39-43.

Pfeifer GP, Hainaut P (2004) The origin of G→T transversions in lung cancer cell lines. Reply to Rodin and Rodin. *Mutation Res.* 545: 145-146.
Pfeifer P *et al.* (1996) Lack of prognostic significance of epidermal growth factor receptor and the oncoprotein p185HER-2 in patients with systemically untreated non-small-cell lung cancer: an immunohistochemical study on cryosections. *Br. J. Cancer* 74: 86-91.
Pfeifer GP *et al.* (1998) p53 mutations, benzo[*a*]pyrene and lung cancer: a reply. *Mutagenesis* 13: 537-538.
Pfeifer GP *et al.* (2002) Tobacco smoke carcinogens, DNA damage and p53 mutations in smoking-associated cancers. *Oncogene* 21: 7435-7451.
Philippson A (1926) Ein Betrag zur Krebsätiologie auf Grund klinischer Beobachtungen. *Klin. Wochenschr.* 33: 1513-1516.
Phillips DH (2002) Smoking-related DNA and protein adducts in human tissues. *Carcinogenesis* 23: 1979-2004.
Pianezza ML *et al.* (1998) Nicotine metabolism defect reduces smoking. *Nature* 393: 750.
Picchio MC *et al.* (2004) Alterations of the tumor suppressor gene parkin in non-small lung cancer. *Clin. Cancer Res.* 10: 2720-2724.
Pick E, Roberts AB (2001) Suppressor and oncogenic roles of transforming growth factor-beta and its signalling pathways in tumorigenesis. *Adv. Cancer Res.* 83: 1-54.
Pierce LM *et al.* (2000) Relationships of *TP53* codon 72 and *HRAS1* polymorphisms with lung cancer risk in an ethnically diverse population. *Cancer Epidemiol. Biomarkers Prev.* 9: 1199-1204.
Pierce DA *et al.* (2003) Joint effects of radiation and smoking on lung cancer risk among atomic bomb survivors. *Radiat. Res.* 159: 511-520.
Pietsch T *et al.* (1998) Expression of the c-kit receptor and its ligand SCF in non-small-cell lung carcinomas. *Int. J. Cancer* 75: 171-175.
Pifarre A *et al.* (1997) Prognostic value of replication errors on chromosomes 2p and 3p in non-small-cell lung cancer. *Br. J. Cancer* 75: 184-189.
Pihan G, Doxsey SJ (2003) Mutations and aneuploidy: co-conspirators in cancer? *Cancer Cell* 4: 89-94.
Pinarbasi H *et al.* (2003) Strong association between the *GSTM1*-null genotype and lung cancer in a Turkish population. *Cancer Genet. Cytogenet.* 146: 125-129.
Pino I *et al.* (2003) Altered patterns of expression of members of the heterogeneous nucleoprotein (hnRNP) family in lung cancer. *Lung Cancer* 41: 131-143.
Pio R *et al.* (2004) αCP-4, encoded by a putative tumor suppressor gene at 3p21, but not its alternative splice variant αCP-4a, is underexpressed in lung cancer. *Cancer Res.* 64: 4171-4179.
Pitterle DM *et al.* (1999) Lung cancer and the human gene for ribonucleotide reductase subunit M1 (RRM1). *Mammalian Genome* 10: 916-922.
Pitti RM *et al.* (1998) Genomic amplification of a decoy receptor for Fas ligand in lung and colon cancer. *Nature* 396: 699-702.
Piyathilake CJ *et al.* (2001) Altered global methylation of DNA: an epigenetic difference in susceptibility for lung cancer is associated with its progression. *Hum. Pathol.* 32: 856-862.
Player A *et al.* (2003) Identification of *TDE2* gene and its expression in non-small cell lung cancer. *Int. J. Cancer* 107: 238-243.
Poisson A *et al.* (2003) Menin interacting proteins as clues toward the understanding of multiple endocrine neoplasia type 1. *Cancer Lett.* 189: 1-10.
Pollan M *et al.* (2003) Clinical value of p53, c-erbB-2, CEA and CA125 regarding relapse, metastasis and death in resectable non-small cell lung cancer. *Int. J. Cancer* 107: 781-790.
Pollock PM *et al.* (1996) Compilation of somatic mutations of the *CDKN2* gene in human cancers: non-random distribution of base substitutions. *Genes Chrom. Cancer* 15: 77-88.
Ponder BAJ (2001) Cancer genetics. *Nature* 411: 336-341.
Ponten I *et al.* (2001) Differences between the mutational consequences of replication of *cis-* and *trans-*opened benzo[*a*]pyrene 7,8-diol 9,10-epoxide-deoxyguanosine adducts in M13mp7L2 constructs. *Chem. Res. Toxicol.* 14: 720-726.
Pope CA III *et al.* (2002) Lung cancer, cardiopulmonary mortality, and long-term exposure to fine particulate air pollution. *JAMA* 287: 1132-1141.
Popescu NC (2000) Comprehensive genetic analysis of cancer cells. *J. Cell Mol. Med.* 4: 151-163.
Potti A *et al.* (2003) CD117 (c-KIT) overexpression in patients with extensive-stage small-cell lung carcinoma. *Ann. Oncol.* 14: 894-897.
Powell CA *et al.* (2003a) Gene expression in lung adenocarcinomas of smokers and non-smokers. *Am. J. Respir. Cell Mol. Biol.* 29: 157-162.
Powell CA *et al.* (2003b) Patterns of allelic loss differ in lung adenocarcinomas of smokers and nonsmokers. *Lung Cancer* 39: 23-29.

Prins J *et al.* (1993) The *myc* family of oncogenes and their presence and importance in small-cell lung carcinoma and other tumour types. *Anticancer Res.* 13: 1373-1386.
Proctor RN (2001) Tobacco and the global lung cancer epidemic. *Nature Rev. Cancer* 1: 82-86.
Prokhortchouk E, Hendrich B (2002) Methyl-CpG binding proteins and cancer: are MeCpGs more important than MBDs? *Oncogene* 21: 5394-5399.
Puisieux A *et al.* (1991) Selective targeting of *p53* gene mutational hotspots in human cancers by etiologically defined carcinogens. *Cancer Res.* 51: 6185-6189.
Pulling LC *et al.* (2003) Promoter hypermethylation of the O(6)-methylguanine-DNA methyltransferase gene: more common in lung adenocarcinomas from never-smokers than smokers and associated with tumour progression. *Cancer Res.* 63: 4842-4848.
Pulling LC *et al.* (2004) Aberrant promoter hypermethylation of the *Death-associated protein kinase* gene is early and frequent in murine lung tumors induced by cigarette smoke and tobacco carcinogens. *Cancer Res.* 64: 3844-3848.
Pylkkanen L *et al.* (2002) Reduced Fhit protein expression and loss of heterozygosity at *FHIT* gene in tumours from smoking and asbestos-exposed lung cancer patients. *Int. J. Oncol.* 20: 285-290.
Qi W, Martinez JD (2003) Reduction of 14-3-3 proteins correlates with increased sensitivity to killing of human lung cancer cells by ionising radiation. *Radiation Res.* 160: 217-223.
Qiu J *et al.* (2004) Development of natural protein microarrays for diagnosing cancer based on an antibody response to tumor antigens. *J. Proteome Res.* 3: 261-267.
Qiuling S *et al.* (2003) Cyclin D1 gene polymorphism and susceptibility to lung cancer in a Chinese population. *Carcinogenesis* 24: 1499-1503.
Quaggin SE *et al.* (1998) Pod-1, a mesoderm-specific basic-helix-loop-helix protein expressed in mesenchymal and glomerular epithelial cells in the developing kidney. *Mech. Dev.* 71: 37-48.
Quinlan DC *et al.* (1992) Accumulation of p53 protein correlates with a poor prognosis in human lung cancer. *Cancer Res.* 52: 4828-4831.
Quinn KA *et al.* (1996) Insulin-like growth factor expression in human cancer cell lines. *J. Biol. Chem.* 271: 11477-11483.
Rachwal WJ *et al.* (1995) Expression and activation of erbB-2 and epidermal growth factor receptor in lung adenocarcinomas. *Br. J. Cancer* 72: 56-64.
Racz A *et al.* (1999) Expression analysis of genes at 3q26-q27 involved in frequent amplification in squamous cell lung carcinoma. *Eur. J. Cancer* 35: 641-646.
Racz A *et al.* (2000) Gene amplification at chromosome 1pter-p33 including the genes *PAX7* and *ENO1* in squamous cell lung carcinoma. *Int. J. Oncol.* 17: 67-73.
Radtke F, Raj K (2003) The role of Notch in tumorigenesis: oncogene or tumour suppressor? *Nature Rev. Cancer* 3: 756-767.
Rahman Q *et al.* (1999) Glutathione redox system in oxidative lung injury. *Crit. Rev. Toxicol.* 29: 543-568.
Rajaee-Behbahani N *et al.* (2001) Altered DNA repair capacity and bleomycin sensitivity as risk markers for non-small cell lung cancer. *Int. J. Cancer* 95: 86-91.
Rajagopalan H *et al.* (2002) Tumorigenesis: RAF/RAS oncogenes and mismatch repair status. *Nature* 418: 934.
Ramaswamy S *et al.* (2003) A molecular signature of metastasis in primary solid tumors. *Nature Genet.* 33: 49-53.
Ramesh R *et al.* (2001) Successful treatment of primary and disseminated human lung cancers by systemic delivery of tumor suppressor gene using an improved liposome vector. *Mol. Ther.* 3: 337-350.
Ramet M *et al.* (1995) p53 protein expression is correlated with benzo[*a*]pyrene-DNA adducts in carcinoma cell lines. *Carcinogenesis* 16: 2117-2124.
Ramnath N *et al.* (2001) MCM2 is an independent predictor of survival in patients with non-small-cell lung cancer. *J. Clin. Oncol.* 19: 4259-4266.
Rampino N *et al.* (1997) Somatic frameshift mutations in the *BAX* gene in colon cancers of the microsatellite mutator phenotype. *Science* 275: 967-969.
Ratnasinghe D *et al.* (2000) Glutathione peroxidase codon 198 polymorphism variant increases lung cancer risk. *Cancer Res.* 60: 6381-6383.
Ratnasinghe D *et al.* (2001a) Polymorphisms of the DNA repair gene *XRCC1* and lung cancer risk. *Cancer Epidemiol. Biomarkers Prev.* 10: 119-123.
Ratnasinghe D *et al.* (2001b) Influence of antioxidants and the CYP1A1 isoleucine to valine polymorphism on the smoking-lung cancer association. *Anticancer Res.* 21(2B): 1295-1299.
Ratnasinghe DL *et al.* (2003) Gene-environment interactions between the codon 194 polymorphism of *XRCC1* and antioxidants influence lung cancer risk. *Anticancer Res.* 23(1B): 627-632.
Ratschiller D *et al.* (2003) Cyclin D1 overexpression in bronchial epithelia of patients with lung cancer is associated with smoking and predicts survival. *J. Clin. Oncol.* 21: 2085-2093.

Raunio H *et al.* (2001) Polymorphisms of CYP2A6 and its practical consequences. *Br. J. Clin. Pharmacol.* 52: 357-363.

Ravi RK *et al.* (1998) Activated Raf-1 causes growth arrest in human small cell lung cancer cells. *J. Clin. Invest.* 101: 153-159.

Rebbeck TR (1997) Molecular epidemiology of the human glutathione *S*-transferase genotypes *GSTM1* and *GSTT1* in cancer susceptibility. *Cancer Epidemiol. Biomarkers Prev.* 6: 733-743.

Rebbeck TR (2004) Biomarkers of inherited susceptibility and cancer. *IARC Sci. Publ.* 157: 91-103.

Reddel RR (2003) Alternative lengthening of telomeres, telomerase, and cancer. *Cancer Lett.* 194: 155-162.

Reeve JG *et al.* (1992) Expression of insulin-like growth factor (IGF) and IGF-binding protein genes in human lung tumor cell lines. *J. Natl. Cancer Inst.* 84: 628-634.

Reeve JG *et al.* (1996) Expression of apoptosis-regulatory genes in lung tumour cell lines: relationship to p53 expression and relevance to acquired drug resistance. *Br. J. Cancer* 73: 1193-1200.

Reichel MB *et al.* (1994) p53 mutations in primary human lung tumors and their metastases. *Mol. Carcinogen.* 9: 105-109.

Reisman DN *et al.* (2003) Loss of BRG1/BRM in human lung cancer cell lines and primary lung cancers: correlation with poor prognosis. *Cancer Res.* 63: 560-566.

Reissmann PT *et al.* (1993) Inactivation of the retinoblastoma susceptibility gene in non-small-cell lung cancer. *Oncogene* 8: 1913-1919.

Reissmann PT *et al.* (1999) Amplification and overexpression of the cyclin D1 and epidermal growth factor receptor genes in non-small-cell lung cancer. *J. Cancer Res. Clin. Oncol.* 125: 61-70.

Reszka E, Wasowicz W (2001) Significance of genetic polymorphisms in glutathione *S*-transferase multigene family and lung cancer risk. *Int. J. Occup. Med. Pub. Health* 14: 99-113.

Reszka E *et al.* (2003) Glutathione *S*-transferase M1 and P1 metabolic polymorphism and lung cancer predisposition. *Neoplasma* 50: 357-362.

Reya T *et al.* (2001) Stem cells, cancer, and cancer stem cells. *Nature* 414: 105-111.

Reynolds SH *et al.* (1991) Activated protooncogenes in human lung tumors from smokers. *Proc. Natl. Acad. Sci. USA* 88: 1085-1089.

Reynolds PR *et al.* (2004) Midkine is regulated by hypoxia and causes pulmonary vascular remodelling. *J. Biol. Chem.* 279: 37124-37132.

Rhee I *et al.* (2002) DNMT1 and DNMT3b cooperate to silence genes in human cancer cells. *Nature* 416: 552-556.

Rhodes DR, Chinnaiyan AM (2004) Bioinformatics strategies for translating genome-wide expression analyses into clinically useful cancer markers. *Ann. N Y Acad. Sci.* 1020: 32-40.

Richardson GE, Johnson BE (1993) The biology of lung cancer. *Semin. Oncol.* 20: 105-127.

Rideout WM *et al.* (1991) 5-methylcytosine as an endogenous mutagen in the p53 tumor suppressor gene. *Princess Takamatsu Symp.* 22: 207-219.

Ried T *et al.* (1994) Mapping of multiple DNA gains and losses in primary small cell lung carcinomas by comparative genomic hybridisation. *Cancer Res.* 54: 1801-1806.

Risch N, Merikangas K (1996) The future of genetic studies of complex human diseases. *Science* 273: 1516-1517.

Roberts CW, Orkin SH (2004) The SWI/SNF complex – chromatin and cancer. *Nature Rev. Cancer* 4: 133-142.

Robles AI *et al.* (2002) The p53 network in lung carcinogenesis. *Oncogene* 21: 6898-6907.

Rocco JW, Sidransky D (2001) p16(MTS-1/*CDKN2*/*INK4a*) in cancer progression. *Exp. Cell Res.* 264: 42-55.

Rodenhuis S *et al.* (1987) Mutational activation of the K-*RAS* oncogene. A possible pathogenetic factor in adenocarcinoma of the lung. *New Engl. J. Med.* 317: 929-935.

Rodenhuis S, Slebos RJ (1992) Clinical significance of *ras* oncogene activation in human lung cancer. *Cancer Res.* 52 (9 Suppl.): 2665s-2669s.

Rodin SN *et al.* (1998) CpG transition strand asymmetry and hitch-hiking mutations as measures of tumorigenic selection in shaping the p53 mutation spectrum. *Int. J. Mol. Med.* 1: 191-199.

Rodin SN *et al.* (2002) Cancerous hyper-mutagenesis in p53 genes is possibly associated with transcriptional bypass of DNA lesions. *Mutation. Res.* 510: 153-168.

Rodin SN, Rodin AS (2000) Human lung cancer and *p53*: the interplay between mutagenesis and selection. *Proc. Natl. Acad. Sci. USA* 97: 12244-12249.

Rodin SN, Rodin AS (2002) On the origin of p53 G:C→T:A transversions in lung cancers. *Mutation Res.* 508: 1-19.

Rodin SN, Rodin AS (2004a) On the excess of G→T transversions in the p53 gene in lung cancer cell lines. Reply to Pfeifer and Hainaut. *Mutation Res.* 545: 141-144.

Rodin SN, Rodin AS (2004b) Response to reply to Pfeifer and Hainaut. *Mutation Res.* 545: 147.

Rodriguez-Salas N *et al.* (2001) Correlation of p53 oncoprotein expression with chemotherapy response in small cell lung carcinomas. *Lung Cancer* 34: 67-74.
Rodriguez-Viciana P *et al.* (1996) Activation of phosphoinositide 3-kinase by interaction with Ras and by point mutation. *EMBO J.* 15: 2442-2451.
Rogalla P *et al.* (1998) HMGIC expression patterns in non-small lung cancer and surrounding tissue. *Anticancer Res.* 18(5A): 3327-3330.
Roh MS, Hong SH (2002) Utility of thyroid transcription factor-1 and cytokeratin 20 in identifying the origin of metastatic carcinomas of cervical lymph nodes. *J. Korean Med. Sci.* 17: 512-517.
Rohr UP *et al.* (2003) Non-small lung cancer cells are prime targets for p53 gene transfer mediated by a recombinant adeno-associated virus type-2 vector. *Cancer Gene Ther.* 10: 898-906.
Rohr UP *et al.* (2004) Expression of the tyrosine kinase c-kit is an independent prognostic factor in patients with small cell lung cancer. *Int. J. Cancer* 111: 259-263.
Rojas M *et al.* (1998) High benzo[*a*]pyrene diol-epoxide DNA adduct levels in lung and blood cells from individuals with combined *CYP1A1 MspI/Msp-GSTM1* *0/*0 genotypes. *Pharmacogenetics* 8: 109-118.
Rojas M *et al.* (2000) Modulation of benzo[*a*]pyrene diolepoxide DNA adduct levels in human white blood cells by *CYP1A1*, *GSTM1* and *GSTT1* polymorphism. *Carcinogenesis* 21: 35-41.
Rojas M *et al.* (2004) High DNA damage by benzo[*a*]pyrene 7,8-diol-9,10-epoxide in bronchial epithelial cells from patients with lung cancer: comparison with lung parenchyma. *Cancer Lett.* 207: 157-163.
Ronai Z *et al.* (1996) K-ras mutation in sputum of patients with or without lung cancer. *J. Cell Biochem. Suppl.* 25: 172-176.
Ronen A, Glickman BW (2001) Human DNA repair genes. *Environ. Mol. Mutagen.* 37: 241-283.
Rosell R *et al.* (1997) Reduced survival in patients with stage-I non-small-cell lung cancer associated with DNA-replication errors. *Int. J. Cancer* 74: 330-334.
Rosell R *et al.* (1999) Genetic susceptibility associated with rare *HRAS1* variable number of tandem repeats alleles in Spanish non-small cell lung cancer patients. *Clin. Cancer Res.* 5: 1849-1854.
Rosell R *et al.* (2001) Predictive molecular markers in non-small cell lung cancer. *Curr. Opin. Oncol.* 13: 101-109.
Rosell R *et al.* (2002) Molecular markers and targeted therapy with novel agents: prospects in the treatment of non-small cell lung cancer. *Lung Cancer* 38 Suppl.: 43-49.
Rosell R *et al.* (2003) Transcripts in pre-treatment biopsies from a three-arm randomised trial in metastatic non-small-cell lung cancer. *Oncogene* 22: 3548-3553.
Rosell R *et al.* (2004a) Molecular predictors of response to chemotherapy in lung cancer. *Semin. Oncol.* 31 Suppl. 1: 20-27.
Rosell R *et al.* (2004b) Gene expression as a predictive marker of outcome in stage IIB-IIIA-IIIB non-small cell lung cancer after induction gemcitabine-based chemotherapy followed by resectional surgery. *Clin. Cancer Res.* 10: 4215S-4219S.
Rosenwald IB *et al.* (2001) Expression of eukaryotic translation initiation factors 4E and 2α is increased frequently in bronchioloalveolar but not in squamous cell carcinomas of the lung. *Cancer* 92: 2164-2171.
Rosenwald IB (2004) The role of translation in neoplastic transformation from a pathologist's point of view. *Oncogene* 23: 3230-3247.
Ross D, Siegel D (2004) NAD(P)H:quinone oxidoreductase 1 (NQO1, DT-diaphorase), functions and pharmacogenetics. *Meth. Enzymol.* 382: 115-144.
Ross DT *et al.* (2000) Systematic variation in gene expression patterns in human cancer cell lines. *Nature Genet.* 24: 227-235.
Roth JA *et al.* (1996) Retrovirus-mediated wild-type p53 gene transfer to tumors of patients with lung cancer. *Nature Medicine* 2: 985-991.
Roth JA, Grammer SF (2004) Gene replacement therapy for non-small cell lung cancer: a review. *Hematol. Oncol. Clin. North Am.* 18: 215-229.
Rottman H (1898) *Über primäre Lungencarcinoma.* Inaugural dissertation, Universität Würzburg.
Rouse J, Jackson SP (2002) Interfaces between the detection, signalling, and repair of DNA damage. *Science* 297: 547-551.
Roz L *et al.* (2002) Restoration of fragile histidine triad (*FHIT*) expression induces apoptosis and suppresses tumorigenicity in lung and cervical cancer cell lines. *Proc. Natl. Acad. Sci. USA* 99: 3615-3620.
Ruano-Ravina A *et al.* (2003) *GSTM1* and *GSTT1* polymorphisms, tobacco and risk of lung cancer: a case-control study from Galicia, Spain. *Anticancer Res.* 23(5b): 4333-4337.
Rubbi CP, Milner J (2003) p53 is a chromatin accessibility factor for nucleotide excision repair of DNA damage. *EMBO J.* 22: 975-986.

Rubinstein DB *et al.* (2004) Receptor for the globular heads of C1q (gC1q-R, p33, hyaluronan-binding protein) is preferentially expressed by adenocarcinoma cells. *Int. J. Cancer* 110: 741-750.
Rudduck C *et al.* (1993) Co-amplification of the gene for parathyroid hormone-related protein (*PTHRP*) and *KRAS2* in a human lung cancer cell line. *Genes Chrom. Cancer* 7: 213-218.
Rüdiger R *et al.* (2001) Disruption of protein phosphatase 2A subunit interaction in human cancers with mutations in the A alpha subunit gene. *Oncogene* 20: 10-15.
Ruffini E *et al.* (2004) The significance of associated pre-invasive lesions in patients resected for primary lung neoplasms. *Eur. J. Cardiothorac. Surg.* 26: 165-172.
Ruivenkamp CAL *et al.* (2002) *Ptprj* is a candidate for the mouse colon-cancer susceptibility locus *Scc1* and is frequently deleted in human cancers. *Nature Genet.* 31: 295-300.
Rusch V *et al.* (1993) Differential expression of the epidermal growth factor receptor and its ligands in primary non-small cell lung cancers and adjacent benign lung. *Cancer Res.* 53: 2379-2385.
Rusch V *et al.* (1995) Aberrant p53 expression predicts clinical resistance to cisplatin-based chemotherapy in locally advanced non-small cell lung cancer. *Cancer Res.* 55: 5038-5047.
Rusch V *et al.* (1997) Overexpression of the epidermal growth factor receptor and its ligand transforming growth factor alpha is frequent in resectable non-small cell lung cancer but does not predict tumor progression. *Clin. Cancer Res.* 3: 515-522.
Rusin MR *et al.* (1996) Intragenic mutations of the p16(INK4), p15(INK4B) and p18 genes in primary non-small-cell lung cancers. *Int. J. Cancer* 65: 734-739.
Rusin M *et al.* (2004) Intronic polymorphism (1541-1542delGT) of the constitutive heat shock protein 70 gene has functional significance and shows evidence of association with lung cancer risk. *Mol. Carcinogen.* 39: 155-163.
Russo G *et al.* (2003a) pRB(2)/p130 target genes in non-small cell lung cancer cells identified by microarray analysis. *Oncogene* 22: 6959-6969.
Russo G *et al.* (2003b) Advantages and limitations of microarray technology in human cancer. *Oncogene* 22: 6497-6507.
Ruttan CC, Glickman BW (2002) Coding variants in human double-strand break DNA repair genes. *Mutation Res.* 509: 175-200.
Ryan KM, Birnie GD (1996) *Myc* oncogenes: the enigmatic family. *Biochem. J.* 314: 713-721.
Ryberg D *et al.* (1990) Ha-*ras*-1 alleles in Norwegian lung cancer patients. *Hum. Genet.* 86: 40-44.
Ryberg D *et al.* (1994) p53 mutations in lung tumors: relationship to putative susceptibility markers for cancer. *Cancer Res.* 54: 1551-1555.
Ryberg D *et al.* (1995) A hereditary genetic marker closely associated with microsatellite instability in lung cancer. *Cancer Res.* 55: 3996-3999.
Ryberg D *et al.* (1997) Genotypes of glutathione transferase M1 and P1 and their significance for lung DNA adduct levels and cancer risk. *Carcinogenesis* 18: 1285-1289.
Rygaard K *et al.* (1993) Expression of the proto-oncogenes c-met and c-kit and their ligands, hepatocyte growth factor/scatter factor and stem cell factor, in SCLC cell lines and xenografts. *Br. J. Cancer* 67: 37-46.
Ryu JS *et al.* (2004) Association between polymorphisms of ERCC1 and XPD and survival in non-small-cell lung cancer patients treated with cisplatin combination chemotherapy. *Lung Cancer* 44: 311-316.
Saad RS *et al.* (2004) Prognostic significance of HER2/neu, p53, and vascular endothelial growth factor expression in early stage conventional adenocarcinoma and bronchioloalveolar carcinoma of the lung. *Mod. Pathol.* In press. [DOI: 10.1038/modpathol.3800171]
Saarikoski ST *et al.* (2000) *CYP2D6* ultrarapid metabolizer genotype as a potential modifier of smoking behaviour. *Pharmacogenetics* 10: 5-10.
Sabol SZ *et al.* (1999) A genetic association for cigarette smoking behavior. *Health Psych.* 18: 7-13.
Sachse R *et al.* (1994) DNA aberrations at the retinoblastoma gene locus in human squamous cell carcinomas of the lung. *Oncogene* 9: 39-47.
Sagawa M *et al.* (1998) K-*ras* point mutation occurs in the early stage of carcinogenesis in lung cancer. *Br. J. Cancer* 77: 720-723.
Sage J *et al.* (2003) Acute mutation of retinoblastoma gene function is sufficient for cell cycle re-entry. *Nature* 424: 223-229.
Saito T *et al.* (2003) Sensitive detection of small cell lung carcinoma cells by reverse transcriptase-polymerase chain reaction for prepro-gastrin-releasing peptide mRNA. *Cancer* 97: 2504-2511.
Saji H *et al.* (2003) Significance of expression of TGFβ in pulmonary metastasis in non-small cell lung cancer tissues. *Ann. Thorac. Cardiovasc. Surg.* 9: 295-300.
Sakaguchi M *et al.* (1996) Inversely correlated expression of p16 and Rb protein in non-small cell lung cancers: an immunohistochemical study. *Int. J. Cancer* 65: 442-445.
Sakatani T, Onyango P (2003) Oncogenomics: prospects for the future. *Expert Rev. Anticancer Ther.* 3: 891-901.

Saksela K *et al.* (1985) Amplification of the c-myc oncogene in a subpopulation of human small cell lung cancer. *Cancer Res.* 45: 1823-1827.
Saksela K *et al.* (1986) Amplification of the N-myc oncogene in an adenocarcinoma of the lung. *J. Cell. Biochem.* 31: 297-304.
Salerno CT *et al.* (1998) Detection of occult micrometastases in non-small cell lung carcinoma by reverse transcriptase-polymerase chain reaction. *Chest* 113: 1526-1532.
Sameshima Y *et al.* (1990) Point mutation of the p53 gene resulting in splicing inhibition in small cell lung carcinoma. *Biochem. Biophys. Res. Commun.* 173: 697-703.
Sameshima Y *et al.* (1992) Alterations of the *p53* gene are common and critical events for the maintenance of malignant phenotypes in small-cell lung carcinoma. *Oncogene* 7: 451-457.
Samuels Y *et al.* (2004) High frequency of mutations of the *PIK3CA* gene in human cancers. *Science* 304: 554.
Sanchez-Cespedes M *et al.* (1998) Detection of chromosome 3p alterations in serum DNA of non-small-cell lung cancer patients. *Ann. Oncol.* 9: 113-116.
Sanchez-Cespedes M *et al.* (1999) Inactivation of the INK4A/ARF locus frequently coexists with *TP53* mutations in non-small cell lung cancer. *Oncogene* 18: 5843-5849.
Sanchez-Cespedes M *et al.* (2001a) Chromosomal alterations in lung adenocarcinomas from smokers and nonsmokers. *Cancer Res.* 61: 1309-1313.
Sanchez-Cespedes M *et al.* (2001b) Increased loss of chromosome 9p21 but not *p16* inactivation in primary non-small cell lung cancer from smokers. *Cancer Res.* 61: 2092-2096.
Sanchez-Cespedes M *et al.* (2001c) Identification of a mononucleotide repeat as a major target for mitochondrial DNA alterations in human tumors. *Cancer Res.* 61: 7015-7019.
Sanchez-Cespedes M *et al.* (2002) Inactivation of *LKB1/STK11* is a common event in adenocarcinomas of the lung. *Cancer Res.* 62: 3659-3662.
Santaraosa M, Ashworth A (2004) Haploinsufficiency for tumour suppressor genes; when you don't need to go all the way. *Biochim. Biophys. Acta* 1654: 105-122.
Santoro MM *et al.* (1996) Constitutive activation of the RON gene promotes invasive growth but not transformation. *Mol. Cell. Biol.* 16: 7072-7083.
Santos E *et al.* (1984) Malignant activation of a K-ras oncogene in lung carcinoma but not in normal tissue of the same patient. *Science* 223: 661-664.
Sanz-Ortega J *et al.* (1999) LOH at the APC/MCC gene (5q21) is frequent in early stages of non-small cell lung cancer. *Pathol. Res. Pract.* 195: 677-680.
Sanz-Ortega J *et al.* (2001) 3p21, 5q21, and 9p21 allelic deletions are frequently found in normal bronchial cells adjacent to non-small-cell lung cancer, while they are unusual in patients with no evidence of malignancy. *J. Pathol.* 195: 429-434.
Sarasin A (2003) An overview of the mechanisms of mutagenesis and carcinogenesis. *Mutation Res.* 544: 99-106.
Sard L *et al.* (1999) The tumour-suppressor gene *FHIT* is involved in the regulation of apoptosis and in cell cycle control. *Proc. Natl. Acad. Sci USA* 96: 8489-8492.
Saretzki G *et al.* (2002) hTERT gene dosage correlates with telomerase activity in human lung cancer cell lines. *Cancer Lett.* 176: 81-91.
Sarries C *et al.* (2002) Pharmacogenomic strategies for developing customized chemotherapy in non-small cell lung cancer. *Pharmacogenomics* 3: 763-780.
Sarvesvaran J *et al.* (1999) Is small cell lung cancer the perfect target for anti-telomerase treatment? *Carcinogenesis* 20: 1649-1651.
Sasaki H *et al.* (1997) Elevated plasma thymosin-alpha1 levels in lung cancer patients. *Eur. J. Cardiothorac. Surg.* 12: 885-891.
Sasaki H *et al.* (2001a) Expression of the sensitive to apoptosis gene, SAG, as a prognostic marker in nonsmall cell lung cancer. *Int. J. Cancer* 95: 375-377.
Sasaki H *et al.* (2001b) Expression of the antiapoptosis gene, AAC-11, as a prognosis marker in non-small cell lung cancer. *Lung Cancer* 34: 53-57.
Sasaki H *et al.* (2001c) Expression of periostin, homologous with an insect cell adhesion molecule, as a prognostic marker in non-small cell lung cancers. *Jpn. J. Cancer Res.* 92: 869-873.
Sasaki H *et al.* (2001d) Expression of the *cdc25B* gene as a prognostic marker in non-small cell lung cancer. *Cancer Lett.* 173: 187-192.
Sasaki H *et al.* (2001e) Overexpression of *Hrad17* gene in non-small cell lung cancers correlated with lymph node metastasis. *Lung Cancer* 34: 47-52.
Sasaki H *et al.* (2001f) Expression of the prothymosin-α gene as a prognostic factor in lung cancer. *Surg. Today* 31: 936-938.
Sasaki H *et al.* (2001 g) Clinical significance of matrix metalloproteinase-7 and Ets-1 gene expression in patients with lung cancer. *J. Surg. Res.* 101: 242-247.

Sasaki H *et al.* (2002a) Beta IGH3, a TGF-beta inducible gene, is overexpressed in lung cancer. *Jpn. J. Clin. Oncol.* 32: 85-89.

Sasaki H *et al.* (2002b) Expression of the *MTA1* mRNA in advanced lung cancer. *Lung Cancer* 35: 149-154.

Sasaki H *et al.* (2002c) Decreased perioxisome proliferator-activated receptor gamma gene expression was correlated with poor prognosis in patients with lung cancer. *Lung Cancer* 36: 71-76.

Sasaki H *et al.* (2003a) *Cten* mRNA expression was correlated with tumor progression in lung cancers. *Lung Cancer* 40: 151-155.

Sasaki H *et al.* (2003b) Molecular processes of chromosome 9p21 deletions in human cancers. *Oncogene* 22: 3792-3798.

Sasaki H *et al.* (2003c) Angioarrestin mRNA expression in early-stage lung cancers. *Eur. J. Surg. Oncol.* 29: 649-653.

Sasaki H *et al.* (2003d) SAGE mRNA expression in advanced-stage lung cancers. *Eur. J. Surg. Oncol.* 29: 900-903.

Sathyanarayana UG *et al.* (2003) Epigenetic inactivation of laminin-5-encoding genes in lung cancers. *Clin. Cancer Res.* 9: 2665-2672.

Sato M *et al.* (1998a) The H-cadherin (*CDH13*) gene is inactivated in human lung cancer. *Hum. Genet.* 103: 96-101.

Sato M *et al.* (1998b) Identification of a 910-kb region of common allelic loss in chromosome bands 16q24.1-q24.2 in human lung cancer. *Genes Chrom. Cancer* 22: 1-8.

Sato H *et al.* (1999) Alternative splicing, but not allelic loss, of the *FHIT* gene increases with development of lung cancer. *Int. J. Oncol.* 15: 81-88.

Sato M *et al.* (2002) The expression of DNA methyltransferases and methyl-CpG-binding proteins is not associated with the methylation status of p14(ARF), p16(INK4a) and RASSF1A in human lung cancer cell lines. *Oncogene* 21: 4822-4829.

Sato M *et al.* (2003) Increased expression and no mutation of the Flap endonuclease (*FEN1*) gene in human lung cancer. *Oncogene* 22: 7243-7246.

Satoh T *et al.* (2002) Activation of peroxisome proliferator-activated receptor-γ stimulates the growth arrest and DNA-damage inducible 153 gene in non-small cell lung carcinoma cells. *Oncogene* 21: 2171-2180.

Sattler M, Salgia R (2003) Molecular and cellular biology of small cell lung cancer. *Semin. Oncol.* 30: 57-71.

Savaraj N *et al.* (2003) Overexpression of mutated MRP4 in cisplatin-resistant small cell lung cancer cell line: collateral sensitivity to azidothymidine. *Int. J. Oncol.* 23: 173-179.

Schabath MB *et al.* (2000) Genetic variants of myeloperoxidase and lung cancer risk. *Carcinogenesis* 21: 1163-1166.

Schabath MB *et al.* (2002) A myeloperoxidase polymorphism associated with reduced risk of lung cancer. *Lung Cancer* 37: 35-40.

Schaefer C *et al.* (2001) A new cancer genome anatomy project web resource for the community. *Cancer J.* 7: 52-60.

Schairer E, Schoniger E (1943) Lungenkrebs und Tabakverbrauch. *Z. Krebsforsch.* 54: 261-269.

Schauer IE *et al.* (1994) Cyclin D1 overexpression vs. retinoblastoma inactivation: implications for growth control evasion in non-small cell and small cell lung cancer. *Proc. Natl. Acad. Sci. USA* 91: 7827-7831.

Scheel J *et al.* (2002) Variability of the human aryl hydrocarbon receptor nuclear translocator (*ARNT*) gene. *J. Hum. Genet.* 47: 217-224.

Schenk S *et al.* (2001) A novel polymorphism in the promoter of the RAGE gene is associated with non-small cell lung cancer. *Lung Cancer* 32: 7-12.

Scherf U *et al.* (2000) A gene expression database for the molecular pharmacology of cancer. *Nature Genet.* 24: 236-244.

Schlegel U *et al.* (1992) p53 gene mutations in primary lung tumors are conserved in brain metastases. *J. Neurooncol.* 14: 93-100.

Schmid M *et al.* (1998) Homozygous deletions of methylthioadenosine phosphorylase (MTAP) are more frequent than p16INK4A (CDKN2) homozygous deletions in primary non-small cell lung cancers (NSCLC). *Oncogene* 17: 2669-2675.

Schmid M *et al.* (2000) A methylthioadenosine phosphorylase (MTAP) fusion transcript identifies a new gene on chromosome 9p21 that is frequently deleted in cancer. *Oncogene* 19: 5747-5754.

Schmidt RA *et al.* (1992) A flow cytometric study of non-small cell lung cancer classified as T1NO. *Cancer* 69: 78-85.

Schmidt B *et al.* (2004) Detection of cell-free nucleic acids in bronchial lavage fluid supernatants from patients with lung cancer. *Eur. J. Cancer* 40: 452-460.

Schneider PM *et al.* (1989) Differential expression of the c-*erbB*-2 gene in human small cell and non-small cell lung cancer. *Cancer Res.* 49: 4968-4971.
Schneider J *et al.* (2003) Pro-gastrin-releasing peptide (ProGRP), neuron specific enolase (NSE), carcinoembryonic antigen (CEA) and cytokeratin 19-fragments (CYFRA 21-1) in patients with lung cancer in comparison to other lung diseases. *Anticancer Res.* 23(2A): 885-893.
Schneider S *et al.* (2004a) Osteopontin but not osteonectin messenger RNA expression is a prognostic marker in curatively resected non-small cell lung cancer. *Clin. Cancer Res.* 10: 1588-1596.
Schneider J *et al.* (2004b) *GSTM1*, *GSTT1*, and *GSTP1* polymorphism and lung cancer risk in relation to tobacco smoking. *Cancer Letts.* 208: 65-74.
Schneider-Stock R *et al.* (2004) Retention of the arginine allele in codon 72 of the p53 gene correlates with poor apoptosis in head and neck cancer. *Am. J. Pathol.* 164: 1233-1241.
Schoket B *et al.* (1998) Smoking-associated bulky DNA adducts in bronchial tissue related to *CYP1A1 Msp*I and *GSTM1* genotypes in lung patients. *Carcinogenesis* 19: 841-846.
Schottenfeld D (2000) Etiology and epidemiology of lung cancer. Chap. 20 in *Lung Cancer. Principles and Practice.* 2nd Ed. HI Pass, JB Mitchell, DH Johnson, AT Turrisi, JD Minna (Eds.). Lippincott Williams & Williams, Philadelphia, pp 367-388.
Schrami P *et al.* (1999) Tissue microarrays for gene amplification surveys in many different tumor types. *Clin. Cancer Res.* 5: 1966-1975.
Schreiber E *et al.* (1992) Human small cell lung cancer expresses the octamer DNA-binding and nervous system-specific transcription factor N-Oct 3 (brain-2). *Cancer Res.* 52: 6121-6124.
Schuler M *et al.* (1998) A phase I study of adenovirus-mediated wild-type p53 gene transfer in patients with advanced non-small cell lung cancer. *Hum. Gene Ther.* 9: 2075-2082.
Schuler M *et al.* (2001) Adenovirus-mediated wild-type p53 gene transfer in patients receiving chemotherapy for advanced non-small-cell lung cancer: results of a multicenter phase II study. *J. Clin. Oncol.* 19: 1750-1758.
Schwab M (1998) Amplification of oncogenes in human cancer cells. *BioEssays* 20: 473-479.
Schwartz AG (2000) Genetic susceptibility to lung cancer. Chap. 21 in *Lung Cancer. Principles and Practice.* 2nd Ed. HI Pass, JB Mitchell, DH Johnson, AT Turrisi, JD Minna (Eds.). Lippincott Williams & Williams, Philadelphia, pp 389-397.
Schwartz AG *et al.* (1996) Familial risk of lung cancer among nonsmokers and their relatives. *Am. J. Epidemiol.* 144: 554-562.
Schwartz AG *et al.* (1999) Increased cancer risk among relatives of non-smoking lung cancer cases. *Genet. Epidemiol.* 17: 1-15.
Schwendel A *et al.* (1997) Primary small-cell lung carcinomas and their metastases are characterized by a recurrent pattern of genetic alterations. *Int. J. Cancer* 74: 86-93.
Scian MJ *et al.* (2004) Tumor-derived p53 mutants induce oncogenesis by transactivating growth-promoting genes. *Oncogene* 23: 4430-4443.
Scolnick DM, Halazonetis TD (2000) Chfr defines a mitotic stress checkpoint that delays entry into metaphase. *Nature* 406: 430-435.
Scott FM *et al.* (1997) High frequency of K-ras codon 12 mutations in bronchoalveolar lavage fluid of patients at high risk for second primary lung cancer. *Clin. Cancer Res.* 3: 479-482.
Seifart C *et al.* (2002) Surfactant protein B gene variations enhance susceptibility to squamous cell carcinoma of the lung in German patients. *Br. J. Cancer* 87: 212-217.
Seike M *et al.* (2002) The promoter region of the human *BUBR1* gene and its expression analysis in lung cancer. *Lung Cancer* 38: 229-234.
Seker H *et al.* (2001) Functional significance of *XPD* polymorphic variants: attenuated apoptosis in human lymphoblastoid cells with the *XPD* 312 Asp/Asp genotype. *Cancer Res.* 61: 7430-7434.
Seki Y *et al.* (2004) STAT3 and MAPK in human lung cancer tissues and suppression of oncogenic growth by JAB and dominant negative STAT3. *Int. J. Oncol.* 24: 931-934.
Sekido Y *et al.* (1991) Preferential expression of *c-kit* protooncogene transcripts in small cell lung cancer. *Cancer Res.* 51: 2416-2419.
Sekido Y *et al.* (1993) Recombinant human stem cell factor mediates chemotaxis of small-cell lung cancer cell lines aberrantly expressing the *c-kit* protooncogene. *Cancer Res.* 53: 1709-1714.
Sekido Y *et al.* (1994) Molecular analysis of the von Hippel-Lindau disease tumor suppressor gene in human lung cancer cell lines. *Oncogene* 9: 1599-1604.
Sekido Y *et al.* (1996) Human semaphorins A(V) and IV reside in the 3p21.3 small cell lung cancer deletion region and demonstrate distinct expression patterns. *Proc. Natl. Acad. Sci. USA* 93: 4120-4125.
Sekido Y *et al.* (2003) Molecular genetics of lung cancer. *Ann. Rev. Med.* 54: 73-87.
Sekimura A *et al.* (2004) Expression of Smac/DIABLO is a novel prognostic marker in lung cancer. *Oncol. Rep.* 11: 797-802.

Sekine I *et al.* (1997) Microsatellite instability in lung cancer patients 40 years of age or younger. *Jpn. J. Cancer Res.* 88: 559-563.
Sekine I *et al.* (1998) Role of epidermal growth factor receptor overexpression, K-ras point mutation and c-myc amplification in the carcinogenesis of non-small cell lung cancer. *Oncol. Rep.* 5: 351-354.
Sellers TA *et al.* (1988) Familial risk of cancer inheritance in the pathogenesis of lung cancer. *Genet. Epidemiol.* 5: 381-391.
Sellers TA *et al.* (1990) Evidence for mendelian inheritance in the pathogenesis of lung cancer. *J. Natl. Cancer Inst.* 82: 1272-1279.
Selvaggi G *et al.* (2002) HER-2/neu overexpression in patients with radically resected nonsmall cell lung carcinoma. Impact on long-term survival. *Cancer* 94: 2669-2674.
Selvaggi G *et al.* (2004) Epidermal growth factor receptor overexpression correlates with a poor prognosis in completely resected non-small-cell lung cancer. *Ann. Oncol.* 15: 28-32.
Sémon M, Duret L (2004) Evidence that functional transcription units cover at least half of the human genome. *Trends Genet.* 20: 229-232.
Sen S *et al.* (2001) A comparative study of telomerase activity in sputum, bronchial washing and biopsy specimens of lung cancer. *Lung Cancer* 33: 41-49.
Senchenko VN *et al.* (2004) Discovery of frequent homozygous deletions in chromosome 3p21.3 LUCA and AP20 regions in renal, lung and breast carcinomas. *Oncogene* 23: 5719-5728.
Seow A *et al.* (1999) NAT2 slow acetylator genotype is associated with increased risk of lung cancer among non-smoking Chinese women in Singapore. *Carcinogenesis* 20: 1877-1881.
Seregni E *et al.* (1996) Pattern of mucin gene expression in normal and neoplastic lung tissues. *Anticancer Res.* 16(4B): 2209-2213.
Shabnam MS *et al.* (2004) Expression of p53 protein and the apoptotic regulatory molecules Bcl-2, Bcl-X(L), and Bax in locally advanced squamous cell carcinoma of the lung. *Lung Cancer* 45: 181-188.
Shan Z *et al.* (2004) Identifying novel homozygous deletions by microsatellite analysis and characterization of tumor suppressor candidate 1 gene, *TUSC1*, on chromsome 9p in human lung cancer. *Oncogene* In press. [DOI: 10.1038/sj.onc.1207857].
Shanley TP *et al.* (2001) The serine/threonine phosphatase, PP2A: endogenous regulator of inflammatory cell signalling. *J. Immunol.* 166: 966-972.
Shapiro GI *et al.* (1995a) Reciprocal Rb inactivation and p16INK4 expression in primary lung cancers and cell lines. *Cancer Res.* 55: 505-509.
Shapiro GI *et al.* (1995b) Multiple mechanisms of p16INK4A inactivation in non-small cell lung cancer cell lines. *Cancer Res.* 55: 6200-6209.
Shaw GL *et al.* (1991) Lung cancer risk associated with cancer in relatives. *J. Clin. Epidemiol.* 44: 429-437.
Shaw GL *et al.* (1998) Genetic polymorphism of *CYP2D6* and lung cancer risk. *Cancer Epidemiol. Biomarkers Prev.* 7: 215-219.
Shay JW, Bacchetti S (1997) A survey of telomerase activity in human cancer. *Eur. J. Cancer* 33: 787-791.
Shay JW *et al.* (2001a) Telomerase and cancer. *Hum. Mol. Genet.* 10: 677-685.
Shay JW *et al.* (2001b) Role of telomeres and telomerase in aging and cancer. Chapter 21 in *Molecular Genetics of Cancer*. JK Cowell (Ed), 2nd Ed. pp 483-508. BIOS Scientific, Oxford.
Shen Y, White E (2001) p53-dependent apoptosis pathways. *Adv. Cancer Res.* 82: 55-84.
Shen JC *et al.* (1992) High frequency mutagenesis by a DNA methyltransferase. *Cell* 71: 1073-1080.
Shen H *et al.* (2002) A novel polymorphism in human cytosine DNA-methyltransferase-3B promoter is associated with an increased risk of lung cancer. *Cancer Res.* 62: 4992-4995.
Shen H *et al.* (2003a) Smoking, DNA repair capacity and risk of nonsmall cell lung cancer. *Int. J. Cancer* 107: 84-88.
Shen H *et al.* (2003b) Dietary folate intake and lung cancer risk in former smokers: a case-control analysis. *Cancer Epidemiol. Biomarkers Prev.* 12: 980-986.
Sherr CJ (1996) Cancer cell cycles. *Science* 274: 1672-1677.
Sherr CJ (2000) Cancer cell cycles revisited. *Cancer Res.* 60: 3689-3695.
Sherr CJ (2004) Principles of tumor suppression. *Cell* 116: 235-246.
Shi D *et al.* (1992) Overexpression of the c-erbB2/neu-encoded p-185 protein in primary lung cancer. *Mol. Carcinogen.* 5: 213-218.
Shi X *et al.* (2003) A single strand conformation polymorphism method by capillary electrophoresis with laser-induced fluorescence for detection of T1151A mutation in hMLH1 gene. *Electrophoresis* 24: 2316-2321.
Shi WX, Chen SQ (2004) Frequencies of poor metabolizers of cytochrome P450 2C19 in esophagus cancer, stomach cancer, lung cancer and bladder cancer. *World J. Gastroenterol.* 10: 1961-1963.

Shibuya K *et al.* (2001) Increased telomerase activity and elevated hTERT mRNA expression during multistage carcinogenesis of squamous cell carcinoma of a lung. *Cancer* 92: 849-855.
Shields PG (2000) Epidemiology of tobacco carcinogenesis. *Curr. Oncol. Rep.* 2: 257-262.
Shields PG (2002) Molecular epidemiology of smoking and lung cancer. *Oncogene* 21: 6870-6876.
Shigemitsu K *et al.* (2001) Genetic alteration of the beta-catenin gene (*CTNNB1*) in human lung cancer and malignant mesothelioma and identification of a new 3p21.3 homozygous deletion. *Oncogene* 20: 4249-4257.
Shih CM *et al.* (2000) Lack of evidence of association of p21WAF1/CIP1 polymorphism with lung cancer susceptibility and prognosis in Taiwan. *Jpn. J. Cancer Res.* 91: 9-15.
Shih J-Y *et al.* (2001) Collapsin response mediator protein-1 and the invasion and metastasis of cancer cells. *J. Natl. Cancer Inst.* 93: 1392-1400.
Shih CM *et al.* (2002) Association of L-myc polymorphism with lung cancer susceptibility and prognosis in relation to age-selected controls and stratified cases. *Lung Cancer* 36: 125-132.
Shih J-Y *et al.* (2003) Collapsin response mediator protein-1: a novel invasion-suppressor gene. *Clin. Exp. Metastasis* 20: 69-76.
Shikada Y *et al.* (2003) Higher expression of RhoC is related to invasiveness in non-small cell lung carcinoma. *Clin. Cancer Res.* 9: 5282-5286.
Shiloh Y (2003) ATM and related protein kinases: safeguarding genome integrity. *Nature Rev. Cancer* 3: 155-168.
Shimakage M *et al.* (2002) Expression of drs mRNA in human lung adenocarcinomas. *Hum. Pathol.* 33: 615-619.
Shimamoto T *et al.* (2004) Hypermethylation of E-cadherin gene is frequent and independent of p16INK4A methylation in non-small cell lung cancer: potential prognostic implication. *Oncol. Rep.* 12: 389-395.
Shimizu T, Sekiya T (1995) Loss of heterozygosity at 9p21 loci and mutations of the MTS1 and MTS2 genes in human lung cancers. *Int. J. Cancer* 63: 616-620.
Shimizu K *et al.* (1983) Structure of the Ki-ras gene of the human lung carcinoma cell line Calu-1. *Nature* 304: 497-500.
Shimizu E *et al.* (1994) RB protein status and clinical correlation from 171 cell lines representing lung cancer, extrapulmonary small cell carcinoma, and mesothelioma. *Oncogene* 9: 2441-2448.
Shimizu T *et al.* (1996) Absence of a mutation of the p21/WAF1 gene in human lung and pancreatic cancers. *Jpn. J. Cancer Res.* 87: 275-278.
Shimizu S *et al.* (2000a) High frequency of clonally related tumors in cases of multiple synchronous lung cancers as revealed by molecular diagnosis. *Clin. Cancer Res.* 6: 3994-3999.
Shimizu Y *et al.* (2000b) Benzo[*a*]pyrene carcinogenicity is lost in mice lacking the aryl hydrocarbon receptor. *Proc. Natl. Acad. Sci USA* 97: 779-782.
Shin MS *et al.* (2002) Alterations of Fas-pathway genes associated with nodal metastasis in non-small cell lung cancer. *Oncogene* 21: 4129-4136.
Shin HS *et al.* (2004) The relationship between the serum intercellular adhesion molecule-1 and the prognosis of the disease in lung cancer. *Korean J. Intern. Med.* 19: 48-52.
Shinmura K, Yokota J (2001) The *OGG1* gene encodes a repair enzyme for oxidatively damaged DNA and is involved in human carcinogenesis. *Antioxid. Redox Signal.* 3: 597-609.
Shinmura K *et al.* (2001) Somatic mutations and single nucleotide polymorphisms of base excision repair genes involved in the repair of 8-hydroxyguanine in damaged DNA. *Cancer Letts.* 166: 65-69.
Shintani Y *et al.* (2003) New prognostic indicator for non-small-cell lung cancer, quantitation of thymidylate synthase by real-time reverse transcription polymerase chain reaction. *Int. J. Cancer* 104: 790-795.
Shintani Y *et al.* (2004) Overexpression of *ADAM9* in non-small cell lung cancer correlates with brain metastasis. *Cancer Res.* 64: 4190-4196.
Shioda T (2004) Application of DNA microarray to toxicological research. *J. Environ. Pathol. Toxicol. Oncol.* 23: 13-31.
Shipman R (1996) Frequent *TP53* gene alterations (mutations, allelic loss, nuclear accumulation) in primary non-small cell lung cancer. *Eur. J. Cancer* 32A: 335-341.
Shiraishi M *et al.* (1989) Amplification of protooncogenes in surgical specimens of human lung carcinomas. *Cancer Res.* 49: 6474-6479.
Shiraishi M *et al.* (2002a) A comprehensive catolog of CpG islands methylated in human lung adenocarcinomas for the identification of tumor suppressor genes. *Oncogene* 21: 3804-3813.
Shiraishi M *et al.* (2002b) *HOX* gene clusters are hotspots of *de novo* methylation in CpG islands of human lung adenocarcinomas. *Oncogene* 21: 3659-3662.
Shirotani Y *et al.* (1994) Alteration in length of telomeric repeats in lung cancer. *Lung Cancer* 11: 29-41.

Shiseki M *et al.* (1996) Comparative allelotype of early and advanced stage non-small cell lung carcinomas. *Genes Chrom. Cancer* 17: 71-77.

Shivapurkar N *et al.* (1999) Deletions of chromosome 4 at multiple sites are frequent in malignant mesothelioma and small cell lung carcinoma. *Clin. Cancer Res.* 5: 17-23.

Shivapurkar N *et al.* (2002a) Loss of expression of death-inducing signalling complex (DISC) components in lung cancer cell lines and the influence of *MYC* amplification. *Oncogene* 21: 8510-8514.

Shivapurkar N *et al.* (2002b) Differential inactivation of caspase-8 in lung cancers. *Cancer Biol. Ther.* 1: 65-69.

Shivapurkar N *et al.* (2003) Apoptosis and lung cancer: a review. *J. Cell. Biochem.* 88: 885-898.

Shoji T *et al.* (2002) Clinical significance of p21 expression in non-small-cell lung cancer. *J. Clin. Oncol.* 20: 3865-3871.

Shridhar R *et al.* (1994) Genetic instability of microsatellite sequences in many non-small cell lung carcinomas. *Cancer Res.* 54: 2084-2087.

Shridhar R *et al.* (1996) Mutations in the arginine-rich protein gene, in lung, breast, and prostate cancers, and in squamous cell carcinoma of the head and neck. *Cancer Res.* 56: 5576-5578.

Shriver SP *et al.* (2000) Sex-specific expression of gastrin-releasing peptide receptor: relationship to smoking history and risk of lung cancer. *J. Natl. Cancer Inst.* 92: 24-33.

Shu W *et al.* (2001) Characterization of a new subfamily of winged-helix/forkhead (Fox) genes that are expressed in the lung and act as transcription repressors. *J. Biol. Chem.* 276: 27488-27497.

Shuey DJ *et al.* (2002) RNAi: gene silencing in therapeutic intervention. *Drug Discov. Today* 7: 1040-1046.

Siddiq F *et al.* (2004) Increased osteonectin expression is associated with malignant transformation and tumor associated fibrosis in the lung. *Lung Cancer* 45: 197-205.

Sieber OM *et al.* (2003) Genomic instability – the engine of tumoigenesis? *Nature Rev. Cancer* 3: 701-708.

Siegfried JM *et al.* (1997a) Prognostic value of specific KRAS mutations in lung adenocarcinomas. *Cancer Epidemiol. Biomarkers Prev.* 6: 841-841.

Siegfried JM *et al.* (1997b) Association of immunoreactive hepatocyte growth factor with poor survival in resectable non-small cell lung cancer. *Cancer Res.* 57: 433-439.

Siegfried JM *et al.* (1999) Evidence for autocrine actions of neuromedin B and gastrin-releasing peptide in non-small cell lung cancer. *Pulm. Pharmacol. Ther.* 12: 291-302.

Siemianowicz K *et al.* (2003) Methylenetetrahydrofolate reductase gene C677T and A1298C polymorphisms in patients with small cell and non-small cell lung cancer. *Oncol. Rep.* 10: 1341-1344.

Sienel W *et al.* (2003) Elevated expression of carcinoembryonic antigen-related cell adhesion molecule 1 promotes progression of non-small cell lung cancer. *Clin. Cancer Res.* 9: 2260-2266.

Sikkink SK *et al.* (2003) In-depth analysis of molecular alterations within normal and tumour tissue from an entire bronchial tree. *Int. J. Oncol.* 22: 589-595.

Sikstrom C *et al.* (1996) Transferrin C3 offers protection against smoking-associated lung cancer? *Carcinogenesis* 17: 1447-1449.

Silini EM *et al.* (1994) K-ras gene mutations: an unfavourable prognostic marker in stage I lung adenocarcinoma. *Virchows Arch.* 424: 367-373.

Silva JM *et al.* (1999) *TP53* gene mutations in plasma DNA of cancer patients. *Genes Chrom. Cancer* 24: 160-161.

Simi L *et al.* (2004) Simultaneous measurement of *MMP9* and *TIMP1* mRNA in human non small cell lung cancers by multiplex real time RT-PCR. *Lung Cancer* 45: 171-179.

Simon GR, Wagner H (2003) Small cell lung cancer. *Chest* 123 1 Suppl.: 259S-271S.

Simon R *et al.* (2003) Tissue microarrays in cancer diagnosis. *Expert Rev. Mol. Diagn.* 3: 421-430.

Singal DP *et al.* (1996) Molecular basis for lack of expression of HLA class I antigens in human small-cell lung carcinoma cell lines. *Int. J. Cancer* 68: 629-636.

Singh B, Arlinghaus RB (1997) Mos and the cell cycle. *Prog. Cell Cycle Res.* 3: 251-259.

Singh B *et al.* (2002) p53 regulates cell survival by inhibiting PIK3CA in squamous cell carcinomas. *Genes Dev.* 16: 984-993.

Singh-Kaw P *et al.* (1995) Stimulatory effects of hepatocyte growth factor on normal and neoplastic human bronchial epithelial cells. *Am. J. Physiol.* 268: 1012-1020.

Singhal S *et al.* (2003a) Alterations in cell cycle genes in early stage lung adenocarcinoma identified by expression profiling. *Cancer Biol. Ther.* 2: 291-298.

Singhal S *et al.* (2003b) Differentially expressed apoptotic genes to early stage lung adenocarcinoma predicted by expression profiling. *Cancer Biol. Ther.* 2: 566-571.

Singleton AB *et al.* (1998) Lack of association between the dopamine D2 receptor gene allele DRD2*A1 and cigarette smoking in a United Kingdom population. *Pharmacogenetics* 8: 125-128.

Siprashvili Z *et al.* (1997) Replacement of Fhit in cancer cells suppresses tumorigenicity. *Proc. Natl. Acad. Sci. USA* 94: 13771-13776.
Sithanandam G *et al.* (2003) Cell cycle activation in lung adenocarcinoma cells by the ErbB3/phosphatidylinositol 3-kinase/Akt pathway. *Carcinogenesis* 24: 1581-1592.
Sjalander A *et al.* (1996) Association between the p21 codon 31 Ai (arg) allele and lung cancer. *Hum. Hered.* 46: 221-225.
Skaug V *et al.* (2000) p53 mutations in defined structural and functional domains are related to poor clinical outcome in non-small cell lung cancer patients. *Clin. Cancer Res.* 6: 1031-1037.
Slebos RJC *et al.* (1989) Cellular protooncogenes are infrequently amplified in untreated non-small lung cancer. *Br. J. Cancer* 59: 76-80.
Slebos RJC *et al.* (1990) K-*ras* oncogene activation as a prognostic marker in adenocarcinoma of the lung. *New Engl. J. Med.* 323: 561-565.
Slebos RJC *et al.* (1991) Relationship between K-*ras* oncogene activation and smoking in adenocarcinoma of the human lung. *J. Natl. Cancer Inst.* 83: 1024-1027.
Slebos RJ *et al.* (1998) p53 alterations in atypical alveolar hyperplasia of the human lung. *Hum. Pathol.* 29: 801-808.
Slonim DK (2002) From patterns to pathways: gene expression data analysis comes of age. *Nature Genet.* 32: 502-508.
Slupphaug G *et al.* (2003) The interacting pathways for prevention and repair of oxidative DNA damage. *Mutation Res.* 531: 231-251.
Smiraglia DJ *et al.* (2001) Excessive CpG island hypermethylation in cancer cell lines versus primary human malignancies. *Hum. Mol. Genet.* 10: 1413-1419.
Smith G *et al.* (1995) Metabolic polymorphisms and cancer susceptibility. *Cancer Surveys* 25: 27-65.
Smith AL *et al.* (1996) Extensive areas of aneuploidy are present in the respiratory epithelium of lung cancer patients. *Br. J. Cancer* 73: 203-209.
Smith ML *et al.* (1997) The p53-regulated cyclin G gene promotes cell growth: p53 downstream effectors cyclin G and Gadd45 exert different effects on cisplatin chemosensitivity. *Exp. Cell Res.* 230: 61-68.
Smith LE *et al.* (2000) Targeting of lung cancer mutational hotspots by polycyclic aromatic hydrocarbons. *J. Natl. Cancer Inst.* 92: 803-811.
Smith SL *et al.* (2003) Maspin – the most commonly-expressed gene of the 18q21.3 serpin cluster in lung cancer – is strongly expressed in preneoplastic bronchial lesions. *Oncogene* 22: 8677-8687.
Smits KM *et al.* (2004) Association of metabolic gene polymorphisms with tobacco consumption in healthy controls. *Int. J. Cancer* 110: 266-270.
Snijders PJ *et al.* (2004) Elevated hTERT mRNA levels: a potential determinant of bronchial squamous cell carcinoma (*in situ*). *Int. J. Cancer* 109: 412-417.
Sobti RC *et al.* (2003) *CYP1A1* and *CYP2D6* polymorphism and risk of lung cancer in a North Indian population. *Biomarkers* 8: 415-428.
Soder AI *et al.* (1997) Amplification, increased dosage and *in situ* expression of the telomerase RNA gene in human cancer. *Oncogene* 14: 1013-1021.
Soini Y *et al.* (2001) Expression of γ-glutamyl cysteine synthetase in nonsmall cell lung carcinoma. *Cancer* 92: 2911-2919.
Solan MJ, Werner-Wasik M (2003) Prognostic factors in non-small cell lung cancer. *Semin. Surg. Oncol.* 21: 64-73.
Sommer S. (1994) Does cancer kill the individual and save the species? *Hum. Mutation* 3: 166-169.
Song S *et al.* (1999) Comparative gene expression profiling in response to p53 in a human lung cancer cell line. *Biochem. Biophys. Res. Commun.* 264: 891-895.
Song N *et al.* (2001) *CYP1A1* polymorphism and risk of lung cancer in relation to tobacco smoking: a case-control study in China. *Carcinogenesis* 22: 11-16.
Song L *et al.* (2003) Activation of Stat3 by receptor tyrosine kinases and cytokines regulates survival in human non-small cell carcinoma cells. *Oncogene* 22: 4150-4165.
Song MS *et al.* (2004) The tumour suppressor RASSF1A regulates mitosis by inhibiting the APC-Cdc20 complex. *Nature Cell Biol.* 6: 129-137.
Sonnenschein C, Soto AM (1999) *The Society of Cells.* BIOS Scientific, Oxford.
Soo Choi Y *et al.* (2003) Prognostic significance of E-cadherin and beta-catenin in resected stage I non-small cell lung cancer. *Eur. J. Cardiothorac. Surg.* 24: 441-449.
Sordella R *et al.* (2004) Gefitinib-sensitizing EGFR mutations in lung cancer activate anti-apoptotic pathways. *Science* 305: 1163-1167.
Sorensen JB *et al.* (1993) Interobserver variability in histopathologic subtyping and grading of pulmonary adenocarcinoma. *Cancer* 71: 2971-2976.
Sorensen M *et al.* (2004) Glutathione *S*-transferase T1 null-genotype is associated with an increased risk of lung cancer. *Int. J. Cancer* 110: 219-224.

Soria J-C *et al.* (2002a) Aberrant promoter methylation of multiple genes in bronchial brush samples from former cigarette smokers. *Cancer Res.* 62: 351-355.

Soria J-C *et al.* (2002b) Lack of *PTEN* expression in non-small cell lung cancer could be related to promoter methylation. *Clin. Cancer Res.* 8: 1178-1184.

Soria JC *et al.* (2003a) Retinoic acid recptor beta and telomerase catalytic subunit expression in bronchial epithelium of heavy smokers. *J. Natl. Cancer Inst.* 95: 165-168.

Soria JC *et al.* (2003b) Lack of interleukin-10 expression could predict poor outcome in patients with stage I non-small cell lung cancer. *Clin. Cancer Res.* 9: 1785-1791.

Soung YH *et al.* (2004) Somatic mutations of *CASP3* gene in human cancers. *Hum. Genet.* 115: 112-115.

Soussi T, Béroud C (2003) Significance of *TP53* mutations in human cancer: a critical analysis of mutations at CpG dinucleotides. *Hum. Mutation* 21: 192-200.

Souza RF *et al.* (1996) Microsatellite instability in the insulin-like growth factor II receptor gene in gastrointestinal tumours. *Nature Genet.* 14: 255-258.

Sozzi G *et al.* (1992) Deletions of 17p and p53 mutations in preneoplastic lesions of the lung. *Cancer Res.* 52: 6079-6082.

Sozzi G *et al.* (1995) Genetic evidence for an independent origin of multiple preneoplastic and neoplastic lung lesions. *Cancer Res.* 55: 135-140.

Sozzi G *et al.* (1996) The *FHIT* gene at 3p14.2 is abnormal in lung cancer. *Cell* 85: 17-26.

Sozzi G *et al.* (1997) Association between cigarette smoking and *FHIT* gene alterations in lung cancer. *Cancer Res.* 57: 2121-2123.

Sozzi G *et al.* (1998) Loss of *FHIT* function in lung cancer and preinvasive bronchial lesions. *Cancer Res.* 58: 5032-5037.

Sozzi G *et al.* (1999) Detection of microsatellite alterations in plasma DNA of non-small cell lung cancer patients: a prospect for early diagnosis. *Clin. Cancer Res.* 5: 2689-2692.

Sozzi G *et al.* (2001) Analysis of circulating tumor DNA in plasma at diagnosis and during follow-up of lung cancer patients. *Cancer Res.* 61: 4675-4678.

Sozzi G *et al.* (2003) Quantification of free circulating DNA as a diagnostic marker in lung cancer. *J. Clin. Oncol.* 21: 3902-3908.

Spankuch-Schmitt B *et al.* (2002) Effect of RNA silencing of polo-like kinase-1 (PLK1) on apoptosis and spindle formation in human cancer cells. *J. Natl. Cancer Inst.* 94: 1863-1877.

Spierings DC *et al.* (2003) Expression of TRAIL and TRAIL death receptors in stage III non-small cell lung cancer tumors. *Clin. Cancer Res.* 9: 397-405.

Spira A, Ettinger DS (2004) Multidisciplinary management of lung cancer. *New Engl. J. Med.* 350: 379-392.

Spira A *et al.* (2004) Effects of cigarette smoke on the human airway epithelial cell transcriptome. *Proc. Natl. Acad. Sci. USA* 101: 10143-10148.

Spitz MR *et al.* (1998) Case-control study of the D2 dopamine receptor gene and smoking status in lung cancer patients. *J. Natl. Cancer Inst.* 90: 358-363.

Spitz MR *et al.* (2001) Modulation of nucleotide excision repair capacity by *XPD* polymorphisms in lung cancer patients. *Cancer Res.* 61: 1354-1357.

Spitz MR *et al.* (2003) Genetic susceptibility to lung cancer: the role of DNA damage and repair. *Cancer Epidemiol. Biomarkers Prev.* 12: 689-698.

Spivack SD *et al.* (1997) The molecular epidemiology of lung cancer. *Crit. Rev. Toxicol.* 27: 319-365.

Spivack SD *et al.* (2003) Phase I and II carcinogen metabolism gene expression in human lung tissue and tumors. *Clin. Cancer Res.* 9: 6002-6011.

Sriuranpong V *et al.* (2001) Notch signalling induces cell cycle arrest in small cell lung cancer cells. *Cancer Res.* 61: 3200-3205.

Sriuranpong V *et al.* (2002) Notch signalling induces rapid degradation of achate-scute homolog 1. *Mol. Cell. Biol.* 22: 3129-3139.

Srivastava SK *et al.* (1985) Effects of two major activating lesions on the structure and conformation of human *ras* oncogene products. *Proc. Natl. Acad. Sci. USA* 82: 38-42.

Stabile LP *et al.* (2002) Human non-small cell lung tumors and cells derived from normal lung express both estrogen receptor α and β and show biological responses to estrogen. *Cancer Res.* 62: 2141-2150.

Stabile LP *et al.* (2004) Inhibition of human non-small cell lung tumors by a c-Met antisense/U6 expression plasmid strategy. *Gene Ther.* 11: 325-335.

Stahlman MT *et al.* (1998) Temporal-spatial distribution of hepatocyte nuclear factor-3beta in developing human lung and other foregut deriviatives. *J. Hisochem. Cytochem.* 46: 955-962.

Staunton JE *et al.* (2001) Chemosensitivity prediction by transcriptional profiling. *Proc. Natl. Acad. Sci. USA* 98: 10787-10792.

Stefanou D *et al.* (2004a) Expression of vascular endothelial growth factor (VEGF) and association with microvessel density in small-cell and non-small-cell lung carcinomas. *Histol. Histopathol.* 19: 37-42.
Stefanou D *et al.* (2004b) Expression of vascular endothelial growth factor and the adhesion molecule E-cadherin in non-small cell lung cancer. *Anticancer Res.* 23(6C): 4715-4720.
Stein CK *et al.* (2002) Direct correlation between FRA3B expression and cigarette smoking. *Genes Chrom. Cancer* 34: 333-340.
Stenhouse G *et al.* (2004) Thyroid transcription factor 1 in pulmonary adenocarcinoma. *J. Clin. Pathol.* 57: 383-387.
Stenson PD *et al.* (2003) Human Gene Mutation Database (HGMD®): 2003 update. *Hum. Mutation* 21: 577-581.
Stieber P *et al.* (1999) Pro-gastrin-releasing peptide (ProGRP) – a useful marker in small cell lung carcinomas. *Anticancer Res.* 19(4A): 2673-2678.
Stokes W (1837) *A Treatise on the Diagnosis and Treatment of Diseases of the Chest.* Hodges & Smith, Dublin.
Storm SM, Rapp UR (1993) Oncogene activation: c-*raf*-1 gene mutations in experimental and naturally occurring tumors. *Toxicol. Letts.* 67: 201-210.
Strange RC *et al.* (2001) Glutathione-*S*-transferase family of enzymes. *Mutation Res.* 482: 21-26.
Straub RE *et al.* (1999) Susceptibility genes for nicotine dependence: a genome scan and followup in an independent sample suggest that regions on chromosomes 2, 4, 10, 16, 17 and 18 merit further study. *Mol. Psych.* 4: 129-144.
Strausberg RL *et al.* (2003) Sequence-based cancer genomics: progress, lessons and opportunities. *Nature Rev. Genet.* 4: 409-418.
Struski S *et al.* (2001) Identification of chromosomal loci associated with non-P-glycoprotein-mediated multidrug resistance to topoisomerase II inhibitor in lung adenocarcinoma cell line by comparative genomic hybridisation. *Genes Chrom. Cancer* 30: 136-142.
Stücker I *et al.* (2002) Genetic polymorphisms of glutathione *S*-transferases as modulators of lung cancer susceptibility. *Carcinogenesis* 23: 1475-1481.
Su LK *et al.* (2000) Inactivation of germline mutant *APC* alleles by attenuated somatic mutations: a molecular genetic mechanism for attenuated familial adenomatous polyposis. *Am. J. Hum. Genet.* 67: 582-590.
Su C *et al.* (2002) Analysis on point mutation of the CDKN2/p16 gene in lung cancer. *Zhonghua Yi Xue Yi Chuan Xue Za Zhi* 19: 37-40.
Su L *et al.* (2003a) p53 (codon 72) and p21 (codon 31) polymorphisms alter *in vivo* mRNA expression of p21. *Lung Cancer* 40: 259-266.
Su L *et al.* (2003b) No association between the p21 codon 31 serine-arginine polymorphism and lung cancer risk. *Cancer Epidemiol. Biomarkers Prev.* 12: 174-175.
Sueoka E *et al.* (1999) Heterogeneous nuclear ribonucleoprotein B1 as a new marker of early detection for human lung cancers. *Cancer Res.* 59: 1404-1407.
Sugimura H *et al.* (1990) Association of rare alleles of the Harvey *ras* protooncogene locus with lung cancer. *Cancer Res.* 50: 1857-1862.
Sugimura H *et al.* (1999) hOGG1 Ser326Cys polymorphism and lung cancer susceptibility. *Cancer Epidemiol. Biomarkers Prev.* 8: 669-674.
Sugio K *et al.* (1992) ras gene mutations as a prognostic marker in adenocarcinoma of the human lung without lymph node metastasis. *Cancer Res.* 52: 2903-2906.
Sugio K *et al.* (1994) K-ras mutations are a relatively late event in the pathogenesis of lung carcinomas. *Cancer Res.* 54: 5811-5815.
Sugio K *et al.* (2001) Clinical significance of the Rb expression in adenocarcinoma of the lung. *Anticancer Res.* 21(3B): 1931-1935.
Sugita M *et al.* (2000) Molecular definition of a small amplification domain within 3q26 in tumors of cervix, ovary, and lung. *Cancer Genet. Cytogenet.* 117: 9-18.
Sugita M *et al.* (2002) Combined use of oligonucleotide and tissue microarrays identifies cancer/testis antigens as biomarkers in lung carcinoma. *Cancer Res.* 62: 3971-3979.
Suh Y-A *et al.* (2002) Loss of retinoic acid receptor β gene expression is linked to aberrant histone H3 acetylation in lung cancer cell lines. *Cancer Res.* 62: 3945-3949.
Sui G *et al.* (2004) Yin Yang 1 is a negative regulator of p53. *Cell* 117: 859-872.
Sullivan DC (2002) Challenges and opportunities for *in vivo* imaging in oncology. *Technol. Cancer Res. Treat.* 1: 419-422.
Sun Z, Yang P (2004) Role of imbalance between neutrophil elastase and alpha 1-antitrypsin in cancer development and progression. *Lancet Oncol.* 5: 182-190.
Sun GF *et al.* (1997) Gene deficiency of glutathione *S*-transferase mu isoform associated with susceptibility to lung cancer in a Chinese population. *Cancer Lett.* 113: 169-172.

Sun SY *et al.* (2001) Overexpression of BCL2 blocks TNF-related apoptosis-inducing ligand (TRAIL)-induced apoptosis in human lung cancer cells. *Biochem. Biophys. Res. Commun.* 280: 788-797.

Sun JL *et al.* (2004a) Expression and structure of *BNIP3L* in lung cancer. *Ai Zheng* 23: 8-14.

Sun W *et al.* (2004b) Identification of differentially expressed genes in human lung squamous cell carcinoma using suppression subtractive hybridization. *Cancer Letts.* 212: 83-93

Sunaga N *et al.* (2001) Constitutive activation of the Wnt signalling pathway by *CTNNB1* (β-catenin) mutations in a subset of human lung adenocarcinoma. *Genes Chrom. Cancer* 30: 316-321.

Sunaga N *et al.* (2002) Contribution of the NQO1 and GSTT1 polymorphisms to lung adenocarcinoma susceptibility. *Cancer Epidemiol. Biomarkers Prev.* 11: 730-738.

Sunaga N *et al.* (2004) Different roles for caveolin-1 in the development of non-small cell lung cancer versus small cell lung cancer. *Cancer Res.* 64: 4277-4285.

Sunahara M *et al.* (1999) Mutational analysis of *p51A/TA63γ*, a *p53* homolog, in non-small cell lung cancer and breast cancer. *Oncogene* 18: 3761-3765.

Sundaresan V *et al.* (1992) p53 and chromosome 3 abnormalities, characteristic of malignant lung tumours, are detectable in preinvasive lesions of the bronchus. *Oncogene* 7: 1989-1997.

Sundaresan V *et al.* (1998) Homozygous deletions at 3p12 in breast and lung cancer. *Oncogene* 17: 1723-1729.

Sundberg K *et al.* (2002) Glutathione conjugation and DNA adduct formation of dibenzo[*a*,l]pyrene and benzo[*a*]pyrene diol epoxides in V79 cells stably expressing different human glutathione transferases. *Chem. Res. Toxicol.* 15: 170-179.

Supino R *et al.* (2001) A role for c-myc in DNA damage-induced apoptosis in a human TP53-mutant small-cell lung cancer cell line. *Eur. J. Cancer* 37: 2247-2256.

Suzuki Y *et al.* (1990) Detection of *ras* gene mutations in human lung cancers by single-strand conformation polymorphism analysis of polymerase chain reaction products. *Oncogene* 5: 1037-1043.

Suzuki H *et al.* (1992) *p53* mutations in non-small cell lung cancer in Japan: association between mutations and smoking. *Cancer Res.* 52: 734-736.

Suzuki H *et al.* (1994) Altered imprinting in lung cancer. *Nature Genet.* 6: 332-333.

Suzuki K *et al.* (1998a) Microsatellite instability in female non-small-cell lung cancer patients with familial clustering of malignancy. *Br. J. Cancer* 77: 1003-1008.

Suzuki K *et al.* (1998b) Loss of heterozygosity in the tuberous sclerosis gene associated regions in adenocarcinoma of the lung accompanied by multiple atypical adenomatous hyperplasia. *Int. J. Cancer* 79: 384-389.

Suzuki C *et al.* (2003a) Identification of COX17 as a therapeutic target for non-small cell lung cancer. *Cancer Res.* 63: 7038-7041.

Suzuki M *et al.* (2003b) Alterations in the mitochondrial displacement loop in lung cancers. *Clin. Cancer Res.* 9: 5636-5641.

Suzuki M *et al.* (2004) RNA interference-mediated knockdown of DNA methyltransferase 1 leads to promoter demethylation and gene re-expression in human lung and breast cancer cells. *Cancer Res.* 64: 3137-3143.

Sweeney C *et al.* (2003) Glutathione *S*-transferase M1, T1, and P1 polymorphisms and survival among lung cancer patients. *Cancer Epidemiol. Biomarkers Prev.* 12: 527-533.

Swinson DE *et al.* (2004) Hypoxia-inducible factor-1α in non small cell lung cancer: relation to growth factor, protease and apoptosis pathways. *Int. J. Cancer* 111: 43-50.

Swisher SG, Roth JA (2000) Gene therapy in lung cancer. *Curr. Oncol. Rep.* 2: 64-70.

Swisher SG, Roth JA (2002) p53 gene therapy for lung cancer. *Curr. Oncol. Rep.* 4: 334-340.

Swisher SG *et al.* (1999) Adenovirus-mediated p53 gene transfer in advanced non-small-cell lung cancer. *J. Natl. Cancer Inst.* 91: 763-771.

Swisher SG *et al.* (2003) Induction of p53-regulated genes and tumor regression in lung cancer patients after intratumoral delivery of adenoviral p53 (INGN 201) and radiation therapy. *Clin. Cancer Res.* 9: 93-101.

Sy SM *et al.* (2003) Genetic alterations of lung adenocarcinoma in relation to smoking and ethnicity. *Lung Cancer* 41: 91-99.

Sy SM *et al.* (2004) Distinct patterns of genetic alterations in adenocarcinoma and squamous cell carcinoma of the lung. *Eur. J. Cancer* 40: 1082-1094.

Szabo E *et al.* (1996) Altered cJUN expression: an early event in human lung carcinogenesis. *Cancer Res.* 56: 305-315.

Szabo E, Kalebic T (2003) Chemoprevention of lung cancer: new directions. *Recent Results Cancer Res.* 163: 172-181.

Szyf M *et al.* (2004) DNA demethylation and cancer: therapeutic implications. *Cancer Letts.* 211: 133-143.

Tabor HK *et al.* (2002) Candidate-gene approaches for studying complex genetic traits: practical considerations. *Nature Rev. Genet.* 3: 1-7.
Taga S *et al.* (1997) Prognostic value of the immunohistochemical detection of p16INK4 expression in nonsmall cell lung carcinoma. *Cancer* 80: 389-395.
Taga S *et al.* (1999) Prognostic impact of telomerase activity in non-small cell lung cancers. *Ann. Surg.* 230: 715-720.
Tagawa K *et al.* (1999) Down-regulation of *KAI1* messenger RNA expression is not associated with loss of heterozygosity of the *KAI1* gene region in lung adenocarcinoma. *Jpn. J. Cancer Res.* 90: 970-976.
Taguchi T *et al.* (1997) Combined chromosome microdissection and comparative genomic hybridization detect multiple sites of amplified DNA in a human lung carcinoma cell line. *Genes Chrom. Cancer* 20: 208-212.
Tai AL *et al.* (2004) Recurrent chromosomal imbalances in nonsmall cell lung carcinoma. *Cancer* 100: 1918-1927.
Taioli E *et al.* (1995) A specific African-American *CYP1A1* polymorphism is associated with adenocarcinoma of the lung. *Cancer Res.* 55: 472-473.
Taioli E *et al.* (2003) Polymorphisms in *CYP1A1*, *GSTM1*, *GSTT1* and lung cancer below the age of 45 years. *Int. J. Epidemiol.* 32: 60-63.
Tajima K *et al.* (2003) Expression of cancer/testis (CT) antigens in lung cancer. *Lung Cancer* 42: 23-33.
Takagi Y *et al.* (1998) p53 mutations in non-small-cell lung cancers occurring in individuals without a past history of active smoking. *Br. J. Cancer* 77: 1568-1572.
Takahama M *et al.* (1999) Enhanced expression of Tie2, its ligand angiopoietin-1, vascular endothelial growth factor, and CD31 in human non-small cell lung carcinomas. *Clin. Cancer Res.* 5: 2506-2510.
Takahashi T *et al.* (1989a) Expression and amplification of *myc* gene family in small cell lung cancer and its relation to biological characteristics. *Cancer Res.* 49: 2683-2688.
Takahashi T *et al.* (1989b) *p53*: a frequent target for genetic abnormalities in lung cancer. *Science* 246: 491-494.
Takahashi T *et al.* (1990) Identification of intronic point mutations as an alternative mechanism for p53 inactivation in lung cancer. J. Clin. Invest. 86: 363-369.
Takahashi T *et al.* (1991) The p53 gene is very frequently mutated in small-cell lung cancer with a distinct nucleotide substitution pattern. *Oncogene* 6: 1775-1778.
Takahashi T *et al.* (1992) Wild-type but not mutant p53 suppresses the growth of human lung cancer cells bearing multiple genetic lesions. *Cancer Res.* 52: 2340-2343.
Takahashi T *et al.* (2002) Increased expression of COX-2 in the development of human lung cancers. *J. Environ. Pathol. Toxicol. Oncol.* 21: 177-181.
Takai D *et al.* (2001) Silencing of *HTR1B* and reduced expression of *EDN1* in human lung cancers, revealed by methylation-sensitive representational difference analysis. *Oncogene* 20: 7505-7513.
Takamizawa J *et al.* (2004) Reduced expression of the *let-7* microRNAs in human lung cancers in association with shortened postoperative survival. *Cancer Res.* 64: 3753-3756.
Takakura S *et al.* (2001) Genetic alterations and expression of the protein phosphatase 1 genes in human cancers. *Int. J. Oncol.* 18: 817-824.
Takamochi K *et al.* (2001) Loss of heterozygosity on chromosomes 9q and 16p in atypical adenomatous hyperplasia concomitant with adenocarcinoma of the lung. *Am. J. Pathol.* 159: 1941-1948.
Takanami I *et al.* (1996) Immunohistochemical detection of basic fibroblast growth factor as a prognostic indicator in pulmonary adenocarcinoma. *Jpn. J. Clin. Oncol.* 26: 293-297.
Takanami I *et al.* (2001a) Significance of autocrine motility factor receptor gene expression as a prognostic factor in non-small-cell lung cancer. *Int. J. Cancer* 95: 384-387.
Takanami I *et al.* (2001b) Expression of ETS-1 is correlated with urokinase-type plasminogen activator and poor prognosis in pulmonary adenocarcinoma. *Tumour Biol.* 22: 205-210.
Takanami I (2003) Overexpression of CCR7 mRNA in nonsmall cell lung cancer: correlation with lymph node metastasis. *Int. J. Cancer* 105: 186-189.
Takanami I, Takeuchi K (2003) Autocrine motility factor-receptor gene expression in lung cancer. *Jpn. J. Thorac. Cardiovasc. Surg.* 51: 368-373.
Takata T *et al.* (2001) Clinical significance of caspase-3 expression in pathologic-stage I, nonsmall-cell lung cancer. *Int. J. Cancer* 96 Suppl.: 54-60.
Takebayashi Y *et al.* (2001) Loss of heterozygosity of nucleotide excision repair factors in sporadic ovarian, colon and lung carcinomas: implication for their roles of carcinogenesis in human solid tumors. *Cancer Letts* 174: 115-125.
Takeda S *et al.* (1993) Detection of *K-ras* mutation in sputum by mutant-allele-specific amplification (MASA). *Hum. Mutation* 2: 112-117.

Takei K *et al.* (1998) A novel tumor suppressor locus on chromosome 18q involved in the development of human lung cancer. *Cancer Res.* 58: 3700-3705.

Takenoshita S *et al.* (1997) Absence of mutations in the transforming growth factor-beta type II receptor in sporadic lung cancers with microsatellite instability and rare H-ras-1 alleles. *Carcinogenesis* 18: 1427-1429.

Takeshima Y *et al.* (1998) p21WAF1/CIP1 expression in primary lung adenocarcinomas: heterogeneous expression in tumor tissues and correlation with p53 expression and proliferative activities. *Carcinogenesis* 19: 1755-1761.

Takeshita H *et al.* (1999) Expression of the *DMBT1* gene is frequently suppressed in human lung cancer. *Jpn. J. Cancer Res.* 90: 903-908.

Takimoto M *et al.* (2002) Frequent expression of new cancer/testis gene D40/AF15q14 in lung cancers of smokers. *Br. J. Cancer* 86: 1757-1762.

Tamura K *et al.* (1997) Deletion of three distinct regions on chromosome 13q in human non-small-cell lung cancer. *Int. J. Cancer* 74: 45-49.

Tamura M *et al.* (1999) PTEN gene and integrin signalling in cancer. *J. Natl. Cancer Inst.* 91: 1820-1828.

Tamura K *et al.* (2002) Homozygous deletion on the chromosomal region 5q12.3 in human lines of small-cell lung cancers. *J. Hum. Genet.* 47: 348-354.

Tan Y *et al.* (1996) IL-2 gene therapy of advanced lung cancer patients. *Anticancer Res.* 16: 1993-1998.

Tan W *et al.* (2001) Frequency of *CYP2A6* gene deletion and its relation to risk of lung and esophageal cancer in the Chinese population. *Int. J. Cancer* 95: 96-101.

Tan D *et al.* (2003a) Loss of Cables protein expression in human non-small cell lung cancer: a tissue microarray study. *Hum. Pathol.* 34: 143-149.

Tan DF *et al.* (2003b) Prognostic significance of expression of p53 oncoprotein in primary (stage I-IIIa) non-small cell lung cancer. *Anticancer Res.* 23(2C): 1665-1672.

Tan D *et al.* (2003c) Thyroid transcription factor-1 expression prevalence and its clinical implications in non-small cell lung cancer: a high-throughput tissue microarray and immunohistochemistry study. *Hum. Pathol.* 34: 597-604.

Tan D *et al.* (2003d) HER-2/neu protein expression and gene alteration in stage I-IIIA non-small-cell lung cancer: a study of 140 cases using a combination of high throughput tissue microarray, immunohistochemistry, and fluorescent *in situ* hybridisation. *Diagn. Mol. Pathol.* 12: 201-211.

Tanabe M *et al.* (2003) Activating transcription factor 4 increases the cisplatin resistance of human cancer cell lines. *Cancer Res.* 63: 8592-8595.

Tanaka H *et al.* (1998) Disruption of the RB pathway and cell-proliferative activity in non-small-cell lung cancers. *Int. J. Cancer* 79: 111-115.

Tanaka F *et al.* (2002) Expression of angiopoietins and its clinical significance in non-small cell lung cancer. *Cancer Res.* 62: 7124-7129.

Tang X *et al.* (2000) Hypermethylation of the death-associated protein (DAP) kinase promoter and aggressiveness in stage I non-small-cell lung cancer. *J. Natl. Cancer Inst.* 92: 1511-1516.

Taniguchi K *et al.* (2002) Polymorphisms in the promoter region of the neutrophil elastase gene are associated with lung cancer development. *Clin. Cancer Res.* 8: 1115-1120.

Taningher M *et al.* (1999) Drug metabolism polymorphisms as modulators of cancer susceptibility. *Mutation Res.* 436: 227-261.

Tani M *et al.* (1997) Infrequent mutations of the transforming growth factor beta-type II receptor gene at chromosome 3p22 in human lung cancers with chromosome 3p deletions. *Carcinogenesis* 18: 1119-1121.

Tani M *et al.* (1999) Mutation and expression of the p51 gene in human lung cancer. *Neoplasia* 1: 71-79.

Tani M *et al.* (2004) Correlation between histone acetylation and expression of the *MYO18B* gene in human lung cancer cells. *Genes. Chrom. Cancer* 40: 146-151.

Taniguchi K *et al.* (2002) Polymorphisms in the promoter region of the neutrophil elastase gene are associated with lung cancer development. *Clin. Cancer Res.* 8: 1115-1120.

Tarafa G *et al.* (2003) Common genetic evolutionary pathways in familial adenomatous polyposis tumours. *Cancer Res.* 63: 5731-5737.

Tarapore P, Fukusawa K (2002) Loss of p53 and centrosome hyperamplification. *Oncogene* 21: 6234-6240.

Tateishi M *et al.* (1990) Immunohistochemical evidence of autocrine growth factors in adenocarcinomas of the human lung. *Cancer Res.* 50: 7077-7080.

Tateishi M *et al.* (1991) Prognostic value of c-erbB-2 protein expression in human lung adenocarcinoma and squamous cell carcinoma. *Eur. J. Cancer* 27: 1372-1375.

Taya Y *et al.* (1984) A novel combination of K-ras and myc amplification accompanied by point mutational activation of K-ras in a human lung cancer. *EMBO J.* 3: 2943-2946.
Tchou-Wong KM *et al.* (2002) Lung-specific expression of dominant-negative mutant p53 in transgenic mice increases spontaneous and benzo(*a*)pyrene-induded lung cancer. *Am. J. Respir. Cell Mol. Biol.* 27: 186-193.
Tefre T *et al.* (1994) Genotyping of the *CYP2D6* gene in Norwegian lung cancer patients and controls. *Pharmacogenetics* 4: 47-57.
Teng DH-F *et al.* (1997) Human mitogen-activated protein kinase kinase 4 as a candidate tumor suppressor. *Cancer Res.* 57: 4177-4182.
Teng DH *et al.* (2001) Mutation analyses of 268 candidate genes in human tumor cell lines. *Genomics* 74: 352-364.
Testa JR *et al.* (1994) Cytogenetic analysis of 63 non-small cell lung carcinomas: recurrent chromosome alterations amid frequent and widespread genomic upheaval. *Genes Chrom. Cancer* 11: 178-194.
Testa JR (1996) Chromosome alterations in human lung cancer. Chap. 2 in *Lung Cancer: Principles and Practice*. Eds. HI Pass, JB Mitchell, DH Johnson, AT Turrisi. Lipincott-Raven, Philadelphia.
Testa JR *et al.* (1997) Advances in the analysis of chromosome alterations in human lung carcinomas. *Cancer Genet. Cytogenet.* 95: 20-32.
Teyssier JR *et al.* (1985) Trisomy in a lung carcinoid tumor: precocious index of malignant transformation? *Cancer Genet. Cytogenet.* 15: 277-282.
Thiagalingam S *et al.* (2002) Loss of heterozygosity as a predictor to map tumor suppressor genes in cancer: molecular basis of its occurrence. *Curr. Opin. Oncol.* 14: 65-72.
Thomas A *et al.* (1999) Familial associations among cancer sites. *Comput. Biomed. Res.* 32: 517-529.
Thompson LH, Schild D (2002) Recombinational DNA repair and human disease. *Mutation Res.* 509: 49-78.
Thorslund T *et al.* (2002) Repair of 8-oxoG is slower in endogenous nuclear genes than in mitochondrial DNA and is without strand bias. *DNA Repair* 1: 261-273.
Thun MJ *et al.* (2002) Tobacco use and cancer: an epidemiologic perspective for geneticists. *Oncogene* 21: 7307-7325.
Tian F *et al.* (2003) Overexpression of COX-2 and its clinical significance in non-small cell lung cancer. *Zhonghua Wai Ke Za Zhi* 41: 407-410.
Timmer T *et al.* (1999) A comparison of genomic structures and expression patterns of two closely related flanking genes in a critical lung cancer region at 3p21.3. *Eur. J. Hum. Genet.* 7: 478-486.
Tischfield JA (1997) Loss of heterozygosity or: how I learned to stop worrying and love mitotic recombination. *Am. J. Hum. Genet.* 61: 995-999.
To-Figueras J *et al.* (2001) Lung cancer susceptibility in relation to combined polymorphisms of microsomal epoxide hydrolase and glutathione *S*-transferase P1. *Cancer Letts.* 173: 155-162.
Tobioka H *et al.* (2004) Expression of occluding, a tight junction-associated protein, in human lung carcinomas. *Virchows Arch.* In press [DOI: 10.1007/s00428-004-1054-9].
Tockman MS (2000) Advances in sputum analysis for screening and early detection of lung cancer. *Cancer Control* 7: 19-24.
Tockman MS, Mulshine JL (2000) The early detection of occult lung cancer. *Chest Surg. Clin. N. Am.* 10: 737-749.
Tockman MS *et al.* (1997) Prospective detection of preclinical lung cancer: results from two studies of heterogeneous nuclear ribonucleoprotein A2/B1 overexpression. *Clin. Cancer Res.* 3: 2237-2246.
Todd S *et al.* (1997) Homozygous deletions of human chromosome 3p in lung tumors. *Cancer Res.* 57: 1344-1352.
Tojo Y *et al.* (2003) Aberrant messenger RNA splicing of the cytokeratin 8 in lung cancer. *Lung Cancer* 42: 153-161.
Tokieda K *et al.* (1999) Surfactant protein B-deficient mice are susceptible to hyperoxic lung injury. *Am. J. Resp. Cell Molec. Biol.* 21: 563-572.
Tokuchi Y *et al.* (1999a) Abnormal *FHIT* transcripts found in both lung cancer and normal lung tissue. *Genes, Chrom. Cancer* 24: 105-111.
Tokuchi Y *et al.* (1999b) The expression of p73 is increased in lung cancer, independent of p53 gene alteration. *Br. J. Cancer* 80: 1623-1629.
Tokuhara T *et al.* (2001a) Clinical significance of *CD151* gene expression in non-small cell lung cancer. *Clin. Cancer Res.* 7: 4109-4114.
Tokuhara T *et al.* (2001b) Neutral endopeptidase/CD10 and aminopeptidase N/CD13 gene expression as a prognostic factor in non-small cell lung cancer. *Jpn. J. Cardiovasc. Surg.* 49: 489-496.
Tokuhata GK, Lilienfeld AM (1963) Familial aggregation of lung cancer in humans. *J. Natl. Cancer Inst.* 30: 289-312.

Toledo G *et al.* (2004) Loss of FHIT protein expression is related to high proliferation, low apoptosis and worse prognosis in non-small-cell lung cancer. *Mod. Pathol.* 17: 440-448.

Tomida S *et al.* (2004) Gene expression-based, individualized outcome prediction for surgically treated lung cancer patients. *Oncogene* 23: 5360-5370.

Tomita M *et al.* (2003) Prognostic significance of bcl-2 expression in resected pN2 non-small cell lung cancer. *Eur. J. Surg. Oncol.* 29: 654-657.

Tomita-Mitchell A *et al.* (2003) The mutational spectrum of the *HPRT* gene from human T cells *in vivo* shares a significant concordant set of hot spots with MNNG-treated human cells. *Cancer Res.* 63: 5793-5798.

Tomiyasu M *et al.* (2002) Quantification of macrophage migration inhibitory factor mRNA expression in non-small cell lung cancer tissues and its clinical significance. *Clin. Cancer Res.* 8: 3755-3760.

Tomizawa Y *et al.* (1998) Identification and characterization of families with aggregation of lung cancer. *Jp. J. Clin. Oncol.* 28: 192-195.

Tomizawa Y *et al.* (1999) Correlation between the status of the p53 gene and survival in patients with stage I non-small cell lung carcinoma. *Oncogene* 18: 1007-1014.

Tomizawa Y *et al.* (2001) Inhibition of lung cancer cell growth and induction of apoptosis after reexpression of 3p21.3 candidate tumor suppressor gene *SEMA3B*. *Proc. Natl. Acad. Sci. USA* 98: 13954-13959.

Tomizawa Y *et al.* (2002) Clinicopathological significance of epigenetic inactivation of *RASSF1A* at 3p21.3 in stage I lung adenocarcinoma. *Clin. Cancer Res.* 8: 2362-2368.

Tomlinson IPM *et al.* (1996) The mutation rate and cancer. *Proc. Natl. Acad. Sci. USA* 93: 14800-14803.

Tomlinson I, Bodmer W (1999) Selection, the mutation rate and cancer: ensuring that the tail does not wag the dog. *Nature Med.* 5: 11-12.

Tomlinson IP *et al.* (2002) Loss of heterozygosity analysis: practically and conceptually flawed? *Genes Chrom. Cancer* 34: 349-353.

Tong X *et al.* (2001a) Cyr61, a member of CCN family, is a tumor suppressor in non-small cell lung cancer. *J. Biol. Chem.* 276: 47709-47714.

Tong AW *et al.* (2001b) Potential clinical application of antioncogene ribozymes for human lung cancer. *Clin. Lung Cancer* 2: 220-226.

Tonini T *et al.* (2003) Molecular basis of angiogenesis and cancer. *Oncogene* 22: 6549-6556.

Top B *et al.* (1995) Comparative analysis of p53 gene mutations and protein accumulation in human non-small-cell lung cancer. *Int. J. Cancer* 64: 83-91.

Topaloglu O *et al.* (2004) Detection of promoter hypermethylation of multiple genes in the tumor and bronchoalveolar lavage of patients with lung cancer. *Clin. Cancer Res.* 10: 2284-2288.

Tosolini A, Testa JR (2000) Genomic imbalances in lung cancer. Chap. 28 in *Lung Cancer. Principles and Practice.* 2nd Ed. HI Pass, JB Mitchell, DH Johnson, AT Turrisi, JD Minna (Eds.). Lippincott Williams & Williams, Philadelphia, pp 502-517.

Townsend D, Tew K (2003) Cancer drugs, genetic variation and the glutathione-*S*-transferase gene family. *Am. J. Pharmacogenomics* 3: 157-172.

Townson JL *et al.* (2003) The role of apoptosis in tumor progression and metastasis. *Curr. Mol. Med.* 3: 631-642.

Toyooka KO *et al.* (2001a) Loss of expression and aberrant methylation of the *CDH13* (H-cadherin) gene in breast and lung carcinomas. *Cancer Res.* 61: 4556-4560.

Toyooka KO *et al.* (2001b) DNA methylation profiles of lung tumours. *Mol. Cancer Ther.* 1: 61-67.

Toyooka S *et al.* (2002) Differential expression of *FEZ1/LZTS1* gene in lung cancers and their cell cultures. *Clin. Cancer Res.* 8: 2292-2297.

Toyooka KO *et al.* (2003a) The *TP53* gene, tobacco exposure and lung cancer. *Hum. Mutation* 21: 229-239.

Toyooka KO *et al.* (2003b) Smoke exposure, histologic type and geography-related differences in the methylation profiles of non-small cell lung cancer. *Int. J. Cancer* 103: 153-160.

Toyooka S *et al.* (2003c) Detection of codon 61 point mutations of the K-ras gene in lung and colorectal cancers by enriched PCR. *Oncol. Rep.* 10: 1455-1459.

Toyoyama H et al. (1999) The reduced expression of E-cadherin, α-catenin and γ-catenin but not β-catenin in human lung cancer. *Oncol. Rep.* 6: 81-85.

Tran YK *et al.* (1999) A novel member of the NF2/ERM/4.1 superfamily with growth suppressing properties in lung cancer. *Cancer Res.* 59: 35-43.

Travis WD *et al.* (2000) Classification, histology, cytology and electron microscopy. Chap. 26 in *Lung Cancer. Principles and Practice.* 2nd Ed. HI Pass, JB Mitchell, DH Johnson, AT Turrisi, JD Minna (Eds.). Lippincott Williams & Williams, Philadelphia, pp 453-495.

Tretyakova N *et al.* (2002) Formation of benzo[*a*]pyrene diol epoxide-DNA adducts at specific guanines within K-ras and p53 gene sequences: stable isotope-labeling mass spectrometry approach. *Biochemistry* 41: 9535-9544.
Tripodis N *et al.* (2001) Complexity of lung cancer modifiers: mapping of thirty genes and twenty-five interactions in half of the mouse genome. *J. Natl. Cancer Inst.* 93: 1484-1491.
Tsai CM *et al.* (1995) Enhanced chemoresistance by elevation of p185neu levels in HER-2/neu-transfected human lung cancer cells. *J. Natl. Cancer Inst.* 87: 682-684.
Tsai YY *et al.* (2003) Genetic susceptibility and dietary patterns in lung cancer. *Lung Cancer* 41: 269-281.
Tsao MS *et al.* (1997) Angiogenesis correlates with vascular endothelial growth factor expression but not with Ki-ras oncogene activation in non-small cell lung carcinoma. *Clin. Cancer Res.* 3: 1807-1814.
Tsao AS *et al.* (2003) Increased phospho-AKT [Ser(473)] expression in bronchial dysplasia: implications for lung cancer prevention studies. *Cancer Epidemiol. Biomarkers Prev.* 12: 660-664.
Tse C *et al.* (2002) Human semaphorin 3B (*SEMA3B*) located at chromosome 3p21.3 suppresses tumor formation in an adenocarcinoma cell line. *Cancer Res.* 62: 542-546.
Tseng JE *et al.* (1999a) Gender differences in *p53* mutational status in small cell lung cancer. *Cancer Res.* 59: 5666-5670.
Tseng JE *et al.* (1999b) Loss of Fhit is frequent in stage I non-small cell lung cancer and in the lungs of chronic smokers. *Cancer Res.* 59: 4798-4803.
Tsoli E *et al.* (2001) Low levels of p27 in association with deregulated p53-pRb protein status enhance tumor proliferation and chromosomal instability in non-small cell lung carcinomas. *Mol. Med.* 7: 418-429.
Tsou JA *et al.* (2002) DNA methylation analysis: a powerful new tool for lung cancer diagnosis. *Oncogene* 21: 5450-5461.
Tsuchiya E *et al.* (1995) High K-ras mutation rates in goblet-cell-type adenocarcinomas of the lungs. *J. Cancer Res. Clin. Oncol.* 121: 577-581.
Tsuji T *et al.* (2001) Antiproliferative activity of REIC/Dkk-3 and its significant down-regulation in non-small-cell lung carcinomas. *Biochem. Biophys. Res. Commun.* 289: 257-263.
Tsukamoto S *et al.* (2001) Reduced expression of cell-cycle regulator p27(Kip1) correlates with a shortened survival in non-small cell lung cancer. *Lung Cancer* 34: 83-90.
Tsurutani J *et al.* (2002) Point mutations in the topoisomerase I gene in patients with non-small cell lung cancer treated with irinotecan. *Lung Cancer* 35: 299-304.
Turken O *et al.* (2003) Prevalence and prognostic value of c-erbB2 expression in non-small cell lung cancer (NSCLC). *Neoplasma* 50: 257-261.
Turney MK *et al.* (2004) Gene expression phenotyping of an ACTH-producing small cell lung cancer line. *Mol. Cell. Endocrinol.* 219: 105-113.
Tuteja N, Tuteja R (2001) Unraveling DNA repair in human: molecular mechanisms and consequences of repair defect. *Crit. Rev. Biochem. Mol. Biol.* 36: 261-290.
Tuveson DA, Jacks T (1999) Modeling human lung cancer in mice: similarities and shortcomings. *Oncogene* 18: 5318-5324.
Tye BK (1999) MCM proteins in DNA replication. *Annu. Rev. Biochem.* 68: 649-686.
Tyndale RF (2003) Genetics of alcohol and tobacco use in humans. *Ann. Med.* 35: 94-121.
Ubagai T *et al.* (2001) Comparative genomic hybridisation analysis suggests a gain of chromosome 7p associated with lymph node metastasis in non-small cell lung cancer. *Oncol. Rep.* 8: 83-88.
Uchida K *et al.* (1996) Somatic *in vivo* alterations of the *JV18-1* gene at 18q21 in human lung cancers. *Cancer Res.* 56: 5583-5585.
Uchida K *et al.* (2002) Expression of progastrin-releasing peptide and gastrin-releasing peptide receptor mRNA transcripts in tumor cells of patients with small cell lung cancer. *J. Cancer Res. Clin. Oncol.* 128: 633-640.
Uchino K *et al.* (2003) Clinical implication and prognostic significance of the tumor suppressor *TSLC1* gene detected in adenocarcinoma of the lung. *Cancer* 98: 1002-1007.
Uchiyama M *et al.* (2003) Loss of heterozygosity of chromosome 12p does not correlate with *KRAS* mutation in non-small cell lung cancer. *Int. J. Cancer* 107: 962-969.
Ueda M *et al.* (2001) Mutations of the beta- and gamma-catenin genes are uncommon in human lung, breast, kidney, cervical and ovarian carcinomas. *Br. J. Cancer* 85: 64-68.
Ueda M *et al.* (2003) Glutathione *S*-transferase GSTM1, GSTT1 and p53 codon 72 polymorphisms in human tumor cells. *Hum. Cell* 16: 241-251.
Uematsu F *et al.* (1991) Association between restriction fragment length polymorphism in the human cytochrome P450IIE1 gene and susceptibility to lung cancer. *Jpn. J. Cancer Res.* 82: 254-256.
Uematsu K *et al.* (2001) Aberrations in the fragile histidine triad (*FHIT*) gene in idiopathic pulmonary fibrosis. *Cancer Res.* 61: 8527-8533.

Uematsu K *et al.* (2003) Activation of the Wnt pathway in non small cell lung cancer: evidence of dishevelled overexpression. *Oncogene* 22: 7218-7221.
Ullmann R *et al.* (2004) Protein expression profiles in adenocarcinoma and squamous cell carcinomas of the lung generated using tissue microarrays. *J. Pathol.* 203: 798-807.
Umiker W (1957) False positive reports in the cytologic diagnosis of lung cancer. *Br. J. Cancer* 11: 391-397.
Uno F *et al.* (2004) Myristoylation of the Fus1 protein is required for tumor suppression in human lung cancer cells. *Cancer Res.* 64: 2969-2976.
Upadhyaya M *et al.* (2004) Characterization of the somatic mutational spectrum of the neurofibromatosis type 1 (*NF1*) gene in neurofibromatosis patients with benign and malignant tumors. *Hum. Mutation* 23: 134-136.
Urban T *et al.* (2000) Detection of codon 12 K-ras mutations in non-neoplastic mucosa from bronchial carina in patients with lung adenocarcinomas. *Br. J. Cancer* 82: 412-417.
U.S. Surgeon General's Advisory Committee on Smoking and Health (1964) Smoking and health. Public Health Service, U.S. Department of Health, Education & Welfare. Wasington DC, p.196.
Usadel H *et al.* (2002) Quantitative adenomatous polyposis coli promoter methylation analysis in tumor tissue, serum, and plasma DNA of patients with lung cancer. *Cancer Res.* 62: 371-375.
Ushijima C *et al.* (2001) High vascularity in the peripheral region of non-small cell lung cancer tissue is associated with tumor progression. *Lung Cancer* 34: 233-241.
Vachtenheim J. (1997) Occurrence of *ras* mutations in human lung cancer. *Neoplasma* 44: 145-149.
Vachtenheim J, Novotna H (1997) Expression of the aromatic L-amino acid decarboxylase mRNA in human tumour cell lines of neuroendocrine and neuroectodermal origin. *Eur. J. Cancer* 33: 2411-2417.
Vachtenheim J *et al.* (1995) Mutations of K-*ras* oncogene and absence of H-*ras* mutations in squamous cell carcinomas of the lung. *Clin. Cancer Res.* 1: 359-365.
Vähäkangas KH *et al.* (2001) p53 and K-ras mutations in lung cancers from former and never-smoking women. *Cancer Res.* 61: 4350-4356.
Vähäkangas K (2003) *TP53* mutations in workers exposed to occupational carcinogens. *Hum. Mutation* 21: 240-251.
van Niekerk CC, Poels LG (1999) Reduced expression of protein tyrosine phosphatase gamma in lung and ovarian tumors. *Cancer Lett.* 137: 61-73.
van Oijen MG, Slootweg PJ (2000) Gain-of-function mutations in the tumor suppressor gene p53. *Clin. Cancer Res.* 6: 2138-2145.
van Rens MT *et al.* (2002) p53 mutation analysis for definite diagnosis of multiple primary lung carcinoma. *Cancer* 94: 188-196.
Van Schooten FJ *et al.* (2004) Myeloperoxidase (*MPO*) –463G→A reduces MPO activity and DNA adduct levels in bronchoalveolar lavages of smokers. *Cancer Epidemiol. Biomarkers Prev.* 13: 828-833.
van der Sijp JR *et al.* (2002) Determination of the molecular relationship between multiple tumors within one patient is of clinical importance. *J. Clin. Oncol.* 20: 1105-1114.
Vandenbergh DJ *et al.* (2002) Smoking status and the human dopamine transporter variable number of tandem repeats (VNTR) polymorphism failure to replicate and finding that never-smokers may be different. *Nicotine Tob. Res.* 4: 333-340.
Varella-Garcia M *et al.* (1998) Chromosomal duplication accompanies allelic loss in non-small cell lung carcinoma. *Cancer Res.* 58: 4701-4707.
Vargas DA *et al.* (2003) Mdm2: a regulator of cell growth and death. *Adv. Cancer Res.* 89: 1-34.
Varley JM (2003) Germline *TP53* mutations and Li-Fraumeni syndrome. *Hum. Mutation* 21: 320.
Veglia F *et al.* (2003) Determinants of exposure to environmental tobacco smoke in 21,588 Italian non-smokers. *Tumori* 89: 665-668.
Veneis P *et al.* (2003) *CYP1A1* T3801C polymorphism and lung cancer: a pooled analysis of 2451 cases and 3358 controls. *Int. J. Cancer* 104: 650-657.
Venter JC *et al.* (2001) The sequence of the human genome. *Science* 281: 1304-1351.
Verona RI *et al.* (2003) Genomic imprinting: intricacies of epigenetic regulation in clusters. *Annu. Rev. Cell Dev. Biol.* 19: 237-259.
Veronese M-L *et al.* (2000) The role of the *FHIT* gene in the pathogenesis of lung cancer. Chap. 8 in *Lung Cancer. Principles and Practice.* 2nd Ed. HI Pass, JB Mitchell, DH Johnson, AT Turrisi, JD Minna (Eds.). Lippincott Williams & Williams, Philadelphia, pp 156-169.
Viard-Leveugle I *et al.* (2003) Frequent loss of Fas expression and function in human lung tumours with overexpression of FasL in small cell lung carcinoma. *J. Pathol.* 201: 268-277.
Vicent S *et al.* (2004) Mitogen-activated protein kinase phosphatase is overexpressed in non-small cell lung cancer and is an independent predictor of outcome in patients. *Clin. Cancer Res.* 10: 3639-3649.

Vineis P *et al.* (2003) *CYP1A1* T3801C polymorphism and lung cancer: a pooled analysis of 2451 cases and 3358 controls. *Int. J. Cancer* 104: 650-657.

Vineis P *et al.* (2004) Tobacco and cancer: recent epidemiological evidence. *J. Natl. Cancer Inst.* 96: 99-106.

Virmani AK *et al.* (1998) Allelotyping demonstrates common and distinct patterns of chromosomal loss in human lung cancer types. *Genes Chrom. Cancer* 21: 308-319.

Virmani AK *et al.* (2000) Promoter methylation and silencing of the retinoic acid receptor-beta gene in lung carcinomas. *J. Natl. Cancer Inst.* 92: 1303-1307.

Virmani AK *et al.* (2001) Aberrant methylation of the adenomatous polyposis coli (*APC*) gene promoter 1A in breast and lung carcinomas. *Clin. Cancer Res.* 7: 1998-2004.

Virmani AK *et al.* (2002) Hierarchical clustering of lung cancer cell lines using DNA methylation markers. *Cancer Epidemiol. Biomarkers Prev.* 11: 291-297.

Virmani A *et al.* (2003a) Aberrant methylation of TMS1 in small cell, non small cell lung cancer and breast cancer. *Int. J. Cancer* 106: 198-204.

Virmani A *et al.* (2003b) Aberrant methylation of the cyclin D2 promoter in primary small cell, nonsmall cell lung and breast cancers. *Int. J. Cancer* 107: 341-345.

Virtanen C *et al.* (2002) Integrated classification of lung tumors and cell lines by expression profiling. *Proc. Natl. Acad. Sci. USA* 99: 12357-12362.

Visbal AL *et al.* (2004) Gender differences in non-small-cell lung cancer survival: an analysis of 4,618 patients diagnosed between 1997 and 2002. *Ann. Thorac. Surg.* 78: 209-215.

Vivanco I, Sawyers CL (2002) The phosphatidylinositol 3-kinase AKT pathway in human cancer. *Nature Rev. Cancer* 2: 489-501.

Vodicka P *et al.* (2004) Genetic polymorphisms in DNA repair genes and possible links with DNA repair rates, chromosomal aberrations and single-strand breaks in DNA. *Carcinogenesis* 25: 757-763.

Vogel U *et al.* (2004) Two regions in chromosome 19q13.2-3 are associated with risk of lung cancer. *Mutation Res.* 546: 65-74.

Vogt U *et al.* (2002) p53 gene status in relation to *ex vivo* chemosensitivity of non-small cell lung cancer. *J. Cancer Res. Clin. Oncol.* 128: 141-147.

Volm M *et al.* (1997) Prognostic value of basic fibroblast growth factor and its receptor (FGFR-1) in patients with non-small cell lung carcinomas. *Eur. J. Cancer* 33: 691-693.

Volm M *et al.* (2002a) Protein expression profiles of non-small cell lung carcinomas: correlation with histological subtype. *Anticancer Res.* 22: 2321-2324.

Volm M *et al.* (2002b) Protein expression profile of primary human squamous cell lung carcinomas indicative of the incidence of metastases. *Clin. Exp. Metastasis* 19: 385-390.

Volm M *et al.* (2002c) Expression profile of genes in non-small cell lung carcinomas from long-term surviving patients. *Clin. Cancer Res.* 8: 1843-1848.

Volm M *et al.* (2002d) Protein expression profiles indicative for drug resistance of non-small cell lung cancer. *Br. J. Cancer* 87: 251-257.

Voorhoeve PM, Agami R (2003) The tumor-suppressive functions of the human *INK4A* locus. *Cancer Cell* 4: 311-316.

Voronov E *et al.* (2003) IL-1 is required for tumour invasiveness and angiogenesis. *Proc. Natl. Acad. Sci. USA* 100: 2645-2650.

Wada C *et al.* (1992) A general transcription initiation factor, human transcription factor IID, overexpressed in human lung and breast carcinoma and rapidly induced with serum stimulation. *Cancer Res.* 52: 307-313.

Walker C *et al.* (1995) Expression of the BCL-2 protein in normal and dysplastic bronchial epithelium and in lung carcinomas. *Br. J. Cancer* 72: 164-169.

Walton R *et al.* (2001) Genetic clues to the molecular basis of tobacco addiction and progress towards personalized therapy. *Trends Mol. Med.* 7: 70-76.

Wan PTC *et al.* (2004) Mechanism of activation of the RAF-ERK signalling pathway by oncogenic mutations of B-RAF. *Cell* 116: 855-867.

Wang XW, Harris CC (1997) *p53* tumor-suppressor gene: clues to molecular carcinogenesis. *J. Cell. Physiol.* 173: 247-255.

Wang Q, Beck WT (1998) Transcriptional suppression of multidrug resistance-associated protein (MRP) gene expression by wild-type p53. *Cancer Res.* 58: 5762-5769.

Wang SS *et al.* (1998a) Alterations of the *PPP2R1B* gene in human lung and colon cancer. *Science* 282: 284-287.

Wang Y *et al.* (1998b) Val384Asp in hMLH1 gene in Chinese, Japanese and German and its etiological role in colorectal cancer. *Zhonghua Yi Xue Yi Chuan Xue Za Zhi* 15: 263-266.

Wang D *et al.* (1998c) Mutagenicity and repair of oxidative DNA damage: insights from studies using defined lesions. *Mutation Res.* 400: 99-115.

Wang SS *et al.* (1999a) Refined mapping of two regions of loss of heterozygosity on chromosome band 11q23 in lung cancer. *Genes Chrom. Cancer* 25: 154-159.
Wang YC *et al.* (1999b) Prognostic significance of p53 codon 72 polymorphism in lung carcinomas. *Eur. J. Cancer* 35: 226-230.
Wang L *et al.* (2000a) C-CAM1, a candidate tumor suppressor gene, is abnormally expressed in primary lung cancers. *Clin. Cancer Res.* 6: 2988-2993.
Wang T *et al.* (2000b) Identification of genes differentially over-expressed in lung squamous cell carcinoma using combination of cDNA subtraction and microarray analysis. *Oncogene* 19: 1519-1528.
Wang X *et al.* (2001a) Human Rad17 is phosphorylated upon DNA damage and also overexpressed in primary non-small cell lung cancer tissues. *Cancer Res.* 61: 7417-7421.
Wang T *et al.* (2001b) L552S, an alternatively spliced isoform of XAGE-1, is over-expressed in lung adenocarcinoma. *Oncogene* 20: 7699-7709.
Wang K-k *et al.* (2002a) Novel candidate tumor marker genes for lung adenocarcinoma. *Oncogene* 21: 7598-7604.
Wang L *et al.* (2002b) hTERT expression is a prognostic factor of survival in patients with stage I non-small cell lung cancer. *Clin. Cancer Res.* 8: 2883-2889.
Wang Y *et al.* (2002c) Sulfotransferase (SULT) 1A1 polymorphism as a predisposition factor for lung cancer: a case-control analysis. *Lung Cancer* 35: 137-142.
Wang H *et al.* (2002d) Antisense oligonucleotide targeted to RIα subunit of cAMP-dependent protein kinase (GEM231) enhances therapeutic effectiveness of cancer chemotherapeutic agent irinotecan in nude mice bearing human cancer xenografts: *in vivo* synergistic activity, pharmacokinetics and host toxicity. *Int. J. Oncol.* 21: 73-80.
Wang D *et al.* (2002e) Linkage disequilibrium mapping of novel lung tumor susceptibility quantitative trait loci in mice. *Oncogene* 21: 6858-6865.
Wang TL *et al.* (2002f) Prevalence of somatic alterations in the colorectal cancer genome. *Proc. Natl. Acad. Sci. USA* 99: 3076-3080.
Wang Y *et al.* (2003a) Association between glutathione *S*-transferase p1 polymorphisms and lung cancer risk in Caucasians: a case-control study. *Lung Cancer* 40: 25-32.
Wang Y-C *et al.* (2003b) Inactivation of *hMLH1* and *hMSH2* by promoter methylation in primary non-small cell lung tumors and matched sputum samples. *J. Clin. Invest.* 111: 887-895.
Wang Y *et al.* (2003c) Mice with alterations in both p53 and Ink4a/Arf display a striking increase in lung tumor multiplicity and progression: differential chemopreventive effect of budesonide in wild-type and mutant A/J mice. *Cancer Res.* 63: 4389-4395.
Wang T *et al.* (2003d) L523S, an RNA-binding protein as a potential therapeutic target for lung cancer. *Br. J. Cancer* 88: 887-894.
Wang LE *et al.* (2003e) Fas A670G polymorphism, apoptotic capacity in lymphocyte cultures, and risk of lung cancer. *Lung Cancer* 42: 1-8.
Wang L *et al.* (2003f) Association of a functional tandem repeats in the downstream of human telomerase gene and lung cancer. *Oncogene* 22: 7123-7129.
Wang Y *et al.* (2003 g) *XRCC3* genetic polymorphism, smoking, and lung carcinoma risk in minority populations. *Cancer* 98: 1701-1706.
Wang H *et al.* (2003h) Substantial reduction in risk of lung adenocarcinoma associated with genetic polymorphism in CYP2A13, the most active cytochrome P450 for the metabolic activation of the tobacco-specific carcinogen NNK. *Cancer Res.* 63: 8057-8061.
Wang BG *et al.* (2004a) A case control study on the impact of CYP450 *Msp*I and GST-M1 polymorphisms on the risk of lung cancer. *Zhonghua Zhong Liu Za Zhi* 26: 93-97.
Wang Z *et al.* (2004b) Mutational analysis of the tyrosine phosphatome in colorectal cancers. *Science* 304: 1164-1166.
Wang J-W *et al.* (2004c) Deregulated expression of LRBA facilitates cancer cell growth. *Oncogene* 23: 4089-4097.
Wang Y *et al.* (2004d) Elevated expression of DNA polymerase kappa in human lung cancer is associated with p53 inactivation: negative regulation of POLK promoter activity by p53. *Int. J. Oncol.* 25: 161-165.
Warburton D *et al.* (2000) The molecular basis of lung morphogenesis. *Mech. Dev.* 92: 55-81.
Warnecke PM, Bestor TH (2000) Cytosine methylation and human cancer. *Curr. Opin. Oncol.* 12: 68-73.
Washimi O *et al.* (1995) *In vivo* occurrence of *p16* (*MTS1*) and *p15* (*MTS2*) alterations preferentially in non-small cell lung cancers. *Cancer Res.* 55: 514-517.
Washio K *et al.* (2003) The effect of epidermal growth factor receptor antisense morpholino oligomer on non-small cell lung cancer cell line. *Oncol. Rep.* 10: 1967-1971.

Watanabe T, Sullenger BA (2000) Induction of wild-type p53 activity in human cancer cells by ribozymes that repair mutant p53 transcripts. *Proc. Natl. Acad. Sci. USA* 97: 8490-8494.
Watanabe J *et al.* (2000) Association of *CYP1B1* genetic polymorphism with incidence to breast and lung cancer. *Pharmacogenetics* 10: 25-33.
Watanabe S *et al.* (2003) Methylated DNA-binding domain 1 and methylpurine-DNA glycosylase link transcriptional repression and DNA repair in chromatin. *Proc. Natl. Acad. Sci. USA* 100: 12859-12864.
Watkins SJ, Norbury CJ (2002) Translation initiation and its deregulation during tumorigenesis. *Br. J. Cancer* 86: 1023-1027.
Watkins DN *et al.* (2003) Hedgehog signalling within airway epithelial progenitors and in small-cell lung cancer. *Nature* 422: 313-317.
Watson MA *et al.* (1998) Human glutathione *S*-transferase P1 polymorphisms: relationship to lung tissue enzyme activity and population frequency distribution. *Carcinogenesis* 19: 275-280.
Weber BL (2002) Cancer genomics. *Cancer Cell* 1: 37-47.
Weber E *et al.* (1999) Retinoic acid-mediated growth inhibition of small cell lung cancer cells is associated with reduced myc and increased p27Kip1 expression. *Int. J. Cancer* 80: 935-943.
Wedlund PJ (2000) The CYP2C19 enzyme polymorphism. *Pharmacology* 61: 174-183.
Wegmann BR *et al.* (1993) Molecular cloning of IGFBP-5 from SCLC cell lines and expression of IGFBP-4, IGFBP-5 and IGFBP-6 in lung cancer cell lines and primary tumours. *Eur. J. Cancer* 29A: 1578-1584.
Wei Q *et al.* (1996) Reduced DNA repair capacity in lung cancer patients. *Cancer Res.* 56: 4103-4107.
Wei Q *et al.* (2000) Repair of tobacco carcinogen-induced DNA adducts and lung cancer risk: a molecular epidemiologic study. *J. Natl. Cancer Inst.* 92: 1764-1772.
Wei X *et al.* (2004) Analysis of the RNA helicase A gene in human lung cancer. *Oncol. Rep.* 11: 253-258.
Weill D *et al.* (2000) Adenoviral-mediated p53 gene transfer to non-small cell lung cancer through endobronchial injection. *Chest* 118: 966-970.
Weiner DB *et al.* (1990) Expression of the *neu* gene-encoded protein ($P185^{neu}$) in human non-small cell carcinomas of the lung. *Cancer Res.* 50: 421-425.
Weinstein EJ *et al.* (2003) URP1: a member of a novel family of PH and FERM domain-containing membrane-associated proteins is significantly over-expressed in lung and colon carcinomas. *Biochim. Biophys Acta* 1637: 207-216.
Weiserbs KF *et al.* (2003) A cross-sectional study of polycyclic aromatic hydrocarbon-DNA adducts and polymorphism of glutathione *S*-transferases among heavy smokers by race/ethnicity. *Biomarkers* 8: 142-155.
Wen Cheng Y, Lee H (2003) Environmental exposure and lung cancer among nonsmokers: an example of Taiwanese female lung cancer. *J. Environ. Sci. Health Part C Environ. Carcinog. Ecotoxicol. Rev.* 21: 1-28.
Wesley UV *et al.* (2004) Role for dipeptidyl peptidase IV in tumor suppression of human non-small cell lung carcinoma cells. *Int. J. Cancer* 109: 855-866.
Wessel I *et al.* (2002) Human small cell lung cancer NYH cells resistant to the bisdioxopiperazine ICRF-187 exhibit a functional dominant Tyr165Ser mutation in the Walker A ATP binding site of topoisomerase IIα. *FEBS Letts* 520:161-166.
West KA *et al.* (2003) Rapid Akt activation by nicotine and a tobacco carcinogen modulates the phenotype of normal human airway epithelial cells. *J. Clin. Invest.* 111: 81-90.
West KA *et al.* (2004) Tobacco carcinogen-induced cellular transformation increases activation of the phosphatidylinositol 3'-kinase/Akt pathway *in vitro* and *in vivo*. *Cancer Res.* 64: 446-451.
Weston A and Godbold JH (1997) Polymorphisms of H-*ras*-1 and *p53* in breast cancer and lung cancer: a meta-analysis. *Envir. Health Persp.* 105: 919-926.
Westra WH *et al.* (1993a) K-*ras* oncogene activation in lung adenocarcinomas from former smokers. *Cancer* 72: 432-438.
Westra WH *et al.* (1993b) Overexpression of the p53 tumor suppressor gene product in primary lung adenocarcinomas is associated with cigarette smoking. *Am. J. Surg. Pathol.* 17: 213-220.
Westra WH *et al.* (1996) K-ras oncogene activation in atypical alveolar hyperplasias of the human lung. *Cancer Res.* 56: 2224-2228.
Whang-Peng J *et al.* (1982a) Specific chromosome defect associated with human small-cell lung cancer; deletion 3p(14-23). *Science* 215: 181-182.
Whang-Peng J *et al.* (1982b) A non-random chromosomal abnormality del 3p(14-23) in human small cell lung cancer (SCLC). *Cancer Genet. Cytogenet.* 6: 119-134.
Whiteside MA *et al.* (2004) A novel time-course cDNA microarray analysis method identifies genes associated with the development of cisplatin resistance. *Oncogene* 23: 744-752.

Whitney JF *et al.* (1995) Transferrin receptor expression in nonsmall cell lung cancer: histopathologic and clinical correlates. *Cancer* 76: 20-25.
Whyatt RM *et al.* (1998) Polycyclic aromatic hydrocarbon-DNA adducts in human placenta and modulation by CYP1A1 induction and genotype. *Carcinogenesis* 19: 1389-1392.
Whyatt RM *et al.* (2001) Biomarkers of polycyclic aromatic hydrocarbon-DNA damage and cigarette smoke exposures in paired maternal and newborn blood samples as a measure of differential susceptibility. *Cancer Epidemiol. Biomarkers Prev.* 10: 581-588.
Wieland I *et al.* (1996) Microsatellite instability and loss of heterozygosity at the *hMLH1* locus on chromosome 3p21 occur in a subset of nonsmall cell lung carcinomas. *Oncology Res.* 8: 1-5.
Wieland I *et al.* (1999) Isolation of *DICE1*: a gene frequently affected by LOH and downregulated in lung carcinomas. *Oncogene* 18: 4530-4537.
Wieland I *et al.* (2000) Molecular characterization of the DICE1 (*DDX26*) tumor suppressor gene in lung carcinoma cells. *Oncol. Res.* 12: 491-500.
Wieland I *et al.* (2004) Ectopic expression of DICE1 suppresses tumor cell growth. *Oncol. Rep.* 12: 207-211.
Wiencke JK *et al.* (1997) Lung cancer in Mexican-Americans and African-Americans is associated with the wild-type genotype of the NAD(P)H: quinone oxidoreductase polymorphism. *Cancer Epidemiol. Biomarkers Prev.* 6: 87-92.
Wiencke JK *et al.* (2002) DNA adduct burden and tobacco carcinogenesis. *Oncogene* 21: 7376-7391.
Wigle DA *et al.* (2002) Molecular profiling of non-small cell lung cancer and correlation with disease-free survival. *Cancer Res.* 62: 3005-3008.
Wikman H *et al.* (2000) hOGG1 polymorphism and loss of heterozygosity (LOH): significance for lung cancer susceptibility in a Caucasian population. *Int. J. Cancer* 88: 932-937.
Wikman H *et al.* (2001) Relevance of *N*-acetyltransferase 1 and 2 (NAT1, NAT2) genetic polymorphisms in non-small cell lung cancer susceptibility. *Pharmacogenetics* 11: 157-168.
Wikman H *et al.* (2002) Identification of differentially expressed genes in pulmonary adenocarcinoma by using cDNA array. *Oncogene* 21: 5804-5813.
Wikman H *et al.* (2004) Caveolins as tumour markers in lung cancer detected by combined use of cDNA and tissue microarrays. *J. Pathol.* 203: 584-593.
Wild CP *et al.* (2002) Molecular epidemiology and cancer: promising areas for future research in the post-genomic era. *Mutation Res.* 499: 3-12.
Willett CG *et al.* (1998) Macrophage-stimulating protein and its receptor in non-small-cell lung tumors: induction of receptor tyrosine phosphorylation and cell migration. *Am. J. Respir. Cell Mol. Biol.* 18: 489-496.
Williams MD, Sandler AB (2001) The epidemiology of lung cancer. *Cancer Treat. Res.* 105: 31-52.
Willis AC *et al.* (2003) p73 can suppress the proliferation of cells that express mutant p53. *Oncogene* 22: 5481-5495.
Willis A *et al.* (2004) Mutant p53 exerts a dominant negative effect by preventing wild-type p53 from binding to the promoter of its target genes. *Oncogene* 23: 2330-2338.
Winn RA *et al.* (2002) γ-Catenin expression is reduced or absent in a subset of human lung cancers and re-expression inhibits transformed cell growth. *Oncogene* 21: 7497-7506.
Wistuba II *et al.* (1999a) Allelic losses at chromosome 8p21-23 are early and frequent events in the pathogeneis of lung cancer. *Cancer Res.* 59: 1973-1979.
Wistuba II *et al.* (1999b) Sequential molecular abnormalities are involved in the multistage development of squamous cell lung carcinoma. *Oncogene* 18: 643-650.
Wistuba II *et al.* (1999c) Comparison of features of human lung cancer cell lines and their corresponding tumors. *Clin. Cancer Res.* 5: 991-1000.
Wistuba II *et al.* (2000) High resolution chromosome 3p allelotyping of human lung cancer and preneoplastic/preinvasive bronchial epithelium reveals multiple, discontinuous sites of 3p allele loss and three regions of frequent breakpoints. *Cancer Res.* 60: 1949-1960.
Wistuba II *et al.* (2002) Smoking molecular damage in bronchial epithelium. *Oncogene* 21: 7298-7306.
Woenckhaus M *et al.* (2003) Microsatellite instability at chromosome 8p in non-small cell lung cancer is associated with lymph node metastasis and squamous differentiation. *Int. J. Oncol.* 23: 1357-1363.
Woerner SM *et al.* (2003) Pathogenesis of DNA repair-deficient cancers: a statistical meta-analysis of putative Real Common Target genes. *Oncogene* 22: 2226-2235.
Wolf G *et al.* (1997) Prognostic significance of polo-like kinase (PLK) expression in non-small cell lung cancer. *Oncogene* 14: 543-549.
Wolf P *et al.* (2001) O(6)-methylguanine-DNA methyltransferase promoter hypermethylation shifts the p53 mutational spectrum in non-small cell lung cancer. *Cancer Res.* 61: 8113-8117.
Wong SF, Lai LC (2001) The role of TGF beta in human cancers. *Pathology* 33: 85-92.

Wong BC, Lo YM (2003) Cell-free DNA and RNA in plasma as new tools for molecular diagnostics. *Expert Rev. Mol. Diagn.* 3: 785-797.

Wong AJ *et al.* (1986) Gene amplification of c-myc and N-myc in small cell carcinoma of the lung. *Science* 233: 461-464.

Wong AKC *et al.* (2000) BRG1, a component of the SWI-SNF complex, is mutated in multiple human tumor cell lines. *Cancer Res.* 60: 6171-6177.

Wong JM *et al.* (2001) Human aryl hydrocarbon receptor polymorphisms that result in loss of CYP1A1 induction. *Biochem. Biophys. Res. Commun.* 288: 990-996.

Wong MP *et al.* (2002) Primary adenocarcinomas of the lung in nonsmokers show a distinct pattern of allelic imbalance. *Cancer Res.* 62: 4464-4468.

Woo IS *et al.* (2002) Infrequent mutations of the activating transcription factor-2 gene in human lung cancer, neuroblastoma and breast cancer. *Int. J. Oncol.* 20: 527-531.

Woo IS *et al.* (2004) Expression of placental growth factor gene in lung cancer. *Tumour Biol.* 25: 1-6.

Wood ME *et al.* (2000) The inherited nature of lung cancer: a pilot study. *Lung Cancer* 30: 135-144.

Wood RD *et al.* (2001) Human DNA repair genes. *Science* 291: 1284-1289.

Woods DB, Vousden KH (2001) Regulation of p53 function. *Exp. Cell Res.* 264: 56-66.

Woolner LB, McDonald JR (1949) Diagnosis of carcinoma of the lung. *JAMA* 139: 8-13.

Wright BE *et al.* (2002) Hypermutable bases in the *p53* cancer gene are at vulnerable positions in DNA secondary structures. *Cancer Res.* 62: 5641-5644.

Wright JG and Christman JW (2003) The role of nuclear factor kappa B in the pathogenesis of pulmonary diseases: implications for therapy. *Am. J. Respir. Med.* 2: 211-219.

Wu X *et al.* (1997) Associations between cytochrome p4502E1 genotype, mutagen sensitivity, cigarette smoking and susceptibility to lung cancer. *Carcinogenesis* 18: 967-973.

Wu X *et al.* (1998) Deletion in poly(ADP-riose)polymerase pseudogene and lung cancer risk. *Carcinogenesis* 19: 93-98.

Wu X *et al.* (1999a) Associations among telomerase activity, p53 protein overexpression, and genetic instability in lung cancer. *Br. J. Cancer* 80: 453-457.

Wu W *et al.* (1999b) Expression of *DMBT1*, a candidate tumor suppressor gene, is frequently lost in lung cancer. *Cancer Res.* 59: 1846-1851.

Wu WG *et al.* (2000) TRAIL-R2 is not correlated with p53 status and is rarely mutated in non-small cell lung cancer. *Anticancer Res.* 20(6B): 4525-4529.

Wu X *et al.* (2001) The association of microsomal epoxide hydrolase polymorphisms and lung cancer risk in African-Americans and Mexican-Americans. *Carcinogenesis* 22: 923-929.

Wu X *et al.* (2002) p53 genotypes and haplotypes associated with lung cancer susceptibility and ethnicity. *J. Natl. Cancer Inst.* 94: 681-690.

Wu X *et al.* (2003a) *XPA* polymorphism associated with reduced lung cancer risk and a modulating effect on nucleotide excision repair capacity. *Carcinogenesis* 24: 505-509.

Wu S *et al.* (2003b) hnRNP B1 protein may be a possible prognostic factor in squamous cell carcinoma of the lung. *Lung Cancer* 41: 179-186.

Wu TC *et al.* (2003c) Loss of telomerase activity may be a potential favourable prognostic marker in lung carcinomas. *Lung Cancer* 41: 163-169.

Wu X *et al.* (2003d) Telomere dysfunction: a potential cancer predisposition factor. *J. Natl. Cancer Inst.* 95: 1211-1218.

Wu XM *et al.* (2003e) Gene polymorphism of myeloperoxidase and genetic susceptibility to lung cancer. *Ai Zheng* 22: 912-915.

Wu PF *et al.* (2004a) Cancer aggregation and complex segregation analysis of families with female non-smoking lung cancer probands in Taiwan. *Eur. J. Cancer* 40: 260-266.

Wu X *et al.* (2004b) Genetic susceptibility to tobacco-related cancer. *Oncogene* 23: 6500-6523.

Wulfkuhle JD *et al.* (2003) Proteomic approaches to the diagnosis, treatment, and monitoring of cancer. *Adv. Exp. Med. Biol.* 532: 59-68.

Wunsch-Filho V *et al.* (2002) Familial cancer aggregation and the risk of lung cancer. *Sao Paulo Med. J.* 120: 38-44.

Wyllie AH (1997) Apoptosis and carcinogenesis. *Eur. J. Cell Biol.* 73: 189-197.

Wynder EL, Graham EA (1950) Tobacco smoking as a possible etiologic factor in bronchiogenic carcinoma. *JAMA* 143: 329-336.

Xi T *et al.* (2004) Many amino acid substitutions variants identified in DNA repair genes during human population screenings are predicted to impact protein function. *Genomics* 83: 970-979.

Xiao S *et al.* (1995) Codeletion of *p15* and *p16* genes in primary non-small cell lung carcinoma. *Cancer Res.* 55: 2968-2971.

Xiao L and Lang W (2000) A dominant role for the c-Jun NH2-terminal kinase in oncogenic ras-induced morphologic transformation of human lung carcinoma cells. *Cancer Res.* 60: 400-408.

Xiao X *et al.* (2004) Development of proteomic patterns for detecting lung cancer. *Dis. Markers* 19: 33-39.

Xie Y *et al.* (2004) Deficiencies in mouse *Myh* and *Ogg1* result in tumor predisposition and G to T mutations in codon 12 of the *K-Ras* oncogene in lung tumors. *Cancer Res.* 64: 3096-3102.

Xinarianos G *et al.* (1999) Telomerase activity in non-small cell lung carcinomas correlates with smoking status. *Int. J. Oncol.* 15: 961-965.

Xinarianos G *et al.* (2000a) *hMLH1* and *hMSH2* expression correlates with allelic imbalance on chromosome 3p in non-small cell lung carcinomas. *Cancer Res.* 60: 4216-4221.

Xinarianos G *et al.* (2000b) Evaluation of telomerase activity in bronchial lavage as a potential diagnostic marker for malignant lung disease. *Lung Cancer* 28: 37-42.

Xing D *et al.* (2002) Polymorphisms of the DNA repair gene *XPD* and risk of lung cancer in a Chinese population. *Lung Cancer* 38: 123-129.

Xing L *et al.* (2003) Expression of angiopoietin-2 gene in non-small cell lung cancer. J. *Huazhong Univ. Sci Technolog. Med. Sci.* 23: 362-364.

Xu H, El-Gewely MR (2003) Differentially expressed downstream genes in cells with normal or mutated p53. *Oncol. Res.* 13: 429-436.

Xu H-J *et al.* (1991) Absence of retinoblastoma protein expression in primary non-small cell lung carcinomas. *Cancer Res.* 51: 2735-2739.

Xu H-J *et al.* (1994) Altered retinoblastoma protein expression and prognosis in early-stage non-small-cell lung carcinoma. *J. Natl. Cancer Inst.* 86: 695-699.

Xu X *et al.* (1996) Cytochrome P450 *CYP1A1 Msp*I polymorphism and lung cancer susceptibility. *Cancer Epidemiol. Biomarkers Prev.* 5: 687-692.

Xu XC *et al.* (1997) Suppression of retinoic acid receptor beta in non-small-cell lung cancer *in vivo*: implications for lung cancer development. *J. Natl. Cancer Inst.* 89: 602-604.

Xu L *et al.* (2000) Microsatellite instability at AAAG repeat sequences in respiratory tract cancers. *Int. J. Cancer* 91: 200-204.

Xu XL *et al.* (2001a) Inactivation of human SRBC, located within the 11p15.5-p15.4 tumor suppressor region, in breast and lung cancers. *Cancer Res.* 61: 7943-7949.

Xu LL *et al.* (2001b) The NAD(P)H:quinone oxidoreductase 1 gene polymorphism and lung cancer: differential susceptibility based on smoking behavior. *Cancer Epidemiol. Biomarkers Prev.* 10: 303-309.

Xu L *et al.* (2001c) Microsatellite instability at AAAG repeat sequences in respiratory tract cancers. *Int. J. Cancer* 91: 200-204.

Xu LL *et al.* (2002a) Counterpoint: the myeloperoxidase –463G-A polymorphism does not decrease lung cancer susceptibility in Caucasians. *Cancer Epidemiol. Biomarkers Prev.* 11: 1555-1559.

Xu C *et al.* (2002b) *CYP2A6* genetic variation and potential consequences. *Adv. Drug Deliv. Rev.* 54: 1245-1256.

Xue Jun H *et al.* (2003) Reduced transcription of the *RB2/p130* gene in human lung cancer. *Mol. Carcinogen.* 38: 124-129.

Yageta M *et al.* (2002) Direct association of TSLC1 and DAL-1, two distinct tumor suppressor proteins in lung cancer. *Cancer Res.* 62: 5129-5133.

Yahata N *et al.* (1998) Telomerase activity in lung cancer cells obtained from bronchial washings. *J. Natl. Cancer Inst.* 90: 684-690.

Yamada T *et al.* (2000) Frequent chromosome 8q gains in human small cell lung carcinoma detected by arbitrarily primed-PCR genomic fingerprinting. *Cancer Genet. Cytogenet.* 120: 11-17.

Yamada KM *et al.* (2001) Tumor suppressor PTEN: modulator of cell signalling, growth, migration and apoptosis. *J. Cell Sci.* 114: 2375-2382.

Yamagata N *et al.* (2003) A training-testing approach to the molecular classification of resected non-small cell lung cancer. *Clin. Cancer Res.* 9: 4695-4704.

Yamamoto F, Perucho M (1984) Activation of a human c-*K-ras* oncogene. *Nucleic Acids Res.* 12: 8873-8885.

Yamane A *et al.* (2003) Suppressive activities of OGG1 and MYH proteins against G:C to T:A mutations caused by 8-hydroxyguanine but not by benzo[*a*]pyrene diol epoxide in human cells *in vivo*. *Carcinogenesis* 24: 1031-1037.

Yamane A *et al.* (2004) Differential ability of polymorphic OGG1 proteins to suppress mutagenesis induced by 8-hydroxyguanine in human cell *in vivo*. *Carcinogenesis* 25: 1689-1694.

Yan J-J *et al.* (1996) Immunohistochemical detection of Bcl-2 protein in small cell carcinomas. *Oncology* 53: 6-11.

Yan H *et al.* (2002a) Small changes in expression affect predisposition to tumorigenesis. *Nature Genet.* 30: 25-26.

Yan H *et al.* (2002b) Allelic variation in human gene epression. *Science* 297: 1143.

Yanagawa N *et al.* (2002) Frequent epigenetic silencing of the *p16* gene in non-small cell lung cancers of tobacco smokers. *Jpn. J. Cancer Res.* 93: 1107-1113.
Yanagawa N *et al.* (2003) Promoter hypermethylation of tumor suppressor and tumor-related genes in non-small cell lung cancers. *Cancer Sci.* 94: 589-592.
Yanagisawa K *et al.* (1996) Molecular analysis of the *FHIT* gene at 3p14.2 in lung cancer cell lines. *Cancer Res.* 56: 5579-5582.
Yanagisawa K *et al.* (2000) Heterogeneities in the biological and biochemical functions of Smad2 and Smad4 mutants naturally occurring in human lung cancers. *Oncogene* 19: 2305-2311.
Yanagisawa K *et al.* (2003a) Proteomic patterns of tumour subsets in non-small-cell lung cancer. *Lancet* 362: 433-439.
Yanagisawa K *et al.* (2003b) Molecular fingerprinting in human lung cancer. *Clin. Lung Cancer* 5: 113-118.
Yanaihara N *et al.* (2001) Physical and transcriptional map of a 311-kb segment of chromosome 18q21, a candidate lung tumor suppressor locus. *Genomics* 72: 169-179.
Yanez L *et al.* (1987) c-K-ras mutations in human carcinomas occur preferentially in codon 12. *Oncogene* 1: 315-318.
Yang P *et al.* (1997) Genetic analysis of families with nonsmoking lung cancer probands. *Genet. Epidemiol.* 14: 181-197.
Yang X *et al.* (1999a) Cloning and characterization of the human BAG-1 gene promoter: up-regulation by tumor-derived TP53 mutants. *Oncogene* 18: 4546-4553.
Yang X *et al.* (1999b) Alpha1-antitrypsin deficiency allele carriers among lung cancer patients. *Cancer Epidemiol. Biomarkers Prev.* 8: 461-465.
Yang HZ *et al.* (2001) Detection of telomerase activity level in human non-small-cell lung cancer. *Hunan Yi Ke Da Xue Xue Bao* 26: 549-550.
Yang YH, Speed T (2002) Design issues for cDNA microarray experiments. *Nature Rev. Genet.* 3: 579-588.
Yang H *et al.* (2002) GATA6 regulates differentiation of distal lung epithelium. *Development* 129: 2233-2246.
Yang HY *et al.* (2003) 14-3-3 sigma positively regulates p53 and suppresses tumor growth. *Mol. Cell Biol.* 23: 7096-7107.
Yang M *et al.* (2004a) Lack of association between Caucasian lung cancer risk and O(6)-methylguanine-DNA methyltransferase-codon 178 genetic polymorphism. *Lung Cancer* 44: 281-286.
Yang P *et al.* (2004b) Glutathione pathway genes and lung cancer risk in young and old age populations. *Carcinogenesis* In press. [DOI: 10.1093/carcin/bgh203]
Yano S *et al.* (2003) Molecular mechanisms of angiogenesis in non-small cell lung cancer, and therapeutics targeting related molecules. *Cancer Sci.* 94: 479-485.
Yao X *et al.* (2004) Epigenetic regulation of the taxol resistance-associated gene *TRAG-3* in human tumors. *Cancer Genet. Cytogenet.* 151: 1-13.
Yap DBS *et al.* (1999) mdm2: a bridge over the two tumour suppressors, p53 and Rb. *Oncogene* 18: 7681-7689.
Yashima K *et al.* (1997) Telomerase expression in respiratory epithelium during the multistage pathogenesis of lung carcinomas. *Cancer Res.* 57: 2373-2377.
Yatabe Y *et al.* (1998) p27KIP1 in human lung cancers: differential changes in small cell and non-small cell carcinomas. *Cancer Res.* 58: 1042-1047.
Yatabe Y *et al.* (2000) Topographical distributions of allelic loss in individual non-small-cell lung cancers. *Am. J. Pathol.* 157: 985-993.
Yatabe Y *et al.* (2002) TTF-1 expression in pulmonary adenocarcinomas. *Am. J. Surg. Pathol.* 26: 767-773.
Yatabe Y *et al.* (2004) Maspin expression in normal lung and non-small-cell lung cancers: cellular property-associated expression under the control of promoter DNA methylation. *Oncogene* 23: 4041-4049.
Ye S (2000) Polymorphism in matrix metalloproteinase gene promoters: implication in regulation of gene expression and susceptibility of various diseases. *Matrix Biol.* 19: 623-629.
Yegen BC (2003) Bombesin-like peptides: candidates as diagnostic and therapeutic tools. *Curr. Pharm. Des.* 9: 1013-1022.
Yeh KT *et al.* (2003) CYP3A5*1 is an inhibitory factor for lung cancer in Taiwanese. *Kaohsiung J. Med. Sci.* 19: 201-217.
Yen L *et al.* (2000) Heregulin selectively upregulates vascular endothelial growth factor secretion in cancer cells and stimulates angiogenesis. *Oncogene* 19: 3460-3469.
Yendamuri S *et al.* (2003) WW domain containing oxidoreductase gene expression is altered in non-small cell lung cancer. *Cancer Res.* 63: 878-881.

Yesner R (2001) Heterogeneity of so-called neuroendocrine lung tumors. *Exp. Mol. Pathol.* 70: 179-182.
Yi ES *et al.* (1997) High c-erbB-3 protein expression is associated with shorter survival in advanced non-small cell lung carcinomas. *Mod. Pathol.* 10: 142-148.
Yin L *et al.* (2001) Genetic polymorphisms of NAD(P)H quinone oxidoreductase, CYP1A1 and microsomal epoxide hydrolase and lung cancer risk in Nanjing, China. *Lung Cancer* 33: 133-141.
Yin Y *et al.* (2002) p53 stability and activity is regulated by Mdm2-mediated induction of alternative p53 translation products. *Nature Cell Biol.* 4: 462-467.
Yin JQ *et al.* (2003) siRNA agents inhibit oncogene expression and attenuate human tumor cell growth. *J. Exp. Ther. Oncol.* 3: 194-204.
Yokoi S *et al.* (2002) A novel target gene, *SKP2*, within the 5p13 amplicon that is frequently detected in small cell lung cancers. *Am. J. Pathol.* 161: 207-216.
Yokoi S *et al.* (2003a) Down-regulation of *SKP2* induces apoptosis in lung cancer cells. *Cancer Sci.* 94: 344-349.
Yokoi S *et al.* (2003b) *TERC* identified as a probable target within the 3q26 amplicon that is detected frequently in non-small cell lung cancers. *Clin. Cancer Res.* 9: 4705-4713.
Yokoi S *et al.* (2004) Amplification and overexpression of *SKP2* are associated with metastasis of non-small cell lung cancers to lymph nodes. *Am. J. Pathol.* 165: 175-180.
Yokomizo A *et al.* (1998) PTEN/MMAC1 mutations identified in small cell, but not in non-small cell lung cancers. *Oncogene* 17: 475-479.
Yokota J *et al.* (1988) Heterogeneity of lung cancer cells with respect to the amplification and rearrangement of *myc* family oncogenes. *Oncogene* 2: 607-611.
Yokota J *et al.* (1989) Multiple genetic alterations in small-cell lung carcinoma. *Princess Takamatsu Symp.* 20: 43-48.
Yokota J *et al.* (2003) Genetic alterations responsible for metastatic phenotypes of lung cancer cells. *Clin. Exp. Metastasis* 20: 189-193.
Yoneda K *et al.* (2001) Development of high density DNA microarray membrane for profiling smoke- and hydrogen peroxide-induced genes in a human bronchial epithelial cell line. *Am. J. Respir. Crit. Care Med.* 164: S85-S89.
Yoneda K *et al.* (2003) Application of high density DNA microarray to study smoke and hydrogen peroxide-induced injury and repair in human bronchial epithelial cells. *J. Am. Soc. Nephrol.* 14 (8 Suppl. 3): S284-S289.
Yoo SH *et al.* (2003) Expression of caveolin-1 is associated with poor prognosis of patients with squamous cell carcinoma of the lung. *Lung Cancer* 42: 195-202.
Yoon JH *et al.* (2001) Methylated CpG dinucleotides are the preferential targets for transversions mutations induced by benzo[*a*]pyrene diol epoxide in mammalian cells: similarities with the p53 mutation spectrum in smoking-associated lung cancers. *Cancer Res.* 61: 7110-7117.
Yoon H *et al.* (2002) Gene expression profiling of isogenic cells with different *TP53* gene dosage reveals numerous genes that are affected by *TP53* dosage and identifies CSPG2 as a direct target of p53. *Proc. Natl. Acad. Sci. USA* 99: 15632-15637.
Yoon SO *et al.* (2003) Roles of matrix metalloproteinases in tumor metastasis and angiogenesis. *J. Biochem. Mol. Biol.* 36: 128-137.
Yoshida T *et al.* (2004) The clinical significance of cyclin B1 and Weel expression in non-small-cell lung cancer. *Ann. Oncol.* 15: 252-256.
Yoshikawa H *et al.* (1999) Mutational analysis of p73 and p53 in human cancer cell lines. *Oncogene* 18: 3415-3421.
Yoshimatsu K *et al.* (2001) Inducible prostaglandin E synthase is overexpressed in non-small cell lung cancer. *Clin. Cancer Res.* 7: 2669-2674.
Yoshimoto K *et al.* (1986a) Rearrangement of the c-myc gene in two giant carcinomas of the lung. *Jpn. J. Cancer Res.* 77: 731-735.
Yoshimoto K *et al.* (1986b) Increased expression of the c-myc gene without gene amplification in human lung cancer and colon cancer cell lines. *Jpn. J. Cancer Res.* 77: 540-545.
Yoshimura A *et al.* (2003) Increased expression of the *LGALS3* (Galectin 3) gene in human non-small-cell lung cancer. *Genes Chrom. Cancer* 37: 159-164.
Yoshino I *et al.* (2003) Detection of loss of heterozygosity by high-resolution fluorescent system in non-small cell lung cancer: association of loss of heterozygosity with smoking and tumor progression. *Chest* 123: 545-550.
Yoshioka N *et al.* (2002) Suppression of anchorage-independent growth of human cancer cell lines by the TRIF52/periostin/OSF-2 gene. *Exp. Cell Res.* 279: 91-99.
Yoshitaka T *et al.* (1996) Mutations of E2F-4 trinucleotide repeats in colorectal cancer with microsatellite instability. *Biochem. Biophys. Res. Commun.* 227: 553-557.

You L *et al.* (2004) Inhibition of Wnt-2-mediated signalling induce programmed cell death in non-small-cell lung cancer cells. *Oncogene* 23: 6170-6174.
Younes M *et al.* (1997) Overexpression of *Glut1* and *Glut3* in stage I nonsmall cell lung carcinoma is associated with poor survival. *Cancer* 80: 1046-1051.
Youssef EM *et al.* (2004) Hypermethylation and silencing of the putative tumor suppressor tazarotene-induced gene 1 in human cancers. *Cancer Res.* 64: 2411-2417.
Yu D *et al.* (1994) c-erbB2/neu overexpression enhances metastatic potential of human lung cancer cells by induction of metastasis-associated properties. *Cancer Res.* 54: 3260-3266.
Yu C *et al.* (2002) Correlation between a single nucleotide polymorphism in the matrix metalloproteinase-2 promoter and risk of lung cancer. *Cancer Res.* 62: 6430-6433.
Yuan A *et al.* (2000) Correlation of total *VEGF* mRNA and protein expression with histologic type, tumor angiogenesis, patient survival and timing of relapse in non-small-cell lung cancer. *Int. J. Cancer* 89: 475-483.
Yuan A *et al.* (2002) Aberrant p53 expression correlates with expression of vascular endothelial growth factor mRNA and interleukin-8 mRNA and neoangiogenesis in non-small-cell lung cancer. *J. Clin. Oncol.* 20: 900-910.
Yuan BZ *et al.* (2004) DLC-1 operates as a tumor suppressor gene in human non-small cell lung carcinomas. *Oncogene* 23: 1405-1411.
Yuasa Y *et al.* (1983) Acquisition of transforming properties by alternative point mutations within c-*bas/has* human proto-oncogene. *Nature* 303: 775-779.
Yuasa Y *et al.* (1984) Mechanism of activation of an N-*ras* oncogene of SW-1271 human lung carcinoma cells. *Proc. Natl. Acad. Sci. USA* 81: 3670-3674.
Zabarovsky ER *et al.* (2002) Tumor suppressor genes on chromosome 3p involved in the pathogenesis of lung and other cancers. *Oncogene* 21: 6915-6935.
Zafer E *et al.* (2004) Detection and typing of human papillomavirus in non-small cell lung cancer. *Respiration* 71: 88-90.
Zajac-Kaye M (2001) Myc oncogene: a key component in cell cycle regulation and its implication for lung cancer. *Lung Cancer* 34 Suppl. 2: S43-S46.
Zangemeister-Wittke U *et al.* (2000) A novel bispecific antisense oligonucleotide inhibiting both bcl-2 and bcl-xL expression efficiently induces apoptosis in tumor cells. *Clin. Cancer Res.* 6: 2547-2555.
Zardo G *et al.* (2002) Integrated genomic and epigenomic analyses pinpoint biallelic gene inactivation in tumors. *Nature Genet.* 32: 453-458.
Zeng X *et al.* (1999) MDM2 suppresses p73 function without promoting p73 degradation. *Mol. Cell. Biol.* 19: 3257-3266.
Zhan S *et al.* (1995) Concordant loss of imprinting of the human insulin-like growth factor II gene promoters in cancer. *J. Biol. Chem.* 270: 27983-27986.
Zhang Y *et al.* (1993) Retroviral vector-mediated transduction of K-ras antisense RNA into human lung cancer cells inhibits expression of the malignant phenotype. *Hum. Gene Ther.* 4: 451-460.
Zhang X-K *et al.* (1994) A specific defect in the retinoic acid response associated with human lung cancer cell lines. *Cancer Res.* 54: 5663-5669.
Zhang CY *et al.* (1995) Point mutation of the somatostatin receptor 2 gene in the human small cell lung cancer cell line. COR-L103. *Biochem. Biophys. Res. Commun.* 210: 805-815.
Zhang ZY *et al.* (1996) Characterization of purified human recombinant cytochrome P4501A1-Ile462 and -Val462: assessment of a role for the rare allele in carcinogenesis. *Cancer Res.* 56: 3926-3933.
Zhang Y *et al.* (1998) ARF promotes MDM2 degradation and stabilizes p53: ARF-INK4a locus deletion impairs both the Rb and p53 tumor suppression pathways. *Cell* 92: 725-734.
Zhang A *et al.* (2000a) Frequent amplification of the telomerase reverse transcriptase gene in human tumors. *Cancer Res.* 60: 6230-6235.
Zhang P *et al.* (2000b) Lack of expression for the suppressor PML in human small cell lung carcinoma. *Int. J. Cancer* 85: 599-605.
Zhang YA *et al.* (2000c) Generation of a ribozyme-adenoviral vector against K-ras mutant human lung cancer cells. *Mol. Biotechnol.* 15: 39-49.
Zhang YA *et al.* (2000d) Anti-tumorigenic effect of a K-ras ribozyme against human lung cancer cell line heterotransplants in nude mice. *Gene Ther.* 7: 2041-2050.
Zhang Z *et al.* (2001) Wildtype *Kras2* can inhibit lung carcinogenesis in mice. *Nature Genet.* 29: 25-30.
Zhang X *et al.* (2002) Genetic polymorphisms of the human *CYP2A13* gene: identification of single-nucleotide polymorphisms and functional characterization of an Arg257Cys variant. *J. Pharmacol. Exp. Ther.* 302: 416-423.

Zhang HT *et al.* (2003a) Mutational analysis of the transforming growth factor beta receptor type I gene in primary non-small cell lung cancer. *Lung Cancer* 40: 281-287.
Zhang LF *et al.* (2003b) Comparison of K-ras gene mutations in tumour and sputum DNA of patients with lung cancer. *Biomarkers* 8: 156-161.
Zhang HT *et al.* (2004a) Defective expression of transforming growth factor beta receptor type II is associated with CpG methylated promoter in primary non-small cell lung cancer. *Clin. Cancer Res.* 10: 2359-2367.
Zhang P *et al.* (2004b) *CHK2* kinase expression is down-regulated due to promoter methylation in non-small cell lung cancer. *Mol. Cancer* 3: 14.
Zhang Z *et al.* (2004c) MDM2 is a negative regulator of p21WAF1/CIP independent of p53. *J. Biol. Chem.* 279: 16000-16006.
Zhao B, Bepler G (2001) Transcript map and complete genomic sequence for the 310 kb region of minimal allele loss on chromosome segment 11p15.5 in non-small-cell lung cancer. *Oncogene* 20: 8154-8164.
Zhao Y *et al.* (1999) p53 gene mutations in non-small cell lung cancer detected by polymerase chain reaction single-strand conformation polymorphism analysis. *Chin. Med. Sci. J.* 14: 134-137.
Zhao B *et al.* (2000) Transgenic mouse models for lung cancer. *Exp. Lung Res.* 26: 567-579.
Zhao H *et al.* (2002) Microsomal epoxide hydrolase polymorphisms and lung cancer risk in non-Hispanic whites. *Mol. Carcinogen.* 33: 99-104.
Zhao X *et al.* (2004) An integrated view of copy number and allelic alterations in the cancer genome using single nucleotide polymorphism arrays. *Cancer Res.* 64: 3060-3071.
Zheng L, Lee WH (2002) Retinoblastoma tumor suppressor and genome stability. *Adv. Cancer Res.* 85: 13-50.
Zhong S *et al.* (1991) Glutathione *S*-transferase mu locus: use of genotyping and phenotyping assays to assess association with lung cancer susceptibility. *Carcinogenesis* 12: 1533-1537.
Zhong Y *et al.* (2001) Loss of H-cadherin protein expression in human non-small cell lung cancer is associated with tumorigenicity. *Clin. Cancer Res.* 7: 1683-1687.
Zhong L *et al.* (2004) Identification of circulating antibodies to tumor-associated proteins for combined use as markers of non-small cell lung cancer. *Proteomics* 4: 1216-1225.
Zhou BB, Elledge SJ (2000) The DNA damage response: putting checkpoints in perspective. *Nature* 408: 433-439.
Zhou XZ, Lu KP (2001) The PIN2/TRF1-interacting protein PinX1 is a potent telomerase inhibitor. *Cell* 107: 347-359.
Zhou X *et al.* (2000) Prognostic implication of microsatellite alteration profiles in early stage non-small cell lung cancer. *Clin. Cancer Res.* 6: 559-565.
Zhou JX *et al.* (2001a) Mechanisms of G1 checkpoint loss in resected early stage non-small cell lung cancer. *Lung Cancer* 32: 27-38.
Zhou JX *et al.* (2001b) A role for p53 in base excision repair. *EMBO J.* 20: 914-923.
Zhou J *et al.* (2001c) Expression of early lung cancer detection marker: hnRNP-A2/B1 and its relation to microsatellite alteration in non-small cell lung cancer. *Lung Cancer* 34: 341-350.
Zhou W *et al.* (2001d) The interaction between microsomal epoxide hydrolase polymorphisms and cumulative cigarette smoking in different histological subtypes of lung cancer. *Cancer Epidemiol. Biomarkers Prev.* 10: 461-466.
Zhou J *et al.* (2002a) Attachment and tension in the spindle assembly checkpoint. *J. Cell Sci.* 115: 3547-3555.
Zhou W *et al.* (2002b) Gene-environment interaction for the *ERCC2* polymorphisms and cumulative cigarette smoking exposure in lung cancer. *Cancer Res.* 62: 1377-1381.
Zhou M *et al.* (2002c) α-Methylacyl-CoA racemase: a novel tumor marker over-expressed in several human cancers and their precursor lesions. *Am. J. Surg. Pathol.* 26: 926-931.
Zhou W *et al.* (2002d) Genetic polymorphisms in *N*-acetyltransferase-2 and microsomal epoxide hydrolase, cumulative cigarette smoking, and lung cancer. *Cancer Epidemiol. Biomarkers Prev.* 11: 15-21.
Zhou W *et al.* (2003a) Polymorphisms in the DNA repair genes XRCC1 and ERCC2, smoking and lung cancer risk. *Cancer Epidemiol. Biomarkers Prev.* 12: 359-365.
Zhou Y *et al.* (2003b) Genome-wide identification of chromosomal regions of increased tumor expression by transcriptome analysis. *Cancer Res.* 63: 5781-5784.
Zhou W *et al.* (2003c) Mapping genetic alterations in tumors with single nucleotide polymorphisms. *Curr. Opin. Oncol.* 15: 50-54.
Zhu Y *et al.* (2001) A single nucleotide polymorphism in the matrix metalloproteinase-1 promoter enhances lung cancer susceptibility. *Cancer Res.* 61: 7825-7829.
Zhu J *et al.* (2002) Gene expression and immunolocalization of 15-lipoxygenase isoenzymes in the airway mucosa of smokers with chronic bronchitis. *Am. J. Respir. Cell Mol. Biol.* 27: 666-677.

Zhu LQ *et al.* (2003a) Relationship between hMSH2 and FHIT gene expression in non-small cell lung cancer. *Ai Zheng* 22: 571-574.
Zhu B-q *et al.* (2003b) Second hand smoke stimulates tumor angiogenesis and growth. *Cancer Cell* 4: 191-196.
Zhu WG *et al.* (2003c) Methylation of adjacent CpG sites affects Sp1/Sp3 binding and activity in the p21 (Cip1) promoter. *Mol. Cell Biol.* 23: 4056-4065.
Zhu CQ *et al.* (2004a) Skp2 gene copy number aberrations are common in non-small cell lung carcinoma, and its overexpression in tumors with ras mutation is a poor prognostic marker. *Clin. Cancer Res.* 10: 1984-1991.
Zhu NL *et al.* (2004b) NKX2.1 regulates transcription of the gene for human bone morphogenetic protein-4 in lung epithelial cells. *Gene* 327: 25-36.
Zhukov TA *et al.* (2003) Discovery of distinct protein profiles specific for lung tumors and pre-malignant lung lesions by SELDI mass spectrometry. *Lung Cancer* 40: 267-279.
Ziegel R *et al.* (2003) *K-ras* gene sequence effects on the formation of 4-(methylnitrosamino)-1-(3-pyridyl)-1-butanone (NNK)-DNA adducts. *Chem. Res. Toxicol.* 16: 541-550.
Ziegel R *et al.* (2004) Endogenous 5-methylcytosine protects neighboring guanines from N7 and O-6 methylation and O6-pyridyloxobutylation by the tobacco carcinogen 4-(methylnitrosamino)-1-(3-pyridyl)-1-butanone. *Biochemistry* 43: 540-549.
Ziegler A *et al.* (1997) Induction of apoptosis in small-cell lung cancer cells by an antisense oligodeoxynucleotide targeting the Bcl-2 coding sequence. *J. Natl. Cancer Inst.* 89: 1027-1036.
Ziegler A *et al.* (2002) Circulating DNA: a new diagnostic gold mine? *Cancer Treat. Rev.* 28: 255-271.
Zienolddiny S *et al.* (1999) DNA mismatch binding in human lung tumor cell lines. *Lung Cancer* 26: 15-25.
Zienolddiny S *et al.* (2001) Loss of heterozygosity is related to *p53* mutations and smoking in lung cancer. *Br. J. Cancer* 84: 226-231.
Zienolddiny S *et al.* (2004) Polymorphisms of the interleukin-1 beta gene are associated with increased risk of non-small cell lung cancer. *Int. J. Cancer* 109: 353-356.
Zimmerman PV *et al* (1987) Ploidy as a prognostic determinant in surgically treated lung cancer. *Lancet* ii: 530-533.
Zimniak P *et al* (1994) Naturally occurring human glutathione *S*-transferase GSTP1-1 isoforms with isoleucine and valine in position 104 differ in enzymic properties. *Eur. J. Biochem.* 224: 893-899.
Zindy F *et al* (1998) Myc signalling via the ARF tumor suppressor regulates p53-dependent apoptosis and immortalization. *Genes Dev.* 12: 2424-2433.
Zingg JM, Jones PA (1997) Genetic and epigenetic aspects of DNA methylation on genome expression, evolution, mutation and carcinogenesis. *Carcinogenesis* 18: 869-882.
Zmirou D *et al* (2000) Personal exposure to atmospheric polycyclic aromatic hydrocarbons in a general adult population and lung cancer risk assessment. *J. Occup. Environ. Med.* 42: 121-126.
Zöchbauer-Müller S *et al.* (2000) Fragile histidine triad (*FHIT*) gene abnormalities in lung cancer. *Clin. Lung Cancer* 2: 141-145.
Zöchbauer-Müller S *et al.* (2001a) 5' CpG island methylation of the *FHIT* gene is correlated with loss of gene expression in lung and breast cancer. *Cancer Res.* 61: 3581-3585.
Zöchbauer-Müller S *et al.* (2001b) Aberrant promoter methylation of multiple genes in non-small cell lung cancers. *Cancer Res.* 61: 249-255.
Zöchbauer-Müller S *et al.* (2002) Aberrant DNA methylation in lung cancer: biological and clinical implications. *Oncologist* 7: 451-457.
Zöchbauer-Müller S *et al.* (2003) Aberrant methylation of multiple genes in the upper aerodigestive tract epithelium of heavy smokers. *Int. J. Cancer* 107: 612-616.
Zojer N *et al.* (2000) Aneuploidy of chromosome 7 can be detected in invasive lung cancer and associated premalignant lesions of the lung by fluorescence *in situ* hybridisation. *Lung Cancer* 28: 225-235.
Zuber J *et al.* (2000) A genome-wide survey of RAS transformation targets. *Nature Genet.* 24: 144-152.
Zurer I *et al.* (2004) The role of p53 in base excision repair following genotoxic stress. *Carcinogenesis* 25: 11-19.

Subject Index

D

G

H

N

T

U

V

Printing: Saladruck, Berlin
Binding: Stein+Lehmann, Berlin